Theory and Algorithms for Cooperative Systems

Series on Computers and Operations Research

Series Editor: P. M. Pardalos *(University of Florida)*

Published

Vol. 1 Optimization and Optimal Control
eds. P. M. Pardalos, I. Tseveendorj and R. Enkhbat

Vol. 2 Supply Chain and Finance
eds. P. M. Pardalos, A. Migdalas and G. Baourakis

Vol. 3 Marketing Trends for Organic Food in the 21st Century
ed. G. Baourakis

Series on Computers and Operations Research | Vol. 4

Theory and Algorithms for Cooperative Systems

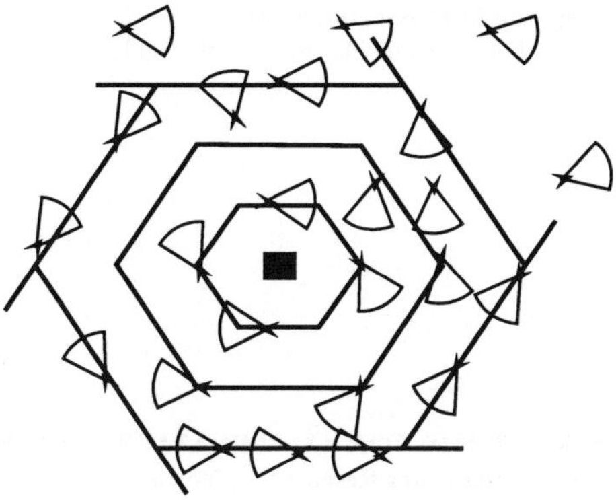

Editors

Don Grundel
Air Force Research Laboratory, USA

Robert Murphey
Air Force Research Laboratory, USA

Panos M Pardalos
University of Florida, USA

World Scientific

NEW JERSEY • LONDON • SINGAPORE • BEIJING • SHANGHAI • HONG KONG • TAIPEI • CHENNAI

Published by

World Scientific Publishing Co. Pte. Ltd.

5 Toh Tuck Link, Singapore 596224

USA office: Suite 202, 1060 Main Street, River Edge, NJ 07661

UK office: 57 Shelton Street, Covent Garden, London WC2H 9HE

British Library Cataloguing-in-Publication Data
A catalogue record for this book is available from the British Library.

THEORY AND ALGORITHMS FOR COOPERATIVE SYSTEMS
Series on Computers and Operations Research — Vol. 4

ISBN 981-256-020-3

Printed in Singapore by World Scientific Printers (S) Pte Ltd

"We must recognize that war is common and strife is justice, and all things happen according to strife and necessity.

"War is the father of all and king of all, who manifested some as gods and some as men, who made some slaves and some freemen."

-Heraclitus (540 BC - 480 BC)

Preface

Over the past several years, cooperative control and optimization have increasingly played a larger and more important role in many aspects of military sciences, biology, communications, robotics, and decision making. At the same time, cooperative systems are notoriously difficult to model, analyze, and solve - while intuitively understood, they are not axiomatically defined in any commonly accepted manner. This is to be expected in what many would consider a most diverse scientific and engineering discipline.

As cooperation may be defined as entities acting together in a coordinated way in the pursuit of shared goals, we find that cooperative systems are pervasive in many areas. In society we find a combination of technology, people and organizations that facilitate the communication and coordination necessary for a group to effectively work together and achieve gain for all its members. In nature, we may consider the obvious coordinated movements of some birds, fish and insects. In a single organism such as the human body, complex systems work together for growth, reproduction, adaptation to the environment, and critical thinking. In man-made systems we find, as examples, robots in manufacturing, unmanned vehicles in search and rescue operations, networked systems for communication, power and transportation.

In studying cooperative systems, several key questions arise. How can local interaction of a number of entities (sub-systems) be integrated to yield some overall result? How does the interaction among many sub-systems yield properties that are absent when these sub-systems are considered in isolation? What are the dynamics and emergent properties of a cooperative system when the number of sub-systems is very large? Regardless of the intense previous research, we are still at the exploration state of understanding the fundamental questions of cooperative systems and how to control and optimize them.

Despite the challenges, researchers and practitioners are increasingly called upon to "solve" cooperative system issues. The works in this volume provide outstanding insight into cooperative control and optimization. They are the result of selected presentations at the fourth Annual Conference on Cooperative Control and Optimization held in Destin, Florida, in November of 2003, and additional invited papers. Although certainly not resolved, we do expect *Theory and Algorithms for Cooperative Systems*, to assist in deeper understanding of these complex problems.

We would like to acknowledge the financial support of the Air Force Research Laboratory and the University of Florida College of Engineering. We are very grateful to the contributing authors, the anonymous referees, and World Scientific for making the publication of this volume possible.

Don Grundel
Robert Murphey
Panos M. Pardalos

Contents

CHAPTER 1

MESH STABILITY IN FORMATION OF DISTRIBUTED SYSTEMS

Chimpalthradi R. Ashokkumar
National Research Council

Robert A. Murphey
AFRL, Munitions Directorate
Eglin AFB, FL

Robert L. Sierakowski
AFRL, Munitions Directorate
Eglin AFB, FL

In formation control of distributed systems, the topology of interconnections is examined in a leader-follower structure. Two fundamental interconnections are introduced. Spatial coordination assists the systems to maintain relative dynamics. When time coordination in relative dynamics is introduced, temporal coordination is defined. Majority of formations is a sequence of spatial and temporal coordinations with a high volume of communication sources. Decomposition technique systematically builds a bridge between an interconnection and a directed graph capable of handling enormous communication sources. A Lyapunov like equation is derived to shape and interpret formation dynamics using a stability-like tool called *mesh stability*. Examples are illustrated.

1. Introduction

Control of distributed aerospace systems is a complex problem. The complexities of the problem can be sometimes made simple provided the structures are explored. The problem is that the distributed systems need to incorporate a certain type of cooperation at lower level as when a system in the formation is directed (by higher level control architecture) to perform a mission operation [22, 32, 13]. Information flow (or data transfer) is a valuable resource in this technology [30, 31]. Intuitively, it is known that unwanted information flow could seriously harm the objectives of a mission operation. Thus a specific information flow to meet the objectives of a mission operation is compared with a high volume of communication useful for the controllability and observability of distributed systems. Following this setting, control of distributed systems in a leader-follower framework is assumed. The goal is to engage and direct all followers in a *formation* so that each follower performs an operation assigned to the leader. Clearly, follower control input generation based on, a) its on-board sensors and b) the information flow from its leader, to shape a formation becomes an important problem.

Conceptually, control of structural systems with miniature transducers embedded in finite elements and control of distributed aerospace systems with an assumed information flow for air superiority share several characteristics. It is well known that the subsystems resulting from a finite element model are naturally connected through a user defined structural degrees of freedom [16, 1]. In distributed systems, we are forced to introduce the topology of interconnections artificially. It is perhaps required to understand control of distributed systems [11, 15, 18, 23, 20, 29]. In our case, the study is motivated to introduce information flow and develop cooperative systems in a formation that can meet the mission objectives. Through trial and error methods, control inputs generated using artificial interconnections have been successfully applied and every system engaged in the formation is carefully controlled. In addition, cooperation is embedded in each system [4]. From the information flow viewpoint, the leader information to the followers is directed in the sense that no information at the leader is received from any of its followers. These assumptions permit to develop an *aggregate system*, a system of systems for which the standard design tools apply [4, 5]. When additional information flow from follower to follower takes place, computer aided control system design tools such as Matlab become obsolete. Some attempts have been made to control distributed systems as well as a structural system by introducing sensor and actuator networking in a reconfigurable architecture [6, 7]. In this architecture, networking capabilities anticipated in centralized, decentralized and distributed architectures are introduced [13].

Assuming a particular data rate, churning information flow from multiple sources in an user defined fashion has shown poor cooperative mission operations [7]. However, it is an important type of computational intelligence required when a higher level control wishes to bifurcate a certain number of systems in an existing formation and assign them to perform a different mission operation. In [7], no latent or computational delay is assumed. Delay can significantly affect the structure of a formation [5]. Further, the observations presented in [7] are applicable for decentralized architectures. The benefits of undirected information flow in a centralized architecture have not been investigated. Centralized architecture is expected to introduce control coupling to assist complex maneuvers. During the development stages, interconnections that couple state vectors of the distributed systems are focused. Synthesis of these interconnections for state and input constrained distributed systems have been recently proposed [8].

The purpose of this chapter is two fold. First, two fundamental interconnections for formation shaping are introduced. In *spatial* coordination mode, an interconnection attempts to generate an information flow based control input that can turn velocity vectors as emphasized in [20], and match slope of each system in phase-plane. In *temporal* coordination mode, while turning the velocity vectors, time coordination among the distributed systems is maintained. It is believed that majority of the mission operations is a sequence of shape and temporal coordinations. The transition from spatial to temporal (or) from temporal to spatial coordination modes on time scale introduces mesh stability [26, 21]. In systems and control, recall stability being used as a tool to interpret and shape trajectories. Likewise, when more than one system is involved, mesh stability attempts to interpret (analysis) and shape (design) formation of the distributed systems. It further provides the time scale behavior of distributed systems, similar to fixed-state terminal control problem [17], but with multiple systems involved. In addition, it directs to shape a formation by constantly modifying the trajectory of one system with respect to the trajectories of other systems.

Secondly, a decomposition technique is introduced to study mesh stability problem in terms of string stability [28], where we attempt to use two systems at any given time and study relative dynamics to interpret and shape formations. To this end, graph theory is used [pages 102-113, 27]. Graph theory is widely applied in formation control of distributed systems [11, 12, 24, 25]. Decomposition technique establishes a bridge between an information flow based interconnection and the graph theory. There are several merits involved in the decomposition technique. Although admissible interconnections are limited, the technique is an excellent procedure to identify cooperative and uncooperative systems in a formation. The cooperative systems are configured using a finite set of leader-follower pairs (Lyapunov systems). The uncooperative systems are the non-communicating follower-

follower pairs. When Lyapunov and uncooperative systems are integrated, we have a *System of Lyapunov Systems.* For Lyapunov systems and System of Lyapunov systems, we present Lyapunov equations to study mesh stability. The difficulty in formation stabilization using string stability is that the distributed systems in regulation, temporal and spatial coordination modes exhibit nonconvex nature. Thus, discrete-time interpretation of a continuous-time Lyapunov equation using its symmetric matrices turns out to be a hard problem [page 85, 9, 10]. When convergence guarantee in the place of stability guarantee is replaced as in Kalman filters [pages 766-772, 14] and in inverse linear quadratic problems [2], the Lyapunov equation presented in this article is expected solve a certain kind of nonconvex problems. In formation sensitivity analysis [3], the velocity vector orientation at each point on follower trajectories is pursued to understand the nonconvex nature of the problem.

This article is written to give a maximal exposition to the above research items contributing to the *Control of Distributed Systems.* Through out our chapter, we assume homogeneous systems with state, input and output vectors having same dimensions. In leader-follower framework, we focus on issues related to mesh stability for trajectories generated from leader information directed to the followers. Although mesh stability is a tool to interpret complex formations of distributed systems with enormous on-off communication sources, the subject is introduced using simple leader to follower information flow structure. The examples are restricted to a rover type of dynamics. Whenever necessary, we either use its first order dynamics with velocity $(\dot{p}_\ell,\ \dot{p}_f)$ as a state variable $(x_\ell,\ x_f)$ or its second order dynamics with velocity and position $(p_\ell,\ p_f)$ as state variables.

2. String Stability of an Interconnection

Consider the leader and follower scalar systems given by,

$$x_\ell(t) = x_\ell(0)\ e^{-t}, \qquad x_\ell(0) = 1,\ t \geq 0 \tag{1}$$

$$x_f(t) = x_f(0)\ e^{-2t}, \qquad x_f(0) = 1,\ t \geq 0. \tag{2}$$

The functions are sign invariant with positive values at all time. The interconnection

$$c(t) = x_\ell(t) - x_f(t) > 0, \quad t \geq 0 \tag{3}$$

is also sign invariant. The string stability of an interconnection $c(t)$ is explained in Figure 1.

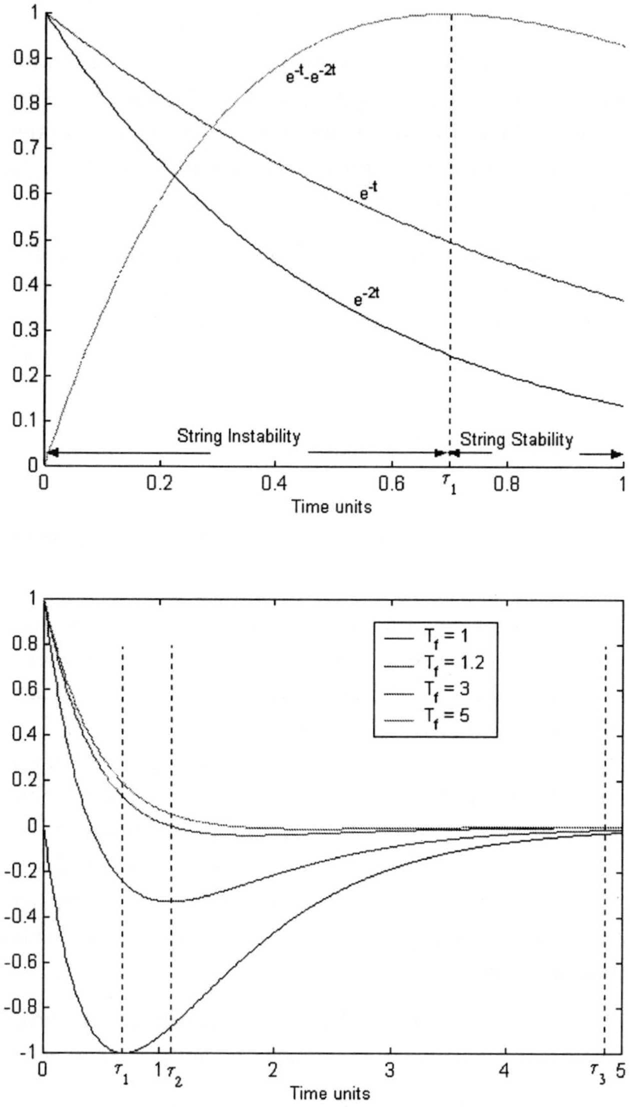

Fig. 1. Top: SS with Relative dynamics Bottom: SS with Interconnections

At $t = 0.7$ units, there is a transition from string instability to string stability. These points on time scale vary as and when the coefficients T_f

of an interconnection,

$$c(t) = T_f x_f(t) - x_\ell(t), \quad T_f \geq 1, \tag{4}$$

are reconfigured. When $T_f = 5$, the response of $c(t)$ is monotonically decreasing. Some shape is preserved during the initial stages, then temporal coordination takes place. In this ad hoc formation, several issues about the cooperative behavior of the distributed systems may be posed. Suppose the string stability is defined as below:[28]

Definition 1: Consider a finite set of distributed systems having an on board processor with a sampling rate Δ time units. An interconnection vector $c(t) \in R^n$ for all $t \in [\tau_j,\ \tau_{j+1}],\ j \in \mathcal{I}$(integer values) is said to be string stable if there exists a sequence $\tau_j + k\Delta,\ k = 1, 2\cdots$ such that,

$$\langle c, c\rangle_{k+1} < \langle c, c\rangle_k, \ \forall\ k = 1, 2\cdots \tag{5}$$

Otherwise it is said to be string unstable. When

$$\langle c, c\rangle_{k+1} = \langle c, c\rangle_k, \ \forall\ k = 1, 2\cdots, \tag{6}$$

the distributed systems in the formation are said to be marginally string stable.

Following the definition, infer that in marginal string stability the distributed systems are brought together as required in cooperative attack type of mission operations. In Figure 1, $T_f = 1$ does not seem to offer this operation in the time duration of interest. Although $T_f = 5$ meets this objective sometime during $t \geq 2$ units, note that the interconnection is such that the leader is five units away from its follower. Such an interconnection may be needed to perform a search and rescue type of mission operations. Likewise, a meaningful details may be drawn from string stability/instability definitions of an interconnection. In order to depict the dependency of string stability on the interconnection parameters, the following theorem is stated.

Theorem 2: String stability of an interconnection with positive systems Consider a sign invariant leader and follower systems

$$\dot{x}_\ell(t) = \sigma_\ell x_\ell(t), \quad x_\ell(t) > 0\ \forall t,\ x_\ell(0) \neq 0,$$
$$\dot{x}_f(t) = \sigma_f x_f(t), \ x_f(t) > 0\ \forall t,\ x_f(0) \neq 0.$$

An interconnection $c(t) = T_f x_f(t) - T_\ell x_\ell(t) \ > 0\ \forall t$, is said to be string stable if and only if,

$$\frac{x_f(t)}{x_\ell(t)} < -\frac{T_\ell \sigma_\ell}{T_f \sigma_f}, \ \forall\ t \in [\tau_j,\ \tau_{j+1}] \tag{7}$$

When the distributed systems are unforced, the dependency of string stability on nonzero initial conditions is inferred from equation (7),

$$\frac{x_f(\tau_j)}{x_\ell(\tau_j)} = \alpha < -\frac{T_\ell \sigma_\ell}{T_f \sigma_f} e^{\sigma t}, \; \forall\, t \in [\tau_j,\ \tau_{j+1}] \tag{8}$$

where $\sigma = \sigma_f - \sigma_\ell$ and $\sigma_f \neq \sigma_\ell$ is the relative stability parameter

The complexities of string stability are best illustrated using Theorem 2. String stability can be defined for almost every system (stable and unstable) with any kind of trajectory (impulse response, step response and so on). In addition, information flow $\pm T_\ell$ and interconnections (or coefficient matrix) $\pm T_f$ with directions ($\pm$) also play an important role. Next, string stability of an interconnection with three scalar systems is analyzed. Consider,

$$c(t) = x_1(t) - x_2(t) - x_\ell(t). \tag{9}$$

In the durations of marginal string stability, let $\langle c, c\rangle_k = 0 \; \forall\, k$. The scenarios are,

$$\begin{aligned} x_\ell(t) &= x_1(t) - x_2(t) \\ x_2(t) &= x_1(t) - x_\ell(t) \\ x_1(t) &= x_\ell(t) + x_2(t) \end{aligned}$$

This is a difficult problem with a directed information flow from the leader. Since no information flow is assumed among the followers, x_1 (or x_2) is not expected to cooperate with x_2 (or x_1). They are however required to cooperate with the leader x_ℓ. The feasible cooperative systems out of this interconnection would be,

$x_\ell(t) \to x_1(t)$		$x_\ell(t) \to -x_2(t)$	Dynamic systems
$x_2(t) \to 0$	(OR)	$x_1(t) \to 0$	Static system

The difficulty begins when the static system becomes dynamic, where analysis problem becomes important. In this discussion, all distributed systems are assumed unforced. In decentralized architecture, interconnections are critical to generate follower control inputs and develop cooperation with the leader [4]. This has been pursued by forcing each follower to pass through a finite set of waypoints. At these waypoints, we may wish to shape a formation by rotating slopes (or velocity vectors) in a phase-plane. This brings in three fundamental operations on time scale, namely, regulation, spatial and temporal coordinations. The decomposition technique attempts to shape a particular follower's formation in a complex interconnection with two or more systems. As and when the spatial, temporal and regulation modes are introduced at each waypoint, the interconnections are accordingly reconfigured and the cooperative control problems are become nonconvex. Typically, mission operations with convexity embedded in each formation is preferred [10]. In the following section, an exposition to the decomposition technique is presented.

3. Decomposition Technique

The leader-follower interconnections for a directed graph are as follows [8]

$$c_j(t) = T_j^{\ell} x_{\ell}(t) + \sum_{k=1}^{j} T_{jk}^{f} x_k(t), \; j \in [1, 2, \cdots N]. \tag{10}$$

The homogeneous distributed systems are $x_{\ell}(t), x_k(t) \in R^n$. The number of systems engaged in the formation is N. The interconnections $c_j(t) \in R^n$ are defined using an information flow $T_j^{\ell} \in R^{n \times n}$ directed to the follower j and an user defined coefficient matrices $T_{jk}^{f} \in R^{n \times n}$. The structure of T_j^{ℓ} reflects the type of information that the leader x_{ℓ} provides to its follower x_j. Synthesis of T_{jk}^{f} for a formation structure is a design problem and T_{jj}^{f} is assumed invertible. For the state and input constrained systems, T_j^{ℓ} and T_{jk}^{f} turn out to be stochastic matrices. Suppose,

$$x(t) = \begin{bmatrix} x_{\ell}(t) \\ x_1(t) \\ \vdots \\ x_N(t) \end{bmatrix} \in R^{Nn}, \tag{11}$$

$$u(t) = \begin{bmatrix} u_{\ell}(t) \\ u_2(t) \\ \vdots \\ u_N(t) \end{bmatrix} \in R^{Nm}, \tag{12}$$

$$\text{and} \;\; u_{\ell}(t), \; u_k(t) \in R^m. \tag{13}$$

The distributed systems with interconnections and information flow can be put into the form,

$$E \, \dot{x}(t) = Ax(t) + Bu(t). \tag{14}$$

Each interconnection can be defined with respect to a vector $w \in R^{Nn}$ and a directed graph $\mathcal{G}(\hat{e}, \hat{n})$ with edges $\hat{e}$ and nodes $\hat{n}$. Let the coefficient vector for an interconnection be $I \in R^{\hat{n}}$, where

$$I = \begin{bmatrix} E'w \\ -A'w \\ -B'w \end{bmatrix}, \tag{15}$$

where the prime denotes transpose. We are now concerned with the mesh stability of a scalar $c(t)$, where

$$c(t) = I'x(t). \tag{16}$$

Consider an edge-node matrix (or incidence matrix) of the graph by,

$$\mathbf{C} \in R^{\hat{e} \times \hat{n}}.$$

Let the row and null spaces of $\mathbf{C}$ be $\mathcal{R}(\mathbf{C})$ and $\mathcal{N}(\mathbf{C})$, respectively. Note that I is spanned by the bases of $\mathcal{R}(\mathbf{C})$ and $\mathcal{N}(\mathbf{C})$. It is known that the basis of $\mathcal{N}(\mathbf{C})$ is a vector $\hat{\mathbf{1}} \in R^{\hat{n}}$ whose components are stacked by 1's.

Theorem 3: Decomposition technique A multi-node interconnection $c(t) = I'x(t)$ for a leader-follower graph $\mathcal{G}(\hat{e}, \hat{n})$ with an edge-node matrix $\mathbf{C}$ is decomposable into a finite set of two-node interconnections if and only if the projection of I on $\mathcal{N}(\mathbf{C})$ is a zero vector.

The proof of this theorem is based on an assumption that a leader-follower graph $\mathcal{G}(\hat{e}, \hat{n})$ has no isolated edges. An isolated edge is an indication that an eigenmode in the interconnection is uncontrollable. Suppose the eigenvalues are complex conjugate, the matrices E, A and B call for composite matrix theory [33]. In the remaining part of this chapter, we focus on each follower $x_f(t)$ where $f \in [1, 2, \cdots N]$ and its cooperation with the leader $x_\ell(t)$. Figure 2 illustrates the decomposition technique.

In the first two graphs, the leader information to followers x_1 and x_2 is used to shape $x_\ell - x_1$ and $x_\ell - x_2$ formations. In the last graph, additional edges are introduced to analyze $x_1 - x_2$ formation. The cooperative and uncooperative edges are shown using solid and dotted lines. The decomposition technique with isolated nodes is beyond the scope of this chapter. In our subsequent discussion, those interconnections that can be decomposed into a finite set of two-node interconnections are considered. Specific issues related to string stability are focused. In particular, we present interconnections for shape and temporal coordinations, and then the Lyapunov functions for the graphs shown in Figure 2 are derived.

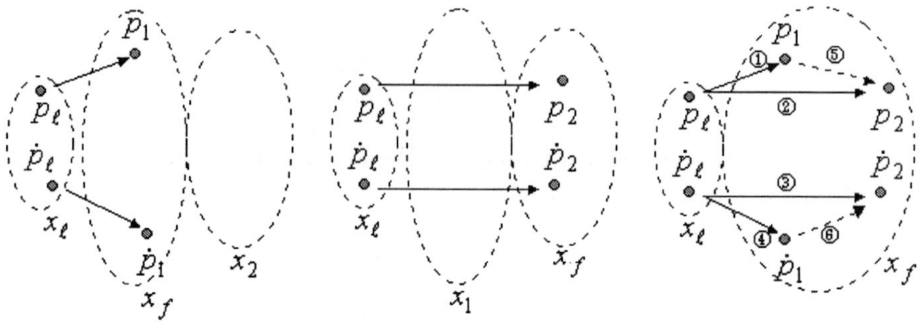

Fig. 2. Left: x_1 with x_ℓ Middle: x_2 with x_ℓ Right: x_1 with x_2 (Integrated Formation)

4. Spatial and Temporal Coordinations

In this section, a procedure to create a formation on leader's phase-plane is presented. First, the follower trajectories are brought to the leader's phase-plane and then each trajectory is shaped with respect to the leader's trajectory. The leader's trajectory is fixed because we assume a setting in which a higher level architecture has already a) identified the leader and b) the leader knows its operation. Therefore, the leader functions independently and a deterministic data to suggest its dynamics is transmitted to the followers. A two-degree of freedom control structure is used for follower control input generation that shapes the string stability based time scale behavior of formations. Spatial coordination attempts to generate control inputs such that the slopes of the leader and follower trajectories are matched at waypoints. If the waypoint is selected at the leader's trajectory, then the follower develops temporal coordination. Figure 3 depicts the principles.

Depending upon the relationship $a < c$ or $a > c$, infer the role of string stability in temporal and spatial coordination modes. In marginal string stability, it is necessary to satisfy additional constraint $a = c$ at all j. In formation regulation, the follower control inputs simply regenerate the leader's trajectory and then shifted. In pursuer evasion problems, leader and follower trajectories intersect with a fine temporal coordination for precision. Here, the slopes need not match. To develop interconnections for spatial and temporal coordinations, the following definitions are used:

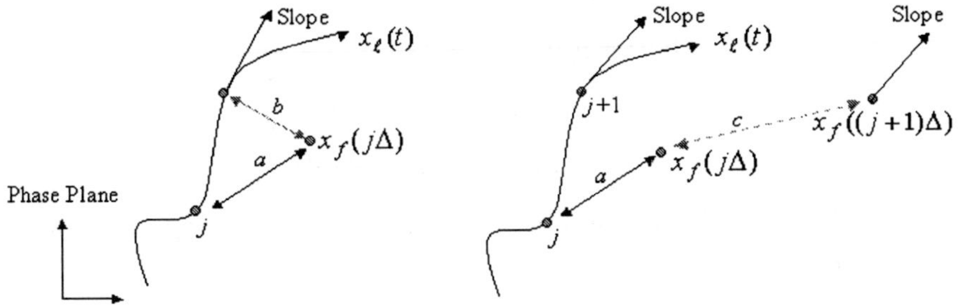

Fig. 3. Follower x_f in temporal (left) and spatial (right) coordination modes

Definition 4: Consider the leader's phase plane and a follower waypoint,

$$\hat{x}_f((j+1)\Delta) = \begin{bmatrix} \hat{p}_{j+1} \\ \dot{\hat{p}}_{j+1} \end{bmatrix}.$$

Denote the leader and follower current positions by,

$$x_\ell(j\Delta) = \begin{bmatrix} p_j \\ \dot{p}_j \end{bmatrix}_\ell \quad \text{and} \quad x_f(j\Delta) = \begin{bmatrix} p_j \\ \dot{p}_j \end{bmatrix}_f.$$

Let the leader's slope at $t = (j+1)\Delta$ be,

$$\beta(j) = \left.\frac{p_{j+1}-p_j}{\dot{p}_{j+1}-\dot{p}_j}\right\}_\ell.$$

If there exists a follower control input such that,

$$\begin{aligned} \beta(j) &= \left.\frac{\hat{p}_{j+1}-p_j}{\dot{\hat{p}}_{j+1}-\dot{p}_j}\right\}_f \\ \dot{\hat{p}}_{j+1}\}_f &= \lim_{\Delta\to 0}\left\{\frac{\hat{p}_{j+2}-\hat{p}_{j+1}}{\Delta}\right\}_f \\ \dot{p}_j\}_f &= \lim_{\Delta\to 0}\left\{\frac{\hat{p}_{j+1}-p_j}{\Delta}\right\}_f, \end{aligned}$$

then the leader-follower *cooperation* is said to be in the *spatial coordination* mode.

Definition 5: Consider the leader's phase plane and a follower waypoint,

$$\hat{x}_f((j+1)\Delta) = \begin{bmatrix} \hat{p}_{j+1} \\ \dot{\hat{p}}_{j+1} \end{bmatrix}.$$

Denote the leader and follower current positions by,

$$x_\ell(j\Delta) = \begin{bmatrix} p_j \\ \dot{p}_j \end{bmatrix}_\ell \quad \text{and} \quad x_f(j\Delta) = \begin{bmatrix} p_j \\ \dot{p}_j \end{bmatrix}_f.$$

If there exists a follower control input such that,

$$
\begin{aligned}
p_j\dot{p}_j\}_\ell + p_j\dot{p}_j\}_f &= 0\\
\hat{\dot{p}}_j\}_f &= \lim_{\Delta\to 0}\left\{\frac{\hat{p}_{j+1}-p_j}{\Delta}\right\}_f\\
\dot{p}_j\}_\ell &= \lim_{\Delta\to 0}\left\{\frac{p_{j+1}-p_j}{\Delta}\right\}_\ell
\end{aligned}
$$

then the leader-follower *cooperation* is said to be in the *temporal coordination* mode.

Note that spatial coordination needs two waypoints, whereas, temporal coordination require one waypoint. The characteristics are similar to the three-point central difference and two-point forward difference schemes. Efforts to synthesize coefficient matrices for formations of a finite set of cooperative aerospace systems is in progress. As and when the waypoints generate follower trajectories, the slope is time varying. Derivation of the coefficient matrices for an interconnection in equation (10) is in progress. Numerical simulation shows that whenever the distributed systems are asymptotically stable, time-invariant coefficient matrices are adequate to shape a formation; however, when uncertainties in initial conditions are present [5], time-invariant coefficient matrices become inadequate. Further, because the formation is made from a sequence of spatial and temporal coordinations, the structure of the coefficient matrices is accordingly reconfigured. Thus in formation control, structural uncertainties in coefficient matrices are unavoidable.

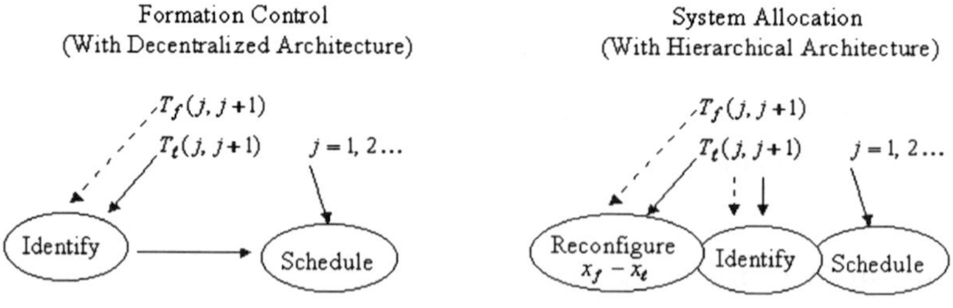

Fig. 4. Interconnections for Formation Control and System Allocation

Figure 4 gives a generic picture on how the information flow and interconnections can affect a) formation control of cooperative systems by a lower level controller and b) system allocation problems by a higher level controller. If formations are parameterized for a fixed leader-follower configuration (i.e., in the absence of a higher level control), note that the problem is equivalent to controlling an aircraft in its flight envelope using conventional gain scheduling algorithm. In such cases, when formation control with structural uncertainties is posed as a convex problem, synthesis of Lyapunov functions with interconnections becomes important. In decentralized platform, it is known that the parameters of string stability do not interact with those parameters at inner loop that preserve stability. Therefore, in all our subsequent discussion, we focus on Lyapunov functions derived with follower outer loop control parameters.

5. System of Lyapunov Systems

In this section, an integrated formation (see Figure 2) for a leader-follower graph $\mathcal{G}(\hat{e}, \hat{n})$ is considered. Recall that in these graphs, the leader information flow to the followers is directed and each follower is expected to develop a cooperation with its leader. Since the information flow from one follower to another follower does not exist, the followers are said to be uncooperative. However, the interconnection in equation (10) suggests that the followers $x_k(t)$, $k \neq j$ can have a formation with respect to $x_j(t)$ through the coefficient matrices T^f_{jk}. Decomposition technique assists to synthesize such interconnections by converting equation (10) into a finite set of two-node edges. These two-node edges offers Lyapunov systems with an information flow $T_\ell \in R^{n\times n}$ and a coefficient matrix $T_f \in R^{n\times n}$, such that an interconnection for the formation of $x_f(t)$ with respect to $x_\ell(t)$ is defined for each $f \in [1, 2, \cdots N]$,

$$c(t) = T_\ell x_\ell(t) + T_f x_f(t). \tag{17}$$

An integrated formation is a formation that we get for graph $\mathcal{G}(\hat{e}, \hat{n})$ with interconnections as in equation (10). A Lyapunov function for the string stability of an interconnection in equation (17) is derived. If the string stability of a particular component (say, position in phase plane) is desired, $c(t)$ becomes a scalar. For brevity, assume $N = 2$ and consider Figure 2. Define the leader and follower dynamics,

$$\begin{aligned} \dot{x}_\ell(t) &= A_\ell x_\ell(t), \quad x_\ell(0) \neq 0 \\ \dot{x}_f(t) &= A_f x_f(t) + B_f K_f x_\ell, \quad x_f(0) \neq 0, \; f \in [1, 2]. \end{aligned} \tag{18}$$

We have assumed m-input systems. Note that the follower control gain $K_f \in R^{m\times n}$ at the outer loop is linear to the leader information $x_\ell(t)$. Various procedures to compute K_f using an aggregate system are presented [4, 7]. To address spatial and temporal coordinations in a formation, two design procedures were adopted:

Design Problem 1: Given nonzero initial conditions $x_\ell(0)$ and $x_f(0)$, find a structurally stable information flow T_ℓ for which a time-invariant coefficient matrix T_f leads to spatial and temporal coordinations as in definitions 2 and 3 through an outer loop control law $u_f(t) = K_f x_\ell(t)$ that minimizes $\langle c(t), c(t)\rangle$. (Here, we attempt to map information flow and formation through interconnections)

Design Problem 2: Given nonzero initial conditions $x_\ell(0)$ and $x_f(0)$, use spatial and temporal coordination definitions to develop an outer loop control law $u_f(t) = K_f x_\ell(t)$. (Here no specific information flow and interconnections are assumed).

Each leader-follower pair leads to a particular type of Lyapunov equation, wherein the follower control parameters are used to shape the formation (similar to the gain matrix used to stabilize a system in Riccati equation). The definition of *Lyapunov System* holds for the fact that the string stability of an interconnection in equation (17) is connected to a Lyapunov equation. Finally, when all Lyapunov systems are integrated for interconnections in equation (10), the follower to follower interconnections are captured. The integrated system is referred as *System of Lyapunov Systems*. The mesh stability of the interconnections in equation (10) is again governed by a Lyapunov equation; however, no follower control parameters are available to shape the integrated formation.

5.1. *Lyapunov Systems*

Consider an interconnection $c(t)$ for a Lyapunov system with leader $x_\ell(t)$ and follower $x_f(t)$. Choose a function $\langle c(t), c(t)\rangle$ for *string* stability and infer that,

$$\begin{aligned}\frac{d}{dt}\langle c(t), c(t)\rangle &= -x'(t)\ Q\ x'(t), \text{ where, } x'(t) = [x'_\ell(t)\ \ x'_f(t)]\\ &= \langle \dot{c}(t), c(t)\rangle + \langle c(t), \dot{c}(t)\rangle. \end{aligned} \tag{19}$$

From equations (17),(18) and (19), we have the Lyapunov equation,

$$\begin{bmatrix} A'_\ell & E' \\ 0 & A'_f \end{bmatrix} \begin{bmatrix} P_\ell & P_c \\ P'_c & P_f \end{bmatrix} + \begin{bmatrix} P_\ell & P_c \\ P'_c & P_f \end{bmatrix} \begin{bmatrix} A_\ell & 0 \\ E & A_f \end{bmatrix} = - \begin{bmatrix} Q_\ell & Q_c \\ Q'_c & Q_f \end{bmatrix} \tag{20}$$

where, $E = B_f K_f$, $P_\ell = T'_\ell T_\ell$, $P_c = T'_\ell T_f$ and $P_f = T'_f T_f$. Rewriting,

$$A'_\ell P_\ell + P_\ell A_\ell = -(E' P'_c + P_c E + Q_\ell) \tag{21}$$

$$A'_\ell P_c + P_c A_f = -(E' P_f + Q_c) \tag{22}$$

$$A'_f P_f + P_f A_f = -Q_f \tag{23}$$

Clearly, the Lyapunov function for string stability at time $t = j\Delta$ is,

$$\begin{aligned} V_j &= x'(t)\ P\ x(t) \\ &= x'(t) \begin{bmatrix} P_\ell & P_c \\ P'_c & P_f \end{bmatrix}_j x(t) \end{aligned} \tag{24}$$

When the coefficient matrices are time-invariant, P and Q are time-invariant for all $j\Delta \in [\tau_j, \tau_{j+1}]$. We will now attempt to redefine string stability using symmetric matrices in equations (20)-(24).

Theorem 6: String Stability An interconnection $c(t)$ with respect to a Lyapunov function $V(x,t) = x'(t)P_j x(t)$ and its derivative $\dot{V}(x,t) = -x'(t)Q_j x(t)$ at a point $t = j\Delta \in [\tau_j, \tau_{j+1}]$ is said to be string stable if and only if $V(x,t) > 0$ and $\dot{V}(x,t) < 0$.

Observe the distinction and similarity of Theorem 2 and Theorem 6. The distinction is that Theorem 6 applies to any kind of systems. The similarity is that both Theorem 2 and Theorem 6 deal with string stability, but the complexity of the problem is visible in Theorem 2. As observed in Theorem 2, the characteristic values of A_ℓ and A_f govern string stability, but not in the sense of stability. In convergence analysis of Kalman filter, this problem has been addressed [page 502, 14] It is also pointed out that in the presence of parametric uncertainties, convergence guarantee is difficult. In our case, the information flow and the coefficient matrices for spatial and temporal coordinations are structurally uncertain. Majority of these issues suggest the nonconvex nature of the problem. Hence, it is difficult to use the characteristic values of P and Q and interpret string stability. The uniqueness of P is however evident from the fact that the distributed systems in decentralized architecture are stable [page 768, 14]. The discrete-time interpretation of the continuous-time Lyapunov equation does offer time-instants τ_j at which the transitions (from string stability to string instability or vice versa) occur. Following the definition of slope in Theorem 6, consider a point $t = j\Delta \in [\tau_j, \tau_{j+1}]$ on the time scale and revisit string stability as follows:

$$\text{If } V((j+1)\Delta) - V(j\Delta) < 0, \text{ then } c(j\Delta) \text{ is string stable} \tag{25}$$

$$\text{If } V((j+1)\Delta) - V(j\Delta) > 0, \text{ then } c(j\Delta) \text{ is string unstable} \tag{26}$$

$$\text{If } V((j+1)\Delta) - V(j\Delta) = 0, \text{ then } c(j\Delta) \text{ is marginally string stable} \tag{27}$$

Two kinds of design parameters that we wish to use and modify the transitions τ_j are a) the follower control gain $K_f(j\Delta)$ and, b) the coefficient matrices $T_f(j\Delta)$ with respect to an information flow T_ℓ. When uncertainty in initial conditions is not present, selection of T_f for a given T_ℓ require synthesis procedure. Conversely, if we choose T_f based on Definitions 4 and 5, we are required to synthesize T_ℓ (all admissible information flow). The constraints imposed while synthesizing T_ℓ and T_f are the time-instants τ_j at which transitions occur for regulation, spatial, and temporal coordination modes. In this process, we may choose $Q_c = P_f E$ and balance Lyapunov equations (20)-(24). The nonconvex nature of the problem may be explained using a rover driven from $x_\ell(0) = [10, 2]'$ to origin by a certain control law. Suppose its follower is at $x_f(0) = [10, -2]'$. Let the discrete-time model be with $\Delta = 1$ unit. The leader and follower inner loops are $[0.89 \pm 0.167i]$ and $[0.854, 0.726]$, respectively. Conceptually, the systems are moving in opposite directions. If we wish to have a formation such that the follower's trajectory is a reflection of leader's, one may use Lyapunov equation to show that K_f does not exist.

5.2. *System of Lyapunov Systems*

In this section, we consider the finite set of Lyapunov systems established in the previous section. Each follower's control is such that a specific formation with respect to its leader is synthesized. When all the followers are stacked in a vector $x_f(t) = [x_1'(t), x_2'(t), \cdots x_N'(t)]'$, an user defined follower to follower interconnections in equation (10)were augmented for analysis. In this case, the coefficient matrices T_ℓ and T_f for the case $N = 2$ will have following structure:

$$T_\ell = \begin{bmatrix} T_1^\ell \\ T_2^\ell \end{bmatrix} \in R^{2n\times n} \text{ and } T_f = \begin{bmatrix} T_{11}^f & 0 \\ T_{12}^f & T_{22}^f \end{bmatrix} \in R^{2n\times 2n}. \tag{28}$$

The decomposition technique is applied to understand the follower-follower relative dynamics. The string stability tool is applied to infer the dynamics of one follower $x_k(t)$ $k \neq j$ with respect to the dynamics of another follower $x_j(t)$ through the coefficient matrices T_{jk}^f. When an interconnection is modified, the Lyapunov systems are modified and the integrated formation of System of Lyapunov system is altered. When a new Lyapunov system is reconfigured, arc uncertainties in the directed graph are introduced. Another scenario would be to preserve the edges of the leader-follower graph and introduce follower to follower information flow. Here, computational complexities are introduced and intelligent data processing is required [7]. Majority of these problems to achieve air superiority are complex in nature. We have presented to a procedure to systematically understand the complexity of a problem by exploring its simplicities.

Edges	Relative	Complex	String Stability	Mesh Stability	Design	Analysis
①	✓		✓		✓	
②	✓		✓		✓	
③	✓		✓		✓	
④	✓		✓		✓	
⑤	✓	✖	✓		?	✓
⑥	✓	✖	✓		?	✓
{①,②,③,④} ∪ {⑤,⑥}		✖		?	?	✖

Table 1: Analysis with Follower to Follower Interconnections

Fig. 5

For the graph in Figure 2, the complexities begin to pop when an analysis problem is seen as a design problem. The table in Figure 5 summarizes these issues.

6. Illustration

The rover models have been extensively used by the authors to understand cooperative systems resulting from a leader-follower graph (Figure 2 is adopted). To understand mesh stability, distributed systems $x_\ell(t),\ x_f(t) \in R^2,\ f \in [1,2]$ are considered. In design setting, the Lyapunov systems $x_\ell(t) - x_1(t)$ and $x_\ell(t) - x_2(t)$ are required to meet either spatial or temporal coordination. In analysis setting with a System of Lyapunov systems, the relative dynamics of the follower $x_1(t) - x_2(t)$ are analyzed.

Figure 6 is a Lyapunov system $x_\ell(t) - x_1(t)$ designed for spatial coordination at all points $j\Delta$. The interconnection is,

$$T_{\ell,f} = \begin{bmatrix} 0 & 0 \\ 0 & 1 \end{bmatrix}.$$

The discrete-time interpretation of the Lyapunov function on right shows the time scale behavior of the rovers in leader-follower framework. When uncertainties in initial conditions are present, the interconnection with time-invariant coefficients becomes inadequate. The structure of T_f in this case is expected to take the role of gain scheduling algorithm. Design problems 1 and 2 are compared in Figure 7. Solutions suggest spatial coordination by Definition 4 can be realized through an interconnection with T_ℓ and T_f.

Similar discussion may be drawn for the temporal coordination problem in Figure 7. The interconnection choice for this case is,

$$T_\ell = \begin{bmatrix} 1 & 0 \\ 0 & 1 \end{bmatrix},\quad T_f = \begin{bmatrix} -1 & 0 \\ & 0\ 1 \end{bmatrix}.$$

Next the System of Lyapunov Systems is presented. Here another Lyapunov system $x_\ell(t) - x_2(t)$ for spatial coordination at all points $j\Delta$ is introduced. A formation of $x_\ell(t) - x_2(t)$ and $x_\ell(t) - x_2(t)$ in leader's phase-plane is shown in Figure 8. In system of systems framework, follower-follower interconnections resulting from $x_1(t) - x_2(t)$ are analyzed. Note that in Figure 8, the interconnection,

$$T_{\ell,f} = \begin{bmatrix} 0 & 0 \\ 0 & 1 \end{bmatrix}$$

is insensitive to $x_2(0) \neq x_1(0)$. It generates control inputs at $x_1(t)$ and $x_2(t)$ and preserves the slopes of the leader. That is, though the inner loops of $x_1(t)$ and $x_2(t)$ are different, formation from $T_{\ell,f}$

becomes sensitive to initial condition $[10, -2]'$ but not to $[-10, -2]'$. Thus sensitivity analysis of interconnections becomes an issue [3]. Information flow T_ℓ, interconnection T_f, follower control K_f, and initial conditions are the parameters contributing to the sensitivity analysis of formations.

The discrete-time interpretation of the Lyapunov function for speed suggests that $\dot{p}_1(t)$ and $\dot{p}_2(t)$ are in perfect coordination with $\dot{p}_\ell(t)$. This

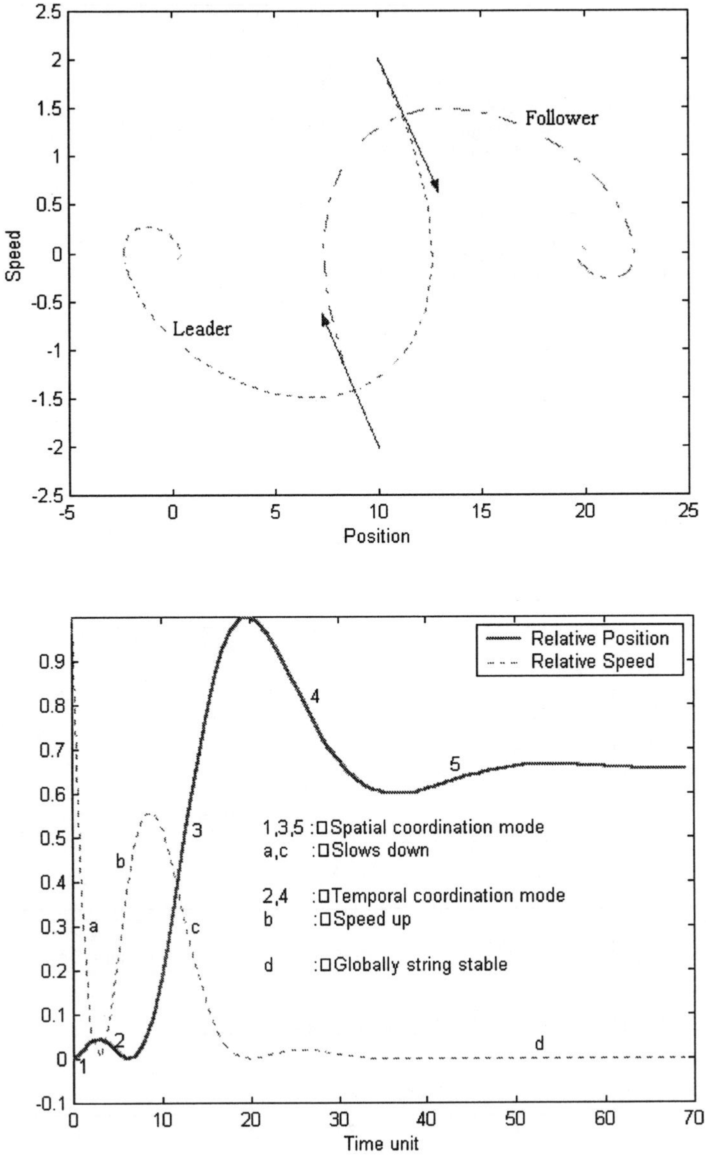

Fig. 6. Top: Spatial Coordination Bottom: Lyapunov Function

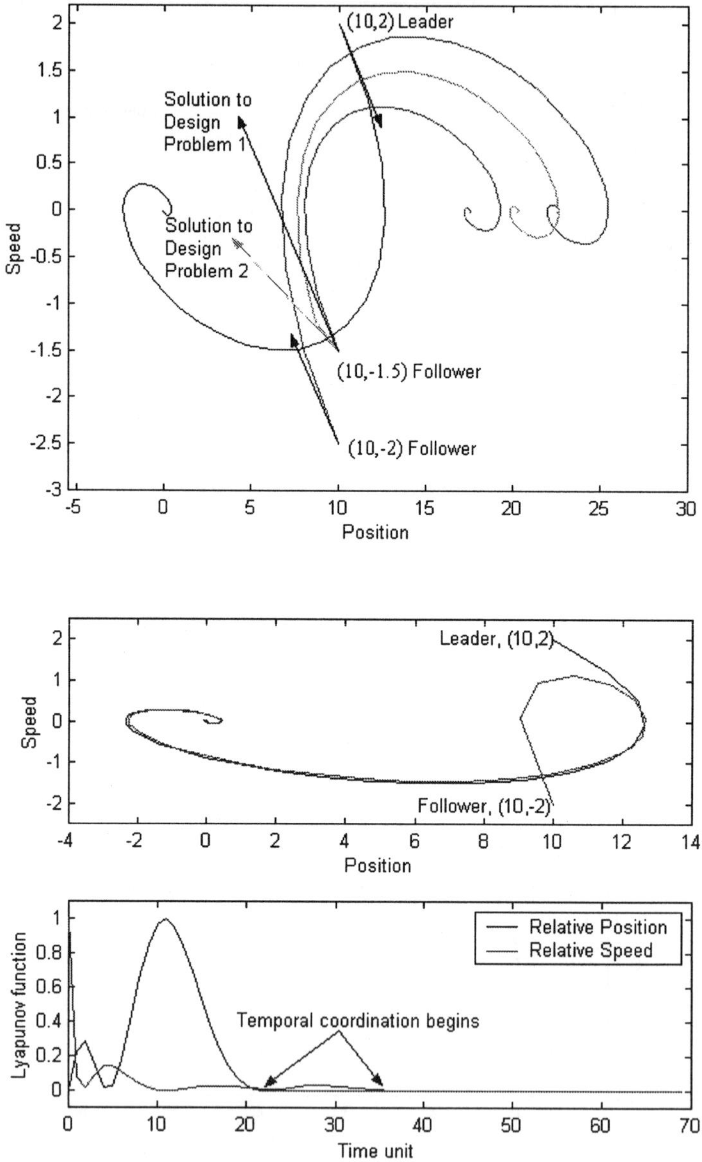

Fig. 7. Top: Uncertainties in initial condition Bottom: Temporal coordination details

is done by using follower control gains K_f shown in Figure 9. Since the follower control is used to preserve coordination with the leader, the coordination of $\dot{p}_1(t)$ with respect to $\dot{p}_2(t)$ is not possible. Observe that the formation of $x_1(t)$ with respect to $x_2(t)$ on leader's phase plane is in regulation mode. String stability suggests that it is difficult for $x_1(t)$ and $x_2(t)$ to have this formation with $x_\ell(t)$ and simultaneously develop an user defined cooperation between $\dot{p}_1(t)$ and $\dot{p}_2(t)$. These problems in control of distributed systems are complex.

7. Conclusions and Future Work

This chapter is concerned with the control distributed systems, where communication (or information flow) becomes a critical data for formation stability. It is a problem associated with the relative dynamics of the distributed systems. In this chapter, a specific communication structure in a directed graph is adopted and a procedure to control distributed homogeneous systems in a decentralized computational platform is successfully demonstrated. The nonconvex properties of formations are summarized in a Lyapunov equation. In these problems, since relative dynamics is difficult to interpret using the positive (-semi) definiteness of symmetric matrices observed in the Lyapunov equation, the Lyapunov functions were used. Suppose the number of systems increase and if several leaders were allocated to perform various mission operations, integer programming with undirected graph becomes an important problem.

References

[1] Ashokkumar, Chimpalthradi R., Curtis, Will J., Murphey, Robert A., and Sierakowski, Robert L., Invariant Interconnected Systems, Cooperative Control Conference, Gainesville, FL 4-6 Dec 2002.

[2] Ashokkumar, C.R., Linear Quadratic Optimality of Infinite Gain Margin Controllers, Journal of Guidance, Control, and Dynamics, Vol. 22, No. 5, pages 720-722, 1999.

[3] Ashokkumar, Chimpalthradi R., and Jeffcoat, David E., Sensitivity Analysis of Distributed Systems in Formation, paper in preparation.

[4] Ashokkumar, Chimpalthradi R., Curtis, Will J., and Murphey, Robert A., Formation Control of Cooperative Systems, Journal of Guidance, Control, and Dynamics, in preparation.

[5] Ashokkumar, Chimpalthradi R., and Jeffcoat, David., Cooperative Systems Under Communication Delay, AIAA Guidance, Control, and Navigation Conference, Austin, TX 11-13 Aug 2003.

[6] Ashokkumar, C.R., and Rao, S.S., Structural Control Using Inverse Optimal

Control Theory, AIAA Journal, Vol. 41, No. 12, pages 2478-2485, December 2003

[7] Ashokkumar, Chimpalthradi R., Curtis, Will J., Murphey, Robert A., Formation Control of Distributed Systems: Computation versus Cooperation, 2nd AIAA Unmanned Unlimited Systems, Technologies, and Operations-Aerospace, Land, and Sea Conference and Workshop & Exhibit, San Diego, California, 15-18 Sep 2003.

[8] Ashokkumar, Chimpalthradi R., and Jeffcoat, David., Dynamic Feasibility of Cooperative Systems, Journal of Guidance, Control, and Dynamics, in preparation.

[9] Barnett, S., *Matrices in Control Theory with Applications to Linear Programming*, Van Nostrand Reinhold Company, London, 1970.

[10] Boyd, S., El Ghaoui, L., Feron, E., and Balakrishnan, V., *Linear Matrix Inequalities and Control Theory*, Siam Studies in Applied Mathematics, Vol. 15, Philadelphia, 1994.

[11] Desai, J.P., Vijay Kumar and Ostrowski, J.P., Control of Changes in Formation for a Team of Mobile Robots, Proceedings of the 1999 IEEE International Conference on Robotics and Automation, Detroit, Michigan, pages 556-1561.

[12] Alexander Fax, J., and Murray, Richard M., Graph Laplacians and Stabilization of Vehicle Formations, The 15th IFAC World Congress, June 2002.

[13] Fowler, Jeffrey M., and D'Andrea, Raffaello, A Formation Flight Experiment, IEEE Control Systems Magazine, pages 35-43, October 2003.

[14] Kailath, T., Sayed, A.H., and Hassibi, B., *Linear Estimation*, Prentice Hall, Upper Saddle River, New Jersey, 07548, 2000.

[15] Kapila, V., Sparks, A.G., Buffington, J.M., and Yan, Q., Spacecraft Formation Flying: Dynamics and Control, Journal of Guidance, Control, and Dynamics, Vol. 23, No. 3, pages 561-564, 2000.

[16] Kurdila, Andrew J., Kamat, Manohar P., Concurrent Multiprocessing for Calculating Nullspace and Range Space Bases for Multibody Simulation, AIAA Journal, Vol. 28, No. 7, pages 1224-1232, 1999.

[17] Lewis, Frank L., and Syrmos, Vassilis L., *Optimal Control*, John Wiley & Sons, Inc., New York, 1995.

[18] Lian, Feng-Li and Richard M. Murray, Real-time Trajectory Generation for the Cooperative Path Planning of Multi-Vehicle Systems, Proc. of the 41st IEEE Conf. on Decision and Control, Las Vegas, Nevada, pages 3766-3769, Dec 2002.

[19] Lyke, James C., Donohoe, Greg W., Karna, Shashi P., Cellular Automata-Based Reconfigurable Systems as Transitional Approach to Gigascale Electronic Architectures, Journal of Spacecraft and Rockets, Vol. 39, No. 4, pages 489-494, 2002.

[20] Moscovitz, Y., and Nicholas, D., Basic Concepts and Methods for Keeping Autonomous Ground Vehicle Formations, Proceedings of the IEEE ISIC/CIRA/ISAS Joint Conference, Gaithersburg, MD, September 14-17, 1998.

[21] Pant, A., Seiler, P., Koo, John T., and Hedrick, K., Mesh Stability of Un-

manned Aerial Vehicle Clusters, Proceedings of the American Control Conference, Arlington, VA, pages 62-68, 25-27 June 2001.

[22] Patcher, M., and Chandler, P.R., Challenges of Autonomous Control, IEEE Control Systems Magazine, pages 93-97, August 1998.

[23] Pledgie, Stephen T., Agrawal, Sunil K., and Robert Murphey, An Integrated Approach to Structured Dynamic Networking, American Control Conference, 2001.

[24] Saber,Reza O., and Murray, Richard M., Distributed Structured Stabilization and Tracking for Formation of Dynamic Multi-Agents, Proceedings of the 41st IEEE Conf. On Decision and Control, Las Vegas, Nevada, pages 209-215, Dec 2002.

[25] Saber,Reza O., and Murray, Richard M., Graph Rigidity and Formation Stabilization of Multi-Vehicle Systems, Proc. of the 41st IEEE Conf. On Decision and Control, Las Vegas, Nevada, pages 2965-2971, Dec 2002.

[26] Pete Seiler, Aniruddha Pant, Hedrick, J.K., Preliminary Investigation of Mesh Stability for Linear Systems, Proceedings of the ASME Dynamic Systems and Control Division, DSC-Vol. 67, pages 359-364, 1999.

[27] Strang, Gilbert, *Linear Algebra and Its Applications*, Saunders College Publishing, Harcourt Brace Jovanovich College Publishers, Fort Worth, 1988.

[28] Swaroop, D., and Hedrick, J.K., String Stability of Interconnected Systems, IEEE Transactions on Automatic Control, Vol. 41, No. 3, pages 349-357, 1996.

[29] Tabuada, P., Pappas, George J., and Lima, P., Feasible Formations of Multi Agent Systems, Proc. of the American Control Conference, Arlington, VA, pages 56-61, 25 June 2001.

[30] Walsh, Gregory C., and Ye Hong, Scheduling of Networked Control Systems, IEEE Control Systems Magazine, pages 57-65, February 2001.

[31] Zang, Wei, Branicky, Michael S., and Phillips, Stephen M., Stability of Networked Control Systems, IEEE Control Systems Magazine, pages 84-99, February 2001.

[32] Linda Wills, Suresh Kannan, San Sander, Murat Guler, Bonnie Heck, Prasad, J.V.R., Daniel Schrage and George Vachtsevanos, An Open Platform for Reconfigurable Control, IEEE Control Systems Magazine, pages 49-63, June 2001.

[33] Yedavalli, Rama K., A Necessary and Sufficient 'Virtual Interior Edge' Solution for Checking Robust Stability of Interval Matrices, Proceedings of the American Control Conference, Philadelphia, Pennsylvania, pages 2309-2313, June 1998.

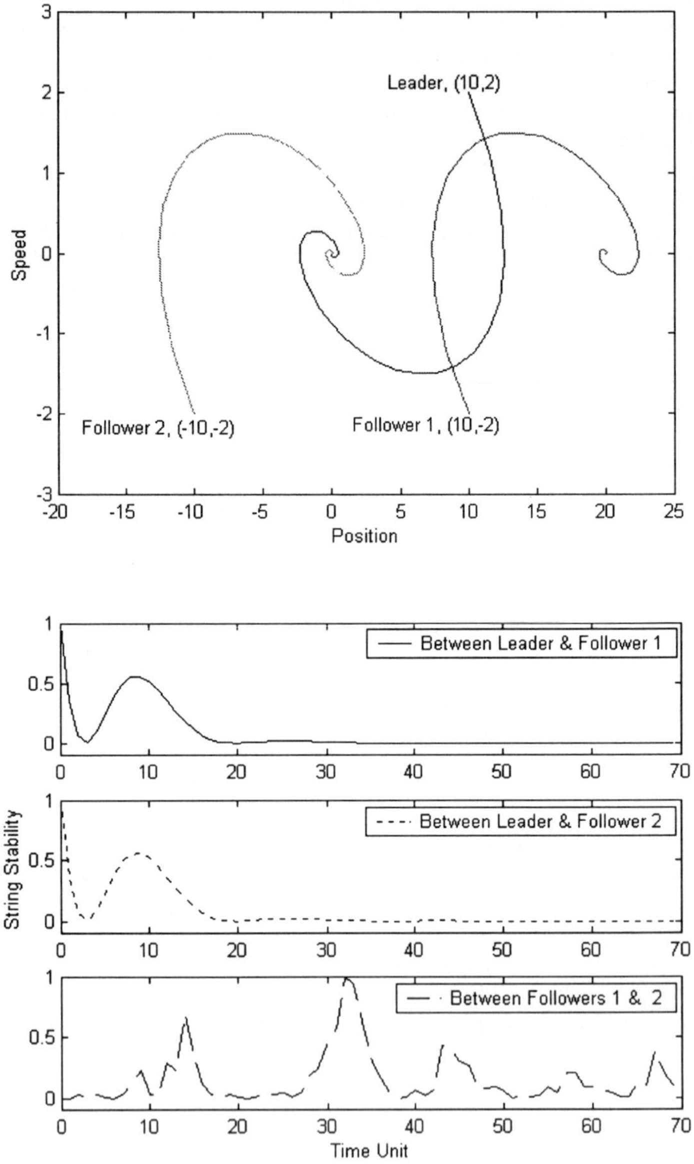

Fig. 8. Top: An integrated formation Bottom: Cooperation in a system of Lyapunov systems

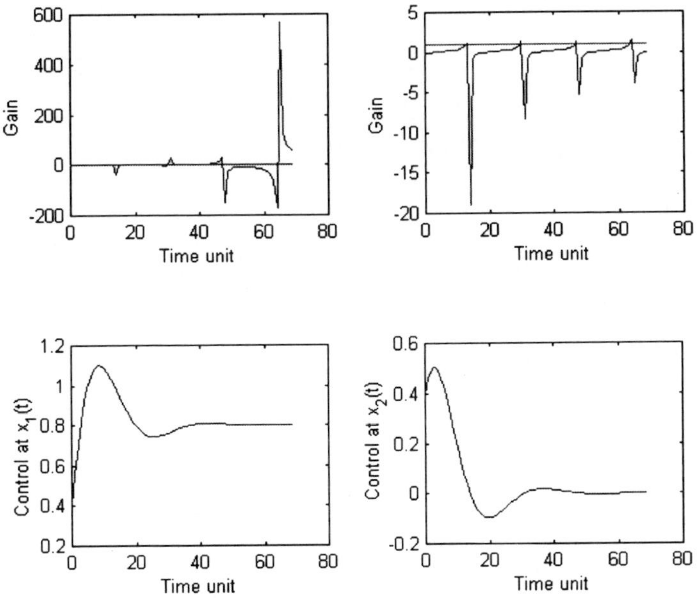

Fig. 9. Follower PD structure based outer loop control

CHAPTER 2

DESIGNING THE CONTROL OF A UAV FLEET WITH MODEL CHECKING

Christopher A. Bohn [a]
Department of Computer Science and Engineering
The Ohio State University
bohn.19@osu.edu

Paolo A.G. Sivilotti
Department of Computer Science and Engineering
The Ohio State University
sivilotti.1@osu.edu

Bruce W. Weide
Department of Computer Science and Engineering
The Ohio State University
weide.1@osu.edu

We model the problem of unmanned aerial vehicles searching for moving targets as a pursuer-evader game in which the pursuers have a speed advantage over the evaders but are incapable of determining an evader's location unless a pursuer occupies the same location as that evader. By treating the players as nondeterministic finite automata, we can model the game and use it as the input for a model checker. By specifying that there is no way to guarantee the pursuer can locate the evader, the model checker will either confirm that this is the case, or it will provide a counterexample showing one search pattern for the pursuer that will guarantee the evader must eventually be caught. As an ongoing effort to reduce the time to find pursuer-winning strategies, we also present heuristics to limit the state space of the game models.

[a]The views expressed in this article are those of the authors and do not necessarily reflect the official policy of the Air Force, the Department of Defense or the US Government.

Keywords:
Model checking, Unmanned aerial vehicle, Pursuer-evader game

1. Introduction

The challenge of an airborne system locating an object on the ground is a common problem for many applications, such as tracking, search and rescue, and destroying enemy targets during hostilities. If the target is not facilitating the search, or is even attempting to foil it by moving to avoid detection, the difficulty of the search effort is greater than when the target aids the search. In recent years, interest in using Unmanned Aerial Vehicles (UAVs) for "the dull, the dirty, and the dangerous" combat and non-combat scenarios has grown, as evidenced by the US Department of Defense's (DoD) investment in UAV development, procurement, and operations: $3 billion in the 1990s, $1 billion from 2000–2002, and a projected $10 billion by 2010 [6]. Their application to the search problem is limited by the size of their sensor footprint – indeed, the DoD has expressed concern over "the soda-straw field of view used by current UAVs that negatively affects their ability to provide broad SA [situational awareness] for themselves, much less for others in a formation" [6].

Where other researchers are addressing the problems of locating stationary targets [1, 7, 9] or of locating moving targets with some probability of success [5], our research is intended to address a technical hurdle for locating moving targets with certainty.

We have abstracted this problem of controlling a fleet of UAVs to meet some search objective into a pursuer-evader game played on a finite grid. The pursuers can move faster than the evaders, but the pursuers cannot ascertain the evaders' locations except by the collocation of a pursuer and evader. Further, not only can the evaders determine the pursuers' past and current locations, they have an oracle providing them with the pursuers' future moves. The pursuers' objective is to locate all evaders eventually, while the evaders' objective is to prevent indefinitely collocation with any pursuer [3, 2]. Part of our contribution is to represent the game as a concurrent system of finite automata and to use symbolic model checking [4] to determine whether all evaders can eventually be located within the model; if so, we can use the model checker's output (a sequence of states that serve as a witness to the satisfaction/dissatisfaction of the model's specification) to generate a search pattern for the UAVs to follow. Using this approach, we are able to address the problem of whether and how the UAVs can be *guaranteed* to locate *all* of a set of *moving* targets.

We demonstrate that model checking can be used to suggest strategies by which a single UAV can locate targets. Moreover, model checking can be used to coordinate the actions of a fleet of UAVs. We also exhibit heuristics that can reduce the size of the models enough to obtain these strategies fast enough for on-the-fly planning.

In this chapter, we present the model of our pursuer-evader game (Section 2), how that model can be used to generate pursuer-winning search strategies when they exist (Section 3), our efforts to reduce the time required to generate search strategies (Section 4), and the directions we intend to take our research (Section 5).

2. Modeling the Game

We begin by describing the model of the pursuer-evader game. After examining the finite automata we use to model the game, we present sufficient conditions to guarantee that the pursuers will locate all evaders.

2.1. *The Model*

In its simplest form, the game is played on a rectilinear grid with one pursuer and one evader taking turns to move. There are no obstacles to either the pursuer or evader, so each can occupy any location and can transition from any location to any adjacent location. Consider Figure 1. In Figure 1(a), the pursuer moves three spaces north and one west on a 6×6 board; this is followed by the evader moving south one space in Figure 1(b). There are four basic variations on the game's simplest form, based on the definition of "adjacent location". In all four variants, the players can move in the four cardinal directions (north, south, east, west); the variations are whether the pursuer, the evader, both, or neither can move diagonally (northwest, northeast, southwest, southeast). These simplifications facilitate an analysis of the required pursuer speed.

An obvious (but, it turns out, inappropriate; see Section 3.2) model for the game is to treat the pursuers and evaders as nondeterministic finite automata (NFAs) with each player's state defined only by its current grid cell. Instead, we define a model that does not explicitly include evaders. Instead, each grid cell has a single boolean state variable CLEARED that indicates whether it is impossible for an undetected evader to occupy that cell. CLEARED is `true` if and only if no undetected evader can occupy that

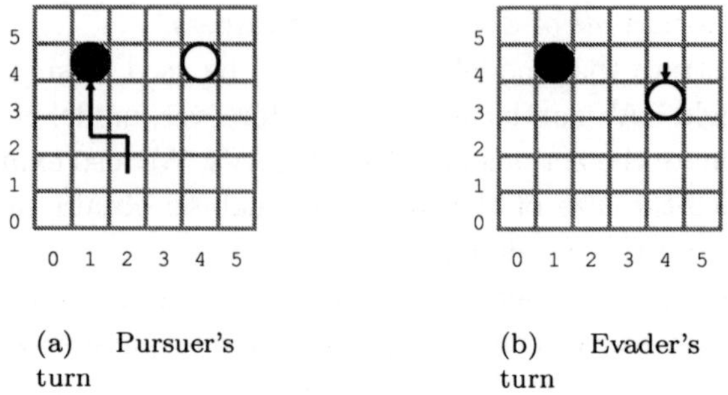

Fig. 1. Examples of movements by the pursuer and the evader. Solid circle is the pursuer; hollow circle is the evader

cell, and CLEARED is `false` if it is *possible* for an undetected evader to occupy that cell. Trivially, cells occupied by pursuers are CLEARED – either there's no evader occupying that cell, or it has been detected. A cell that is not CLEARED becomes CLEARED when and only when a pursuer occupies it. A CLEARED cell ceases to be CLEARED when and only when it is adjacent to an unCLEARED cell during the evaders' turn to move; if all its neighboring cells are CLEARED then it remains CLEARED.

Compare Figure 2 with Figure 1. In this hypothetical scenario, the pursuer has CLEARED a region of the southwest corner of the grid, as shown by the shaded portion of Figure 2(a) and can conclude the evader must be outside that region. As in Figure 1(a), the pursuer moves four spaces north and west in Figure 2(b), increasing the CLEARED region by three cells. Since the pursuer does not know where the evader is located, the CLEARED region must shrink in accordance with the union of all possible moves by the evader. A move by the evader south from the northeasternmost corner would not cause the evader to enter a previously-CLEARED cell, but Figure 2(c) shows there are six ways the evader *could* move from an unCLEARED cell into a CLEARED cell, and the five CLEARED cells that could now be occupied by the evader may no longer be considered CLEARED.

By modeling the evaders' possible locations instead of their actual locations, the model does not need to be modified when additional evaders are incorporated. The same model is used whether there are one, ten, or (importantly in many practical situations) an unknown number of evaders. Henceforth, we talk of "the evaders" without saying how many there are.

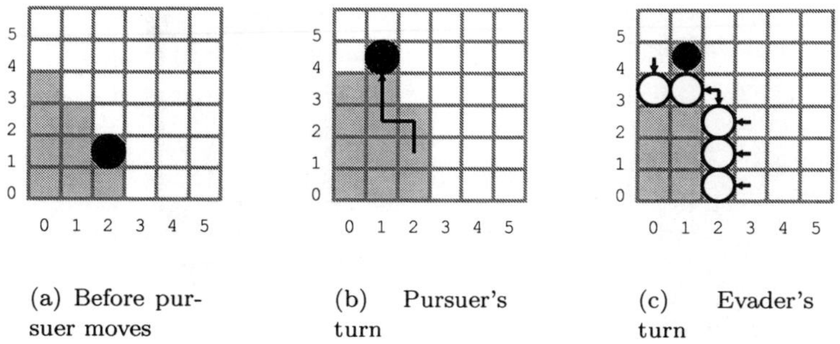

Fig. 2. Examples of changes in the possible locations for the evader. Evader is known to be in unshaded region

2.2. *Sufficient Pursuer Qualities for Simple Game Variants*

We wish to characterize the necessary and sufficient pursuer qualities for pursuer victories in the myriad variations of the game. While we have proven sufficient conditions for a pursuer win in general, we have yet to establish the necessary conditions except for specific game instances.

2.2.1. *One Pursuer*

We have established the sufficient pursuer qualities to guarantee a pursuer win for the simpler forms of the game. When the game is played with one pursuer and with evaders that move 1 space/turn and do not move diagonally on a rectilinear grid with the shorter dimension being n cells, we have shown the existence of a search pattern that will guarantee the pursuer's victory if the pursuer can move n spaces per turn. If the evaders can move diagonally, then such a pursuer-winning strategy exists if the pursuer can move $n+1$ spaces per turn [2].

The formal proofs can be found in [2] however we shall provide here an informal operational argument. Consider the case in which the evaders do not move diagonally and the pursuer can move n spaces/turn. As shown in Figure 3, the pursuer can cause the number of CLEARED cells in a column to increase, up to the point at which all cells in that column are CLEARED. The pursuer can then shift into the initial position for the next column (while preserving the CLEARED cells in the previous column). By repeatedly applying this process, eventually every cell in each column will be CLEARED.

We can demonstrate that a pursuer speed of $n-1$ spaces/turn is insufficient in general. Consider a 2×2 grid, and for simplicity's sake assume the

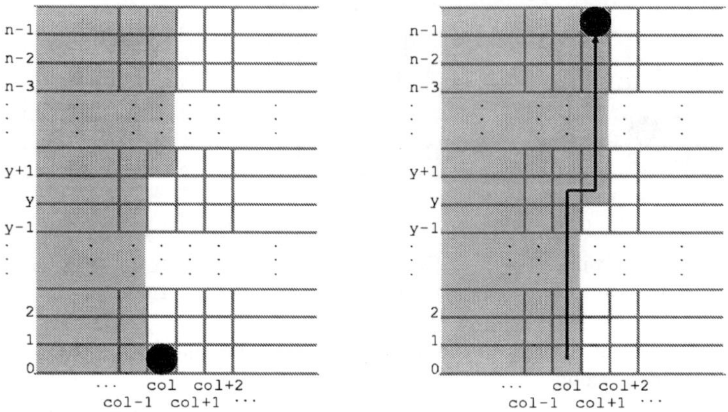

(a) Initial conditions: cells $y+1..n-1$ are CLEARED in this column

(b) Pursuer moves n spaces

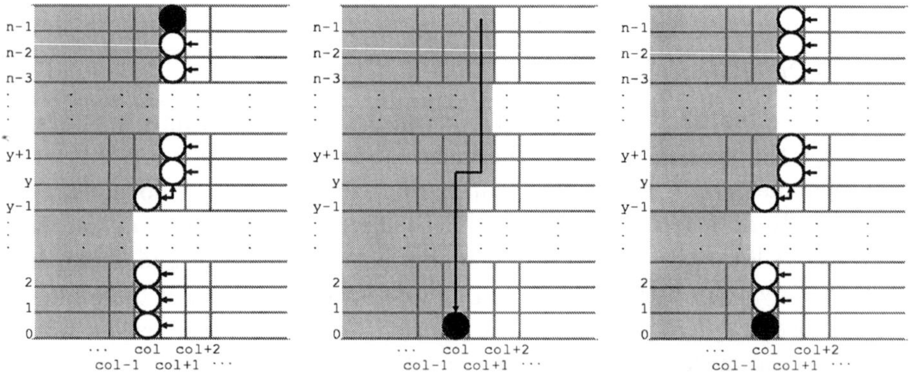

(c) Evaders move, causing cells on the frontier to be unCLEARED

(d) Pursuer moves n spaces

(e) Evaders move, causing cells on the frontier to be unCLEARED; cells $y..n-1$ are CLEARED in this column

Fig. 3. Execution of the "clear-cell" algorithm when the evaders cannot move diagonally

pursuer does not move diagonally either. The pursuer is not guaranteed to find even a single evader if the pursuer's speed is $n-1=1$ spaces/turn.

The evaders can avoid detection indefinitely as follows: after each move by the pursuer, move to (or stay in) the cell diagonally opposite the pursuer. Because the pursuer cannot move 2 spaces/turn, it cannot move to a diagonally opposite corner in a single move; because the evaders can always view the pursuer, they can always move to the diagonally opposite corner.

Further, we believe (but have not proved) that for all values of $n \geq 2$, $n-1$ spaces/turn is an insufficient speed to assure the pursuer's victory. This is based on the observation that the evaders have less freedom of movement in the boundary cells (and even less still in the corner cells), and so evaders have increasing freedom of movement as the ratio of non-border cells to border cells increases. And so we believe that if the pursuer is not certain to locate the evaders at $n-1$ spaces/turn when all cells are corner cells, then the pursuer will not be able to locate the evaders at $n-1$ spaces/turn when there are non-border cells.

2.2.2. *Multiple Pursuers*

We have also turned our attention to characterizing the necessary and sufficient pursuer qualities for a fleet of pursuers on a grid with shorter dimension n. Clearly, if at least one pursuer moves n spaces/turn when the evaders do not move diagonally, the pursuers can win – but we can do better. Through model checking, we have shown that there are configurations in which all but one pursuer can remain stationary, and the single moving pursuer can detect all evaders even when its speed is less than n spaces/turn. More generally, if all pursuers move at least 1 space/turn (the same speed as the evaders), then a pursuer-winning strategy exists for k pursuers if:

$$\sum_{p \text{ is a pursuer}} p.speed \geq n - k + 1 \tag{1}$$

It may not be practical – or even possible – to characterize analytically the necessary pursuer qualities for games in which there are arbitrary restrictions on players' movements. Instead, it may be more practical to use the model checker, as described in Section 3, to determine whether the proposed pursuer qualities are sufficient. As an example of the difficulty in characterizing the necessary and sufficient conditions of a game variant in general, our analysis of even the "simple" game variants was aided by model checking: informed by tool-generated search strategies for specific game instances of a game variant, we reanalyzed the model of that game variant and showed the sufficiency of weaker pursuer characteristics than were indicated by our previous analysis. In short, we learned something from the model checker.

3. Generating Search Strategies

Now that we have covered the model of the pursuer-evader game, we shall explain how we obtain pursuer-winning search strategies by using model checking. We first offer an overview of symbolic model checking. We follow this with the manner in which we use a symbolic model checker to obtain a sequence of computation states that correspond to a series of moves the pursuers can make to detect every evader.

3.1. *Symbolic Model Checking*

Model checking is a technique for verifying that finite state concurrent systems satisfy certain properties expressed in a temporal logic. Typical examples of its use are in verifying properties of complex digital circuit designs and communication protocols [4].

Symbolic model checking has three features that are particularly desirable. First is that for certain logics, such as Computational Tree Logic (CTL), a model can be checked in time linear in the number of the model's states; more precisely, checking a specification f expressed as a predicate in CTL requires $\mathcal{O}\left((|S| + |R|) \cdot |f|\right)$ time, where S is the set of states and R is the set of transitions between states in the model.

The second feature of symbolic model checking is that the model's states are not represented explicitly; rather, the model checker uses binary decision diagrams. Because of this, symbolic model checkers use less memory than explicit-state model checkers for a given model (or, for the same memory footprint, symbolic model checkers can check larger models). When first introduced, symbolic model checking could verify models with up to $10^{20}(\sim 2^{66})$ states; subsequent refinements have permitted the checking of models with up to $10^{120}(\sim 2^{400})$ states [4].

The third feature is that if the specification being checked is not satisfied by the model, the model checker can provide a specific counterexample showing why the property does not hold. We make use of this last capability by modeling the pursuer-evader game as a nondeterministic finite automaton and instructing the model checker to prove that no matter how the pursuer moves, some evader can successfully avoid the pursuer. If the model checker indicates this property is true, then for the given game conditions there is no guarantee the pursuer and an evader will be collocated.

On the other hand, if the model checker indicates the property is false, then as a counterexample, it will indicate the model's state changes that, when properly interpreted, correspond to a sequence of moves the pursuer can make to guarantee it will locate all the evaders.

3.2. *Model Checking the Game*

Had we modeled all players as NFAs with state defined only by the occupied grid cell, then we would check whether the evaders can always avoid detection. The model checker would promptly offer the counterexample of the pursuer moving immediately to the evaders' locations; this is not a useful result.

Instead, we model each grid cell as CLEARED or unCLEARED, and the property to check against this model can be described using CTL:

$$\mathsf{AG} \bigvee_{\substack{0 \leq r < n \\ 0 \leq c < m}} \neg\text{CLEARED}(r, c) \tag{2}$$

That is, along every computation path ("A") beginning at the initial state, every state ("G") includes at least one unCLEARED cell. If the specification is not satisfied, then the counterexample can be used to generate a search pattern that will cause every cell to be CLEARED. When all cells are CLEARED, there can be no undetected evaders.

Example models coded in the model description language for the SMV[b] model checker can be found in [3].

As reported in Section 2.2, we have determined mathematically the sufficient pursuer speed to guarantee a pursuer victory for the simple forms of the game on an arbitrarily-sized grid. As a sanity test, we note that the results of model checking specific instances of these simple forms match the analytical results: when the pursuer speed is set to what we believe to be the minimally-sufficient speed, the pursuer has a winning search strategy; when the pursuer speed is less, the pursuer does not have a winning search strategy. Interestingly, the winning search strategies suggested by the model checker do not correspond to those used in the proofs of our analysis. So, there are winning strategies other than that of Figure 3.

Clearly, a single pursuer travelling over a rectilinear grid with no obstacles is not a faithful model of the real world, though it is useful for investigating some of the issues in using a model checker for our purpose.

---

[b]http://www.cs.cmu.edu/~modelcheck/smv.html

For this reason, we have demonstrated that we can also use the model checker with a hexagonal grid, with multiple pursuers, when the pursuer's sensor footprint is not simply the cell it currently occupies, or when there is terrain (modeled as cells the evaders cannot enter and cell boundaries the evaders cannot cross).

For example, consider the terrain in Figure 4, and assume the evaders cannot occupy the central lake, cannot traverse the sheer western wall of the crevice, and can traverse the oblique eastern wall of the crevice in only one direction. If we model this terrain on a 4×6 grid, then by Equation 1, we know the pursuers can detect all evaders if the sum of the k pursuers' speeds is at least $5 - k$ spaces/turn. Because of the restrictions on the evaders' movements, however, it is possible for a single pursuer to locate all evaders when the pursuer's speed is only 3 spaces/turn.

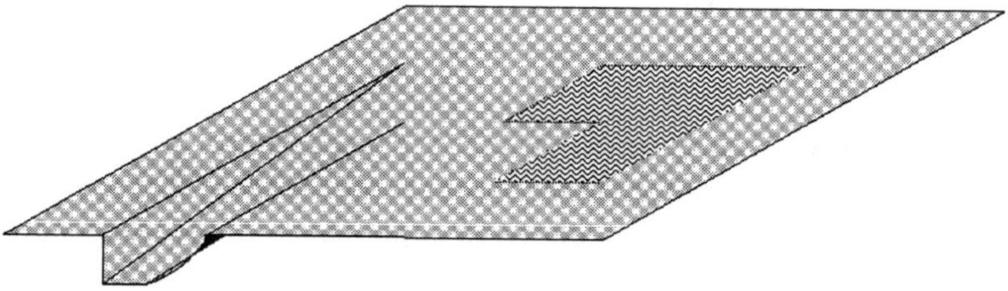

Fig. 4. Example of terrain with crevice and lake

4. Performance

In this section we consider the performance of the technique described in Section 3.2: how much time is required to obtain a winning search strategy? After demonstrating why the method is intractable, we present three heuristics, two of which we have implemented.

4.1. *Model Checking Complexity*

As stated in Section 3.1, CTL has the desirable characteristic that CTL properties can be model checked in time linear in the product of the size of the model and the size of the specification [4]. Unfortunately, the model we have described has a number of states that is exponential in the size of the grid. Specifically, for an $m \times n$ grid with k pursuers, the fastest of which moves s spaces/turn, the number of states is:

$$\underbrace{(mn)^k}_{\substack{\text{pursuers'} \\ \text{locations}}} \cdot \underbrace{2^{mn}}_{\substack{\text{cells} \\ \text{CLEARED?}}} \cdot \underbrace{(s+1)}_{\substack{\text{"clock"} \\ \text{artifact}}} \tag{3}$$

Since the number of states is exponential in the number of grid cells, the execution time is exponential in the number of grid cells. We cannot attain a faster time without running the risk of failing to find a legitimate pursuer-winning strategy. From [8], we can conclude that even if we were to use a different temporal logic, the execution time would still be exponential in the number of grid cells.

Table 1 shows the effect on the execution time and memory displacement to run SMV on a 933MHz Pentium III system for different-sized grids with one pursuer and movement only in the four cardinal directions. The attentive reader will notice that the number of states listed in the table for each configuration is exactly twice the number of states described by Equation 3; this is due to an extra boolean variable necessitated by our choice of model checker. The number of *reachable* states reported by SMV is of interest because of an optimization in SMV that reduces the execution time to a function of the number of reachable states. The last two rows of Table 1 are incomplete, not because checking a model with $2^{43.7}$ states is impossible, but because checking a model with $2^{43.7}$ states required more time than we were willing to allot for our initial experiments.

Table 1. Time needed to obtain a pursuer search strategy.

Board Dimensions	Number of Locations	Pursuer Speed (moves/turn)	Number of States	Reachable States	Time
2×2	4	2	$2^{8.6}$	$2^{6.6}$	$0.01s$
2×4	8	2	$2^{13.6}$	$2^{11.1}$	$0.04s$
3×3	9	3	$2^{15.2}$	$2^{13.3}$	$0.08s$
3×5	15	3	$2^{21.9}$	$2^{17.6}$	$0.66s$
4×4	16	4	$2^{23.3}$	$2^{19.4}$	$5.39s$
4×6	24	4	$2^{31.9}$	$2^{23.3}$	$7m\ 26s$
5×5	25	5	$2^{33.2}$	$2^{25.7}$	$2h\ 4m\ 3s$
5×7	35	5	$2^{43.7}$	??	$> 24h$
6×6	36	6	$2^{45.0}$	??	$> 24h$

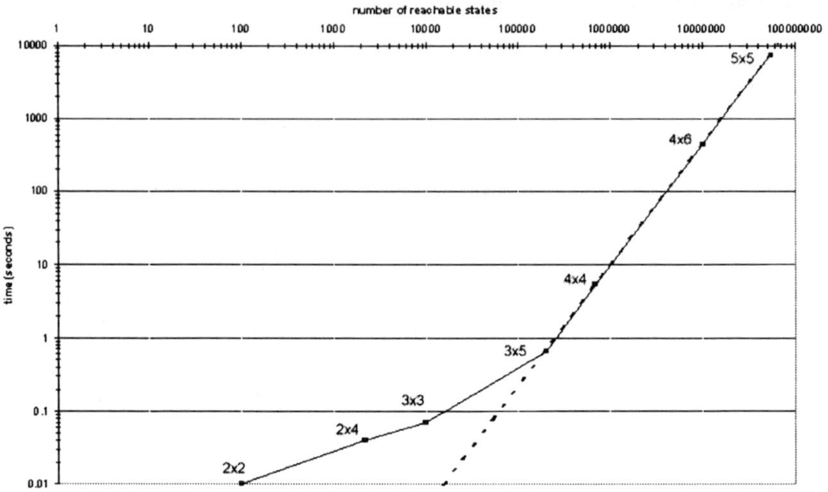

Fig. 5. Execution time vs. number of reachable states

4.2. *Heuristics to Reduce the State Space*

We are considering heuristics that reduce the size of the state space, albeit at the cost of potentially excluding pursuer-winning search patterns. We consider this to be an acceptable trade-off. The benefit is that we can dramatically reduce the execution time. If the model checker indicates it cannot find a pursuer-winning search strategy, then one may or may not exist for the conditions checked; however, if such a search strategy is found, then we know it is a valid search strategy.

4.2.1. *The "Sweep" Heuristic*

The first heuristic we present is to break the problem of clearing the grid into the smaller problem of clearing one column and ending up positioned to clear the next column, without permitting any undetected evaders to pass into previously-CLEARED columns; see Figure 6. This corresponds to the algorithm in the proof sketch of Section 2.2.1. If it is ever possible for the evader to enter the westernmost region, then the technique of clearing columns will not compose. However, if it is possible to accomplish this feat, repeated applications of this "clear-column" procedure can be composed to clear the whole grid by sweeping from one side of the grid to the other. Now, if we only need to model $w \times n$ cells explicitly (where $w \ll m$), then the number of states is:

$$(wn)^k \cdot 2^{wn+1} \cdot (s+1) \tag{4}$$

and the property to check is:

$$\neg \mathsf{E} \left[\text{CLEARED}(west) \mathsf{U} \begin{pmatrix} \bigwedge_{0 \le r < n} \text{CLEARED}(r, 0) \wedge \text{CLEARED}(west) \wedge \\ clock = 0 \wedge \text{position}(pursuer) = (0, 1) \end{pmatrix} \right] \tag{5}$$

That is, there does not exist a computation path ("E") in which the western region remains CLEARED until ("U") all cells in the first explicitly-modeled column are CLEARED along with the western region, and the pursuer is positioned in the southern cell of the second explicitly-modeled column at the start of a turn. If the specification is not satisfied, then the counterexample can be used to generate a search pattern that will cause the first modeled column to be CLEARED and the pursuer positioned to clear the next column at the beginning of its turn. Repeated application of this search pattern will eventually sweep the entire grid.

We have, in fact, used such models with a model checker. Using this heuristic, we obtained a pursuer-winning strategy for one pursuer moving 6 spaces/turn on a "$\infty \times 6$" grid ($w = 2$) in 76 seconds on a 933 MHz Pentium III [3]. Contrast this with the performance for the 6×6 grid in Table 1.

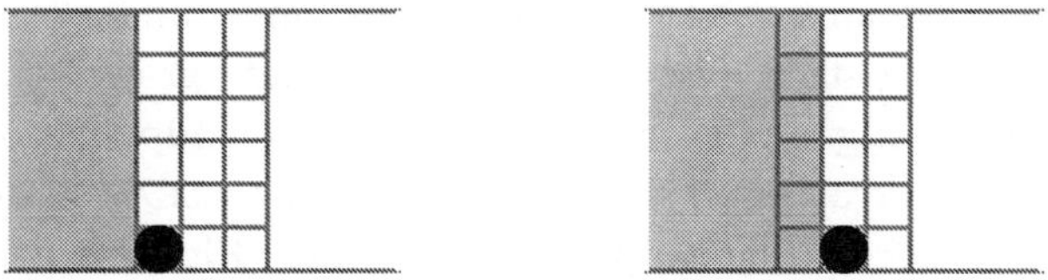

(a) Before column is cleared

(b) After column is cleared

Fig. 6. Abstraction of grid unbounded along the horizontal axis

4.2.2. *The "Cleared-Bars" Heuristic*

Besides composing subsolutions, we also consider changes to the manner in which we model the game. For example, many possible states do not bring the pursuer any closer to winning the game. One such state is the "7-10 split" in Figure 7(a). Unless the pursuer can move diagonally 8 spaces/turn (or cardinally 14 spaces/turn), then more cells will become unCLEARED in the next turn than the pursuer could make CLEARED. We believe that the resulting state could have been achieved without passing through this state first. Similarly, the "checkerboard" in Figure 7(b) is not a useful state, and it not even reachable without diagonal moves.

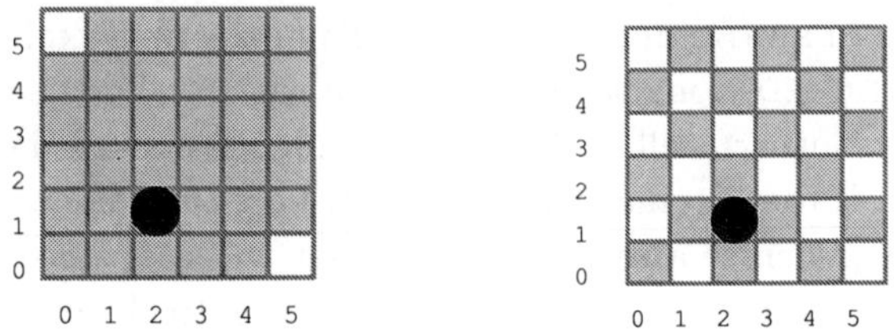

(a) All cells CLEARED except two at opposite corners.

(b) Alternating CLEARED and unCLEARED cells.

Fig. 7. Examples of undesirable states

So instead of considering whether each cell is CLEARED, we instead can define sets of contiguous CLEARED cells. For example, under the belief that if a pursuer-winning strategy exists, one exists that "grows" the CLEARED area as a set of contiguous bars, we can define the endpoints of CLEARED cells in each row (or column) and require that the CLEARED cells in each row be contiguous from one endpoint to the other (Figure 8(a)). With this model, the number of states is:

$$\underbrace{(mn)^{2k}}_{\text{pursuers' locations}} \cdot \underbrace{(m+1)^{2n}}_{\text{endpoints of bars}} \cdot \underbrace{(s+1)}_{\text{"clock" artifact}} \tag{6}$$

The first term is raised to the power of $2k$ instead of k because there are conditions in which the pursuers' current and last locations are needed to update the bars correctly. The middle term is $m+1$ instead of m to provide for "endpoints" when there are no CLEARED cells in a given row.

Table 2 contrasts the execution time without a heuristic to that with this heuristic.

4.2.3. *The "Cleared-Regions" Heuristic*

Alternatively, we might instead define the CLEARED regions geometrically by possibly-overlapping convex polygons. Figure 8(b) shows how the CLEARED area in Figure 2(a) can be described using three rectangles. While this will dramatically increase the complexity of the model description, we believe it will also dramatically decrease the number of states in the model. Consider, for example, a game with k pursuers moving s spaces/turn, with r rectangles describing the CLEARED regions. The size of the state space would be:

$$\underbrace{(mn)^k}_{\substack{\text{pursuers'}\\ \text{locations}}} \cdot \underbrace{(mn)^{2r}}_{\substack{\text{diagonal}\\ \text{corners of}\\ \text{rectangles}}} \cdot \underbrace{(s+1)}_{\substack{\text{"clock"}\\ \text{artifact}}} \tag{7}$$

Under the belief that if a pursuer-winning strategy exists, one exists when r is a small constant, this heuristic should offer substantial runtime improvement.

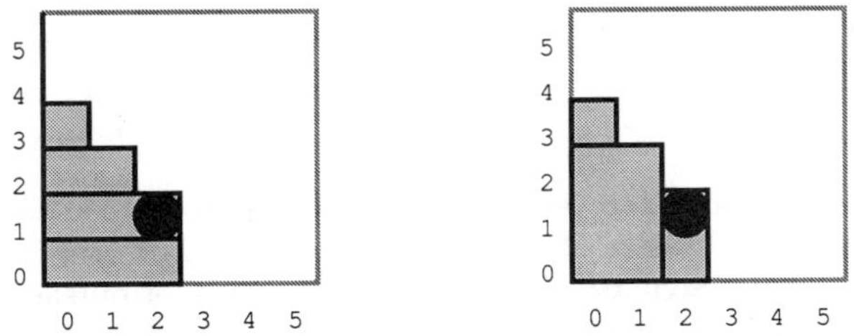

(a) Using CLEARED bars

(b) Using CLEARED regions.

Fig. 8. Alternate ways to describe the configuration of Figure 2(a)

Table 2. Time benefit of "Cleared-Bars" Heuristic.

Board Dimensions	Pursuer Speed (moves/turn)	Time Cleared-Cells	Time Cleared-Bars
4×4	4	$5.39s$	$1.69s$
5×5	5	$2h\ 4m\ 3s$	$32.27s$
6×6	6	$> 24h$	$32m\ 56s$

5. Conclusions and Future Work

We have shown how we can model the problem of UAVs searching for ground targets as a pursuer-evader game, and we have shown how we can use model checking to generate pursuer-winning search strategies when they exist. After considering the complication that the execution time is exponential in the size of the grid, we presented three heuristics to reduce the time to obtain search strategies, and the two that have been implemented have provided significant reductions in the execution time, and the third promises an even greater savings.

We plan to add additional heuristics to our arsenal. Once we have the state space growth managed, our work will address improving the fidelity of the model, such as by increasing the size of the grids, by permitting concurrent movement of the pursuers and evaders, and by placing restrictions on the pursuers' maneuverability.

Acknowledgments

The authors gratefully thank the US Air Force and the Air Force Institute of Technology for their direct support of the primary author. This work was supported by the AFRL/VA and AFOSR Collaborative Center of Control Science (Grant F33615-01-2-3154).

References

[1] Baum, M., A search-theoretic approach to cooperative control for uninhabited air vehicles. Master's thesis, The Ohio State University, 2002.

[2] Bohn, C. A. and Sivilotti, P. A. G., Upper bounds for pursuer speed in rectilinear grids. Technical Report OSU–CISRC–1/04–TR01, The Ohio State University, 2004.

[3] Bohn, C. A., Sivilotti, P. A. G., and Weide, B. W., Using model checking to find a hidden evader. In *Proceedings, Workshop on Agent/Swarm Programming*, pages 1–7, 2003.

[4] Clarke, Jr., E. M., Grumberg, O., and Peled, D. A., *Model Checking*. The MIT Press, 1999.

[5] Kanchanavally, S., Ordóñez, R., and Layne, J., Mobile target tracking by networked uninhabited autonomous vehicles via hospitability maps. Submitted to *American Control Conference*, 2004.

[6] Office of the Secretary of Defense, Unmanned aerial vehicles roadmap: 2002–2027, 2002.

[7] Oğraş, Ü. Y., Cooperative search strategies for multi-vehicle teams. Master's thesis, The Ohio State University, 2002.

[8] Reif, J. H., The complexity of two-player games of incomplete information. *Journal of Computer and System Sciences*, 29(2):207-301, 1984

[9] Zhang, C. and Ordóñez, R., Decentralized adaptive coordination and control of uninhabited autonomous vehicles via surrogate optimization. In *Proceedings, American Control Conference*, pages 2205-2210, 2003.

CHAPTER 3

APPLYING SIMULATED ANNEALING TO THE MULTIDIMENSIONAL ASSIGNMENT PROBLEM

Wilson K. Clemons, Don A. Grundel and David E. Jeffcoat

Air Force Research Laboratory, Munitions Directorate

Eglin AFB, FL

{wilson.clemons, don.grundel, david.jeffcoat}@eglin.af.mil

The multidimensional assignment problem (MAP) is a combinatorial optimization problem that is known to be NP-hard, and therefore, heuristic methods are generally used to find good solutions to it. The problem has many recognized applications such as multi-agent path planning, data association, and multi-searcher problems. Simulated annealing has proven to be effective in solving many combinatorial optimization problems, but we find no references in the literature in which simulated annealing is applied to the MAP. In this chapter, we evaluate a simulated annealing algorithm for solving the MAP and report experimental results using several controlling factors in the algorithm. These factors include the cooling schedule and initial temperature, the neighborhood definition, and the method of finding a starting solution. A design of experiments approach is used to find the most effective controlling factors in the algorithm. Algorithm performance measures include time to solution and quality of solution. For a small problem, the algorithm finds the optimal solution in every case tested. For a large problem, the algorithm finds results that average 1.2 units from the optimal solution. The results show that simulated annealing is an effective method for solving the MAP.

Keywords:

Multidimensional assignment problem, simulated annealing

1. Introduction

The multidimensional assignment problem (MAP) is an intuitive problem structure for many current combinatorial optimization problems, but it is known to be NP-hard. Other heuristic methods have been utilized in solving this problem, but there is much room for improvement in solution quality and run time. Our basic objective is to apply simulated annealing to the MAP and evaluate the results.

1.1. *Simulated Annealing*

Simulated annealing can be thought of as an evolution of a local search (or neighborhood search). The fundamental difference is that simulated annealing randomly accepts worse solutions, sometimes called 'uphill moves.' These 'uphill moves' often result in escaping an undesirable local minimum.

'Simulated annealing' refers to the annealing process used in materials science, typically for metals. Annealing is used to improve the final properties of a solid by changing the temperature profile during the cooling of molten materials. Instead of simply reducing the temperature of the material at a steady rate, the temperature is raised periodically using precise criteria in order to produce specific physical qualities in the material. These temperature fluctuations improve the characteristics of the final material. The improvement may be attributed to the fact that the imperfections in the crystal structure can be worked out during this period of temperature fluctuation, thus allowing the molecules to form a better structure. Similarly, simulated annealing randomly accepts solutions that are farther away from the final state throughout the procedure. This allows the algorithm to escape undesirable local optima and converge to a better final solution.

Kirkpatrick, Gelatt, and Vecchi [7] first proposed simulated annealing in 1983. Their article, "Optimization by Simulated Annealing," describes and applies simulated annealing as a method for combinatorial optimization. The authors describe this algorithm in a detailed comparison to the actual properties of molecules in the physical annealing process. They also discuss applications of simulated annealing in specific areas, and then compare this method to another heuristic method. The central message of this article is the connection between statistical mechanics (the behavior of systems with many degrees of freedom in thermal equilibrium at a finite temperature) and multivariate and combinatorial optimization.

However, much of the literature has expressed skepticism about the relevance of the physical annealing process to combinatorial optimization. In this chapter, we emphasize the experimental evaluation of an algorithm's performance, rather than constraining the algorithm structure by focusing on the relationship with the physical annealing process. Johnson, et al. [6] explicitly state that there is no need to be overly concerned with the analogy between the physical annealing process and this algorithm. They are skeptical about the significance of this relationship. They mention that the simulated annealing algorithm is very similar to the much older 'Metropolis Algorithm,' a simulation procedure developed in the late 1940s to model the behavior of gas particles. They point out that simulated annealing simply uses this older technique as its inner core.

Many other studies have been done on simulated annealing. Collins, et al. [2] in 1988 provide an annotated bibliography of simulated annealing. Ogbu and Smith [9] in 1990, Jeffcoat and Bulfin [5] in 1990, and Van Laarhoven [14] et al., in 1992, apply simulated annealing to specific job shop scheduling problems. Since these studies, a great deal of research has been conducted in this area, but there appears to be no cases of applying simulated annealing to the MAP.

1.2. *Multidimensional Assignment Problem*

The MAP is simply a higher dimensional version of the standard 2-dimensional or linear assignment problem. The MAP is known to be NP-hard. Some general studies on the MAP include Peirskalla [12] in 1968 and Pardolos and Pitsoulis [10] in 2000 with a survey on the MAP. In Figure 1, we represent a hypothetical MAP in terms of a manager, location, and a product. Each of these items represents a dimension. Specifically, manager is the first dimension, location is the second dimension, and product is the third dimension. This is a three-dimensional MAP with four elements in each dimension. So, each number in the left side of Figure 1 represents an element in its respective dimension. In the figure, manager one is assigned to location one, and both of these are assigned to product one. The same goes for manager two, location two, product two, and so forth. We can model this as a minimization problem where the objective function represents a cost. Each assignment has a corresponding cost coefficient, c_{ijk}, where i, j, and k represent the element numbers of an assignment. In other words, it will cost a certain amount to assign manager one to location one, and assign these to product one. We would represent this cost coefficient as c_{111}. So, the total cost of this solution is given by $C = c_{111} + c_{222} + c_{333} + c_{444}$. Using set notation, we can represent the solution of the left side of Figure 1 as $\{111, 222, 333, 444\}$. In this particular instance of the problem, $c_{111} = 12$, $c_{222} = 17$, $c_{333} = 24$, and $c_{444} = 10$, yielding a total cost of 63. However, if we exchange some of the assignments to construct the optimal solution of this problem instance as shown on the right side of Figure 1, the solution becomes $\{133, 214, 322, 411\}$, and the total cost $C = c_{133} + c_{214} + c_{322} + c_{411}$. Replacing these variables with cost values, we get $C = 6 + 12 + 17 + 2$, or 37, an improvement of 26 units from the initial solution.

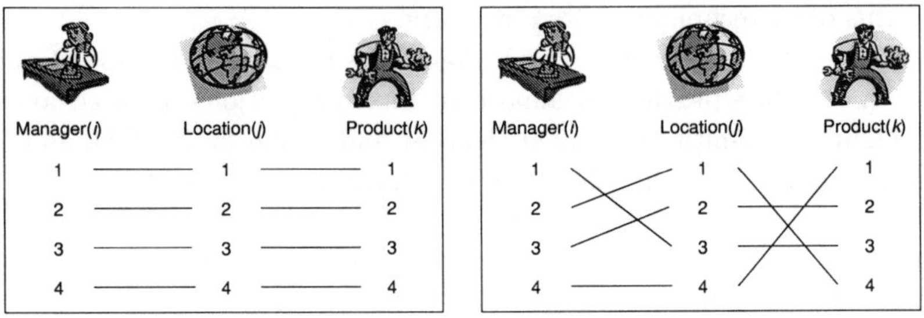

Fig. 1. Left: Starting solution of some random instance has cost 63. Right: Optimal solution of the same instance has cost 37.

The number of cost coefficients depends on problem size. The total number of cost coefficients for a fully dense problem is given in expression (1), where n_i is the number of elements in dimension i of D dimensions.

$$\prod_{i=1}^{D} n_i \tag{1}$$

Similarly, as the number of elements in each dimension increase or as D increases, the number of feasible solutions increases exponentially. The number of feasible solutions is given by expression (2).

$$Number\ of\ Feasible\ Solutions = \prod_{i=2}^{D} \frac{n_i!}{(n_i - n_1)!} \tag{2}$$

The MAP can be expressed as an integer program. In the following formulation, the objective is to minimize the sum of the cost coefficients subject to constraints that insure that each element of the first dimension has exactly one assignment and that each element of any additional dimensions has at most one assignment. The formulation can be expressed as

$$
\begin{aligned}
\min \sum_{i_1=1}^{n_1} \cdots \sum_{i_D=1}^{n_D} c_{i_1 \cdots i_D} x_{i_1 \cdots i_D} & \\
\text{s.t.} \sum_{i_2=1}^{n_2} \cdots \sum_{i_D=1}^{n_D} x_{i_1 \cdots i_D} &= 1 \ \forall \ i_1 = 1, 2, \ldots, n_1, \\
\sum_{i_1=1}^{n_1} \cdots \sum_{i_{k-1}=1}^{n_{k-1}} \sum_{i_{k+1}=1}^{n_{k+1}} \cdots \sum_{i_D=1}^{n_D} x_{i_1 \cdots i_D} &\leq 1 \\
\forall \ k = 2, \ldots, D_1, \text{ and } \forall \ i_k = 1, 2, \ldots, n_k, & \\
\sum_{i_2=1}^{n_2} \cdots \sum_{i_{D-1}=1}^{n_{D-1}} x_{i_1 \cdots i_D} &\leq 1 \ \forall \ i_D = 1, 2, \ldots, n_D, \\
x_{i_1 \cdots i_D} \in \{0, 1\} \ \forall \ i_1, i_2, \ldots, i_D, & \\
n_1 \leq n_2 \leq \cdots \leq n_D, &
\end{aligned}
$$

Also, in the formulation, the assignment variables have a value of either zero or one. If the assignment variable has a value of zero, this means that we have not currently chosen the particular assignment. Likewise, if the assignment variable has a value of one, then we have currently selected the particular assignment. Several heuristic methods have been applied to the MAP in recent years. Poore and Rijavec [13] applied Lagrangian relaxation to the MAP in 1993. Murphey, Pardolos, and Pitsoulis [8] applied GRASP to the multitarget multisensor tracking problem in 1998. Aiex et al. [1] used GRASP with path relinking in 2001. Pasiliao [11] developed an exact solution method, using branch and bound in 2003.

2. The Algorithm

The simulated annealing algorithm is a template for solving optimization problems, rather than an all-encompassing formula that can be applied to any problem. It is important to carefully select parameters and settings for each individual procedure in order to most effectively solve a specific problem. This includes setting the starting temperature, cooling schedule, number of iterations per phase, method for terminating the algorithm, and type of local search. In our case, some of these parameters are selected based on previous research. Those parameters do not change throughout our investigation. Other parameters are determined using a design of experiments approach. One of the "set" parameters is the type of local search that is used.

2.1. *Two-Exchange Local Search*

The two-exchange local search has proven to be an effective local search for the MAP in the past [11]. We demonstrate this procedure in Figure 2. The initial solution has a cost of sixty-three and is the combination of the first element of the first dimension with the first element of the second and third dimension. Similarly, the second elements from each dimension are assigned together, and so forth. After we determine an initial solution, we perform the local search. The two-exchange local search is as follows:

(1) Randomly select a column (dimension).
(2) Randomly select two elements from that column.
(3) Exchange the assignment for the two elements.
(4) If the new solution has a lower cost than the current solution, accept the new solution. Otherwise, go to step 1. Continue the algorithm until no better solution can be found in the neighborhood.

In this specific case, the first solution with a lower cost is found by exchanging elements two and three from column one. The cost is 47, an improvement of 13 from the initial solution. We perform this procedure again and obtain a cost of 38 by exchanging the first and fourth elements in column three. Finally, we exchange elements one and three from column one and obtain a value of 37. At this point, no neighboring solutions yield an objective function value less than 37, so we terminate the algorithm and accept this solution, which is a local minimum. The global optimal solution is 30, so it is clear that we are trapped in a local optimal solution. With simulated annealing, we hope to escape this local optimum and find a solution much closer to the global optimum.

Along with the type of local search, we use a set method for finding an initial solution. In the annealing algorithm, we use the same method that is shown in the top left of Figure 2. That is, we simply start by assigning each element in the first dimension with the corresponding elements from each additional dimension. So, our first assignment is $\{111, 222, 333, 444\}$. Although the names of the cost variables are similar for the starting solutions of each problem, the cost coefficients are randomly generated, and therefore these initial costs are independent from problem to problem. Our method for generating problems is discussed in [3].

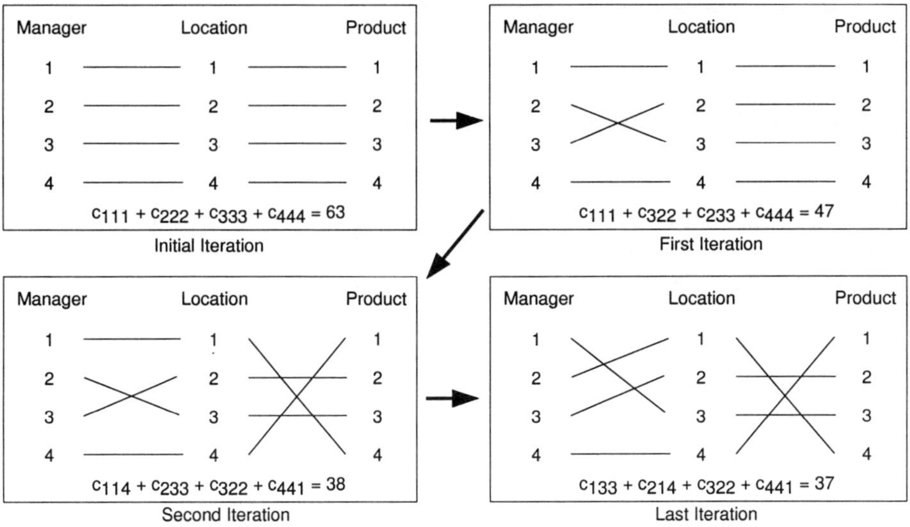

Fig. 2. Walking through iterations of simple local search.

2.2. *The Simulated Annealing Algorithm*

Just as we predetermine the type of local search and the method for generating an initial solution, we also predetermined the type of cooling schedule and the method for terminating the algorithm. Our cooling schedule is relatively simple compared with some other cooling schedules in the literature. The schedule is as follows: multiply the current temperature T by the cooling ratio R after each phase of the algorithm is completed. So, the initial temperature and the cooling ratio strictly determine the temperature during each phase.

The method for terminating the algorithm is determined analytically by investigating the specific problem. The idea is to continue with the algorithm until virtually no uphill moves are accepted for some period of time. Ogbu and Smith [9] use this concept by continuing the algorithm until T becomes very low. We use this same idea.

We define the variable Δ as,

$$\Delta = cost(S_{new}) - cost(S_{current}) \tag{3}$$

This is simply the difference between the proposed solution and the current solution at any given step of the algorithm. We consider a Δ of one a small value for our problem. The determination of what constitutes a small or large change in the objective function requires specific knowledge of the problem and will obviously vary from one problem to another. Further insight into these problems can be found in [3]. After making this estimate, we set a preliminary stopping criterion of,

$$e^{-\Delta/T} < 0.0001 \tag{4}$$

The exponential corresponds to the probability of accepting an uphill move, so that equation (4) means that we want the algorithm to terminate whenever the probability of accepting an uphill move is less than 0.0001. Starting from equation (4), the stopping criterion can then be expressed in terms of T. Using a small cost difference ($\Delta = 1$), we get,

$$T < \frac{-1}{ln(0.0001)} = 0.1086 \tag{5}$$

For the sake of simplicity, this threshold value of T is truncated to 0.10. Therefore, we terminate the algorithm whenever T becomes less than 0.10. The input parameters for the algorithm are initial temperature T, cooling ratio R, and the number of iterations per phase NUM. We show the basic algorithm along with parameters and variables in the following:

Input Parameters:
T_{start}: Initial Temperature
R: Cooling ratio used for reducing T before each new phase starts
NUM_{max}: Maximum number of iterations per phase

Other Variables:
NUM: Iteration number. Initialized to zero at the beginning of each phase

T: Current temperature. Updated before each new phase starts S_{best}: Best solution found in the algorithm - updated whenever a better solution is found

$S_{current}$: Solution that is currently accepted in the algorithm
S_{new}: Solution that is a candidate for becoming $S_{current}$

The Algorithm:

Step 1. Get the initial feasible solution, $S_{initial}$, and initialize $S_{current}$ and S_{best} to $S_{initial}$

$$S_{current} = S_{best} = S_{initial}.$$

Step 2. Randomly select a neighbor, S_{new}. If $cost(S_{new}) < cost(S_{best})$, then $S_{best} = S_{new}$.

Step 3. Accept or reject S_{new} as $S_{current}$ by

(a) Let $\Delta = cost(S_{new}) - cost(S_{current})$.

(b) If $\Delta \leq 0$, then $S_{current} = S_{new}$.

(c) Otherwise, generate uniform random number in [0,1], $u \sim U(0,1)$. If $u \leq e^{\Delta/T}$, then $S_{current} = S_{new}$, otherwise, no change to $S_{current}$.

Step 4. Repeat Steps 2 though 4 until $NUM > NUM_{max}$

(a) Set $NUM = NUM + 1$.

(b) If $NUM > NUM_{max}$, then proceed to **Step 5**. Otherwise, go back to **Step 2**.

Step 5. End this phase of the algorithm.

(a) If $T > 0.1$, then $T = T \cdot R$ and go to **Step 2**.

(b) Otherwise, terminate the algorithm and return S_{best}.

3. Design of Experiments

We use a Design of Experiments (DOE) approach for investigating the best values for certain parameters in the algorithm. Table 1 summarizes this model. It includes five factors, each at two levels. The factors in the experiment are initial temperature (T), cooling ratio (R), number of iterations per phase (NUM), number of dimensions in the MAP (D), and number of elements in each dimension (N). T, R, and NUM are direct parameters of the simulated annealing algorithm. The number of dimensions (D) and the number of elements per dimension (N), on the other hand, are related to the problem size. The smallest problem has three dimensions with five elements in each dimension and the largest problem has six dimensions with ten elements in each dimension. These two problems correspond to a relatively small problem and a relatively large problem, respectively. Also, all of the problems that we run are balanced problems. That is, there are an equal number of elements in each dimension, $n_1 = n_2 = \cdots = n_D = N$.

Table 1. Design of Experiments. * - j represents the number of neighbors of a feasible solution

	Levels	
Factors	**Low**	**High**
T	1	10
R	0.9	0.99
NUM	1j*	10j*
D	3	6
N	5	10

Also, notice that in Table 1, the third factor, *NUM*, is dependent upon the size of the neighborhood j. The size of a neighborhood when performing a two-exchange local search is given by $j = D\binom{N}{2}$. This shows that j is a function of D and N. Specifically, j increases when either D or N increases. Further, the low level of NUM and high level of NUM are determined by $NUM = j$ and $NUM = 10j$, respectively.

This shows that *NUM*, like j, is also a function of D and N. The low level of *NUM* is simply the size of a neighborhood for the two-exchange local search, while the high level is ten times the size of a neighborhood.

The major response variable for this experiment is the optimality gap. This is the difference between the objective function value of the optimal solution and the objective function value of the solution determined by the simulated annealing algorithm. The optimal solution is determined during the generation of the problem, which allows us to compare the solution found using simulated annealing to the optimal solution. Since these problems are known to be NP-hard, this is a very helpful tool. Also, the problems are fully dense and the cost variables for individual problems are approximately normally distributed. The method for generating sample problems is discussed in [3].

In addition to the optimality gap, runtime is a response variable for this model. However, it is secondary to the optimality gap since our primary objective is to obtain better solutions to these problems. Nevertheless, the trade-off between runtime and quality of solution is an interesting area for future research.

Figures 3 and 4 show some of the results of this experiment. Figure 3 shows that the algorithm generally yields a smaller optimality gap when the starting temperature (T) is 10. Figure 4 shows the interaction between the cooling ratio (R) and the number of iterations per phase (NUM). The top line shows the results for NUM at its low level and the bottom line is at the high level. Similarly, the low level of R is the first x-axis value, and the high level is the second value. The optimality gap is better (smaller) at the high level of *NUM*, 10 times the size of the neighborhood. Also, the optimality gap is better at the high level of R, 0.99.

This experiment tells us that the high levels of T, R, and NUM yield the best optimality gap. Specifically, the best results are generally found when the starting temperature (T) is 10, the cooling ratio (R) is 0.99, and the number of iterations per phase (NUM) is 10 times the size of the neighborhood.

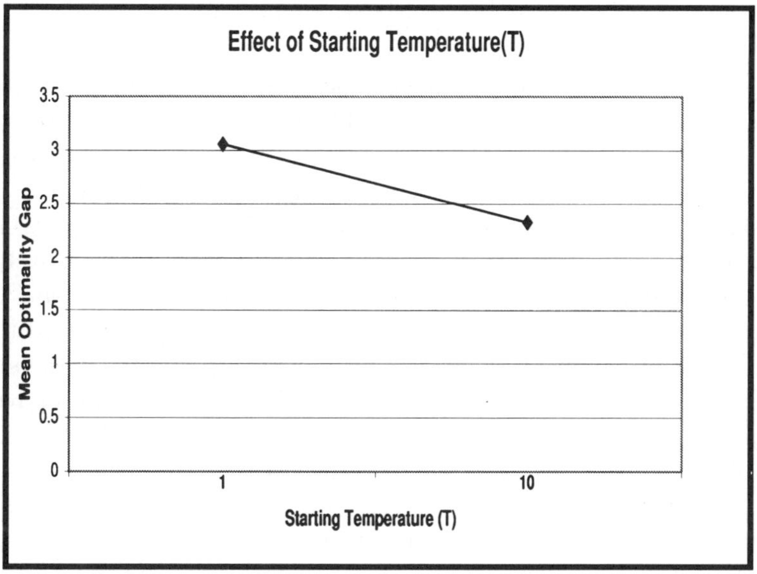

Fig. 3. Effect of starting temperature.

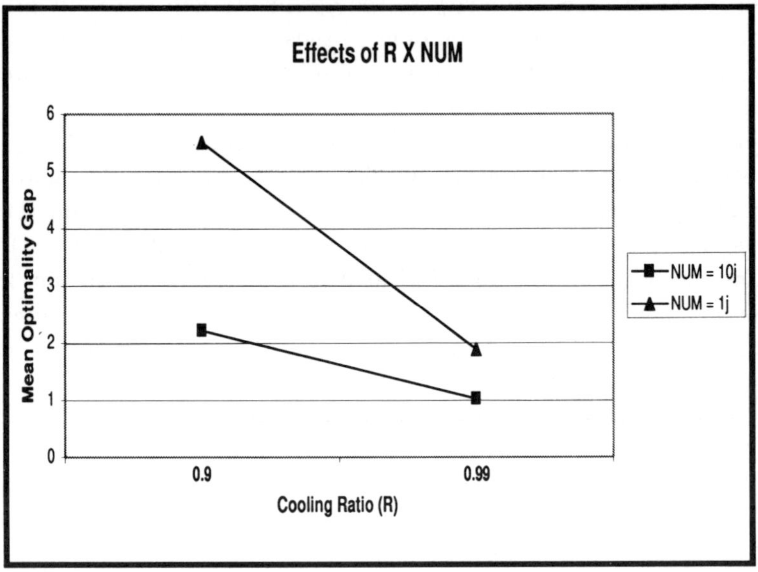

Fig. 4. Effects of $R \times NUM$.

4. Results

Using the best algorithm parameters determined by the previous experiment, we run five replications of four different problems. Again, the problems are determined by the high and low levels of the number of dimensions and number of elements in the DOE. The low level for the number of dimensions (D) is 3 and the high level of D is 6. The low and high levels for the number of elements per dimension (N) are 5 and 10, respectively. As mentioned earlier, the optimal solution of these problems is known. Table 2 shows the output for all four problems. The algorithm converged to the optimal solution in every replication for the small problem. As the size of the problem increases, the optimality gap tends to increase. One interesting point is that the mean value for the second problem ($D = 3$; $N = 10$) is actually larger than the mean value for the last problem ($D = 6$; $N = 10$). This indicates that our optimality gap does not necessarily increase as the number of dimensions in the problem increases. However, since the number of iterations per phase increases with the number of dimensions, this could simply mean that we spend more time finding the solution to the larger problems, with similar final results. Nevertheless, this could be a valuable area of future research.

Table 2. Output values for each replication, four different problems

Replication Number	D	N	Initial Value	S.A. Value	Optimal Value	Runtime (sec)	Optimality Gap
1	3	5	53	15	15	0.741	0
2	3	5	53	15	15	0.771	0
3	3	5	53	15	15	0.811	0
4	3	5	53	15	15	0.791	0
5	3	5	53	15	15	0.792	0
Mean	3	5	53	15	15	0.7812	0
1	3	10	149	53	53	3.585	0
2	3	10	149	56	53	3.425	3
3	3	10	149	56	53	3.526	3
4	3	10	149	56	53	3.515	3
5	3	10	149	53	53	3.515	0
Mean	3	10	149	54.8	57	3.5132	1.8
1	6	5	69	26	25	1.733	1
2	6	5	69	26	25	1.723	1
3	6	5	69	26	25	1.712	1
4	6	5	69	26	25	1.713	1
5	6	5	69	26	25	1.712	1
Mean	6	5	69	26	57	1.7186	1
1	6	10	139	59	57	8.403	2
2	6	10	139	58	57	8.403	1
3	6	10	139	58	57	8.432	1
4	6	10	139	58	57	8.443	1
5	6	10	139	58	57	8.453	1
Mean	6	10	139	58.2	57	8.4268	1.2

Figures 5 and 6 summarize solution value results using the best values for T, R, and NUM. Figure 5 shows solution results of five replications of the largest problem ($D = 6$ and $N = 10$). The black bar represents the objective function value of the initial solution. The gray bar represents the objective function value of the solution found using simulated annealing, and the white bar represents the value of the optimal solution. This gives an interesting perspective on the proportions of these values. Simulated annealing yields solution values that are within two units of the optimal solution values in every case. The average difference is 1.2.

Figure 6 shows the average results for all four problem sizes. The last problem in this figure summarizes results in Figure 5 and compares this problem size to the other problem sizes. Simulated annealing finds the optimal solution for all five replications of the smallest problem ($D = 3$, $N = 5$), since the average solution value is identical to the value of the optimal solution.

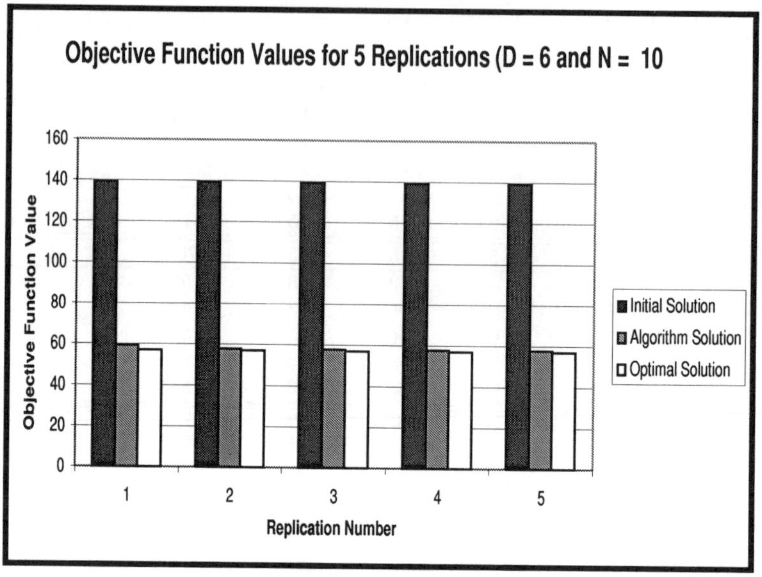

Fig. 5. Objective Function Values for $D = 6$ and $N = 10$.

Figure 7 provides an overview of runtimes for the four problem sizes. As expected, larger problems take longer to solve.

5. Conclusions

The results of this study are encouraging. In investigating the simulated annealing algorithm, we predetermined several algorithm settings, and then experimentally determined other parameter values using a DOE approach. The DOE provided a very efficient framework for determining effective tuning parameters. The final algorithm used a two-exchange local search, an initial temperature of 10, and iterations per phase of ten times the size of the neighborhood. Also, the cooling schedule was very simple: multiply the temperature by the cooling ratio, .99, after each phase is completed.

This combination of factors in the algorithm yielded the best results of any combination that we ran and generally worked very well in solving our particular MAP instances. In the smallest problem ($D = 3$, $N = 5$), the algorithm found the optimal solution in every instance. In the largest problem ($D = 6$ and $N = 10$), it was within two units of the solution for every instance and averaged 1.2 units from the optimal solution. In general, we found the simulated annealing algorithm is effective in finding

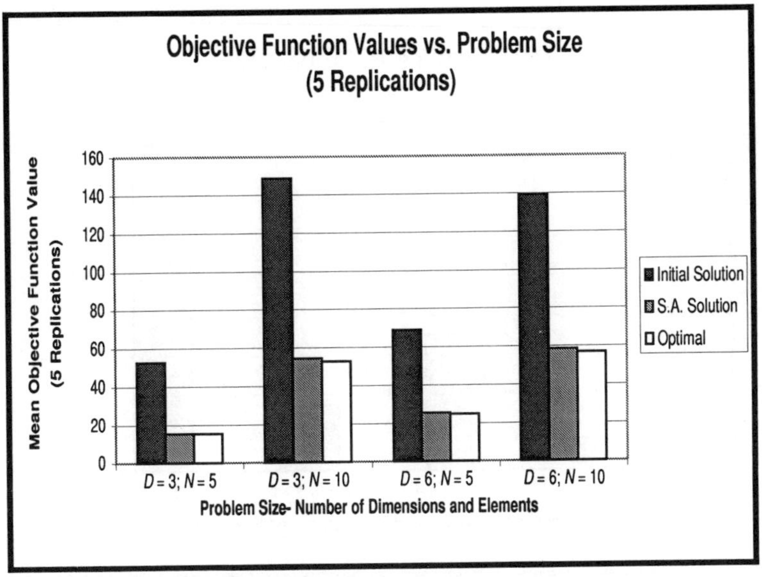

Fig. 6. Mean Objective Function Values for each problem size.

good solutions for the MAP.

6. Future Work

In this introductory study, we accomplish two primary objectives. The first is to demonstrate the viability of using simulated annealing to solve the MAP. However, the second is to gain insight into which aspects of this methodology warrant further study. The following describes some of the most critical of these aspects.

One possible area deals with the effect of D (number of dimensions) on the optimality gap. These effects are interesting in that our experience shows the optimality gap does not necessarily increase when D increases. This is obviously interrelated with the runtime, since our runtime is dependent on the number of dimensions. However, it is an interesting area for future investigation.

Also, the algorithm has longer runtimes when using the higher values of the parameters. Although our study considers solution quality to be the most important response variable, the trade-off between solution quality and run time is a major issue. Future work in this direction could be very valuable.

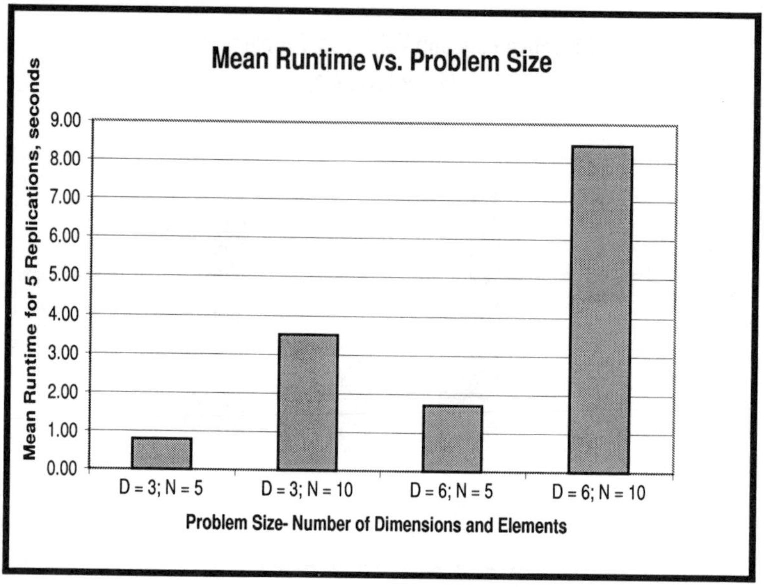

Fig. 7. Mean runtimes for each problem size.

Another interesting area for future work is to investigate how this heuristic method compares with other heuristic methods. The results of this study indicate that the simulated annealing algorithm should compare favorably with other methods.

Finally, the ultimate reason for studying these heuristic methods is to eventually use them to solve real-life problems. The use of this method for certain practical applications of the MAP is a valuable future endeavor. Simulated annealing could be applied to problems such as multiagent path planning or the multi-sensor, multi-target tracking data association problem.

References

[1] Aiex, R. M., M. G. C. Resende, P. M. Pardalos, G. Toraldo. "GRASP with Path Relinking for the Three-Index Assignment Problem." Technical report, AT&T Labs Research, Florham Park, NJ, 2001.

[2] Collins, N. E., R. W. Eglese, B. L. Golden. "Simulated Annealing: An Annotated Bibliography." American Journal of Mathematical and Management Sciences, 8 : pages 209-307, 1988.

[3] Grundel, Don A., Panos M. Pardolos. "Test Problem Generator for the Mul-

tidimensional Assignment Problem." To appear in Computational Optimization and Applications.

[4] Gosavi A. Simulation-Based Optimization. Dordrecht, Neth.: Kluwer Academic Publishers, 2003.

[5] Jeffcoat, D. E., R. L. Bulfin. "Simulated Annealing for Resource-Constrained Scheduling." European Journal of Operational Research, 70 : pages 43-51, 1993.

[6] Johnson, David S., Cecilia R. Aragon, Lyle A. McGeoch, Catherine Schevon. "Optimization by Simulated Annealing: An Experimental Evaluation; Part I, Graph Partitioning." Operations Research, 37 : pages 867-892, 1989.

[7] Kirkpatrick S., C. D. Gelatt, M. P. Vecchi. "Optimization by Simulated Annealing." Science, 220 : pages 671-680, 1983.

[8] Murphey, R., P. Pardalos, L. Pitsoulis. "A Greedy Randomized Adaptive Search Procedure for the Multitarget Multisensor Tracking Problem." DIMACS Series in Discrete Mathematics and Theoretical Computer Science, American Mathematical Society, 40 : pages 277-302, 1998.

[9] Ogbu, F. A., D. K. Smith. "The Application of the Simulated Annealing Algorithm to the Solution of the n/m/Cmax Flowshop Problem." Computers & Operations Research, 17 : pages 243-253, 1990.

[10] Pardalos, P., Pitsoulis, L. (eds.). *Nonlinear Assignment Problems, Algorithms and Applications*, Kluwer, Dordrecht, pp. 1-12, 2000.

[11] Pasiliao, E. L. "Algorithms for Multidimensional Assignment Problems." Ph.D. Dissertation, University of Florida, 2003.

[12] Pierskalla, W. "The Multidimensional Assignment Problem." Operations Research, 16 : pages 422-431, 1968.

[13] Poore, A., Rijavec, N. "A Lagrangian Relaxation Algorithm for Multidimensional Assignment Problems Arising from Multitarget Tracking." Society of Industrial and Applied Mathematics Journal on Optimization, 3 : pages 544-563, 1993.

[14] Van Laarhoven, Peter J. M., Emile H. L. Aarts, Jan Karel Lenstra. "Job Shop Scheduling by Simulated Annealing." Operations Research, 40 : 113-125, 1992.

CHAPTER 4

ON THE PERFORMANCE OF HEURISTICS FOR BROADCAST SCHEDULING

Clayton W. Commander
Department of Industrial and Systems Engineering,
University of Florida, Gainesville, FL
clayton8@ufl.edu

Sergiy I. Butenko
Department of Industrial Engineering
Texas A&M University, College Station, TX
butenko@tamu.edu

Panos M. Pardalos
Center for Applied Optimization
Department of Industrial and Systems Engineering
University of Florida, Gainesville, FL
pardalos@ufl.edu

In the Broadcast Scheduling Problem (BSP), a finite set of stations are to be scheduled in a time division multiple access (TDMA) frame. In a TDMA frame, time is divided into equal length transmission slots. Unconstrained message transmission can result in a collision of messages, rendering them useless. Therefore, the objective of the BSP is to provide a collision free broadcast schedule which minimizes the total frame length or maximizes the slot utilization within the frame. In this chapter, we introduce the BSP as a NP-complete combinatorial optimization problem and discuss several heuristics which have been applied to the problem. The heuristics are tested on over 60 networks of varying sizes and densities and the results compared.

Keywords:

Broadcast scheduling problem, ad-hoc networks, combinatorial optimization, NP-complete, heuristics

1. Introduction

In recent years, ad-hoc networks have become increasingly popular among many researchers. These networks are used to apply a packet switching technique over a shared radio channel and provide high-speed communications between a large number of potentially mobile stations which may be geographically disbursed. Each station is capable of both sending and receiving messages over the network. Hence, if a message is sent and the intended station is not able to receive it from the transmitting station, other intermediate stations may serve as relays by forwarding the message to the intended recipient.

Since all stations in the network share the transmission channel, stations must be scheduled to transmit messages in such a manner that prevents message collisions [33]. There are two types of message collision. The first, herein referred to as *direct collision*, occurs when two neighboring stations transmit during the same time slot. The second, which will be called *hidden collision* is a result of two non-neighboring stations transmitting simultaneously to a station that can receive messages from both transmitting stations. What is desired is a broadcast schedule that is guaranteed to produce collision free transmissions.

In [17], it is shown that the time division multiple access (TDMA) protocol can be used to provide a collision free scheduling procedure. In a TDMA network, time is divided into frames, each frame consisting of a number of unit length slots. The goal is to schedule as many stations as possible in the same time slot provided that simultaneous broadcasts will not cause any collision, either hidden or direct. Specific transmission criteria will defined in Section 2.

In general, two optimization problems are the focus of most broadcast scheduling research [33]. The more common problem attempts to minimize the total number of slots in each frame. The other problem optimizes the schedule by maximizing the slot usage per frame, which will maximize the throughput [31, 33]. Both of these problems are *NP*-complete [5, 9] and a motivational argument will be given in Section 3 of this chapter. It is interesting to note however, that a schedule which prevents only direct collisions can be implemented in polynomial time [12].

In this chapter, we introduce and formally define the BSP. Then we will briefly discuss complexity issues with the problem. Section 4 will be a review of several heuristics which have been applied to the problem. Finally, some concluding remarks will be given in Section 5.

2. The Broadcast Scheduling Problem

An ad-hoc TDMA network can be described as a graph $G = (V, E)$ where the vertices $V = \{1, \ldots, n\}$ represent the stations in the network and the edge set E represents the set of transmission links between the vertices. Two stations $i, j \in V$ are said to be *one-hop neighbors* if and only if (iff) they can directly communicate. That is, stations i and j are one-hop neighbors iff there exists an undirected edge $(i, j) \in E$. One-hop neighbors which transmit in the same time slot will cause a direct collision. Now suppose that $(i, j) \notin E$ but there exists some intermediate node $k \in V$ such that $(i, k) \in E$ and $(k, j) \in E$, then we will say that stations i and j are *two-hop neighbors*. Two-hop neighboring stations transmitting in the same slot will result in a hidden collision.

The topology of an ad-hoc network can be described by an $N \times N$ symmetric binary matrix C, where $N = \mid V \mid$. $C = \{c_{ij}\}$ is known as the *adjacency matrix* or *connectivity matrix* and is defined as follows:

$$c_{ij} = \begin{cases} 1, & \text{if } (i,j) \in E \text{ and } i \neq j, \\ 0, & \text{otherwise.} \end{cases}$$

We assume that there are M time slots per frame, and that one packet length in time is equal to the time required to transmit one packet of information. Furthermore, we assume that packets are transmitted at the beginning of each time slot, and that each packet is received during the same slot in which it is sent. With these assumptions, we now represent the broadcast schedule as a binary $M \times N$ matrix $S = \{s_{mn}\}$, where

$$s_{mn} = \begin{cases} 1, & \text{if station } n \text{ is scheduled tor transmit in slot } m, \\ 0, & \text{otherwise.} \end{cases}$$

It is helpful when performing analysis on the efficiency of a broadcast schedule to know what percentage of the available slots are utilized in each frame. Let ρ_n be the slot utilization for station n. Then,

$$\begin{aligned} \rho_n &= \frac{\text{the number of slots assigned to station } n}{\text{frame length}} \\ &= \frac{\sum_{m=1}^{M} s_{mn}}{M}. \end{aligned}$$

Hence, the total slot utilization of the entire network, ρ, is given by

$$\rho = \frac{\sum_{n=1}^{N} \rho_n}{N} = \frac{\sum_{m=1}^{M} \sum_{n=1}^{N} s_{mn}}{NM}.$$

We now define the Broadcast Scheduling Problem as follows:

Minimize M

subject to:

$$\sum_{m=1}^{M} s_{mn} \geq 1, \qquad \forall\, n, \tag{1}$$

$$c_{ij} + s_{mi} + s_{mj} \leq 2, \qquad \forall\, i, \forall\, j, \text{and } \forall\, m, i \neq j, \tag{2}$$

$$c_{ik}s_{mi} + c_{kj}s_{mj} \leq 1, \qquad \forall\, i, \forall\, j, \forall\, k, \text{and } \forall\, m, i \neq j, j \neq k, k \neq i. \tag{3}$$

The first constraint allows each station to broadcast at least once per frame. Constraint (2) ensures that one-hop neighbors do not transmit in the same slot. Lastly, constraint (3) ensures that no two-hop neighbors transmit in the same slot [33].

2.1. *Lower Bounds on the Frame Length*

Establishing a lower bound on the frame length M is possible by means of a few lemmata provided in the literature. Three of which will be described here. In [31], Wang and Ansari propose a lower bounding lemma based on the degrees of vertices in the graph. Specifically, for a given network, $G = (V, E)$, define the *degree* of a given vertex $v \in V$, denoted $\deg(v)$, to be the number of edges incident to v. Then, the frame length M satisfies the following inequality:

$$M \geq \delta(G) + 1, \tag{4}$$

where $\delta(G) = \max_{v \in V} \deg(v)$.

Though the bound in (4) is relatively simple to calculate, it does not provide a tight lower bound on M. In [15], another bounding lemma is given which produces tighter bounds on M. Consider the network $G = (V, E)$ as we have previously defined. That is, V is the set of stations, and E is the set of direct collisions. Suppose that the edge set E is expanded to also include hidden collisions, and we call this new set E'. We now have a new graph, namely $G' = (V, E')$ which is known as the *augmented* graph of G.

Recall that a *clique* in a graph $G' = (V, E')$ is a subset of V such that any two vertices in that subset are adjacent. Then in [15], Jungnickel shows that

$$M \geq \omega(G'), \tag{5}$$

where $\omega(G')$ is the maximal cardinality of a clique in G'. The latter provides excellent bounds on the frame length; however, calculating the maximum clique of a graph is *NP*-hard in general.

The third and final bound to be discussed is a method based on semidefinite programming and is the value of the so-called Lovász theta function. Lovász first introduced the θ-function in [21]. Given a graph $G = (V, E)$, a subset of vertices $V' \subseteq V$ is said to be an *independent set* if no two members of V' are connected by an edge in E. Lovász showed that the θ-function has the property that

$$MIS(G) \leq \theta(G) \leq \bar{\chi}(G), \tag{6}$$

where $MIS(G)$ is the maximum independent set in G and $\bar{\chi}(G)$ is the chromatic number of the complement of G. The θ-function can be used to provide good approximations for $MIS(G)$ and can be computed in polynomial time with arbitrary accuracy [11] whereas computing $MIS(G)$ is *NP*-hard in general. As is usually the case with calculating bounds, tight bounds in general are computationally more expensive to calculate. This holds true with the lemmata given here.

2.2. *Computational Complexity*

The BSP belongs to a class of computationally difficult problems known as *NP*-complete [9]. Belonging to this class suggests that an algorithm which solves the problem to optimality in polynomial time is unlikely to exist. It was first proven that BSP is *NP*-complete by Wang & Ansari in [31]. It was shown again by Ephremides & Truong in [4].

3. Heuristics

In this section, we will introduce and discuss four heuristics which have been applied to the BSP with varying degrees of success. The algorithms are as follows:

◇ Greedy Randomized Adaptive Search Procedure (GRASP) [3],
◇ Sequential Vertex Coloring (SVC) [33],
◇ Mean Field Annealing (MFA) [31], and
◇ Mixed Neural-Genetic Algorithm (HNN-GA) [30].

3.1. *GRASP*

Greedy Randomized Adaptive Search Procedure (GRASP), originally introduced by Feo and Resende in [6], is a two-phase iterative metaheuristic for combinatorial optimization [7, 8, 28]. In the first phase, referred to as the construction phase, a greedy randomized initial feasible solution is created. Then in the second phase, the initial solution is improved by the application of a local search procedure. The solution which is best out of all GRASP iterations is returned. GRASP has been applied to many combinatorial problems such as quadratic assignment [20, 22], job shop scheduling [2, 1], private virtual circuit routing [26], and satisfiability [29]. GRASP was also successfully applied to the BSP by the current authors in [3].

3.1.1. *Construction Phase*

The construction phase for the GRASP constructs a solution iteratively from a partial broadcast schedule which is initially empty. The stations are first sorted in descending order of the number of one-hop and two-hop neighbors. Then, a station is randomly selected greedily and assigned in the first slot. This selection is called the greedy randomized choice and is so named because of its bias towards stations with higher numbers of neighbors.

Next, a so-called *Restricted Candidate List* (RCL) is created and consists of those stations which may simultaneously broadcast with the greedy assigned station. From this RCL a station is randomly selected and assigned in the same slot as the greedy assigned station. A new RCL is created and another station is randomly selected. This process continues until RCL $= \emptyset$, at which time the slot number is incremented and a new station is assigned by the greedy function. Note that the greedy selection is biased towards those stations which have not been selected at random from an RCL. This bias helps to ensure a minimum frame length is attained [3].

3.1.2. *Local Search*

The local search phase used is a swap-based procedure which is adapted from a similar method for graph coloring implemented by Laguna and Martí in [18]. First, the two slots with the fewest number of scheduled transmissions are combined and the total number of slots is now given as $k = m - 1$. Denote the new broadcast schedule as $s_{m',n}$. Now, let the function $f(s) = \sum_{i=1}^{k} E(m'_i)$, where $E(m'_i)$ is the set of collisions in slot m'_i. $f(s)$ is then minimized by the application of a local search procedure as follows.

A colliding station in the combined slot is chosen randomly and every attempt is made to swap this station with another from the remaining $k-1$ slots. After a swap is made, $f(s)$ is re-evaluated. If the result is better, that is if $f(s)$ has a lower value than before the swap, the swap is kept and the process repeated with the remaining colliding stations.

If after every attempt to swap a colliding station the result is unimproved, a new colliding station is chosen and the swap routine is attempted. This continues until either a successful swap is made or for some specified number of iterations. If a solution is improved such that $f(s) = 0$, then the frame length has been successfully decreased by one slot. The value of k is then decremented and the process is repeated beginning with the combination of the two "smallest" slots. If the procedure ends with $f(s) > 0$, then no improved solution was found.

3.2. *Sequential Vertex Coloring*

In [33], Yeo et al. take a multi-objective optimization approach to solving the BSP. They implement a two-phase heuristic based on the idea of sequential vertex coloring (SVC). In the first phase, they only consider the problem of minimizing the frame length. Then in phase 2, the frame length is fixed with the solution from phase 1 and the utilization within the frame is maximized.

3.2.1. *Frame Length Minimization*

For this phase, a relaxed BSP is solved by replacing constraint (1) with $\sum_{m=1}^{M} s_{mn} = 1$ $(\forall i = 1 \ldots, N)$. The result is the frame length minimization problem which is analogous to the vertex coloring problem described above. An algorithm based on the sequential vertex ordering method is used to solve phase 1.

This is done by first ordering the stations in descending order of the number of one-hop and two-hop neighbors. The first vertex is colored and the list of the other $N - 1$ vertices are scanned downward. The remaining vertices are colored with the smallest color which has not already been assigned to a one-hop neighboring station. The process is continued until all vertices have been assigned a color. The number of colors, known as the *chromatic number* of the graph is returned as the minimum frame length.

3.2.2. *Utilization Maximization*

Beginning with this initial schedule, phase 2 attempts to maximize the throughput in the TDMA frame. To maximize the utilization within the frame whose length was determined in phase 1, an ordering method of the sequential vertex coloring algorithm is applied. The stations are now ordered in ascending order of the the number of one-hop and two-hop neighbors. The first ordered station is then assigned to any slots in which it can simultaneously broadcast with the previously assigned stations. This process is repeated with every station in the ordered list.

3.3. *Mean Field Annealing*

In 1997, Wang and Ansari [31] proposed a heuristic for the BSP based on Mean Field Annealing (MFA). In statistical mechanics, the physical process of annealing is used to relax a system to the state of minimal energy. This is done by heating the solid until it melts and then cooling it slowly so that at each temperature the particles randomly arrange themselves until reaching thermal equilibrium.

In [16], Kirkpatrick et al. introduced a new algorithm for combinatorial problems known as *simulated annealing*. Based on the theory of the physical process, simulated annealing was shown to asymptotically converge to the global minimum after performing a number of so-called transitions at decreasing temperatures.

Though simulated annealing is guaranteed to converge to the global optimal solution, this process is quite often computationally expensive. Mean field annealing, a heuristic which mimics the idea of mean field approximation from statistical physics [24] can be employed instead. In MFA, the stochastic process in simulated annealing is replaced by a set of deterministic equations. Though MFA does not guarantee convergence to a global optimal solution, it can provide an excellent approximation to an optimal solution and is much less expensive computationally.

3.4. *Mixed Neural-Genetic Algorithm*

As in the algorithm presented by Yeo et al. in [33] Salcedo-Sanz et al. [30] introduced a two-phase heuristic based combining both Hopfield neural networks [14] and genetic algorithms as in [32]. As with the vertex coloring algorithm, phase one of the mixed neural-genetic algorithm establishes a minimum framelength and phase two attempts to maximize the utilization within the slot.

3.4.1. *Frame Length Minimization*

The frame length minimization problem presented in [30] is the same as described above. For the solution, a discrete-time binary Hopfield neural network (HNN) is used. As described in [30], the HNN can be described as a graph whose vertices are the neurons (stations) and whose edges are the direct collisions. The graph is then mapped to the schedule matrix S as defined in Section 2 above. The neurons are updated one at a time after a randomized initialization until the system converges, thus yielding the minimum frame length.

3.4.2. *Utilization Maximization*

In this phase, a genetic algorithm is used to maximize the channel utilization within the frame length that was determined in phase one. A HNN is also used to ensure that all constraints are satisfied. Genetic algorithms receive their name from the an explanation of the way the behave. Not surprisingly, they are based on Darwin's Theory of Natural Selection. Genetic algorithms store a set of solutions and then work to replace these solutions with better ones based on some fitness criterion such as the objective function value.

4. Computational Results

With the exception of the algorithm presented in [4] all the aforementioned heuristics were tested using three examples introduced by Wang and Ansari in [31] which have become the de facto test cases for broadcast scheduling algorithms. These examples include 15, 30, and 40 station networks with varying densities. The graphs of these networks can be seen in Figure 1.

Though these examples provide reasonable insight into the quality of a given heuristic, they are relatively trivial networks. Notice in Figure 3 that the lower bound calculations are trivial for these test cases. Therefore, in order to get a better comparison of the various heuristics, the authors generated 60 random unit disc graphs having varying radii and stations. Twenty graphs of networks having 50, 75, and 100 stations were generated with each broken up into four sets of five graphs with radii of 20, 30, 40, and 50 units. Unit disc graphs are popular models. In such a graph, each node has the same transmission range equal to the radius of a disc centered at the node.

4.1. *Average Time Delay*

Evaluating the average time delay of packets as the utilization of the network increases is an effective means of evaluating a broadcast scheduling heuristic. The following assumptions were made before deriving the average time delay:

(1) Packets have a fixed length, and the length of a time slot is equal to the time required to transmit a packet.
(2) The interarrival time for each station i is statistically independent from other stations, and packets arrive according to a *Poisson* process with a rate of λ_i (packets/slot). The total traffic in station i consists of its own traffic and the traffic incoming from other stations. Packets are stored in buffers in each station and the buffer size is infinite.
(3) The probability distribution of the service time of station i is deterministic. Define the service rate of station i to be μ_i (packets/slot).
(4) Packets may be transmitted only at the beginning of each time slot.

Using the above assumptions, an ad-hoc network can be modeled as a system of N M/D/1 queues, where N is the number of stations in the network. The *Pollaczek-Khinchin (P-K) formula* [13] is used to determine the average time delay of each queue. Letting D_i represent the average time delay for each station i, then by P-K, we have:

$$D_i = \frac{1}{\mu_i} + \frac{\lambda_i/\mu_i^2}{2(1-\rho_i)}, \tag{7}$$

where $\mu_i = \sum_{m=1}^{M} s_{mi}/M$, and $\rho_i = \lambda_i/\mu_i$. Thus, the total time delay is given by

$$D = \frac{\sum_{i=1}^{N} \lambda_i D_i}{\sum_{i=1}^{N} \lambda_i}. \tag{8}$$

Notice that the average time delay is most affected by the number of time slots in the frame. The second dominating factor is the total utilization within the frame. As expected, those heuristics which result in schedules

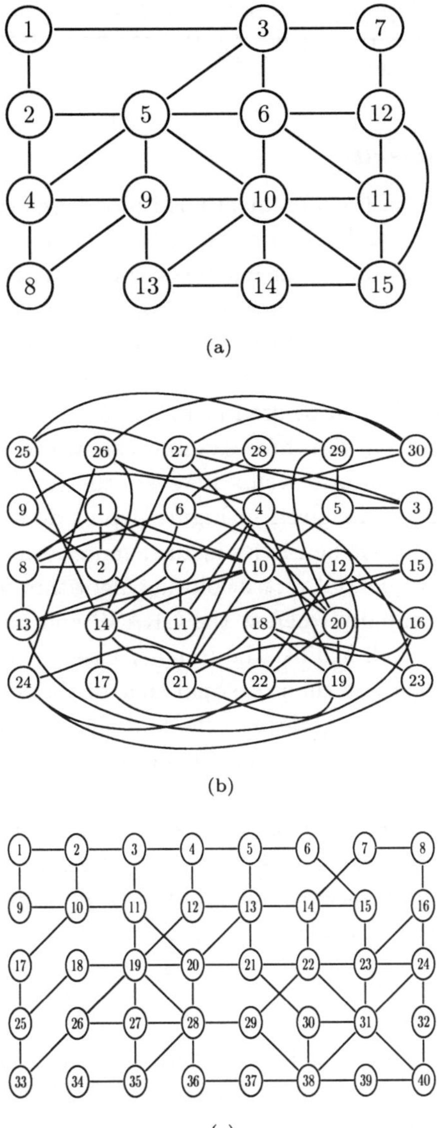

Fig. 1. (a) 15 station network. (b) 30 station network. (c) 40 station network.

with fewer slots and greater throughput experience a less significant delay as the arrival rate increases. Note that some delay plots my be indistinguishable due to the overlapping with others with similar slots and utilizations. Graphs of the average time delay for the 15, 30 and 40 station networks can be seen in Figure 2.

4.2. *Numerical Results*

Comparative results for the aforementioned broadcast scheduling heuristics are shown in Figures 3 and 4. Due to lack of availability, the MFA algorithm was not considered for the larger instances who results are given in Figure 4. Note that in the 15, 30, and 40 station examples, GRASP and the mixed neural-genetic algorithm all achieve schedules with optimal frame lengths of 8, 10, and 8 respectively. Examples of typical broadcast schedules for these example networks can be seen in Figure 5 [3].

With the results from the three examples shown in Figure 3 it appears as if all the routines perform well. However, the results of the randomly generated unit disc graphs given in Figure 4 provide a better insight into the abilities of the various heuristics against more formidable instances.

Notice that the sequential vertex coloring algorithm tends to decrease in performance rather rapidly whereas the GRASP and mixed neural-genetic algorithm fare quite well against the tougher cases. For all cases, GRASP attains a frame length less than or equal that of the other heuristics with quite comparable utilizations despite the fact that the objective is only to minimize the frame length. The mixed neural-genetic algorithm also yields good results with respect to utilization and frame length.

Recall that it is the total number of slots that most greatly affects the average time delay of packets in the network. Thus, if one were considering the frame length minimization version of the BSP, it appears as if GRASP would be the heuristic of choice.

5. Conclusion

In this chapter, we introduced the Broadcast Scheduling Problem as an important *NP*-complete combinatorial problem. We formally defined the problem and discussed several algorithms which have been applied to the BSP, all with competitive results. Over 60 different networks of varying size and densities were tested and the results compared. Some heuristics approached the problem of minimizing the total slots per frame, while others partitioned the BSP into two problems and used varying methods to arrive at a final answer. As research on ad-hoc networks increases, so too will applications of the BSP and other exciting problems which are constantly arising in research on optimization in telecommunication.

Acknowledgments

The authors would like to thank Dr. Sancho Salcedo-Sanz for donating the source code from the mixed neural-genetic algorithm [14] described in Section 3.4 to be tested on the unit disk graphs described previously. The SVC algorithm was coded by the current authors for testing.

References

[1] R.M. Aiex, S. Binato, and M.G.C. Resende. Parallel GRASP with path-relinking for job shop scheduling, *Parallel Computing*, vol. 29, Pages 393-430, 2003.

[2] S. Binato, W.J. Hery, D.M. Loewenstern, and M.G.C. Resende. A GRASP for job shop scheduling, *Essays and Surveys on Metaheuristics*, C.C. Ribeiro and P. Hansen, Eds., Kluwer Academic Publishers, Pages 58-79, 2002.

[3] C.W. Commander, S.I. Butenko, and P.M. Pardalos. A Greedy Randomized Adaptive Search Procedure for the Broadcast Scheduling Problem, in progress.

[4] A. Ephremides and T.V. Truong. Scheduling broadcast in multihop radio networks. *IEEE Transactions on Communications*, vol. 38, Pages 456-460, 1990.

[5] S. Even, O. Goldreich, S. Moran, and P. Tong. On the NP-completeness of Certain Network Testing Problems. *Networks*, 14, 1984.

[6] T.A. Feo and M.G.C. Resende. A Probabilistic Heuristic for a Computationally Difficult Set Covering Problem. *Operations Research Letters*, 8, Pages 67-71, 1989.

[7] T.A. Feo and M.G.C. Resende. Greedy Randomized Adaptive Search Procedures. *Journal of Global Optimization*, 6, Pages 109-133, 1995.

[8] P. Festa and M.G.C. Resende. GRASP: An Annotated Bibliography. In C.C. Ribeiro and P. Hansen, editors, *Essays and Surveys on Metaheuristics*, Pages 325-367. Kluwer Academic Publishers, 2001.

[9] M.R. Garey and D.S. Johnson. *Computers and Intractability: A Guide to the Theory of NP-Completeness.* New York: W.H. Freeman, 1979.

[10] F. Glover. Tabu Search and Adaptive Memory Programming - Advances, Applications, and Challenges. In R.S. Barr, R.V. Helgason, and J.L. Kenngington, editors, *Interfaces in Computer Science and Operations Research*, Pages 1-75. Kluwer Academic Publishers, 1996.

[11] M. Grotschel, L. Lovász, and A. Schrijver. Geometric Algorithms and Combinatorial Optimization. *Springer-Verlag, Berlin*, 1987

[12] B. Hajek and G. Sasaki. Link Scheduling in Polynomial Time. *IEEE Transactions on Information Theory*, 34, Pages 910-918, 1988.

[13] F.S. Hillier and G.J. Lieberman. *Introduction to Operations Research.* New York: McGraw Hill, 2001.

[14] J.J. Hopfield and D.W. Tank. Neural Networks and Physical Systems with

Emergent Collective Computational Abilities. *Proceedings of the National Academy of Science*, Pages 2541-2554, 1982.

[15] D. Jungnickel, *Graphs, Networks and Algorithms.* Berlin, Germany: Springer-Verlag, 1999.

[16] S. Kirkpatrick, C. Gelatt, and M. Vecchi. Optimization by Simulated Annealing. *Science*, 220, Pages 671-680, 1983.

[17] L. Kleinrock, and J. Silvester. Spatial Reuse in Multihop Packet Radio Networks. *Proceedings of the IEEE 75*, 1987.

[18] M. Laguna and R. Martí. A GRASP for Coloring Sparse Graphs. *Computational Optimization and Applications*, vol. 19, no. 2, Pages 165-178, 2001.

[19] M. Laguna and R. Martí. GRASP and Path Relinking for 2-Layer Straight Line Crossing Minimization. INFORMS Journal on Computing, 11, Pages 44-52, 1999.

[20] Y. Li, P.M. Pardalos, and M.G.C. Resende. A Greedy Randomized Adaptive Search Procedure for the Quadratic Assignment Problem, *Quadratic Assignment and Related Problems*, P.M. Pardalos and H. Wolkowicz, eds., *DIMACS Series on Discrete Mathematics and Theoretical Computer Science*, vol. 16, Pages 237-261, 1994.

[21] L. Lovász. On the Shannon Capacity of a Graph. *IEEE Transactions on Information Theory*, vol. IT-25, no. 1, Pages 1-7, 1979.

[22] C.A.S. Oliveira, P.M. Pardalos, and M.G.C. Resende. GRASP with Path-Relinking for the QAP, *5th Metaheuristics International Conference*, Pages 57.1-57.6, 2003.

[23] P.M. Pardalos, L.S. Pitsoulis, and M.G.C. Resende. A Parallel GRASP Implementation for the Quadratic Assignment Problem, *Parallel Algorithms for Irregular Problems*, A. Ferreira and J. Rolim, eds, Kluwer Academic Publishers, Pages 111-130, 1995.

[24] C. Peterson and B. Soderberg. A New Method for Mapping Optmization Problems onto Neural Networks, *International Journal of Neural Systems*, vol. 1, no. 1, Pages 3-22, 1989.

[25] L.S. Pitsoulis, P.M. Pardalos, and M.G.C. Resende. A Parallel GRASP for MAX-SAT, *Lecture Notes in Computer Science*, vol. 1180, Pages 575-585, 1996.

[26] M.G.C. Resende and C.C. Ribeiro. A GRASP with Path-relinking for Private Virtual Circuit Routing, *Networks*, vol. 41, Pages 104-114, 2003.

[27] M.G.C. Resende and C.C. Ribeiro. GRASP with Path-relinking: Recent Advances and Applications, *Metaheuristics: Progress as Real Problem Solvers*, T. Ibaraki, K. Nonobe and M. Yagiura, (Eds.), Kluwer Academic Publishers, 2005.

[28] M.G.C. Resende and C.C. Ribeiro. Greedy randomized adaptive search procedures, in *Handbook of Metaheuristics*, F. Glover and G. Kochenberger, eds., Kluwer Academic Publishers, Pages 219-249, 2003.

[29] M.G.C. Resende and T.A. Feo. A GRASP for Satisfiability. *Cliques, Coloring, and Satisfiability: Second DIMACS Implementation Challenge*, David S. Johnson and Michael A. Trick ,Eds., *DIMACS Series on Discrete Mathematics and Theoretical Computer Science*, vol. 26, Pages 499-520, American

Mathematical Society, 1996.

[30] S. Salcedo-Sanz, C. Busoño-Calzón, and A.R. Figueiral-Vidal. A Mixed Neural-Genetic Algorithm for the Broadcast Scheduling Problem, *IEEE Transactions on Wireless Communications*, vol. 2, no. 2, 2003.

[31] G. Wang and N. Ansari. Optimial Broadcast Scheduling in Packet Radio Networks Using Mean Field Annealing. *IEEE Journal on Selected Areas in Communications*, 15(2), 1997.

[32] Y. Watanabe, N. Mizuguchi, and Y. Fujii. Solving Optimization Problems by Using a Hopfield Neural Network and Genetic Algorithm Combination. *Syst. Comput. Japan* vol. 29, no. 10, Pages 68-73, 1998.

[33] J. Yeo, H. Lee, and S. Kim. An Efficient Broadcast Scheduling Algorithm for TDMA Ad-hoc Networks. *Computers and Operations Research*, 29, Pages 1793-1806, 2002.

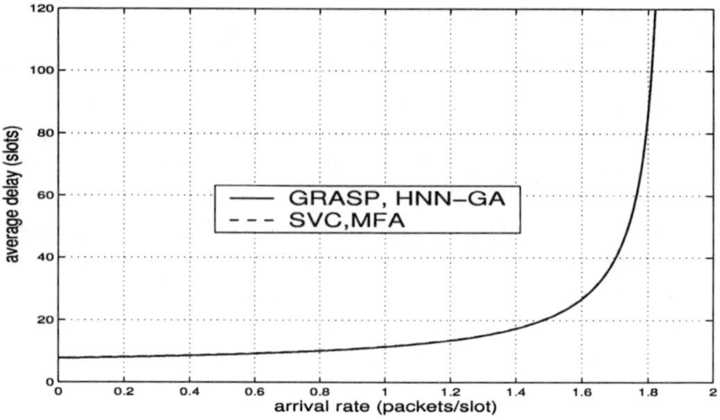

(a)

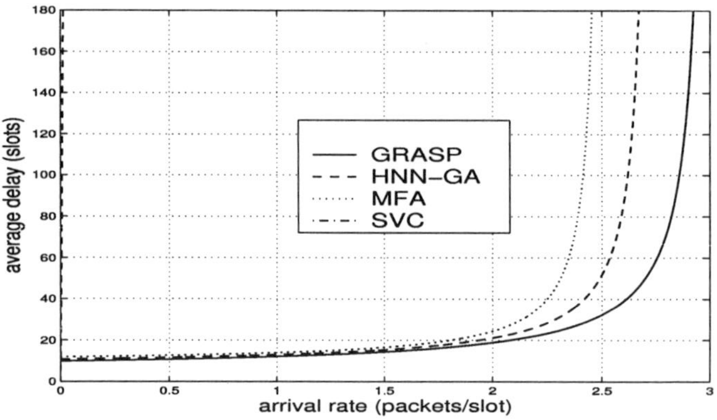

(b)

Fig. 2. Comparison of average time delays for example networks: (a) 15 station network, (b) 30 station network, (c) 40 station network.

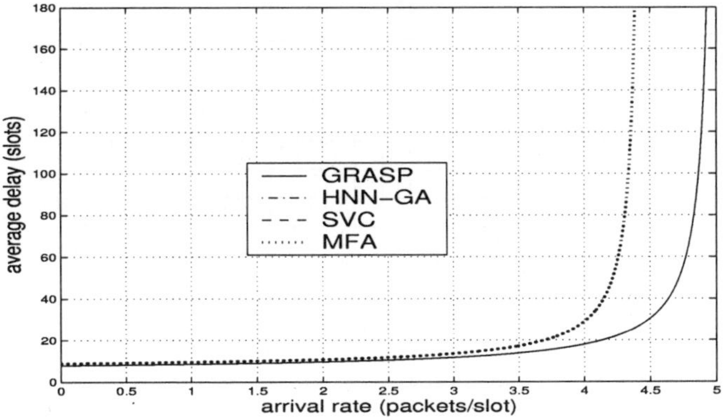

(c)

Fig. 2. (*Continued*)

Stations	LB	Frame Length				Channel Utilization			
		GRASP	HNN-GA	SVC	MFA	GRASP	HNN-GA	SVC	MFA
15	8	8	8	8	8	0.167	0.167	0.15	0.15
30	10	10	10	11	12	0.120	0.117	0.112	0.108
40	8	8	8	8	9	0.203	0.209	0.188	0.197

Fig. 3. In this table, LB represents the lower bounds of M which were derived using the method in [15]. The results compare the heuristics discussed in Section 4.

		GRASP	HNN-GA	SVC
	r = 20	n = 7, ρ = .288	n = 7, ρ = .2926	n = 7.4, ρ = .1997
n = 50	r = 30	n = 8.8, ρ = .1709	n = 9, ρ = .1985	n = 10, ρ = .1416
	r = 40	n = 13.2, ρ = .1142	n = 13.2, ρ = .1142	n = 15.2, ρ = .0876
	r = 50	n = 15.8, ρ = .0835	n = 16, ρ = .0845	n = 17.8, ρ = .0701
	r = 20	n = 8, ρ = .209	n = 8, ρ = .2187	n = 9.6, ρ = .1764
n = 75	r = 30	n = 13, ρ = .1213	n = 13, ρ = .1273	n = 13.8, ρ = .0964
	r = 40	n = 18.4, ρ = .0786	n = 18.4, ρ = .0803	n = 20.6, ρ = .0576
	r = 50	n = 25.2, ρ = .0564	n = 26, ρ = .0583	n = 28.2, ρ = .0523
	r = 20	n = 10.8, ρ = .1628	n = 10.8, ρ = .1966	n = 11.6, ρ = .1126
n = 100	r = 30	n = 16.8, ρ = .0899	n = 16.8, ρ = .0938	n = 17, ρ = .0769
	r = 40	n = 24.8, ρ = .0613	n = 24.8, ρ = .0640	n = 26.6, ρ = .0515
	r = 50	n = 33.2, ρ = .0425	n = 34, ρ = .0435	n = 35.6, ρ = .0360

Fig. 4. The average frame length and utilization are given for the randomly generated networks consisting of 50, 75, and 100 stations with radii of 20, 30, 40, and 50 units.

slot \ station	1	2	3	4	5	6	7	8	9	10	11	12	13	14	15
1					B										
2						B		B							
3							B			B					
4	B								B			B			
5			B					B							B
6	B							B			B				
7				B			B							B	
8		B					B						B		

(a)

slot \ station	1	2	3	4	5	6	7	8	9	10	11	12	13	14	15	16	17	18	19	20	21	22	23	24	25	26	27	28	29	30
1										B	B																			B
2						B			B					B																
3						B	B													B										
4				B									B												B					
5		B										B											B				B			
6					B														B							B				
7		B	B													B					B									
8								B														B						B		
9	B																	B					B							B
10									B				B		B		B							B					B	

(b)

slot \ station	1	2	3	4	5	6	7	8	9	10	11	12	13	14	15	16	17	18	19	20
1		B						B							B					B
2	B				B		B										B		B	
3				B						B				B		B				
4	B					B		B			B									
5		B				B		B				B								
6		B				B	B											B		
7	B							B					B				B			
8			B						B						B					

slot \ station	21	22	23	24	25	26	27	28	29	30	31	32	33	34	35	36	37	38	39	40
1												B	B	B				B		
2			B											B					B	
3								B		B			B							
4											B		B	B		B				
5		B			B							B			B		B			
6				B					B					B						
7						B			B					B						B
8	B						B										B			

(c)

Fig. 5. Example GRASP broadcast schedules for the networks given in Figure 1: (a) 15 station network, (b) 30 station network, (c) 40 station network [3].

CHAPTER 5

NATURAL LANGUAGE PROCESSING IN CONTROL OF UNMANNED AERIAL VEHICLES

Emily M. Craparo

Laboratory for Information and Decision Systems
Massachusetts Institute of Technology
emily1@mit.edu

Felix Sheng-Ho Chang and Jae W. Lee

Computer Science and Artificial Intelligence Laboratory
Massachusetts Institute of Technology

Robert Berwick and Eric Feron

Laboratory for Information and Decision Systems
Massachusetts Institute of Technology

Unmanned aerial vehicles (UAVs) are becoming increasingly useful in military, commercial, and scientific applications. Currently, UAVs can be found performing surveillance and reconnaissance missions for the military, collecting scientific data in areas that are inhospitable or inaccessible to humans, and furthering commercial and agricultural enterprises. One of the primary needs of military and civilian users is to develop an interface for a single human operator to coordinate multiple UAVs with the same ease that air traffic controllers coordinate multiple aircraft.

This chapter develops the framework for a natural language interface to a UAV. We apply our expertise in air traffic control and airport operations, combined with existing natural language processing techniques, to achieve a higher recognition success rate than traditional natural language processing endeavors in a more general domain of discourse typically do. Because there already exists a structured, yet intuitive, language for air traffic control, this chapter takes advantage of the phraseology already developed for this purpose. A corpus of air traffic control commands was gathered from recorded exchanges between pilots and controllers at Boston's Logan Airport, as well as Hanscom Field in Bedford, MA, and these were used as the "target language" for this chapter.

The implementation of a language processing system that operates according to this language is described. We believe that this is the first attempt at formalizing air traffic control phraseology for use in an unmanned system.

Keywords:

Natural language interface, language processing, dialog system, unmanned air traffic control

1. Introduction

Natural language processing (NLP) is making its way into more and more areas of everyday life. Current applications of natural language processing include automated telephone customer service aids with speech processing capabilities; a variety of translation and language tools available via search engine sites such as Google and AltaVista; automated weather reporting programs that accept raw data and pass information on to the public in plain, unedited English [2]; and grammar and style checkers in commercially available word processing software. Indeed, it is not surprising that this should be so, for people already regard computers as social entities. For example, studies show that a human is more apt to rate a computer's performance positively if the computer has recently said something flattering about the human. Additionally, the human is more likely to rate a computer's performance positively if that computer asks for the rating than if a different computer asks the same questions [15]. Given this predisposition toward treating computers as one would treat other people, it is only fitting that human-computer interfaces mirror existing human communication systems.

Current interfaces between humans and unmanned aerial vehicles (UAVs), however, do little to recognize the human affinity for verbal communication. Existing control schemes, such as datalink and radio control, severely restrict the extent to which UAVs can be successfully integrated into the existing civil and military air traffic control system. This need not be, however - both the highly structured nature of air traffic control phraseology and the algorithmic and goal-driven nature of flight make air traffic control an ideal venue for the application and development of natural language processing technology.

Churcher et al. intended to use speech recognition technology to automatically transcribe certain, essential parts of transmissions between the air traffic control (ATC) and airborne pilots [3]. They claimed that they could use this transcript for ATC training purposes or for relaying this information back to the pilot real-time to help maneuvering the aircraft. They used IBM's off-the-shelf commercial continuous speech recognizer, and it gave them only a modest accuracy of recognition (around 30%) in its base form. However, when the device was augmented with other knowledge sources and higher levels of linguistics such as contextual information and context-free syntax, the accuracy could be greatly improved to over 70% even in noisy environments.

An important example of such high-level knowledge is the structure of a discourse. Discourse is defined as a collocated, related group of sentences, and a group of sentences must be coherent in order to make a discourse [12]. There are at least two different approaches to coherence. One is an informational approach, where the relation between sentences impose constraints such as result, explanation, parallel, elaboration, and occasion on the information in the utterances [10]. Historically, this approach has been applied predominantly to monologues between a speaker and hearers. The other approach is called intentional approach, where utterances are understood as actions, requiring that the listeners infer the underlying goal [9]. This intentional approach has been applied mostly to dialogues.

The notion of intentional coherence in discourse plays a significant role in ATC system, because the high level tasks of landing, takeoff, and maneuvering around the airport must be coordinated by a group of sentences, not just individual sentences. Even human pilots are urged to take advantage of the predictable nature of flight discourse; a student guide to voice communications published by the United States Navy gives this advice [16]:

> Know what to **expect**. As you progress through each flight, you should know what is expected to happen. If you know what is to be said ahead of time, responding correctly will be much easier.

Thus, pilots are expected to rely on their knowledge of the established structure of both flight and communications to form their responses to ATC commands. The manual continues:

> Every conversation with a controlling agency or service follows a specific progression...proper communication involves the realization of which progression phase you are in and making correct and timely responses.

This aspect of communication is not lost on computational linguists. Grosz argued that a discourse could be represented as a composite of three interacting components: a linguistic structure, an intentional structure, and an attentional state [9]. Grosz also pointed out that task-oriented dialogues have a linguistic structure that closely parallels the intentional structure of the task being performed [8]. The fundamental idea is that a discourse has an underlying purpose which it aims to achieve, called the discourse purpose (DP), and that each segment of a discourse also has a finer-grained purpose, called discourse segment purpose (DSP). Then they are organized in a tree-like hierarchy with two coherence relations: dominance and satisfaction-precedence. This structure helps a discourse management system understanding the intention of a speaker.

Figure 1 illustrates the way in which a natural language processing module might fit into a feedback control loop: the human controller compares the UAV's current state with the desired state in much the same way as an air traffic controller. The controller then issues a command to the UAV, which is processed by the natural language processing module to produce a "machine language" command. This command may then be sent to a planning or optimization module, which then sends an exact trajectory into the low-level control system, from which appropriate commands are sent to the actuators and control surfaces of the aircraft.

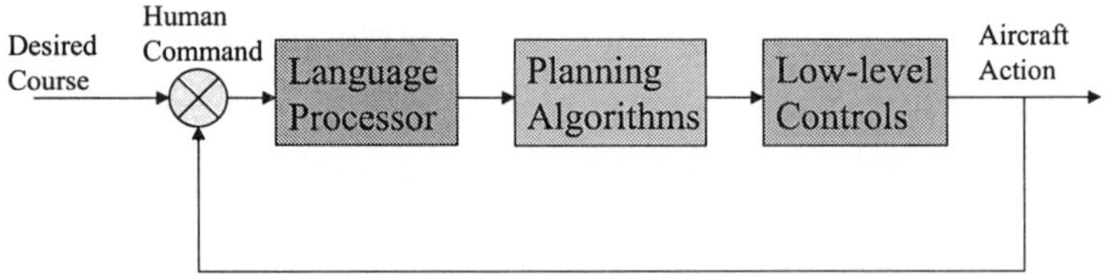

Fig. 1. Incorporation of a natural language processor into a feedback loop.

2. Overview of Example Implementation

The remainder of this chapter describes the implementation of a basic natural language processing system built to interpret air traffic control commands and simulate appropriate aircraft responses. It was constructed using a corpus of air traffic control transmissions collected from Boston's Logan International Airport and the Laurence G. Hanscom Field in Bedford, MA.

The system is comprised of specialized modules as shown in Figure 2. Ground and air traffic controllers can issue commands to the preprocessor, which then passes their edited sentences to the sentence parser. The sentence parser recognizes a set of sentence structures, and converts sentences into standardized verb templates. These verb templates are then passed to the discourse manager. In the discourse manager, consecutive verb templates are analyzed for feasibility and inconsistencies (for example, if an aircraft was told two conflicting destinations). Inconsistencies are reported to the human controller in the same way a pilot would ask a controller for clarifications. If no inconsistencies are found, the verb templates are transformed into low-level commands for the aircraft. For this simulation, a database contains the current states of the airport and the various aircraft in the system. It is consulted when the discourse manager checks for inconsistencies, and it is updated by the discourse manager when new commands arrive. Finally, the graphical simulator displays the airport and aircraft states, and it maintains the simulation clock.

3. Detailed Description

3.1. *Parser*

Our parser is based on the existing Earley parser [5], with significant modifications to existing dictionaries and syntax rules. The dictionary contains the verbs relevant to air traffic control, as well as new classes of words for airplane names, taxiway names, units of measurement, and numbers that conform to conventional air traffic control phraseology. The syntax rules were modified to handle the rather truncated grammar of air traffic control, in which most sentences are imperatives and "unnecessary" words such as prepositions are often (but not always) omitted. The parser filters out semantically empty phrases such as "I want you to [*command*]" or "I'm going to have you [*command*]."

The verb frames generated by our parser are rather specialized - they output information with relevant headings based on the verb being parsed.

For example, the input sentence "*taxi to runway three via zulu*" gener-

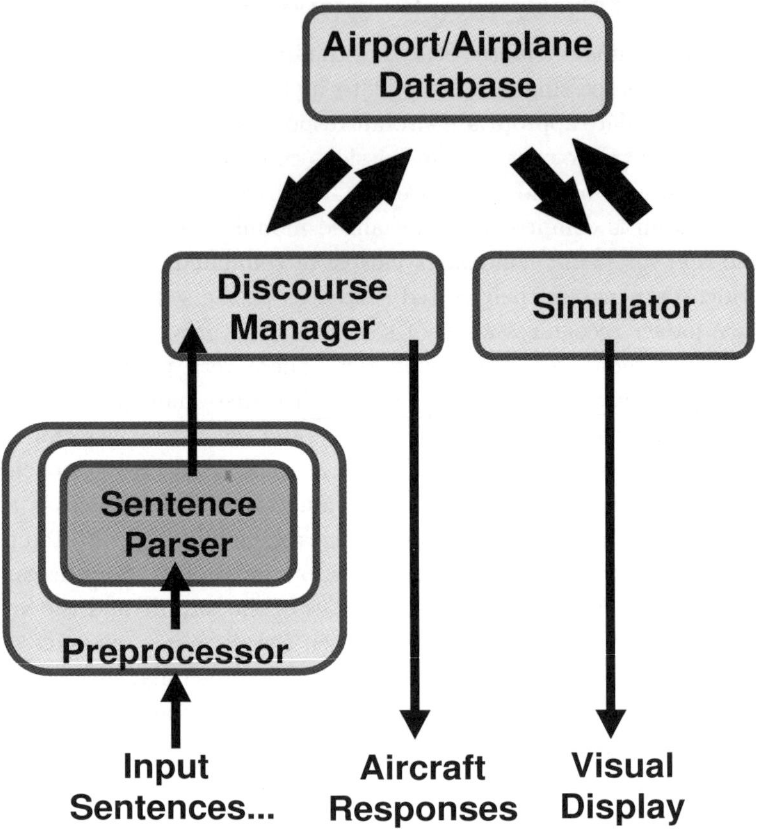

Fig. 2. System layout.

ates
`[OUTPUT](go :on (runway :num 3) :via (taxiway :num zulu) :agent you)`
while the sentence "*turn three*" generates
`[OUTPUT] (turn :heading (heading :to 3) :agent you)`

See the appendix for further examples.

3.2. *Discourse Manager*

A major issue with our parser is that of overgeneration. Since the context-free grammar rules used by the parser were intentionally made as general as possible, many nonsensical sentences are parsed, i.e. "follow ramp," which could result in unexpected aircraft behavior. It was to alleviate this problem that the discourse manager was created.

When the discourse manager (shown in Figure 3) receives a verb template from the parser, it performs semantic interpretation to determine what action to take and updates the states of aircraft in the system database accordingly. If, however, the discourse manager finds the command to be nonsensical or inconsistent with previous commands, it generates a response message back to the requester of the action. It takes the context of the system into account, and rejects a message requesting an action in conflict with the context even if the message is a syntactically and semantically valid one at the sentence level.

As shown in Figure 3, this module has three internal stages: a preprocessor, dispatchers, and handlers. The first stage is the entry point of the entire discourse manager, which takes a verb template from the parser and delivers a string of message back to the caller function. The second stage consists of a group of dispatch functions. The preprocessor transforms the arguments of a given verb template into a data structure containing (record, value) pairs, and invokes an appropriate dispatcher depending on the verb. We have over a dozen verb entries in our dictionary, such as "climb," "descend," "taxi," "remain," and"maintain." Each verb entry has its own dispatch function in the form of dispatch_*(). For example, any command containing verb "maintain" is processed by dispatch_maintain().

It is the responsibility of these dispatch functions to further parse their arguments, and invoke the appropriate actions (such as "change the altitude"). At this stage, these functions can detect and reject messages that are semantically valid in the sentence level, but are not coherent in the greater context.

First in the dispatch functions' analysis of a command is the notion of history. For example, there exist commands that nullify the effects of the previous command, such as "Cancel that", "Never mind", and "Let's not do that." In order to resolve the meanings of "that", we need to keep track of the history of dialogues. Under our constrained situation, we assume that we can undo only the preceding operation, and if there is nothing cancellable, the system cannot search further back into history. For example,

```
TOWER: Delta, clear to taxi to two-niner.
PLANE: Roger.
TOWER: Cancel that.
PLANE: Roger. Please specify a new destination.
```

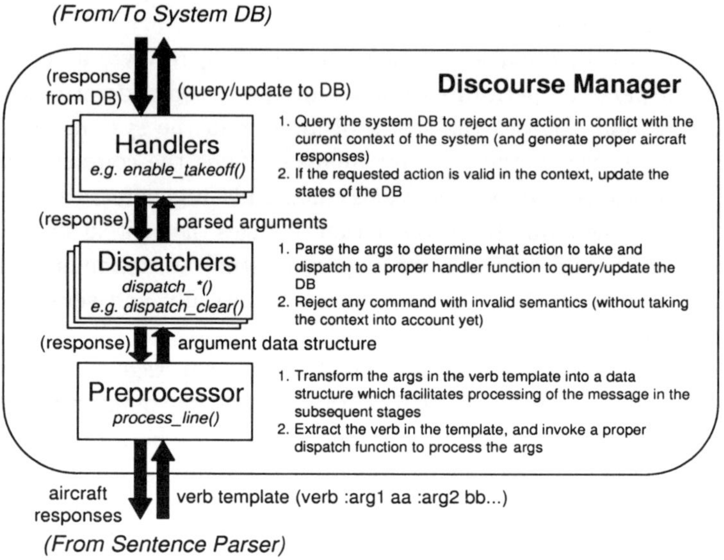

Fig. 3. The discourse manager.

A second stage of analysis deals with implicit references. For example, there are commands such as "remain this frequency" that require contextual knowledge for complete comprehension. Thus, the dispatcher can query the airport/aircraft database for clarification of some commands.

The notions set forth by Grosz et al. [9] for dealing with the structure of a discourse are particularly relevant in the design of the discourse manager. In our preliminary ATC system, we define two dominant intentions: takeoff and landing. These intentional structures determine the response of the system given an input utterance. That is, the utterance is mapped to a corresponding (intermediate) intention, and handlers look up the database to see whether or not all the prerequisites have been met. For example, it doesn't make sense to issue ground-specific commands to an aircraft in mid-air, and vice versa. Nonsensical or inapplicable commands are rejected and the user is requested to make a new command.

The last stage has a group of handlers that actually update the states of the system. Typically, a more sophisticated discourse manager is needed to resolve references and to determine the intentional structure of dialogues. However, for this project we can constrain the input sentences to the predefined corpus for ATC, and this enables us to interpret the human commands within the specific domain of air control. Without this constraint, the problem becomes much more complicated and unpredictable, and the discourse manager needs to be much more robust [9, 13, 1, 14, 4].

3.3. *Database*

The airport is modeled as a set of points on the ground. Each point has a specific (X,Y) coordinate, as well as an adjacency matrix that describes which other points are connected to this point. Figure 4 shows the points and the segments (in blue) that we model for the Laurence G. Hanscom airport in Bedford, Massachusetts.

To aid in the simulation, we implemented a variety of ground navigation algorithms. For example, we implemented Dijkstra's Shortest Path with a slight modification: we heavily penalized any segment that is designated as a runway. Although an airplane will not enter a runway unless it has been given clearance to do so, we still wanted the airplanes to minimize their time on the runway. This constraint, which mirrors safety procedures practiced by human pilots, reduces the number of runway incursions that are likely to occur.

Other airport states that we need are a landing priority queue, and a take off priority queue. In a typical airport, multiple aircraft would often want to take off at about the same time. Airport controllers resolve the conflict by assigning a priority number for them: when a lower priority aircraft is ready to take off, if there are high priority aircraft that are still getting read, the lower priority aircraft can not go ahead. Instead, it has to wait.

4. Results

The system described above has been tested on a number of realistic situations. Some example scenarios follow.

The first scenario involves an airplane (taking off). Here, the controller initially intends for the aircraft to go to one runway, and subsequently decides to change to another runway. The natural language processing system on the aircraft understands the commands, waits for new orders, and proceeds to perform aircraft run-up, lineup, position, and takeoff on its own. Transcript of exchange:

```
ALASKA: Ready to taxi.
```

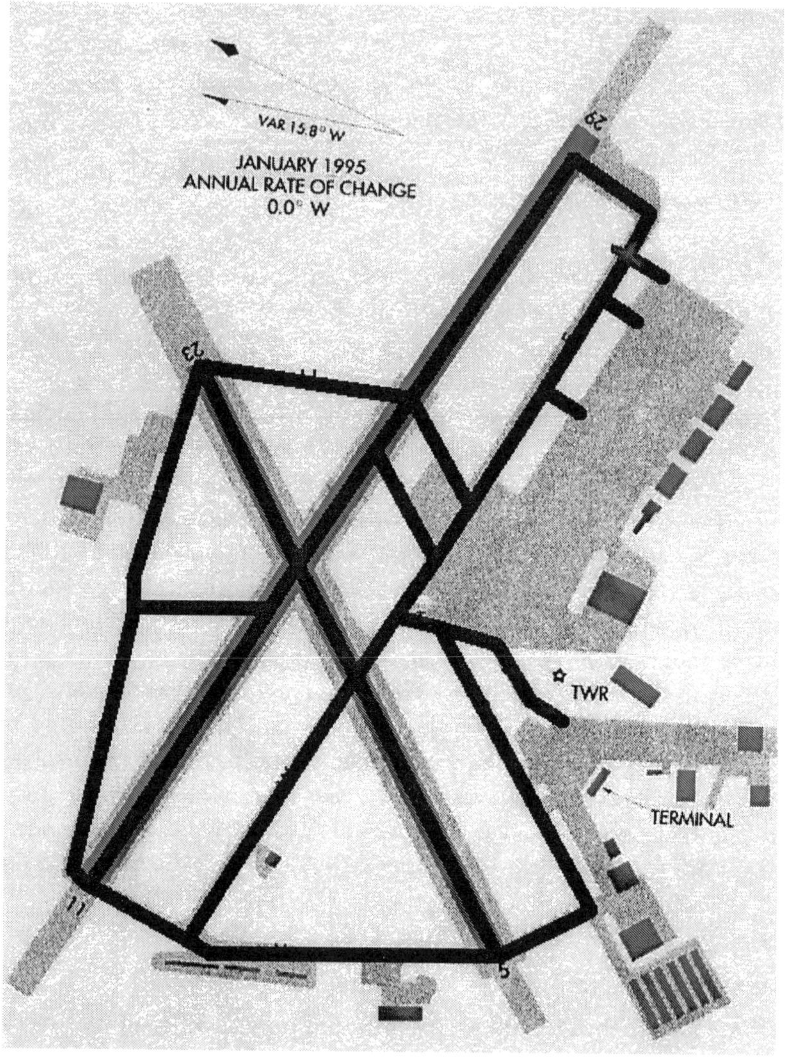

Fig. 4. Laurence G. Hanscom airfield [6].

```
CONTROLLER: Alaska, clear to taxi to runway 5.
ALASKA: Roger. [Begins to taxi]
CONTROLLER: Cancel that.
ALASKA: Roger. Awaiting a new destination.
CONTROLLER: Taxi to twoniner via echo.
ALASKA: Roger. [Taxis to runway 29 via echo]
```

```
ALASKA: Runup completed. Ready for departure.
CONTROLLER: Alaska, hold short.
ALASKA: Roger.
CONTROLLER: Alaska, position and hold.
ALASKA: Position and hold. [Taxis to position]
    Ready for departure.
CONTROLLER: Clear for takeoff.
ALASKA: Roger. [Takes off]
```

The second scenario involves two aircraft (Horizon and United) who both want to land. The controller told Horizon that it should wait behind United. Landing clearance is granted for United first, and then Horizon lands subsequently. Transcript of exchange:

```
HORIZON: Inbound
UNITED: Inbound
CONTROLLER:
Horizon, you're number two for landing behing a United.
United, you're clear for landing runway 29.
UNITED: Roger. Preparing to land on runway 29. [Horizon does
not prepare.]
CONTROLLER: Horizon, clear to land runway 29.
HORIZON: Roger. Preparing to land on runway 29.
```

The third scenario involves two aircraft (Alaska and Boeing) who both need to cross the same intersection. The controller had previously told Alaska to hold for Boeing (meaning Alaska has a lower priority). Transcript of exchange:

```
CONTROLLER: Alaska, hold for the Boeing.
ALASKA: Roger.
```

If they should both reach the intersection at about the same time, Boeing (shown here in orange) will wait until Alaska passes.

In the fourth scenario, Alaska is heading toward runway 29 when the controller dictates a particular taxiway to use. However, the controller says the wrong runway number. The discourse manager catches that, and the aircraft asks for clarification. Once the controller issues the new commands, the aircraft is able to perform its run-up and take off. Transcript of exchange:

```
ALASKA: Ready to taxi.
```

```
CONTROLLER: Alaska, clear to runway 29.
ALASKA: Roger. [Begins to taxi to 29]
CONTROLLER: Taxi to runway 10 via echo.
ALASKA: I do not think there is such a runway.
CONTROLLER: Taxi to runway 29 via echo.
ALASKA: Roger.
```

5. Future Work

Future additions and improvements to this work might include the integration of optimal path planning algorithms for aircraft in flight; a multiple-aircraft interface that allows the controller to address all or part of a fleet of aircraft, as well as an individual craft; a database manager capable of incorporating information extracted from such sources as automated weather and airport advisories and transmissions between other aircraft ("party-line information") as well as transmissions to that aircraft; cooperative strategies for multiple-aircraft operations; and of course further additions to the corpus of known sentences and a more sophisticated discourse manager capable of time-sensitive discourse and improved and expanded intentional inference.

Acknowledgments

This research was supported by AFOSR-DARPA MURI "Complex Adaptive Networks for Cooperative Control," University of Illinois, Subaward number 03-132 and by a Department of Defense NDSEG Graduate Fellowship.

References

[1] Carberry, S. (1990). *Plan Recognition in Natural Language Dialog*, MIT press.
[2] Chandioux, J. (1976). "METEO: un systeme operationnel pour la traduction automatique des bulletins meteorologiques destines au grand public." *Meta*, 21, pages 127-133.
[3] Churcher, Gavin E.; Ateall, Eric S. and Souter, Clive, "Dialogues in Air Traffic Control", *Proceedings of the 11th Twente Workshop on Language Technology*, 1996
[4] Doherty, Patrick; Granlund, Gosta; Kuchcinski, Krzystof; Sandewall, Erik; Nordberg, Klas; Skarman, Erik and Wiklund, Johan, "The WITAS Unmanned Aerial Vehicle Project", *Proceedings ECAI*, 2000.
[5] Earley, J, "An efficient Context-Free Parsing Algorithm", *Communications of the ACM*, 6(8), pages 451-455, 1970.

[6] *Airport/Facility Directory, Northeastern U.S.*, National Aeronautical Charting Office, Federal Aviation Administration, U.S. Department of Transportation, 2002.

[7] *2002 Federal Aviation Regulations and Aeronautical Information Manual*, Federal Aviation Administration, U.S. Department of Transportation.

[8] Grosz, B., "The Representation and Use of Focus in Dialogue Understanding", Ph.D. thesis, University of California, Berkeley, 1977.

[9] Grosz, Barbara and Sidner, Candance L., "Attention, Intentions, and the Structure of Discourse", *Computational Linguistics*, 12(3), pages 175-204, 1986.

[10] Hobbs, J.R. "Coherence and Coreference", *Cognitive Science*, 3, pages 67-90.

[11] *Private Pilot Manual*, Jeppenson Sanderson, Inc., Englewood, CO, 80112-5498, 2001.

[12] Jurafsky, D. and Martin, J.H., *Speech and Language Processing*, Prentice Hall, 2000.

[13] Litman, D.J. and Allen, J.F., "A Plan Recognition Model for Subdialogues in Conversation", *Cognitive Science*, 11, pages 163-200, 1987.

[14] Passonneau, R. and Litman, D.J., "Intention-based Segmentation: Human Reliability and Correlation with Linguistic Cues", ACL 93, Columbus, Ohio, 1993.

[15] Reeves, B. and Nass, C., *The Media Equation: How People Treat Computers, Television, and New Media Like Real People and Places.* Cambridge, University Press, Cambridge, 1996.

[16] United States Navy, Naval Air Training Command, *Voice Communications: Student Guide.* CNATRA P-806 (REV. 4-98) PAT. NAS Corpus Christi, TX, 1998.

[17] Wickens, Christopher D.; Mavor, Anne S.; Parasuraman, Raja and McGee, James P., "The Future of Air Traffic Control: Human Operators and Automation", Panel on Human Factors in Air Traffic Control Automation, National Research Council, 1998.

[18] Yangarber, Roman; Grishman, Ralph; Tapanainen, Pasi and Huttunen, Silja, "Automatic Acquisition of Domain Knowledge for Information Extraction", *Proceedings of the 18th International Conference on Computational Linguistics*, 2000.

Appendix: Examples of Parsed Outputs

```
sentence: ''altimeter six''
   [OUTPUT] (altimeter :value (6) :agent you)
   sentence: ''cancel that''
   [OUTPUT] (undo :agent you)
   sentence: ''change of plans''
   [OUTPUT] (undo :agent you)
   sentence: ''change speed to six''
   [OUTPUT] (setspeed :goal (6) :agent you)
```

```
sentence: ``change speed to six knots''
[OUTPUT] (setspeed :goal (6) :agent you)
sentence: ``checkin at five o'clock''
[OUTPUT] (checkin :time (timevalue :at 5) :agent you)
sentence: ``checkin in five minutes''
[OUTPUT] (checkin :time (timevalue :in 5) :agent you)
sentence: ``clear for takeoff''
[OUTPUT] (enable-takeoff :agent you)
sentence: ``clear the runway''
[OUTPUT] (leave :goal (runway) :agent you)
sentence: ``clear the taxiway''
[OUTPUT] (leave :goal (taxiway) :agent you)
sentence: ``clear to land''
[OUTPUT] (land :agent you)
sentence: ``clear to land runway five''
[OUTPUT] (land :on (runway :num 5) :agent you)
sentence: ``clear to land runway five left''
[OUTPUT] (land :left (runway :num 5) :agent you)
sentence: ``clear to takeoff runway two-niner''
[OUTPUT] (takeoff :on (runway :num 29) :agent you)
sentence: ``clear for landing runway two-niner''
[OUTPUT] (land :on (runway :num 29) :agent you)
sentence: ``clear to runway five''
[OUTPUT] (go :on (runway :num 5) :agent you)
sentence: ``clear to taxi to two''
[OUTPUT] (go :on (runway :num 2) :agent you)
sentence: ``cleared direct to runway five''
[OUTPUT] (go :on (runway :num 5) :agent you)
sentence: ``cleared to runway five''
[OUTPUT] (go :on (runway :num 5) :agent you)
sentence: ``climb and maintain three''
[OUTPUT] (climb-maintain :alt (3) :agent you)
sentence: ``climb and maintain three feet''
[OUTPUT] (climb-maintain :alt (3) :agent you)
sentence: ``contact departure''
[OUTPUT] (setfreq :goal (speaker :name departure) :agent
you)
sentence: ``contact departure at three''
[OUTPUT] (setfreq :goal (speaker :name departure :freq 3)
```

```
:agent you)
   sentence: ``contact ground''
   [OUTPUT] (setfreq :goal (speaker :name ground) :agent you)
   sentence: ``contact ground on five''
   [OUTPUT] (setfreq :goal (speaker :name ground :freq 5)
:agent you)
   sentence: ``contact ramp''
   [OUTPUT] (setfreq :goal (speaker :name ramp) :agent you)
   sentence: ``contact ramp on ten''
   [OUTPUT] (setfreq :goal (speaker :name ramp :freq 10)
:agent you)
   sentence: ``contact three''
   [OUTPUT] (setfreq :goal (speaker :freq 3) :agent you)
   sentence: ``contact tower''
   [OUTPUT] (setfreq :goal (speaker :name tower) :agent you)
   sentence: ``contact tower on eight''
   [OUTPUT] (setfreq :goal (speaker :name tower :freq 8)
:agent you)
   sentence: ``continue on tango''
   [OUTPUT] (go :on (taxiway :num tango) :agent you)
   sentence: ``continue on zulu until runway two''
   [OUTPUT] (go :on (taxiway :num zulu) :until (runway :num 2)
:agent you)
   sentence: ``cross runway three at zulu''
   [OUTPUT] (cross :road (runway :num 3) :at (taxiway :num
zulu) :agent you)
   sentence: ``cross runway two''
   [OUTPUT] (cross :road (runway :num 2) :agent you)
   sentence: ``cross two''
   [OUTPUT] (cross :road (runway :num 2) :agent you)
   sentence: ``cross two at five knots''
   [OUTPUT] (cross :fix 2 :at-speed 5 :agent you)
   sentence: ``cross two at three feet''
   [OUTPUT] (cross :fix 2 :at-altitude 3 :agent you)
   sentence: ``descend and maintain five''
   [OUTPUT] (descend-maintain :alt (5) :agent you)
   sentence: ``descend and maintain five feet''
   [OUTPUT] (descend-maintain :alt (5) :agent you)
   sentence: ``exit on taxiway tango''
```

```
   [OUTPUT] (go :on (taxiway :num tango) :agent you)
   sentence: ''exit the ramp''
   [OUTPUT] (leave :goal ramp :agent you)
   sentence: ''exit the ramp and follow alaska''
   [OUTPUT] (leave :goal ramp :agent you)
   [OUTPUT] (behind :goal (alaska) :agent you)
   sentence: ''exit the ramp behind continental''
   [OUTPUT] (leave :goal ramp :agent you)
   [OUTPUT] (behind :goal (continental) :agent you)
   sentence: ''expect boeing on tango''
   [OUTPUT] (expect :plane (boeing) :on (taxiway :num tango)
:agent you)
   sentence: ''expect three feet five minutes after
departure''
   [OUTPUT] (lock-altitude :alt 3 :time 5 :agent you)
   sentence: ''expect traffic on november''
   [OUTPUT] (expect :on (taxiway :num november) :agent you)
   sentence: ''fall in behind the alaska''
   [OUTPUT] (behind :goal (alaska) :agent you)
   sentence: ''follow a continental that is behind a boeing''
   [OUTPUT] (behind :goal (continental :behind boeing) :agent
you)
   sentence: ''follow in behind that alaska''
   [OUTPUT] (behind :goal (alaska) :agent you)
   sentence: ''follow that boeing''
   [OUTPUT] (behind :goal (boeing) :agent you)
   sentence: ''follow that boeing ahead to the runway''
   [OUTPUT] (behind :goal (boeing) :agent you)
   [OUTPUT] (go :on (runway) :agent you)
   sentence: ''follow that continental directly ahead of you''
   [OUTPUT] (behind :goal (continental :ahead you) :agent you)
   sentence: ''follow the boeing from your left''
   [OUTPUT] (behind :goal (boeing :left you) :agent you)
   sentence: ''get behind the continental''
   [OUTPUT] (behind :goal (continental) :agent you)
   sentence: ''give way to boeing''
   [OUTPUT] (behind :goal (boeing) :agent you)
   sentence: ''give way to the boeing''
   [OUTPUT] (behind :goal (boeing) :agent you)
```

```
sentence: ``go november''
[OUTPUT] (go :on (taxiway :num november) :agent you)
sentence: ``go straight down tango''
[OUTPUT] (go :on (taxiway :num tango) :agent you)
sentence: ``hold five''
[OUTPUT] (hold-heading :heading (heading :to 5) :agent you)
sentence: ``hold for the continental''
[OUTPUT] (behind :goal (continental) :agent you)
sentence: ``hold heading five''
[OUTPUT] (hold-heading :heading (heading :to 5) :agent you)
sentence: ``hold short''
[OUTPUT] (hold-short :agent you)
sentence: ``hold short of runway three on tango''
[OUTPUT] (hold-short :of (runway :num 3) :on (taxiway :num
tango) :agent you)
sentence: ``hold short of taxiway zulu''
[OUTPUT] (hold-short :of (taxiway :num zulu) :agent you)
sentence: ``hold short of zulu for spacing''
[OUTPUT] (hold-short :of (taxiway :num zulu) :agent you)
sentence: ``hold short of zulu for the continental''
[OUTPUT] (hold-short :of (taxiway :num zulu) :for
(continental) :agent you)
sentence: ``intercept five''
[OUTPUT] (intercept :patient 5 :agent you)
sentence: ``let the boeing turn in front of you''
[OUTPUT] (behind :goal (boeing) :agent you)
sentence: ``maintain four feet''
[OUTPUT] (maintain :alt (4) :agent you)
sentence: ``maintain four feet at departure''
[OUTPUT] (maintain :alt (4) :when on-departure :agent you)
sentence: ``maintain heading of four''
[OUTPUT] (maintain :heading 4 :agent you)
sentence: ``maintain this frequency''
[OUTPUT] (nop :agent you)
sentence: ``monitor four''
[OUTPUT] (setfreq :goal (speaker :freq 4) :agent you)
sentence: ``monitor ground''
[OUTPUT] (setfreq :goal (speaker :name ground) :agent you)
sentence: ``monitor ramp''
```

```
[OUTPUT] (setfreq :goal (speaker :name ramp) :agent you)
sentence: ``monitor tower''
[OUTPUT] (setfreq :goal (speaker :name tower) :agent you)
sentence: ``move ahead''
[OUTPUT] (move :agent you)
sentence: ``move ahead before the boeing''
[OUTPUT] (move :agent you)
[OUTPUT] (ahead :goal (boeing) :agent you)
sentence: ``move as soon as you can''
[OUTPUT] (move :agent you)
sentence: ``on departure maintain five''
[OUTPUT] (maintain :alt (5) :when on-departure :agent you)
sentence: ``on departure maintain five feet''
[OUTPUT] (maintain :alt (5) :when on-departure :agent you)
sentence: ``position and hold''
[OUTPUT] (position :agent you)
[OUTPUT] (hold :agent you)
sentence: ``remain on ten''
[OUTPUT] (setfreq :goal (speaker :freq 10) :agent you)
sentence: ``remain this frequency''
[OUTPUT] (nop :agent you)
sentence: ``right in front of the continental''
[OUTPUT] (move :agent you)
[OUTPUT] (ahead :goal (continental) :agent you)
sentence: ``straight ahead on zulu''
[OUTPUT] (go :on (taxiway :num zulu) :agent you)
sentence: ``straight on to runway three''
[OUTPUT] (go :on (runway :num 3) :agent you)
sentence: ``straight on to zulu''
[OUTPUT] (go :on (taxiway :num zulu) :agent you)
sentence: ``taxi ahead''
[OUTPUT] (move :agent you)
sentence: ``taxi ahead on zulu''
[OUTPUT] (go :on (taxiway :num zulu) :agent you)
sentence: ``taxi into position and hold''
[OUTPUT] (position :agent you)
[OUTPUT] (hold :agent you)
sentence: ``taxi quebec''
[OUTPUT] (go :on (taxiway :num quebec) :agent you)
```

```
   sentence: ‘‘taxi to position and hold’’
   [OUTPUT] (position :agent you)
   [OUTPUT] (hold :agent you)
   sentence: ‘‘taxi to runway three via zulu’’
   [OUTPUT] (go :on (runway :num 3) :via (taxiway :num zulu)
:agent you)
   sentence: ‘‘taxi via taxiway tango’’
   [OUTPUT] (go :via (taxiway :num tango) :agent you)
   sentence: ‘‘the boeing is going in front of you’’
   [OUTPUT] (behind :goal (boeing) :agent you)
   sentence: ‘‘then runway three’’
   [OUTPUT] (go :on (runway :num 3) :agent you)
   sentence: ‘‘then zulu’’
   [OUTPUT] (go :on (taxiway :num zulu) :agent you)
   sentence: ‘‘turn heading three’’
   [OUTPUT] (turn :heading (heading :to 3) :agent you)
   sentence: ‘‘turn left five degrees’’
   [OUTPUT] (turn :heading (heading :left 5) :agent you)
   sentence: ‘‘turn left five degrees to heading ten’’
   [OUTPUT] (turn :heading (heading :left 5 :to 10) :agent
you)
   sentence: ‘‘turn left to heading three’’
   [OUTPUT] (turn :heading (heading :left unknown :to 3)
:agent you)
   sentence: ‘‘turn three’’
   [OUTPUT] (turn :heading (heading :to 3) :agent you)
   sentence: ‘‘turn to heading three’’
   [OUTPUT] (turn :heading (heading :to 3) :agent you)
   sentence: ‘‘you are following a continental’’
   [OUTPUT] (behind :goal (continental) :agent you)
   sentence: ‘‘you are going to encounter a alaska’’
   [OUTPUT] (expect :plane (alaska) :agent you)
   sentence: ‘‘you are going to see an alaska behind a
boeing’’
   [OUTPUT] (expect :plane (alaska :behind boeing) :agent you)
   sentence: ‘‘you are going to see an alaska’’
   [OUTPUT] (expect :plane (alaska) :agent you)
   sentence: ‘‘you are number five’’
   [OUTPUT] (set-priority :new (priority :num 5 :event
```

```
unknown) :agent you)
   sentence: ''you are number five behind a cessna''
   [OUTPUT] (set-priority :new (priority :num 5 :event unknown
:behind cessna) :agent you)
   sentence: ''you are number five for landing''
   [OUTPUT] (set-priority :new (priority :num 5 :event
landing) :agent you)
   sentence: ''you are number five for landing behind a
boeing''
   [OUTPUT] (set-priority :new (priority :num 5 :event landing
:behind boeing) :agent you)
   sentence: ''you are number five for takeoff''
   [OUTPUT] (set-priority :new (priority :num 5 :event
takeoff) :agent you)
   sentence: ''you are number five for takeoff behind a
cessna''
   [OUTPUT] (set-priority :new (priority :num 5 :event takeoff
:behind cessna) :agent you)
   sentence: ''you are number five to go''
   [OUTPUT] (set-priority :new (priority :num 5 :event
unknown) :agent you)
```

CHAPTER 6

LYAPUNOV-BASED PARTIAL STABILIZATION OF A NONHOLONOMIC UAV MODEL WITH POLYTOPIC INPUT CONSTRAINTS

J. Willard Curtis

Air Force Research Laboratory, Munitions Directorate
Eglin AFB, FL
jess.curtis@eglin.af.mil

This chapter presents a practical method for regulating the position of a nonholonomic system (the well-studied unicycle model) to the origin in finite time. The technique is based on a (weak) bi-partite control Lyapunov function (clf), and it produces a family of piece-wise continuous stabilizing control laws. This clf-based approach guarantees regulation to the origin even when the system is subject to rectangular or polytopic actuator constraints, such as minimum forward velocity and maximum turn rates. It also offers the designer the flexibility to incorporate performance objectives other than stabilization via a point-wise control value selection. Specifically, we employ the vertex enumeration algorithm to derive a complete parameterization of a constrained stabilizing set of input commands at any particular state.

1. Introduction

Ensuring good performance, via state-dependent feedback control, for the class of nonlinear systems is a challenging problem. Some effective methods for the design of such controllers are linearization [p. 485 – 488, 10], gain scheduling [12], feedback linearization [9], recursive integrator backstepping [16], state-dependent Riccati control [5], and Lyapunov-based methods. Strategies based on Lyapunov functions are particularly valuable because they offer analytical guarantees such as global asymptotic stability, inverse-optimality, and robustness to uncertainty [8].

Most Lyapunov-based methods for nonlinear systems can be interpreted in a control Lyapunov function [17, 1] (clf) framework. Thus, clfs have become a powerful tool for the feedback stabilization of nonlinear systems even though a method of finding a clf for a general nonlinear system remains an open problem (although there are such clf-building techniques for systems in cascade form [11, 16] and systems which are amenable to feedback linearization). There exist several 'universal formulas' that, given a known clf, generate almost smooth control laws which are guaranteed to globally asymptotically stabilize the closed-loop system such as Sontag's formula [18] and Freeman and Kokotovic's point-wise min-norm formula [8].

An important extension to the theory of clf-based control is emerging for nonlinear systems that are subject to input constraints [13, 14, 15, 21, 20]. This is a deserving area of research due to the fact that the vast majority of real feedback systems have saturating actuators, and the design of stabilizing controllers for constrained systems is a long standing problem even for linear systems (see [19] for some recent contributions). In [14] the authors introduce a universal formula for the case of a scalar input which is constrained to be positive and possibly bounded, and in [15] a clf-based formula is derived for inputs that lie within a Minkowski ball. An interesting technique is presented in [21] where it is shown how control Lyapunov functions can be used in a receding horizon formulation to stabilize systems with input constraints, but this receding-horizon approach is computationally intensive and may not be amenable to a real-time implementation. Another promising result is found in [20] where the authors construct a one-parameterized family of universal formulas for systems with very general input constraints.

This chapter extends those constrained-clf results by presenting a *complete* parameterization of all universal formulas that stabilize a nonlinear system under polytopic input constraints with respect to a known clf. In particular we build upon the approach presented in [6, 7] where it is shown that Lyapunov stability can be interpreted as a point-wise one-dimensional constraint on control values and where a complete parameterization of unconstrained universal formulas is developed.

We show that the point-wise stability requirement, $\dot{V} < 0$, can be interpreted as adding another bounding hyperplane to the polytope of input constraints. From this half-space representation we exploit the *vertex enumeration* algorithm introduced in [3, 2] to find a vertex representation of the polytope. It is then shown how a vector of weights can be used to completely parameterize the set of feasible stabilizing control values at any given state.

The benefit of this parameterization is that it can be combined with an optimization routine to improve closed-loop performance, while not requiring the intense computational load associated with receding-horizon control. Thus, the technique in this chapter represents a practical method for constructing high-performance stabilizing controllers for nonlinear affine systems with polytopic input constraints.

2. Control Lyapunov Functions

Consider the following affine nonlinear system with multiple inputs, i.e.,

$$\dot{x} = f(x) + g(x)u, \tag{1}$$

where $x \in \mathbb{R}^n$, $u \in \mathcal{U} \subset \mathbb{R}^m$ and $f(0) = 0$. We will assume throughout the chapter that f and g are locally Lipschitz functions, and the goal will be to regulate the state x to the origin.

Definition 1: A **polytope**, $P \subset \mathbb{R}^m$, is the bounded intersection of a finite number of closed half-spaces:

$$P \triangleq \left\{z \in \mathbb{R}^m : \; z^T h_i \leq b_i, 2m \leq i \leq k < \infty\right\}$$

We assume that the control values are constrained to lie in some polytope, $\mathcal{U}$. This input constraint can be written compactly as follows:

$$\mathcal{U} = \left\{u \in \mathbb{R}^m : \; Mu \leq b\right\}, \tag{2}$$

where the matrix $M \in \mathbb{R}^{k \times m}$ and the vector $b \in \mathbb{R}^k$ define the polytope (k being the dimension of vector b). For the ubiquitous special case when $\mathcal{U}$ is a hyper-box, $M \in \mathbb{R}^{2m \times m}$ can be represented as

$$M \triangleq \begin{bmatrix} I \\ -I \end{bmatrix},$$

and the vector b is of length $2m$ with the ith element of b containing the magnitude of the control constraint along the ith semi-axis. (In other words $\mathcal{U}$ is a hyper-box whose faces are perpendicular to the axes. The ith face of $\mathcal{U}$ lies a distance $b(i)$ from the origin.)

Definition 2: a C^1 function $V(x) : \mathbb{R}^n \to \mathbb{R}$ is said to be a **control Lyapunov function (clf)** for system (1) if $V(x)$ is positive definite, radially unbounded, and if

$$\inf_{u \in \mathcal{U}} V_x^T (f + gu) < 0,$$

for all $x \neq 0$.

We assume that a clf for the constrained system is known *a priori*. Also, it is assumed that there is a minimum desired rate of decrease, $\dot{V} = V_x^T(f + gu) \leq -\epsilon(\|x\|)$, where $\epsilon(\|x\|)$ is a positive-definite function and $V_x(x)$ is the gradient of V with respect to x; this assumption will ensure a closed control value set.

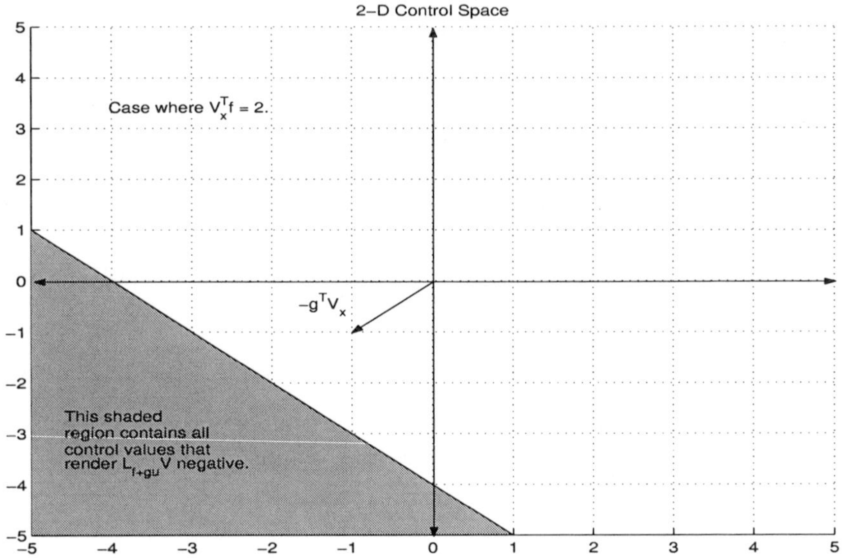

Fig. 1. Stability Constraint

2.1. *The Stabilization Constraint*

Definition 3: The **Stabilizing Set**, denoted $\mathcal{S}(x)$ is a state-dependent control value set containing all the points in $\mathbb{R}^m$ which satisfy:

$$\mathcal{S}(x) \triangleq \{u \in \mathcal{U}\colon V_x^T(f + gu) \leq -\epsilon(\|x\|)\}, \tag{3}$$

Definition 2 guarantees that $\mathcal{S}(x)$ is always non-empty if $\epsilon(\|x\|)$ is chosen to be sufficiently small. S is thus a closed state-dependent half-space in re^m.

Definition 4: Given a constraint polytope, $\mathcal{U} \subset \mathbb{R}^m$, A control law $u(t)$ is **feasible** if $u(\cdot) \in \mathcal{L}^\infty$ and $u(t) \in \mathcal{U}$ for all $t \geq t_0$.

It can be trivially shown that any continuous selection from S will be a feasible control and will render the closed loop system asymptotically stable (since $\dot{V} < 0 \;\; \forall \;\; x \neq 0$ and $u \in S \Longrightarrow u \in \mathcal{U}$).

The next theorem shows how the stability constraint embodied in (3) can be folded into the input constraints described by (2).

Theorem 5: $\mathcal{S}(x) = \left\{u \in \mathbb{R}^m : \overline{M}u < \overline{b}\right\}$, where $\overline{M}$ and $\overline{b}$ are defined as follows:

$$\overline{M} = \begin{bmatrix} M \\ V_x^T g \end{bmatrix},$$

$$\overline{b} = \begin{bmatrix} b \\ -V_x^T f - \epsilon(\|x\|) \end{bmatrix}.$$

Proof: The matrix inequality

$$\overline{M}u \leq \overline{b} \tag{4}$$

represents a system of inequalities. The first k rows ensure that $u \in \mathcal{U}$. The last $(k+1)$ row, $V_x^T gu < -V_x T f - \epsilon(\|x\|)$ is simply a restatement of $V_x^T f + V_x^T gu < -\epsilon(\|x\|)$.

■

Note that Theorem (5) reveals an important result: asymptotic stability at any fixed state can be viewed as a *linear constraint* on the input (see Figure 1). This reduces the design of a feasible, stabilizing control to performing a (continuous in the state) point-wise selection from the well-defined set $\mathcal{S}(x)$.

Figure 2 shows a two dimensional example of the the state-dependent set S at some state where $V_x^T f + \epsilon(\|x\|) = 2$, $V_x^T g = (1 \;\; 1)^T$, and $\mathcal{U}$ is rectangular. Note that S is always closed, always convex, and always a polytope.

3. Removing Redundancy and Vertex Enumeration

Before delving into a parameterization of $\mathcal{S}$, we note that it is possible that the addition of the stability constraint, as outlined in Theorem 5, will render one or more of the input constraints redundant or that the stability constraint will itself be redundant. For example, it is possible that any control value in $\mathcal{U}$ will stabilize the system. In this case, the new system of inequalities described by (4) has an inactive constraint because the last row of $\overline{M}u \leq \overline{b}$ ($V_x^T gu < -V_x T f - \epsilon(\|x\|)$) is redundant.

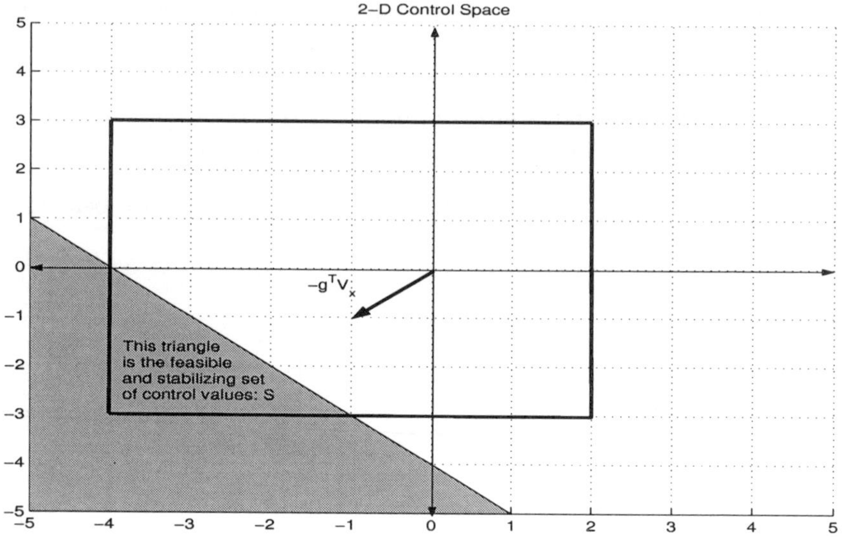

Fig. 2. Stabilizing Polytope

Other than needlessly enlarging the size of the inequality constraints, this redundancy can impair our parameterization and optimization algorithms. Fortunately, we can remove this redundancy with a simple linear program. To accomplish this consider the following system of inequalities:

$$A\xi \leq d,$$
$$s^T \xi \leq t.$$

We can determine whether $s^T \xi \leq t$ is redundant by performing the following linear program:

$$\begin{aligned} f^* = \max \quad & s^T \xi, \\ \text{subject to } & A\xi \leq d \\ & s^T \xi \leq t + m \|b\|_\infty . \end{aligned}$$

Then $s^T \xi \leq t$ is redundant if $f^* \leq t$. Each row in (4) is sequentially tested and removed if redundant. This process is repeated until no redundancy is found.

In order to put the constraints in a form amenable to a convenient parameterization, it will first be necessary to find the vertices of the convex polytope defined by (4). This problem can be solved via the *vertex enumeration* algorithm introduced in [3]. We will not repeat the details of this algorithm, but its function is explained below.

The Minkowski-Weyl Theorem [p. 29, 22] states that every polytope $\mathcal{P}$ can be described as the intersection of a finite set of half-spaces, $\mathcal{P} = \{\xi \in \mathbb{R}^m : A\xi \leq d\}$, called the $\mathcal{H}-representation$, or described equivalently in terms of the convex hull of its vertices $\mathcal{V} = \{v_1, v_2, ...v_N\}, \quad v_i \in \mathbb{R}^m, \quad$ and $N < \infty$, termed its $\mathcal{V} - representation$.

The vertex enumeration algorithm transforms a polytope's $\mathcal{H}-repre-sentation$ into its $\mathcal{V} - representation$. More specifically, given (4) which has been pruned of its redundant inequalities, the algorithm generates a set of points $\mathcal{V} = \{v_1, v_2, ...v_N\}, \quad v_i \in \mathbb{R}^m$ corresponding to the (exterior) vertices of the polytope $S \subset \mathbb{R}^m$.

4. Parameterizing $\mathcal{S}$

The constrained stabilizing control value set, $\mathcal{S}(x)$, can be completely parameterized in terms of its vertices.

Definition 6: Given a polytope's $\mathcal{V}-representation$, we define the $m \times N$ **vertex matrix** V_m and the $N \times 1$ **weighting vector** w as follows:

$$V_m = \begin{bmatrix} v_1 \; v_2 \; \cdots \; v_N \end{bmatrix},$$
$$w^T = \begin{bmatrix} w_1 \; w_2 \; \cdots \; w_N \end{bmatrix},$$

where $\sum_{i=1}^{N} w_i = 1$ and $w_i \geq 0$.

Theorem 7:

$$\forall \; u \in \mathcal{S} \; \exists \; w \text{ such that } u = Vw,$$

where w and V are defined by Definition 6.

Proof: Follows directly from the fact that the S is a closed, convex polytope and from the fact that Vw is a convex weighting of points on the surface of S.

■

The weighting vector w therefore defines a particular convex combination of the vertices, resulting in a point on the interior or on the surface of $\mathcal{S}$. Thus, if we know the $\mathcal{V} - representation$ of $\mathcal{S}$ at some state, then S is completely parameterized by the weighting vector w at that state. We have now reduced the problem of designing a stabilizing controller in the presence of input constraints to the point-wise choice of an $N-$dimensional weighting vector.

Since we will now explore optimization routines for choosing the weighting vector, a remark on the size of N is in order: when the input constraints are rectangular, the polytope S will have $2m$ faces and 2^m vertices.

5. Optimizing the Weighting Vector

5.1. *Min-Norm Control*

Freeman and Kokotovic's min-norm formula (MNF) is based on the reasonable notion that control effort is a valuable resource. In [8] the MNF is defined point-wise as the control value of minimum norm that produces a desired rate of decrease (α_V) in the clf. Given the our assumptions and notation, the MNF control value at every state can be described as

$$u_{MNF} = \min_{u \in \mathcal{S}} \|u\| \,, \tag{5}$$

where $\epsilon \triangleq \alpha_V$. Under our parameterization, the MNF can be easily implemented via a quadratic program, by choosing the weighting vector such that $\|u\| = w^T V^T V w$ is minimized at every state.

5.2. *Max-Rate Control*

The control law which is developed in [20] is defined point-wise as the control value which lies (almost on) the boundary of the constraint set. Furthermore, this value is chosen such that the maximum (point-wise) rate of decrease in V is achieved. Given our notation and assumptions, their control law would be written as:

$$u_{MR} = \max_{u \in \mathcal{S}} -\dot{V}. \tag{6}$$

Under our parameterization, this universal formula can be solved via a simple linear program: maximize $-V_x^T g V w$ over w at every state, or more simply, by testing each vertex and choosing the one maximizes this measure. (Proving that at least one vertex achieves this maximum is a simple exercise left to the reader.)

5.3. *Receding Horizon Control*

Implementing a receding-horizon strategy similar to the one proposed in [21] can now be easily accomplished. First, we discretize the time index ($x_k \approx x(k\delta t)$), then we define an Euler approximation to the system dynamics, and we define a discrete cost function $J(u, x)$:

$$\begin{aligned} x_0 &= x(0), \\ x_k &= x_{k-1} + \big(f(x_{k-1}) + g(x_{k-1})V(x_{k-1})w(i)\big) \cdot \delta t, \\ J(u, x_k) &= \sum_{k}^{k+T} x_k^T Q x_k, \end{aligned}$$

where T is some integer that controls the length of the objective function horizon. The optimization problem can now be posed as follows. Find a sequence of weighting vectors $\big(w(0),\ w(1),\ \cdots,\ w(i),\ \cdots,\ w(i_{max})\big)$ such that J is minimized. The number of time steps for which each $w(i)$ is held constant, the length of the time horizon, the number of distinct weighting vectors (i_{max}) and the size of the time step are all parameters that can be tuned in order to trade off accuracy and computational load. Having found such an optimal sequence, the first element, w_k, is used as our control law, and the optimization is repeated at regular intervals. This is the standard receding horizon formulation, but note that this is an any-time algorithm: the optimization can be halted at any point and the current weighting vector used to build the control law. Also, any valid weighting vector can be used to initialize the optimization.

6. Unicycle Model System

When considering the motion of an Unmanned Aerial Vehicle (UAV) in the lateral plane, its motion can be modelled kinematically as follows:

$$\begin{aligned} \dot{p}_x &= v \sin(\theta), \\ \dot{p}_y &= v \cos(\theta), \\ \dot{\theta}_x &= \omega; \end{aligned} \tag{7}$$

where p_x and p_y are the spatial coordinates, θ is the vehicle's heading angle, and where the control variables are v the linear velocity and ω the angular velocity. We will assume that v is constrained to lie within some closed set of values:

$$0 < \underline{v} \leq v \leq \bar{v} < \infty,$$

and we note the constraint that $v > 0$ renders many of the published algorithms for controlling this system (e.g. [4]) obsolete. Additionally, the angular rate of the vehicle is limited as follows:

$$|\omega| \leq \bar{\omega}.$$

Asymptotic stabilization in the usual sense is impossible, however our motivation in studying this problem is related to a real-world need to autonomously control teams of UAVs. High-level control schemes exist for the purpose of coordinating the actions of these vehicles and for generating way-points to which the vehicles must be regulated. The regulation problem can thus be posed as follows: we desire to drive the spatial state of the UAV to the coordinates of a target way point in finite-time.

We denote this objective as "partial stabilization" since we are not concerned with controlling the angular orientation of the vehicle. Note that we are also unconcerned with the state of the vehicle as time goes to infinity; rather, we assume that after the vehicle moves to within some ϵ-ball of the way-point's coordinates a new way-point will be supplied by the high-level controller or (by default) the vehicle will regulate again to its previous way point.

To aid in the analysis of this problem we first perform the following change of variables:

$$\begin{aligned} x_1 &= p_x \sin\phi - p_y \cos\phi, \\ x_2 &= p_x \cos\phi + p_y \sin\phi, \\ x_3 &= \phi. \end{aligned}$$

Under these coordinates the origin corresponds to the location of the UAV which is always oriented down the negative x_2 semi-axis. The target way-point's coordinates correspond to x_1 and x_2. Under this transformation our new state equation becomes:

$$\begin{pmatrix} \dot{x}_1 \\ \dot{x}_2 \\ \dot{x}_3 \end{pmatrix} = \begin{pmatrix} \omega x_2 \\ -\omega x_1 + v \\ \omega \end{pmatrix} \tag{8}$$

Equation (8) has decoupled θ from the position coordinates, and the evolution of the position coordinates can be written as:

$$\begin{aligned} \dot{x} &= g(x)u. \qquad (9) \\ x &\triangleq \begin{pmatrix} x_1 \\ x_2 \end{pmatrix}, \\ g(x) &\triangleq \begin{pmatrix} \omega x_2 \\ -\omega x_1 + v \end{pmatrix}, \\ u &\triangleq \begin{pmatrix} v \\ \omega \end{pmatrix}. \end{aligned}$$

This system's input constraints are clearly rectangular and can be written in the form of (2) as follows:

$$Mu \le b;$$
$$M \triangleq \begin{bmatrix} 1 & 0 \\ 0 & 1 \\ -1 & 0 \\ 0 & -1 \end{bmatrix}, \qquad b \triangleq \begin{bmatrix} \bar{v} \\ \bar{\omega} \\ -\underline{v} \\ \bar{\omega} \end{bmatrix}.$$

It remains to find a control Lyapunov function for system (9) in order to use the control-vertex technique for stabilization. Since (9) is nonholonomic, we will search for a non-smooth control Lyapunov-like function, whose purpose will be similar to that of the clf: at every state it will indicate a direction in the control space, $V_x g$, along which increasing magnitudes result in increasingly quick stabilization.

First we note that the input constraints result in a minimum turn radius, r, which can be calculated as $r = \frac{\underline{v}}{\bar{\omega}}$. Now, consider the following bi-partite scalar-valued function:

$$V(x) = \begin{cases} -x_2 + r, & if \;\; x \in \Omega \\ \phi^2 + \alpha x_2^2, & elsewhere, \end{cases} \tag{10}$$

where $\phi \in (\pi, -\pi]$ is the angle between x and the negative x_2 semi-axis, and where Ω is the union of two circular disks of radius r centered at the points $-r$ and r on the x_1 axis:

$$\Omega \triangleq \Big\{ x \in \mathbb{R}^2 : \\ (x_1 - r)^2 + x_2^2 \le r^2 \; \cup \; (x_1 + r)^2 + x_2^2 \le r^2 \Big\} \tag{11}$$

Theorem 8: Assuming that V_x is appropriately defined (in the sense of Dini) along the boundary of Ω, at every state there exists a control value, $u \in \mathcal{U}$, such that $\dot{V} = V_x^T g u < 0$.

We will sketch the proof as follows. When $x \in \Omega$ any control value such $\omega = 0$ will make $\dot{V}$ negative:

$$\dot{V} = V_x^T g u = \begin{bmatrix} 0 & -1 \end{bmatrix} \begin{bmatrix} 0 & x_2 \\ 1 & -x_1 \end{bmatrix} u.$$

When $x \in \mathbb{R}^2 \setminus \Omega$ the target waypoint is outside of the minimum turn radius areas. This means that executing a hard turn ($\omega = \pm\bar{\omega}$) will drive x to the negative x_2 semi-axis in finite time. For the case when $x_1 < 0$ a hard right turn will increase ϕ (which is negative for $x_1 < 0$), while ϕ is decreased by a hard left turn when $x_1 > 0$. Thus there will always be a control value that will drive x to the negative x_2 semi-axis. Once on this semi-axis any control value such that $\omega = 0$ results in regulation to the origin in finite time.

One should note that the choice of $\epsilon(\|x\|)$ is constrained by the system's input constraints and initial conditions, but ϵ can always be chosen small enough such that $u \in \mathcal{U}$ renders $\dot{V} < 0$.

Theorem 9: The condition that $\dot{V} < -\epsilon$ at every state is sufficient to ensure that the target waypoint x is regulated to the origin in finite time so long as $x \in \mathbb{R}^2 \setminus \Omega$ is guaranteed to never enter Ω.

We sketch the proof as follows. When $x \in \Omega$, it must exit Ω in finite time; this follows directly from $\dot{V} < \epsilon$ and the fact that $V(x)$ contains no critical points inside Ω. Assuming that re-entry into Ω is impossible, it is clear that the state will be regulated to the negative x_2 semi-axis in finite-time, and from thence to the origin in finite time being aided by the system's drift, $v > 0$.

There remains the difficulty of ensuring that forward trajectories of $x(t_0) \in \mathbb{R}^2 \setminus \Omega$ will never enter Ω. We first remark that the point where x exits Ω is on a surface ($\partial\Omega^+$) defined by the union of two semi-circles:

$$\partial\Omega^+ \triangleq \{x \in \partial\Omega : x_2 > 0\},$$

where $\partial\Omega$ indicates the boundary of Ω. This is a direct result of the vehicle's minimum turn radius and our construction of V inside Ω.

Further, we note that no control value can drive x in the opposite direction across $\partial\Omega^+$: this can be seen by supposing that $x : x_2 > 0$ is at some point very close to $\partial\Omega^+$ but is not inside Ω. Every control value where $\omega = 0$ results in positive x_1 motion at a rate at least equal to $\underline{v}$ due to the system's drift term. It can be shown that executing a hard turn in either direction will never decrease the distance between x and $\partial\Omega^+$. Finally, any other control value will increase the distance between x and $\partial\Omega^+$. This shows that once the state leaves Ω it can only re-enter Ω on the set $\partial\Omega^-$, where $\partial\Omega^-$ is defined as follows:

$$\partial\Omega^- \triangleq \{x \in \partial\Omega : x_2 \leq 0\}.$$

Note that entry into Ω across this boundary is easy because the system's drift term naturally pushes the state in this direction whenever $x : x2 < 0,\ |x1| \leq 2r$. Also, it can be seen that x may cross $\partial\Omega^-$ into Ω even when the control value makes $\dot{V} < 0$: when $x = [r, \; -r + \epsilon]^T$ for example, the control value $u = [\bar{v}, \; 0]$ makes $\dot{V} < 0$ because of the αx_2^2 term in V.

We remedy this problem by imposing an additional constraint on the final control law: whenever $x \in \partial\Omega^-$ we set

$$u = \begin{bmatrix} \underline{v} \\ \pm\bar{\omega} \end{bmatrix}, \tag{12}$$

with $-\bar{\omega}$ when $x_1 < 0$ and $\bar{\omega}$ when $x_1 > 0$. This forces the state to follow the curve of $\partial\Omega^-$ directly to the origin whenever x moves onto $\partial\Omega^-$.

We can now cast this partial stabilization problem into our vertex parameterization framework: we first test whether $x \in \partial\Omega^-$ and if this is true then we define the vertex matrix as

$$V_m = \begin{bmatrix} \underline{v} \\ \pm\bar{\omega} \end{bmatrix}.$$

Otherwise, we derive V_m in the usual manner via the stabilizing control value set, S, and the vertex enumeration algorithm.

7. Discussion and Conclusion

This chapter introduced a method for algorithmically parameterizing stabilizing control laws that obey polytopic input constraints, given a known clf. This technique is a general method, being appropriate for the class of smooth nonlinear systems which are affine in the control. Our approach relies on a fast (polynomial time) algorithm to generate the vertices of a state-dependent polytope, and it is amenable to real-time implementation. In particular we have demonstrated the following:

- Lyapunov stability is equivalent to a point-wise inequality constraint on the input, and this constraint is expressed in a form where it can be easily folded into rectangular or polytopic input constraints.
- The set of simultaneously feasible and stabilizing controls is a polytope in $\mathbb{R}^m$ that can be completely parameterized by a weighting of its vertices.
- Any universal formula can be represented via our parameterization.

Finally, we have constructed a clf-like function that is suitable for the partial regulation of a unicycle-model system to the origin in finite time with constrained actuation, and we have shown how to use our novel vertex-enumeration with this clf-like function to solve the way-point regulation problem. The resulting control strategy is very flexible: instead of producing a single control law, it produces a closed set that evolves point-wise in the control space. Secondary desiderata, such as cooperative-control mode commands, can then be used to choose a specific control value at every state – leading to a family of possible control signals, each of which may be useful in different mission-level contexts.

References

[1] Z. Artstein. Stabilization with relaxed controls. *Nonlinear Analysis, Theory, Methods, and Applications*, 7(11):1163–1173, 1983.

[2] David Avis, David Bremner, and Raimund Seidel. How good are convex hull algorithms. *Computational Geometry*, 7:265–301, 1997.

[3] David Avis and Komei Fukudo. Reverse search for enumeration. *Discrete Applied Mathematics*, 65:21–46, 1996.

[4] C. Canudas de Wit and O. Sordalen. Exponential stabilization of mobile robots with nonholonomic constraints. *IEEE Transactions on Automatic Control*, 37(11):1791–1797, November 1992.

[5] James R. Cloutier. Adaptive matched augmentation proportional navigation. In *Proceedings of the AIAA Missile Sciences Conferenence*, Monterey, California, Nov 1994.

[6] Jess W. Curtis and Randal W. Beard. A graphical understanding of lyapunov-based nonlinear control. In *Proceedings of the IEEE Conference on Decision and Control*, Las Vegas, NV, December 2002.

[7] Jess W. Curtis and Randal W. Beard. Satisficing: A new approach to constructive nonlinear control. *IEEE Transactions on Automatic Control*, to appear 2004.

[8] R. A. Freeman and P. V. Kokotovic. *Robust Nonlinear Control Design: State-Space and Lyapunov Techniques.* Systems and Control: Foundations and Applications. Birkhauser, 1996.

[9] Alberto Isidori. *Nonlinear Control Systems.* Communication and Control Engineering. Springer Verlag, New York, New York, second edition, 1989.

[10] Hassan K. Khalil. *Nonlinear Systems.* Macmillan Publishing Company, New York, New York, 1996.

[11] M. Krstic, I. Kanellakopoulos, and P. V. Kokotovic. *Nonlinear and Adaptive Control Design.* Wiley, 1995.

[12] D. A. Lawrence and W. J. Rugh. Gain scheduling dynamic linear controllers for a nonlinear plant. *Automatica*, 31:380–390, 1995.

[13] Y. Lin and E. Sontag. A universal formula for stabilization with bounded controls. *Systems Control Letters*, 16:393–397, 1991.

[14] Y. Lin and E. Sontag. Control-Lyapunov universal formulas for restricted inputs. *Control-Theory and Advanced Technology*, 10:1981–2004, December 1995.

[15] M. Malisoff and E. Sontag. Universal formulas for CLF's with respect to Minkowski balls. In *Proceedings of the European Control Conference*, Brus-

sels, Belgium, July 1997.

[16] R. Sepulchre, M. Jankovic, and P. Kokotovic. *Constructive Nonlinear Control.* Communication and Control Engineering Series. Springer-Verlag, 1997.

[17] E. D. Sontag. A Lyapunov-like characterization of asymptotic controllability. *SIAM Journal on Control and Optimization*, 21:462–471, May 1983.

[18] E. D. Sontag. Smooth stabilization implies coprime factorization. *IEEE Transactions on Automatic Control*, 34:435–443, 1989.

[19] A. Stoorvogel and A. Saberi. Editors, Special issue on control of systems with bounded control. *Int. J. Robust and Nonlinear Control*, 9, 1999.

[20] Rodolfo Suarez, Julio Solis-Daun, and Baltazar Aguirre. Global clf stabilization for systems with compact convex control value sets. In *Proceedings of the IEEE Conference on Decision and Control*, Orlando, FL, December 2001.

[21] Mario Sznaier and Rodolfo Suarez. Suboptimal control of constrained nonlinear systems via receding horizon state dependent Riccati equations. In *Proceedings of the IEEE Conference on Decision and Control*, Orlando, FL, December 2001.

[22] G. M. Ziegler. *Lectures on Polytopes.* Springer, 1996.

CHAPTER 7

COOPERATIVE OPTIMIZATION FOR SOLVING LARGE SCALE COMBINATORIAL PROBLEMS

Xiaofei Huang
AirPrism, Inc.
Redwood City, CA 94065, U.S.A.
huangxiaofei@ieee.org

This chapter presents a cooperative system for the minimization of energy functions in a general form. The system consists of a number of agents working together in a cooperative way to achieve a certain objective. A novel cooperation scheme is presented which has two parameters to control the cooperation of agents from two different perspectives; the first is used for controlling the level of influence among agents in decision-making, the second is used for controlling the rate of information exchanges among agents. Different settings of the parameters could lead to completely different computational behaviors of the system. When the influence level is balanced with the exchange rate, the system always has a unique equilibrium and it reaches the equilibrium regardless of initial conditions. The equilibrium is also the global optimum of the system if a consensus is reached among agents in this case. When the influence level is at the strongest, the system can always reach the Nash equilibrium, a strategic equilibrium in game theory, which formally studies conflict and cooperation in a system of agents. To demonstrate its power, two case studies are provided where the number of variables of the problems ranges from 10,000 to 100,000. Using the evaluation framework for stereo matching provided by Middlebury College, we show that the solutions found by the cooperative system are significantly better than simulated annealing. Furthermore, the operations of the system are simple and inherently parallel. Our computer simulation has suggested that if the system is implemented in parallel, it can find the stereo matching solution in less than 0.5 milliseconds.

Keywords:
Combinatorial optimization, cooperative optimization, NP problems

1. Introduction

The general methods for combinatorial optimization [9, 10] are 1) local search [9, 10], 2) Simulated Annealing [6], 3) genetic algorithms [3], 4) tabu search, 5) Branch-and-Bound [7, 5, 10], and 6) and Dynamic Programming [5, 10]. The first four methods are classified as local optimization. Many optimization problems of computer vision, image processing, and other fields, have to deal with problems that are nonlinear in nature and very large in scale. Often times, they have local optima in a number exponentially growing with the size of the problem. This would defy the effectiveness in practice of the first four methods due to the local optimum problem. Furthermore, these problems have to deal with thousands to millions of variables, which are beyond the capability of the last two methods in terms of time and space complexity.

This chapter presents a cooperative system for solving large-scale combinatorial problems in practice. The system consists of multiple dynamic agents. These agents may be people, neurons, computers, firms, airplanes, or any combination of these. At first, a problem is decomposed into a number of sub-problems with manageable complexities and each one is assigned to an agent. Then, those agents work together in a cooperative way, instead of independently, to solve those sub-problems.

A formal definition of a cooperative system for optimization is presented. The theoretical foundations of the system are also laid out. The computational capability of the system is determined by its cooperative scheme among the agents of the system. A novel cooperation scheme is presented which determines the computational behaviors of the system. It has two parameters to control the cooperation of agents from two different perspectives. The first one is used for controlling the level of influence among agents in decision-making. The second one is used for controlling the rate of information exchanges among agents. Different settings of the parameters could lead to completely different computational behaviors of the system. Some of those are directly related to the search of global optima and many of them are not possessed by conventional optimization methods. They are presented in the theoretical foundations section.

The binary constraint-based optimization problem is used in this chapter as an example to show the principle of decomposing a complex combinatorial optimization problem into a set of sub-problems of manageable complexities. Many problems in computer vision and image processing have been formalized as this problem. Also, the famous traveling salesman problem is formalized as a special case of the problem in section 2. To demonstrate its power, we show the successful applications of the cooperative system in solving hard, large-scale optimization problems from DNA image analysis as well as stereo matching from computer vision, where the number of variables varies from 10,000 to 100,000. Using the evaluation framework for stereo matching provided by Middlebury College, we show that the cooperative system is much better than simulated annealing in terms of the quality of solutions.

2. The Cooperative System for Optimization

2.1. *The System*

A cooperative system for optimization consists of a number of agents working together in a cooperative way to achieve certain objective. These agents may be people, neurons, computers, firms, airplanes, or any combination of these.

Definition 1: A cooperative system for optimization consists of a set of agents: $A = \{a_1, a_2, \ldots, a_n\}$; and for each agent i,

- a set of options : $D_i = \{o_1, o_2, \ldots, o_{m(i)}\}$,
- an objective function: $E_i : D_1 \times D_2 \times \ldots \times D_n \to \Re$,
- and a cooperation scheme for making the choice: $S_i : D_1 \times D_2 \times \ldots \times D_n \to D_i$.

The objective function of the system is $\sum_i E_i(x)$, denoted as $E(x)$, where $x \in D_1 \times D_2 \times \ldots \times D_n$.

The choice of agent i is denoted as $\tilde{x}_i \in D_i$. All of them together forms the choice of the system, denoted as $\tilde{x}$,

$$\tilde{x} = \{\tilde{x}_1, \tilde{x}_2, \ldots, \tilde{x}_n\}$$

Let D be the Cardesian product of D_is

$$D = D_1 \times D_2 \times \ldots \times D_n \ .$$

Obviously, $\tilde{x} \in D$. Sometimes, the objective function of a system is also called the global objective of the system and the objective functions of agents are called the sub-objectives of the system.

An optimal solution of the system is denoted as $(x_1^*, \ldots, x_n^*)$ or simply x^*, where $x^* = (x_1^*, \ldots, x_n^*)$. It is also called the global optimum of

$E(x)$. The minimum value of the global objective function is denoted as E^*. Obviously, $E^* = E(x^*)$.

Because of the interindependence among the sub-objectives, there is no efficient algorithm that can guarantee to find the global optimum for the global objective within a polynomial time. Different cooperation schemes can lead to substantially different computational behaviors in optimization.

For example, we can have a simple cooperation scheme by letting each agent make a choice in a manner which minimizes its own objective function. However, a system with this cooperation scheme can hardly find the optimal solution for itself.

Our cooperation scheme is to let the agents compromise with each other in their decision makings. It makes an analogy with team playing, where the team members work together to achieve the best for the team, but not necessarily the best for each member. When an agent tries to optimize its own objective function, it communicates with other agents to consider their choices. Its own choice is made as a result of compromising between the choice of itself and those of other agents in an attempt to resolve conflicts in choosing options.

The theoretical investigation that follows shows that all agents operating in this manner together makes a better choice for the system than the simple scheme. This is an iterative process, in which the most important operation is the option discarding. The option discarding is a process where each agent discards certain options from its option set which are unlikely to be chosen in a solution for the system. As the iteration proceeds, we can expect more and more options be discarded from the option set for each agent. After some iteration steps, if there is only one option left for each agent, then a solution is found for the system. The reason for us doing that is based on the computational properties of such a system which will be shown in the following sections. Specifically, there are necessary conditions for the system to decide if an option can be in the optimal solution.

2.2. *The Problem*

The cooperative system can be applied to minimize the energy functions from computer vision, such as stereo matching and shape from shading. They have the following general form:

$$E(x_1, x_2, \ldots, x_n) = \sum_i C_i(x_i) + \sum_{i,j,i \neq j} C_{ij}(x_i, x_j). \tag{1}$$

The minimization of an energy function of this form is called the binary constraint-based optimization. C_i is called a unary constraint on variable x_i and C_{ij} is called a binary constraint on variable x_i and x_j. The optimization of (1) is NP-hard.

The famous Traveling Salesman Problem (TSP) can also be formalized as the minimization of energy functions of the above form. In an instance of the TSP, we are given an integer $n > 0$ and the distance between every pair of n cities in the form of any $n \times n$ matrix $(d_{ij})_{n\times n}$, where $d_{ij} \in \Re^+$. A tour is a closed path that visits every city exactly once. The problem is to find a tour with minimal total length.

Let x_i be the ith city in a tour and

$$D_i = \{city_1, city_2, \ldots, city_n\}, \qquad \text{for } i = 1, 2, \ldots, n.$$

Obviously, $x_i \in D_i$. Let $x_{a(i)}$ be the adjacent city of city x_i in a tour, then

$$a(i) \in \{(i + n - 1)\ \%\ n, (i + n + 1)\ \%\ n\}\ .$$

where % is the modulus operator. Let

$$C_i(x_i) = 0, \qquad \text{for } i = 1, 2, \ldots, n,$$

and

$$C_{ij}(x_i, x_j) = \begin{cases} \infty & \text{if } x_i = x_j, \\ d_{x_i x_j}/2 & \text{if } j = a(i) \text{ and } x_i \neq x_j \\ 0 & \text{if } j \neq a(i) \text{ and } x_i \neq x_j \end{cases}$$

With those choices, the optimal solution x^* to (1) is the shortest tour with a length of E^*.

2.3. *Applying the System to Solve the Problem*

To use the cooperative system defined in Definition 1 to solve the minimization problem (1), we can decompose the objective function (1) as the summation of the following n sub-objective functions

$$C_i(x_i) + \sum_{j,\ j\neq i} C_{ij}(x_i, x_j), \qquad \text{for } i = 1, 2, \ldots, n, \tag{2}$$

and set the ith one to be the objective function for agent i,

$$E_i(x) = C_i(x_i) + \sum_{j,\ j\neq i} C_{ij}(x_i, x_j).$$

The cooperation scheme S_i for agent i is defined as making a choice in a manner which minimizes the following function

$$(1 - \lambda_k)\, E_i(x) + \lambda_k \sum w_{ij} c_j^{(k-1)}(x_j)\ , \tag{3}$$

which is called the modified objective function for agent i, denoted as $\tilde{E}_i^{(k)}(x)$. Here k is the iteration step. $E_i(x)$ is the objective function of agent i. w_{ij} are non-negative real values and $(w_{ij})_{n\times n}$ should be a propagation matrix defined as follows:

Definition 2: A propagation matrix $(w_{ij})_{n\times n}$ is a irreducible, nonnegative, real-valued square matrix and satisfies

$$\sum_{i=1}^{n} w_{ij} = 1, \qquad \text{for } 1 \le j \le n\ .$$

Its ith row is denoted as w_i, and it is simply denoted as W.

A matrix W is called reducible if there exists a permutation matrix P such that PWP^T has the block form

$$\begin{pmatrix} A & B \\ O & C \end{pmatrix} .$$

$c_j^{(k-1)}(x_j)$ in (3) is an unary constraint introduced by the system on the variable x_j, called the assignment constraint. It stores the intermediate solution in the minimization of the modified objective function $\tilde{E}_i^{(k)}$ defined in (3). We can rewrite

$$\min \tilde{E}_i^{(k)}$$

as

$$\min_{x_i} \min_{x_j \in X \backslash x_i} \tilde{E}_i^{(k)}(x)\ .$$

where $X \backslash x_i$ denotes the set X_i minuses $\{x_i\}$. The inner optimization result is defined as $c_i^{(k)}(x_i)$,

$$c_i^{(k)}(x_i) = \min_{x_j \in X \backslash x_i} \tilde{E}_i^{(k)}(x)\ . \tag{4}$$

Equivalently, we can rewrite it as a different equation of $c_i^{(k)}(x_i)$ by substituting $\tilde{E}_i^{(k)}(x)$ using (3),

$$c_i^{(k)}(x_i) = \min_{x_j \in X \backslash x_i} \left((1 - \lambda_k)\, E_i(x) + \lambda_k \sum_j w_{ij} c_j^{(k-1)}(x_j) \right) . \tag{5}$$

Let the choice of agent i be x_i, $c_i^{(k)}(x_i)$ is the minimal value of $\tilde{E}_i^{(k)}$ under the choice. To minimize $\tilde{E}_i^{(k)}$, those values of x_i which have smaller function values $c_i(x_i)$ are preferred more than those which have higher ones. Therefore, $c_i^{(k)}(x_i)$ defines the preference of options for agent i.

Adding $c_j^{(k)}(x_j)$ together with $E_i(x)$ in $\tilde{E}_i^{(k)}(x)$ (see (3)) lets agent i compromise its choice with those of others. Parameter λ_k in (3) controls the level of the cooperation at iteration k. It is called the cooperation strength satisfying $0 \leq \lambda_k < 1$. A higher value for λ_k in (5) will let agent i weigh the choices of the other agents more than the one of itself. As a consequence, a stronger cooperation in the optimization is reached in this case. It was found that a stronger cooperation increases the chance for the system to find the optimal solution.

When agent i makes a choice to minimize its modified objective function $\tilde{E}_i^{(k)}(x)$, it also makes suggestions for the choices of other agents. Let $\tilde{x}_j(\tilde{E}_i)$ be the jth component in the solution for $\tilde{E}_i^{(k)}(x)$, that value is the suggestion from agent i for the choice of agent j. Although we can increase the cooperation strength λ_k in (3) to let agent i compromise more with others, it is still not guaranteed that the value is the same as the choice of agent j, $\tilde{x}_j(\tilde{E}_j)$.

If the suggested choice for agent j from agent i, $\tilde{x}_j(\tilde{E}_i)$, is the same as the choice of agent j, $\tilde{x}_j(\tilde{E}_j)$ for all i, it is said that the cooperative system has reached a consensus for agent j. If the consensus is reached for all the agents, the cooperative system is called to have found a consensus solution. It was found that if the system converges to a consensus solution, it must also be the optimal solution of the objective function $E(x)$.

Definition 3: The system is called reaching a consensus solution if, for any j, $\tilde{x}_j(\tilde{E}_i) = \tilde{x}_j(\tilde{E}_j)$, for any $\tilde{E}_i(x)$ containing the variable x_j.

At each iteration, each agent refines its assignment constraint $c^{(k)}(x_i)$ using (5). If none of the agents can refine its assignment constraint any further at certain iteration, then the system has reached an equilibrium.

Definition 4: The system is called reaching an equilibrium if, for any agent i, it can not refine its assignment constraint,

$$c^{(k)}(x_i) = c^{(k-1)}(x_i).$$

The cooperative optimization also offers the necessary conditions at each iteration to discard variable values. That is, for any option x_i, if it is in an optimal solution, then it should satisfy

$$c_i^{(k)}(x_i) \leq c_i^{(k)}(\tilde{x}_i) + t_i^{(k)} . \tag{6}$$

Any option that doesn't satisfy the above inequality can be discarded. For the exact form of $t_i^{(k)}$, please see (30) in the following section.

2.4. *The Propagation Matrix*

The propagation matrix W defines the neighborhood relations among the agents where agent i is the neighbor of agent j only if w_{ij} is not zero. In the optimization process (5), agents only communicate with their neighbors. Another way to understand the role of w_{ij} in the optimization process (5) is to treat it as a propagation process for $c_i^{(0)}(x_i)$. To make it clear, we can simplify the process by dropping the minimization operator and set $\lambda_k = 1$,

$$\begin{pmatrix} c_1^{(k)}(x_1) \\ c_2^{(k)}(x_2) \\ \vdots \\ c_n^{(k)}(x_n) \end{pmatrix} = \begin{pmatrix} w_{11} & w_{12} & \dots & w_{1n} \\ w_{21} & w_{22} & \dots & w_{2n} \\ \vdots & \vdots & \ddots & \vdots \\ w_{n1} & w_{n2} & \dots & w_{nn} \end{pmatrix} \begin{pmatrix} c_1^{(k-1)}(x_1) \\ c_2^{(k-1)}(x_2) \\ \vdots \\ c_n^{(k-1)}(x_n) \end{pmatrix} \tag{7}$$

Or equivalently,

$$\begin{pmatrix} c_1^{(k)}(x_1) \\ c_2^{(k)}(x_2) \\ \vdots \\ c_n^{(k)}(x_n) \end{pmatrix} = \begin{pmatrix} w_{11} & w_{12} & \dots & w_{1n} \\ w_{21} & w_{22} & \dots & w_{2n} \\ \vdots & \vdots & \ddots & \vdots \\ w_{n1} & w_{n2} & \dots & w_{nn} \end{pmatrix}^k \begin{pmatrix} c_1^{(0)}(x_1) \\ c_2^{(0)}(x_2) \\ \vdots \\ c_n^{(0)}(x_n) \end{pmatrix} \tag{8}$$

The process (8) can uniformly propagate the assignment constraints $c_i^{(0)}(x_i)$ for any choice of w_{ij} as long as $\sum_i w_{ij} = 1$. That is

$$\lim_{k\to\infty} \begin{pmatrix} c_1^{(k)}(x_1) \\ c_2^{(k)}(x_2) \\ \vdots \\ c_n^{(k)}(x_n) \end{pmatrix} = \begin{pmatrix} \mu_1 & \mu_1 & \dots & \mu_1 \\ \mu_2 & \mu_2 & \dots & \mu_2 \\ \vdots & \vdots & \ddots & \vdots \\ \mu_n & \mu_n & \dots & \mu_n \end{pmatrix} \begin{pmatrix} c_1^{(0)}(x_1) \\ c_2^{(0)}(x_2) \\ \vdots \\ c_n^{(0)}(x_n) \end{pmatrix} \tag{9}$$

where

$$\mu_i = \frac{1}{n} \sum_j w_{ij}. \tag{10}$$

In other words, the process (8) achieves the uniformness in propagation to each assignment constraint $c_i^{(k)}(x_i)$,

$$c_i^{(k)}(x_i) = \frac{\mu_i}{n} \sum_i c_i^{(0)}(x_i), \quad \text{when } k \to \infty.$$

If the propagation matrix is also symmetrical, i.e. for any i and j,

$$w_{ij} = w_{ji},$$

then

$$\lim_{k \to \infty} \begin{pmatrix} c_1^{(k)}(x_1) \\ c_2^{(k)}(x_2) \\ \vdots \\ c_n^{(k)}(x_n) \end{pmatrix} = \frac{1}{n} \begin{pmatrix} 1 & 1 & \dots & 1 \\ 1 & 1 & \dots & 1 \\ \vdots & \vdots & \ddots & \vdots \\ 1 & 1 & \dots & 1 \end{pmatrix} \begin{pmatrix} c_1^{(0)}(x_1) \\ c_2^{(0)}(x_2) \\ \vdots \\ c_n^{(0)}(x_n) \end{pmatrix} \tag{11}$$

In this case, the process (8) also achieves the uniformness in propagation to all the assignment constraints,

$$c_1^{(k)}(x_1) = c_2^{(k)}(x_2) = \dots = c_n^{(k)}(x_n) = \frac{1}{n} \sum_i c_i^{(0)}(x_i), \quad \text{when } k \to \infty. \tag{12}$$

From those investigations, we know that the results of the propagation process will make the assignment constraint $c_i^{(k)}(x_i)$ contain not only the optimization results from the ith agent, i.e. $c_i^{(0)}(x_i)$, but also those of others, i.e. $c_j^{(0)}(x_j)$ for any $j \neq i$. As the iteration proceeds, the assignment constraint $c_i^{(k)}(x_i)$ tends to have the optimization results of all the agents in a uniform way (see Equation (10) and Equation (12)), independent of the choice of w_{ij}. This is another way to understand why the optimization process (5) can find the global optimum through the cooperative optimization of agents in a system. In this chapter, we call the matrix $(w_{ij})_{n \times n}$ the propagation matrix.

2.5. *The System in a General Form*

The previous subsection has shown that the cooperative system can be used to solve the minimization of an objective function in a general form (1). The objective function of the famous TSP has also been given using this form.

In fact, the optimization method the system defines is more general than we have explained. It doesn't impose any restrictions on the arity of the constraints in the objective function (1). It can contain constraints of arities besides unary and binary. It has no restrictions on the objective functions for the agents as long as $\sum_i E_i(x) = E(x)$. Therefore, the system can be used to minimize an objective function as long as the function can be decomposed into the summation of a set of sub-objective functions.

There is no assumption about the independence of the sub-objective functions. Otherwise, the original minimization problem becomes trivial to solve. Because of the interdependence of the sub-objective functions, as in the case of the binary constraint-based optimization, an optimization problem becomes NP-hard most of the time.

Definition 5: $\{E_i(X_i)\}$ is called a good decomposition of an objective function $E(x)$ if it satisfies the following three conditions:

- $E_i(x)$ contains x_i,
- $\sum_i E_i(x) = E(x)$,
- decrease in $E_i(x)$ leads to decrease in $E(x)$, for any $x_i \in D_i$.

If it satisfies only the first two, it is called a decomposition of $E(x)$.

Formula (2) provides a simple decomposition of an objective function in the general form (1). A general good decomposition is provided below:

$$C_i(x_i)/2 + \sum_j (C_{ij}(x_i, x_j) + w_{ij} C_j(x_j)/2) \ , \tag{13}$$

where w_{ij} can be any real value as long as $\sum_i w_{ij} = 1$.

In its general form, the system only needs a composition of $E(x)$. The objection function for agent i is $E_i(x)$. The cooperation scheme S_i for agent i is still chosen as picking the $x_i \in D_i$ to minimize the modified objective function (3), which may contain constraints of arities higher than two. The update function for the assignment constraints $c^{(k)}(x_i)$ remains the same as (5). Equivalently, the choice of agent i at iteration k, $\tilde{x}_i^{(k)}$, is the x_i which minimizes $c_i^{(k)}(x_i)$:

$$c_i^{(k)}(\tilde{x}_i^{(k)}) = \min_{x_i \in D_i} c_i^{(k)}(x_i).$$

That is

$$S_i = \{\tilde{x}_i^{(k)} \mid c_i^{(k)}(\tilde{x}_i^{(k)}) = \min_{x_i \in D_i} c_i^{(k)}(x_i)\} \ . \tag{14}$$

Let $c^{(k)} = (c_1^{(k)}, c_2^{(k)}, \ldots, c_n^{(k)})$, then the difference equation (5) for cooperative optimization can be simplified to

$$c_i^{(k)}(x_i) = \min_{x_j \in X \backslash x_i} \left((1 - \lambda_k) E_i + \lambda_k (w_i, c^{(k-1)}) \right) \tag{15}$$

where $(w_i, c^{(k-1)})$ stands for the dot product of w_i and $c^{(k-1)}$. w_i is the i-th row of the propagation matrix defined in Definition 2.

The above difference equation is the parallel version for updating $c_i^{(k)}(x_i)$. That is, all agents update their assignment constraints synchronously. It also has a sequential version where agents update their assignment constraints asynchronously. That is, at time k, there is only one agent i doing (15), while for other agents $j, j \neq i$,

$$c_j^{(k)}(x_j) = c_j^{(k-1)}(x_j).$$

Such a cooperation scheme guarantees that the objective function of the system is $\sum_i E_i(x)$. Any other cooperation scheme will lead to a different computational behavior of the system. For example, any change in the form of the modified objective function (3) such as $\sum_i w_{ij} \neq 1$ or $\sum_j w_{ij} = 1$ (summation over j instead of i), will lead to a different objective function for the system or make it hard to investigate its computational properties.

2.6. *The Framework*

The cooperation scheme (14) of the cooperative system depends on the assignment constraint $c_i^{(k)}(x_i)$, which is updated iteratively based on Equation (5). This update function for the assignment constraint $c_i^{(k)}(x_i)$ has parameter λ_k to control the level of cooperation among agents. It can be generalized further by breaking the update function (5) into two steps. First, for each option x_i, finding the solution, denoted as $\tilde{x}(\tilde{E}_i(x_i))$, for the modified objective function on the right side of the function,

$$\min_{x_j \in X \setminus x_i} \left((1 - \lambda_k)\, E_i(x) + \lambda_k \sum_j w_{ij} c_j^{(k-1)}(x_j) \right) , \tag{16}$$

and then, updating the assignment constraint using the solution:

$$c_i^{(k)}(x_i) = (1 - \mu_k) E_i(\tilde{x}(\tilde{E}_i(x_i))) + \mu_k \Big(\sum_{j, j \neq i} w_{ij} c_j^{(k-1)} \tilde{x}_j(\tilde{E}_i) + w_{ii} c_i^{(k-1)}(x_i) \Big) , \tag{17}$$

where λ_k in (5) is replaced by μ_k. If $\mu_k = \lambda_k$, then the update function falls back to the original form.

The cooperation scheme defined by (16) and (17) is the parallel version. That is, all agents do the solution finding following the assignment constraint updating synchronously. It also has a sequential version where agents do these tasks asynchronously. That is, at time k, there is only one agent i doing (16) and (17), while for other agents $j, j \neq i$,

$$c_j^{(k)}(x_j) = c_j^{(k-1)}(x_j).$$

λ_k and μ_k are two parameters used by the generalized cooperation scheme. λ_k decides the weight of the global information $c_j^{(k-1)}(x_j)$ versus the local information E_i in minimizing the modified objective function. if $\lambda_k \to 1$, then each agent makes decisions that are the best only for others and completely sacrifices itself. Therefore, λ_k is termed the influence level of the system.

Parameter μ_k decides the weight of the global information $c_j^{(k-1)}(x_j)$ versus the local information E_i in updating the assignment constraint for agent i. A higher value of μ_k leads to faster information flow among the agents. If $\mu_k \to 1$, then the assignment constraint for each agent is overwhelmed by the global information $c_j^{(k-1)}(x_j)$. Therefore, it controls the information flow rate among the agents in the system, and is termed the information exchange rate of the system.

Those two parameters together control the cooperation among the agents in two different perspectives. The system has new emerging computational properties as the collective behavior of those agents working together under this generalized cooperation scheme. Different scheme instances can be defined by using different settings of the two parameters. As a consequence, the system defines different optimization algorithms of different computational properties. Therefore, the generalized cooperative scheme offers a framework for defining different optimization algorithms.

We will show in the following section that the optimization defined by the system degrades to the conventional local optimization when the influence level is at its strongest point. In this case, a consensus solution among all the agents can always be reached, but the decisions can hardly be the best for the system as a whole due to the local optimum problem inherited in the local optimization paradigm. When the level of cooperation is just balanced, i.e. $\lambda_k = \mu_k$, then any consensus solution the system converges to is also the optimal solution of the system. In this case, it has the desirable behavior of a cooperative optimization. Therefore, local optimization and cooperative optimization are unified under this framework as special cases using different settings of the generalized cooperation scheme.

We will also show that the convergence process is slow with a higher exchange rate, but the system is more tolerant to noise and has a higher chance to reach a consensus among the agents. On the other hand, a lower exchange rate leads to a faster convergence process, but the system is less tolerant to noise and has a lower chance to reach a consensus among the agents.

3. Theoretical Foundations

This section will show several important properties of the cooperative system. All proofs of the theorems in this chapter are provided in the appendix. It will show that the system with a balanced cooperation has a unique equilibrium. The system always converges to an equilibrium with an exponential rate with any initial condition, insensitive to the perturbation to its intermediate solutions. There are sufficient conditions for the system to identify global optima and there are necessary conditions for the system to discard options to reduce search spaces.

Without loss of generality, we assume all energy functions are nonnegative functions throughout this chapter. First, we will show the computational properties of the system when it operates in parallel, and the influence level and the exchange rate are balanced, i.e., $\lambda_k = \mu_k$ for any k.

3.1. *General Properties*

The following theorem shows that $c_i^{(k)}(x_i)$ for $x_i \in D_i$ have a direct relationship to the lower bound on the optimal cost E^*.

Theorem 6: Given any propagation matrix W and the general initial condition $c^{(0)} = 0$ or $\lambda_1 = 0$, then $\sum_i c_i^{(k)}(x_i)$ is a lower bound on $E(x_1, \ldots, x_n)$, that is

$$\sum_i c_i^{(k)}(x_i) \leq E(x_1, x_2, \ldots, x_n), \qquad \text{for any } k \geq 1 \ . \tag{18}$$

In particular, let $E_-^{(k)} = \sum c_i^{(k)}(\tilde{x}_i)$, then $E_-^{(k)}$ is a lower bound on the optimal cost E^*, that is

$$E_-^{(k)} \leq E^* \ . \tag{19}$$

Here, subscript "-" in $E_-^{(k)}$ indicates that it is a lower bound on E^*.

This theorem tells us that $\sum c_i^{(k)}(\tilde{x}_i)$ provides a lower bound on the energy function E. We will show in the next theorem that this lower bound is guaranteed to be improved as the iteration proceeds.

Theorem 7: Given any propagation matrix W, a constant cooperation strength λ, and the general condition $c^{(0)} = 0$, then $\{E_-^{(k)} | k \geq 0\}$ is a non-decreasing sequence with upper bound E^*.

If a consensus solution is found at some step or steps, then we can find out the closeness between the consensus solution and the global optimum in cost. If the algorithm converges to a consensus solution, then it must be the global optimum also. The following theorem makes those points clearer.

Theorem 8: Given any propagation matrix W, and the general initial condition $c^{(0)} = 0$ or $\lambda_1 = 0$, if a consensus solution $\tilde{x}$ is found at iteration step k_1 and remains the same from step k_1 to step k_2, then the closeness between the cost of $\tilde{x}$, $E(\tilde{x})$, and the optimal cost, E^*, satisfies the following inequality,

$$0 \leq E(\tilde{x}) - E^* \leq \left(\prod_{k=k_1}^{k_2} \lambda_k \right) \left(E(\tilde{x}) - E_-^{(k_1 - 1)} \right), \tag{20}$$

and

$$0 \leq E(\tilde{x}) - E^* \leq \frac{\prod_{k=k_1}^{k_2} \lambda_k}{1 - \prod_{k=k_1}^{k_2} \lambda_k} (E^* - E_-^{(k_1 - 1)}) \,, \tag{21}$$

where $(E^* - E_-^{(k_1-1)})$ is the difference between the optimal cost E^* and the lower bound on the optimal cost, $E_-^{(k_1-1)}$, obtained at step $k_1 - 1$. When $k_2 - k_1 \to \infty$ and $1 - \lambda_k \geq \epsilon > 0$ for $k_1 \leq k \leq k_2$,

$$E(\tilde{x}) \to E^* \,.$$

3.2. *Convergence Properties*

The behavior of the cooperative system depends on the dynamic behavior of the difference equations (5). Its convergence properties are revealed in the following two theorems. The first one shows that, given any propagation matrix and a constant cooperation strength, then there does exist a solution to satisfy the difference equations (5). The second part shows that the cooperative system converges linearly to that solution.

Theorem 9: Given any symmetric propagation matrix W and a constant cooperation strength λ, the Difference Equations (5) have one and only one solution, denoted as $(c_i^{(\infty)}(x_i))$ or simply $c^{(\infty)}$.

Theorem 10: Given any symmetric propagation matrix W and a constant cooperation strength λ, then the cooperative system, with any choice of the initial condition $c^{(0)}$, converges to $c^{(\infty)}$ with linear ("exponential" in other contexts) convergence of rate λ. That is

$$\|c^{(k)} - c^{(\infty)}\|_\infty \leq \lambda^k \|c^{(0)} - c^{(\infty)}\|_\infty \,. \tag{22}$$

This theorem is called the convergence theorem. It indicates that the cooperative system is stable and has a unique attractor, $c^{(\infty)}$. Hence, the evolution of the cooperative system is robust, insensitive to perturbations, and its final solution is independent of initial conditions. In contrast, conventional algorithms based on iterative local improvement have many local attractors due to the local minimum problem. The evolutions of these algorithms are sensitive to perturbations, and their final solutions are dependent on initial conditions.

3.3. *Sufficient Conditions*

In this subsection, we shall provide three sufficient conditions for recognizing global optimums and two necessary conditions for reducing search space and ambiguity in decision-making.

Theorem 11: (Sufficient Condition 1) If a consensus, $\tilde{x}$, is found at some step with the choice of $\lambda = 0$, then the consensus is also a global optimum.

This sufficient condition is, therefore, a weak sufficient condition since the possibility of finding a consensus without cooperation ($\lambda = 0$) is quite low in dealing with complex problems.

Theorem 12: (Sufficient Condition 2) Given a propagation matrix W and the general initial condition $c^{(0)} = 0$ or $\lambda_1 = 0$. If $E_-^{(k+1)} \leq E_-^{(k)}$ at some step k, then a consensus solution found at that step is also a global optimum.

The above theorem provides us the second sufficient condition for recognizing a global optimum. This sufficient condition does not restrict the choice of cooperation strength λ. The whole range of the cooperation strength can be exploited to increase the chance of finding a consensus solution.

The second sufficient condition is stronger than the first one. Given any problem, if a global optimum can be found under the first sufficient condition, it can also be found under the second sufficient condition. At the same time, there exist some problems whose global optima can be found under the second sufficient condition only. Intuitively, the possibility of finding the consensus solution is much higher for the cooperative system with cooperation ($\lambda > 0$) than without cooperation ($\lambda = 0$).

Theorem 13: Sufficient Condition 3 Given the propagation matrix $W = (1/n)_{n\times n}$, and the general initial condition $c^{(0)} = 0$ or $\lambda_1 = 0$. If a consensus $\tilde{x}$ is found at each iteration from step k_1 to step k_2 with λ having a fixed value, and the second minimum value of the variable assignment constraint $c^{(k_2)}(\hat{x}_i)$ satisfies the following inequality:

$$c^{(k_2)}(\hat{x}_i) > \lambda^{(k_2-k_1)}(E(\tilde{x}) - E_-^{(k_1)}) + \lambda E(\tilde{x})/n + (1-\lambda)E_i(\tilde{x}), \qquad (23)$$

for all i, then $\tilde{x}$ is a global optimum.

This sufficient condition does not restrict the choice of cooperation strength λ. The whole range of the cooperation strength can be exploited to increase the chance of finding a consensus.

3.4. *Necessary Conditions*

The following theorem provides us the first necessary condition for an option to be in the global optimum.

Theorem 14: (Necessary Condition 1) Given a propagation matrix W, and the general initial condition $c^{(0)} = 0$ or $\lambda_1 = 0$. If option x_i^* ($x_i^* \in D_i$) is in the global optimum, then $c_i^{(k)}(x_i^*)$, for any $k \geq 1$, must satisfy the following inequality,

$$c_i^{(k)}(x_i^*) \leq (E^* - E_-^{(k)}) + c_i^{(k)}(\tilde{x}_i^{(k)}) \qquad (24)$$

where $E_-^{(k)}$ is, as defined before, a lower bound on E^* obtained by the cooperative system at step k.

Theorem 15: (Necessary Condition 2) Given a symmetric propagation matrix W and the general initial condition $c^{(0)} = 0$ or $\lambda_1 = 0$. If option x_i^* ($x_i^* \in D_i$) is in the global optimum, then $c_i^{(k)}(x_i^*)$ must satisfy the following inequality,

$$c_i^{(k)}(x_i^*) \leq \frac{E^*}{n} + \sqrt{\frac{n-1}{n}}|\alpha_2^{(k)}|E^*. \qquad (25)$$

Here $\alpha_2^{(k)}$ is computed by the following recursive function:

$$\begin{cases} \alpha_2^{(1)} = \lambda_1 \alpha_2 + (1 - \lambda_1) \\ \alpha_2^{(k)} = \lambda_k \alpha_2 \alpha_2^{(k-1)} + (1 - \lambda_k) \end{cases}$$

where α_2 is the second largest eigenvalue of the propagation matrix W.

For the particular choice of W=$\frac{1}{n}(1)_{n \times n}$,

$$\alpha_2^{(k)} = (1 - \lambda_k)$$

and

$$c_i^{(k)}(x_i^*) \leq \frac{E^*}{n} + \sqrt{\frac{n-1}{n}}(1 - \lambda_k)E^*. \tag{26}$$

Inequality (24) and Inequality (25) provide two criteria for checking if an option can be in some global optimum. If either of them is not satisfied, the option can be discarded from the option set to reduce the search space.

Both thresholds in (24) and (25) become tighter and tighter as the iteration proceeds. Therefore, more and more options can be discarded and the search space can be reduced. With the choice of the general initial condition $c^{(0)} = 0$, the right hand side of (24) decreases as the iteration proceeds because of the property of $E_-^{(k)}$ revealed by Theorem 8. With the choice of a constant cooperation strength λ, and supposing $W \neq \frac{1}{n}(1)_{n \times n}$, then $\alpha_2 > 0$ and $\{\alpha_2^{(k)} | k \geq 1\}$ is a monotonic decreasing sequence satisfying

$$\frac{1 - \lambda}{1 - \lambda \alpha_2} < \alpha_2^{(k)} \leq (1 - \lambda) + \lambda \alpha_2. \tag{27}$$

This implies that the right hand side of (25) monotonically decreases as the iteration proceeds.

Based on Theorem 14 and Theorem 15, an ambiguity reduction rule is given as follows:

Ambiguity Reduction Rule: Let E^+ be an upper bound of E^*, $E^+ \geq E^*$. For any $x_i \in D_i$ $(1 \leq i \leq n)$, if $c_i^{(k)}(x_i)$, at some step $k \geq 1$, satisfies

$$c_i^{(k)}(x_i) > (E^+ - E_-^{(k)}) + c_i^{(k)}(\tilde{x}_i^{(k)}), \tag{28}$$

or

$$c_i^{(k)}(x_i) > \frac{E^+}{n} + \sqrt{\frac{n-1}{n}}|\alpha_2^{(k)}|E^+, \tag{29}$$

then the option x_i can be discarded from domain D_i to reduce the search space and the ambiguity in decision making for agent i.

In the above rule, we use an upper bound E^+ on the optimal cost E^* instead of itself in (24) and (25), because an upper bound can be obtained more easily than the optimal cost E^* in most cases. $E(\tilde{x}^{(k)})$ provided by the algorithm, for example, can be used here as E^+.

The application of the above ambiguity reduction rule guarantees the retaining of options in any global optimum. If all options but one for each agent are discarded, then the ambiguity for each agent in decision-making is eliminated, and the global optimum is found.

Rule (28) and Rule (29) provides the theoretical basis for us to choose the threshold $t_i^{(k)}$ in the option discarding process (6):

$$t_i^{(k)} = \min((E(\tilde{x}^{(k)}) - E_-^{(k)}) + c_i^{(k)}(\tilde{x}_i^{(k)}), \frac{E(\tilde{x}^{(k)})}{n} + \sqrt{\frac{n-1}{n}}|\alpha_2^{(k)}|E(\tilde{x}^{(k)})). \tag{30}$$

3.5. *Strong Cooperation*

When the cooperative system operates sequentially and the influence level is the strongest, then the optimization defined by the system falls back to the conventional local search.

Theorem 16: Given a good decomposition $\{E_i\}$, $\lambda_k = 1 - \epsilon$ (ϵ is an positive infinitesimal value), $\mu = 0$, if the cooperative system operates sequentially, then the optimization it defines is equivalent to conventional local search. Given $0 < \mu < 1$, then the optimization it defines is equivalent to conventional local search with a lazy style.

If we view the system as a game where the objective function E_i for agent i is treated as the utility function for the agent, then it is not hard to find out that the equilibrium of the system in this case is also a pure strategy Nash equilibrium, a strategic equilibrium in game theory, which formally studies conflict and cooperation in a system of agents. In this case, the list of choices, one for each agent, has the property that no agent can unilaterally change his choice and get a better payoff. In our definition, without loss of generality, we let each agent minimize its utility function (the objective function) instead of maximize it.

4. Experiments and Results

4.1. *Stereo Matching*

Stereo matching is one of the most active research areas in computer vision [11, 2, 13, 8]. Like many other problems from computer vision it can be formulated as the global optimization of multivariate energy functions, which is NP-hard [1] in computational complexity in discrete space. They have the general binary constraint form (1) that the TSP can be formalized too as described in the previous section.

We have successfully applied the cooperative system with the balanced cooperation to minimize the energy functions from stereo matching. To choose the propagation matrix, we can have all the options as long as the matrix is square, irreducible, and with non-negative elements as defined in Definition 2. Since each site i in an image has four neighbors and is associated with one agent, we set $w_{ij} = 0.25$ if site j is the neighbor of site i. Otherwise, it is zero.

In the iterative function (5), the parameter λ_k is updated as

$$\lambda_k = (k-1)/k, \qquad \text{where } k \geq 1.$$

Hence, the cooperation becomes stronger as the iteration proceeds.

In the experiments, we reduce the threshold $t_i^{(k)}$ in (6) exponentially with the increase of iteration step k:

$$t_i^{(k)} = 100 * 0.92^k \qquad \text{for any } i \text{ at step } k.$$

Hence, more and more options will be discarded as the iteration proceeds. Eventually, there should be only one option left for each agent. The improvement of the cooperative optimization in this chapter over the one in [4] is on the adjustment of the threshold. In [4], the threshold reduces strictly with iteration step k. In this chapter, it will remain unchanged for the next iteration if we see more than 0.1% options are discarded at the current iteration.

However, by doing that, the final solution is not guaranteed to be the global optimum because this threshold could be tighter than the one suggested by theory and the optimal options in the global optimum could be discarded.

The script we use for evaluation under the Middlebury College framework is based on script exp6_gc.txt. The other settings come from the default values in the framework.

The following four tables show the performance of the cooperative system (upper rows in a table) and the simulated annealing algorithm offered by the framework (lower rows in a table) over the four test image sets. From the tables we can see that the former is significantly better than the latter in terms of the quality of solutions.

image = Map (variables = 61344)						
	ALL	NON OCCL	OCCL	TEXTRD	TEXTRLS	D_DISCNT
Disparity Error	4.08	1.12	16.08	1.13	0.47	3.69
Bad Pixels	5.91%	0.53%	90.76%	0.52%	0.95%	5.15%
Disparity Error	5.08	3.94	13.73	3.94	3.73	6.97
Bad Pixels	18.85%	14.30%	90.70%	14.06%	41.43%	23.97%
image = Sawtooth (variables = 164920)						
	ALL	NON OCCL	OCCL	TEXTRD	TEXTRLS	D_DISCNT
Disparity Error	1.40	0.68	7.31	0.71	0.42	1.62
Bad Pixels	4.41%	1.86%	92.39%	1.95%	0.99%	6.56%
Disparity Error	2.24	1.92	7.11	1.79	2.95	2.17
Bad Pixels	12.23%	9.85%	94.40%	8.70%	21.26%	14.84%
image = Tsukuba (variables = 110592)						
	ALL	NON OCCL	OCCL	TEXTRD	TEXTRLS	D_DISCNT
Disparity Error	1.18	0.81	5.43	0.95	0.55	1.67
Bad Pixels	4.03%	1.75%	90.21%	2.54%	0.68%	8.11%
Disparity Error	2.36	2.22	5.50	1.41	2.99	2.39
Bad Pixels	19.94%	18.14%	88.04%	6.65%	33.86%	18.39%
image = Venus (variables = 166222)						
	ALL	NON OCCL	OCCL	TEXTRD	TEXTRLS	D_DISCNT
Disparity Error	1.48	1.02	7.92	0.88	1.25	1.42
Bad Pixels	4.40%	2.77%	91.40%	2.38%	3.57%	9.68%
Disparity Error	3.55	3.41	8.03	2.61	4.63	2.32
Bad Pixels	26.29%	25.04%	92.74%	13.81%	47.96%	21.45%

For the Tsukuba images of the ground truth shown in Figure 1, the recovered depth images of the two stereo algorithms are shown in Figure 2 and Figure 3. We can see by comparing these three images that the result of the cooperative system is much better than that of the simulated annealing in all types of areas.

Our computer simulation has suggested that the system, in a parallel implementation, can find the solutions for the four test image pairs in 0.187,

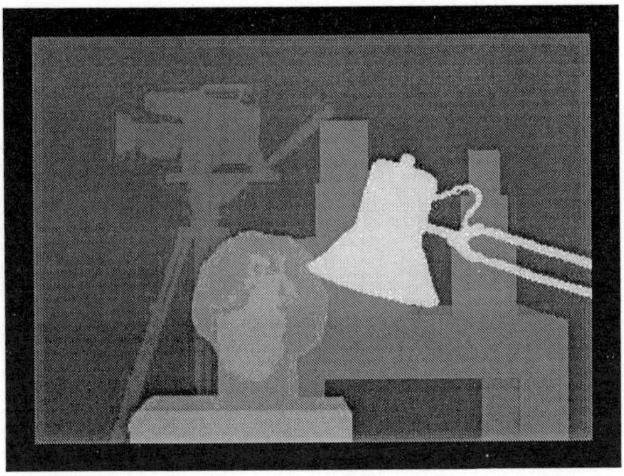

Fig. 1. The ground truth.

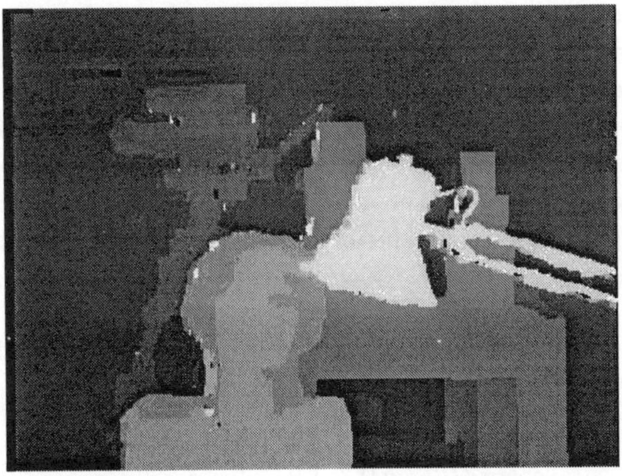

Fig. 2. The depth image recovered by the cooperative system from the Tsukuba images.

0.311, 0.362, and 0.446 milliseconds, respectively.

4.2. *DNA Image Analysis*

DNA image analysis is used to find gene spots in a gene chip. A gene chip may contain thousands or more gene spots, each spot is used for detecting the expression level of the gene printed at the spot. Any living thing

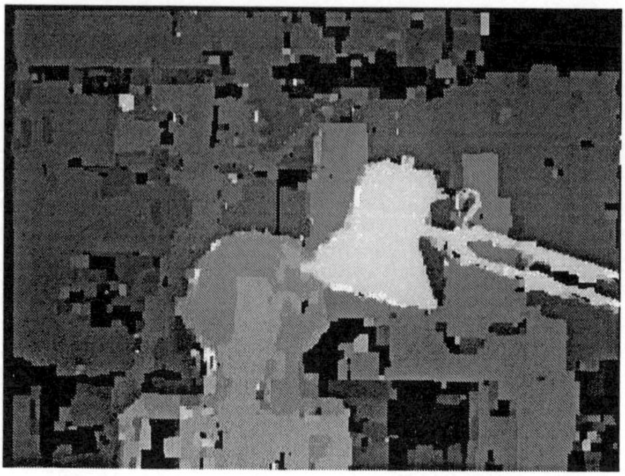

Fig. 3. The depth image recovered by simulated annealing algorithm from the Tsukuba images.

is controlled by a number of genes. A human body, for example, is controlled by around 30,000 genes. Each gene is turned on or off, known as the expression level, due to its reaction to medicines, diseases, or the growing stages. To obtain the gene expression levels, it is not only costly, but also time-consuming. It is desirable to use computers to help people in finding the correct locations of the genes in a chip. Like many other problems from image processing, it can be formulated as the global optimization of a multivariate energy function, which has the general binary constraint form (1).

The unary constraint $C_i(x_i)$ in (1) is defined as the dis-likelihood of the occurrence of a gene spot i at site x_i. We use a circular shape of the following form as the mask to compute $C_i(x_i)$:

$$\cos(\pi d/(2 * \sigma)) + 1.0, \quad \text{for } d < 2 * \sigma \ ,$$

where d is the radius distance from a site to the site i, σ is the radius of the circular mask, and we set it to be 5 in our experiments.

The binary constraint $C_{i,j}(x_i, x_j)$ in (1) is defined as the difference of the expected distance of two gene spots, i and j, from the detected gene spots.

$$C_{ij}(x_i, x_j) = \begin{cases} 0 & \text{if } \Delta d_{ij} < \varepsilon, \\ \Delta d_{ij} & \text{otherwise} \end{cases}$$

where Δd_{ij} is the difference of the expected distance from the detected distance of gene spot i and gene spot j. ε is a parameter set to 5 in our experiments.

With a couple of DNA images randomly selected from a pool of thousands, where each image has thousands of gene spots, we found that the cooperative system successfully found all genes, resisting interference from dust speckles, high backgrounds, or missing gene spot rows. Figure 4 shows two blocks of genes detected successfully by the cooperative system from a gene chip containing 4, 602 gene spots. Figure 5 shows an area in a block of genes detected successfully by the cooperative system when there are dust speckles.

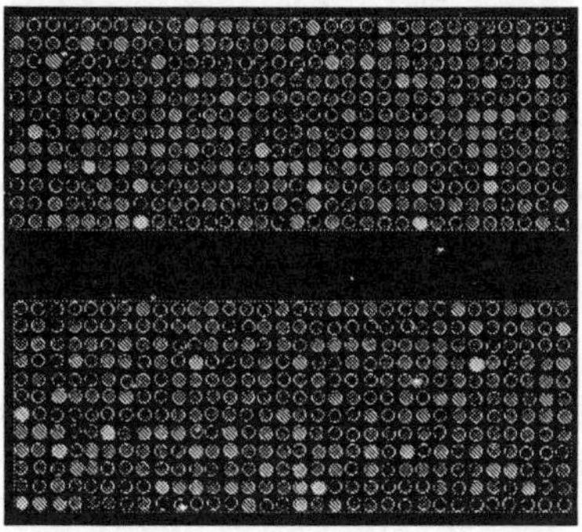

Fig. 4. Two blocks of genes detected successfully by the cooperative system from a gene chip containing 4, 602 gene spots.

5. Conclusions

A formal description of a cooperative system for optimization has been presented. To demonstrate its power, its applications to stereo matching problems from computer vision and the DNA image analysis are provided. Using a common evaluation framework provided by Middlebury College, the system has shown a much better overall performance in terms of solution quality than that of simulated annealing. Furthermore, the operations of the system are simple and inherently parallel. Our computer simulation has suggested that if the system is implemented in parallel, it can find the solution for any stereo matching problem from the framework in less than 0.5 millisecond.

The optimization of the system in the balanced case is based on a cooperation process where one of the key operations is option discarding. Such a

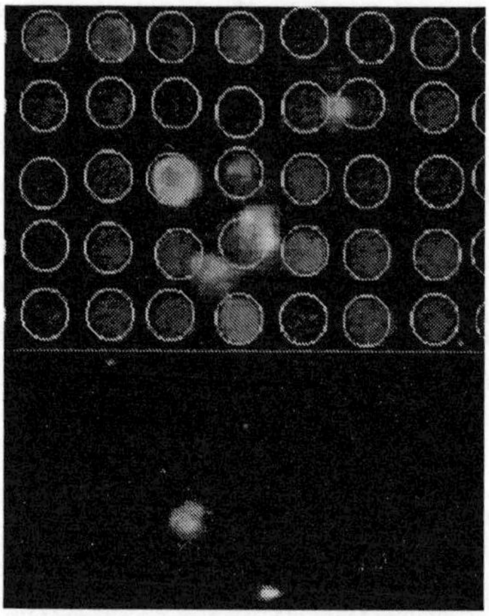

Fig. 5. A area in a block of genes detected successfully by the cooperative system from a gene chip containing 4,602 gene spots where there are dust speckles.

process is the same in principle as the cooperation process used by Marr and Poggio in [8], and Lawrence and Kanade in [13], where "option" is termed "unit" and "discarding" is termed "inhibition." The cooperation optimization is fundamentally different from most known optimization methods. It has many interesting computational properties not possessed by conventional ones. They could help us in understanding cooperative computation possibly used by human brains in solving early vision problems.

The cooperative principle opens a completely new way for attacking hard optimization problems. The influence level and the information exchange rate defined in the cooperation scheme open new dimensions in discovering optimization algorithms, just like the temperature used in simulated annealing. Different settings of these two parameters lead to completely different computational behaviors of the system. When the influence level is balanced with the exchange rate, the system always has a unique equilibrium and it is guaranteed to reach the equilibrium with any initial conditions. The equilibrium is also the global optimum of the system if a consensus is reached among agents in this case. When the influence level is at the strongest, the system can always reach an equilibrium which is also a Nash equilibrium, a strategic equilibrium in game theory.

Further investigation on this new optimization paradigm is desirable both from the theoretical perspective in understanding the tractability of NP-hard problems and from the practical perspective in a wide range of applications in operations research, engineering, biological sciences, and computer science.

References

[1] Atkinson, K. (1989). *Computers and Intractability.* Kluwer Academic Publishers, San Francisco, U.S.A.

[2] Boykov, Y., Veksler, O., and Zabih, R. (2001). Fast approximate energy minimization via graph cut. *IEEE TPAMI*, 23(11):1222–1239.

[3] Hinton, G., Sejnowski, T., and Ackley, D. (1992). Genetic algorithms. *Cognitive Science*, pages 66–72.

[4] Huang, X. (2004). A general global optimization algorithm for energy minimization from stereo matching. In *ACCV*, Korea.

[5] Jr., E. G. C., editor (1976). *Computer and Job-Shop Scheduling.* Wiley-Interscience, New York.

[6] Kirkpatrick, Gelatt, C., and Vecchi, M. (1983). Optimization by simulated annealing. *Science*, 220:671–680.

[7] Lawler, E. L. and Wood, D. E. (1966). Brand-and-bound methods: A survey. *OR*, 14:699–719.

[8] Marr, D. and Poggio, T. (1976). Cooperative computation of stereo disparity. *Science*, 194:209–236.

[9] Michalewicz, Z. and Fogel, D. (2002). *How to Solve It: Modern Heuristics.* Springer-Verlag, New York.

[10] Papadimitriou, C. H. and Steiglitz, K., editors (1998). *Combinatorial Optimization.* Dover Publications, Inc.

[11] Scharstein, D. and Szeliski, R. (2002). A taxonomy and evaluation of dense two-frame stereo correspondence algorithms. *IJCV*, 47:7–42.

[12] Varga, R., editor (1962). *Matrix iterative Analysis.* :Prentice-Hall, Elglewood Cliffs, N.J.

[13] Zitnick, C. L. and Kanade, T. (2000). A cooperative algorithm for stereo matching and occlusion detection. *IEEE TPAMI*, 2(7).

Appendix: Proofs of Theorems

The Properties of Propagation Matrix

Property 5.1: If W is a symmetric propagation matrix, then its has n real eigenvalues, $\alpha_1, \alpha_2, \ldots, \alpha_n$, not necessarily distinct, which satisfy

$$1 = \alpha_1 \geq |\alpha_2| \geq \cdots \geq |\alpha_n| \geq 0$$

Proof. We have assumed the propagation matrix has nonnegative elements and the sum of each row is equal to 1. Obviously, 1 is an eigenvalue and its corresponding eigenvector is $(1)_{n\times 1}$.

According to the Principal Axes Theorem in linear algebra, if W is symmetric and real square matrix, then W has n real eigenvalues.

From the theory of linear algebra,

$$r_\sigma(W) \leq \max_i \sum_j |w_{ij}|$$

where r_σ is called the spectral radius of W. r_σ is the maximum size of eigenvalues of W,

$$r_\sigma(W) = \max_i |\alpha_i|,$$

Since $\sum_j w_{ij} = 1$ and $w_{ij} \geq 0$, we have

$$r_\sigma(W) \leq 1$$

This implies

$$|\alpha_i| \leq 1$$

Hence we have

$$1 = \alpha_1 \geq |\alpha_2| \geq \cdots \geq |\alpha_n| \geq 0. \qquad \S$$

Property 5.2: If W is a symmetric, irreducible propagation matrix, then 1 is a simple eigenvalue of W.

Proof: According to the Perron-Frobenius Theorem [12], 1 is a simple eigenvalue of W provided that W is irreducible. §

Property 5.3: If W is an irreducible propagation matrix, then

$$\lim_{k\to\infty} W^k = \frac{1}{n}(1)_{n\times n}$$

This property directly follows the previous property.

Property 5.4: If we have the following difference equation,

$$c^{(k+1)} = Wc^{(k)}, \qquad \text{for } k \geq 0,$$

where W is an irreducible propagation matrix, then

$$\lim_{k\to\infty} c_i^{(k)} = \frac{1}{n}\sum_j c_j^{(0)}, \qquad \text{for any } i.$$

This property directly follows the previous property.

Proof of Theorem 6

To prove Theorem 6 we use the principle of mathematical induction. Let $k = 1$.

Case 1: Choose $c^{(0)} = 0$. From (15),

$$\begin{aligned}\sum_i c_i^{(1)}(x_i) &= \sum_i \min_{x_j, j \neq i} ((1-\lambda_1)E_i) \\ &\leq (1-\lambda_1)\sum_i E_i = (1-\lambda_1)E(x_1, x_2, \ldots, x_n) \\ &\leq E(x_1, x_2, \ldots, x_n) \ .\end{aligned}$$

Case 2: Choose $\lambda_1 = 0$. From (15),

$$\sum_i c_i^{(1)}(x_i) = \sum_i \min_{x_j, j \neq i} E_i \leq \sum_i E_i = E(x_1, x_2, \ldots, x_n) \ .$$

Hence, the inequality (18) is correct when $k = 1$.

Assume that, for some $k \geq 1$,

$$\sum_i c_i^{(k)}(x_i) \leq E(x_1, x_2, \ldots, x_n) \ . \tag{31}$$

From (15),

$$\begin{aligned}\sum_i c_i^{(k+1)}(x_i) &= \sum_i \min_{x_j, j \neq i} \left(\lambda_{k+1}(w_i, c^{(k)}) + (1-\lambda_{k+1})E_i\right) \\ &\leq \sum_i \left(\lambda_{k+1}(w_i, c^{(k)}) + (1-\lambda_{k+1})E_i\right) \\ &= \lambda_{k+1}\sum_i \sum_j w_{ij} c_j^{(k)}(x_j) + (1-\lambda_{k+1})E(x_1, x_2, \ldots, x_n) \\ &= \lambda_{k+1}\sum_i c_i^{(k)}(x_i) + (1-\lambda_{k+1})E(x_1, x_2, \ldots, x_n) \ ,\end{aligned}$$

since $\sum_i w_{ij} = 1$.

Combining the above result with the assumption (31), we get

$$\sum_i c_i^{(k+1)}(x_i) \leq E(x_1, x_2, \ldots, x_n) \ . \tag{32}$$

This proves the inequality (18) for any $k \geq 1$. Now we prove inequality (19).

For any $k \geq 1$,

$$E_{-}^{(k)} = \sum_i \min_{x_i} c^{(k)}(x_i)$$
$$\leq \sum_i c^{(k)}(x_i^*) \leq E(x_1^*, x_2^*, \ldots, x_n^*) = E^* . \quad \S$$

Proof of Theorem 7

The proof of Theorem 7 needs the following lemma.

Lemma 17: *Choose a propagation matrix W, a constant cooperation strength λ, and the general initial condition $c^{(0)} = 0$, then $\{c_i^{(k)}(x_i)|k \geq 0\}$ is a non-decreasing sequence for any $x_i \in D_i$ and $1 \leq i \leq n$.*

Proof: We prove this lemma by the principle of mathematical induction.

Let $k = 0$.

From (15), we have

$$c_i^{(1)}(x_i) = \min_{x_j, j \neq i} (\lambda(w_i, c^{(0)}) + (1 - \lambda)E_i) .$$

Using the condition $c^{(0)} = 0$ and $E_i \geq 0$ by the assumption of nonnegative constraints,

$$c_i^{(1)}(x_i) = (1 - \lambda)\tilde{E}_i^{(0)} \geq c_i^{(0)}(x_i) .$$

Assume that, for some $k - 1 \geq 0$,

$$c_i^{(k)}(x_i) \geq c_i^{(k-1)}(x_i), \qquad \text{for any } x_i \in D_i \text{ and } 1 \leq i \leq n , \tag{33}$$

then

$$(w_i, c^{(k)}) \geq (w_i, c^{(k-1)}) ,$$

since $w_{ij} \geq 0$. Thus,

$$(\lambda(w_i, c^{(k)}) + (1 - \lambda)E_i) \geq (\lambda(w_i, c^{(k-1)}) + (1 - \lambda)E_i) .$$

This implies

$$\min_{x_j, j \neq i} (\lambda(w_i, c^{(k)}) + (1 - \lambda)E_i) \geq \min_{x_j, j \neq i} (\lambda(w_i, c^{(k-1)}) + (1 - \lambda)E_i) .$$

That is

$$c_i^{(k+1)}(x_i) \geq c_i^{(k)}(x_i), \qquad \text{for any } x_i \in D_i \text{ and } 1 \leq i \leq n ,$$

from the definition of $c_i^{(k+1)}(x_i)$ in (15).

Hence, $c_i^{(k+1)}(x_i) \geq c_i^{(k)}(x_i)$ holds for any $k \geq 0$, any $x_i \in D_i$, and $1 \leq i \leq n$. That is, $\{c_i^{(k)}(x_i)|k \geq 0\}$ is a non-decreasing sequence for any $x_i \in D_i$ and $1 \leq i \leq n$. §

Proof: From Lemma 17, for any $k \geq 0$,

$$\sum_i c_i^{(k+1)}(\tilde{x}_i^{(k+1)}) \geq \sum_i c_i^{(k)}(\tilde{x}_i^{(k+1)}) \geq \sum_i c_i^{(k)}(\tilde{x}_i^{(k)}) .$$

This implies

$$E_-^{(k+1)} \geq E_-^{(k)}, \qquad \text{for } k \geq 0 .$$

Hence, $\{E_-^{(k)}|k \geq 0\}$ is a non-decreasing sequence.

According to Theorem 6,

$$E_-^{(k)} \leq E^* .$$

Hence, $\{E_-^{(k)}|k \geq 0\}$ is up-bounded by E^*. §

Proof of Theorem 8

Assume $\tilde{x}$ is a consensus solution found at step k. From (15), we have

$$\begin{aligned}\sum_i c^{(k)}(\tilde{x}_i) &= \sum_i \min_{x_j, j\neq i} (\lambda_k(w_i, c^{(k-1)}) + (1-\lambda_k)E_i) \\ &= \lambda_k \sum_i \sum_j w_{ij} c_j^{(k-1)}(\tilde{x}_j) + (1-\lambda_k)\sum_i \tilde{E}_i \\ &= \lambda_k \sum_i c_i^{(k-1)}(\tilde{x}_i) + (1-\lambda_k)E(\tilde{x}) .\end{aligned}$$

since $\sum_i w_{ij} = 1$. Based on the condition that $\tilde{x}$ is a consensus solution found from step k_1 to step k_2, the above equation holds for $k_1 \leq k \leq k_2$. Combining these results yields the following equation

$$\sum_i c_i^{(k_2)}(\tilde{x}_i) = A \sum_i c_i^{(k_1-1)}(\tilde{x}_i) + (1-A)E(\tilde{x}) , \tag{34}$$

where $A = \prod_{k=k_1}^{k_2} \lambda_k$.

From (34) and the lower bound theorem, we have

$$A \sum_i c_i^{(k_1-1)}(\tilde{x}_i) + (1-A)E(\tilde{x}) = \sum_i c_i^{(k_2)}(\tilde{x}_i) = E_-^{(k_2)} \leq E^* . \tag{35}$$

Rewrite (35), we have

$$E(\tilde{x}) - E^* \leq A(E(\tilde{x}) - \sum_i c_i^{(k_1-1)}(\tilde{x}_i)) \leq A(E(\tilde{x}) - E_-^{(k_1-1)}) . \tag{36}$$

That proves the first inequality.

Using Difference Equations (15), we have

$$\sum_i c^{(k)}(\tilde{x}_i) \le \lambda_k \sum_i \sum_j w_{ij} c_j^{(k-1)}(x_j^*) + (1-\lambda_k) \sum_i E_i^*$$
$$= \lambda_k \sum_i c_i^{(k-1)}(x_i^*) + (1-\lambda_k) E^* \,.$$

Since the above inequalities holds for $k_1 \le k \le k_2$, we have

$$\sum_i c_i^{(k_2)}(\tilde{x}_i) \le A \sum_i c_i^{(k_1-1)}(x_i^*) + (1-A)E^* \,. \tag{37}$$

Combine (34) and (37),

$$A \sum_i c_i^{(k_1-1)}(\tilde{x}_i) + (1-A)E(\tilde{x}) \le A \sum_i c_i^{(k_1-1)}(x_i^*) + (1-A)E^* \,. \tag{38}$$

This implies

$$E(\tilde{x}) \le E^* + \frac{\prod_{k=k_1}^{k_2} \lambda_k}{1 - \prod_{k=k_1}^{k_2} \lambda_k} \left(\sum_i c_i^{(k_1-1)}(x_i^*) - \sum_i c_i^{(k_1-1)}(\tilde{x}_i) \right) . \tag{39}$$

According to Theorem 6, with the general initial condition $c^{(0)} = 0$ or $\lambda_1 = 0$,

$$\sum_i c_i^{(k_1-1)}(x_i^*) \le E^* \,,$$

and by definition,

$$E_-^{(k_1-1)} = \sum_i c_i^{(k_1-1)}(\tilde{x}_i^{(k_1-1)}) \le \sum_i c_i^{(k_1-1)}(\tilde{x}_i) \,.$$

Hence,

$$E(\tilde{x}) \le E^* + \frac{\prod_{k=k_1}^{k_2} \lambda_k}{1 - \prod_{k=k_1}^{k_2} \lambda_k} (E^* - E_-^{(k_1-1)}) \,. \tag{40}$$

From condition $1 - \lambda_k \ge \epsilon > 0$ for $k_1 \le k \le k_2$,

$$\prod_{k=k_1}^{k_2} \lambda_k \le (1-\epsilon)^{k_2-k_1} \,.$$

From (40),

$$E(\tilde{x}) \le E^* + \frac{B}{1-B}(E^* - E_-^{(k_1-1)}) \,,$$

where $B = (1-\epsilon)^{k_2-k_1}$.

When $k_2 - k_1 \to \infty$, $B \to 1$.

Hence $E(\tilde{x}) \to E^*$, when $k_2 - k_1 \to \infty$, since $E(\tilde{x}) \geq E^*$. §

Proof of Theorem 9

Here we only prove that the difference equations have at least one solution. The uniqueness property will be proved after the proof of Theorem 10.

Proof: According to Lemma 17, when we choose $c^{(0)} = 0$, $\{c_i^{(k)}(x_i)|k \geq 0\}$ is a nondecreasing sequence for any $x_i \in D_i$ and $1 \leq i \leq n$. According to Theorem 6, we know $\sum_i c_i^{(k)}(x_i)$ is bounded above. Because of that and the nonnegative property of $c_i^{(k)}(x_i)$, $c_i^{(k)}(x_i)$ for any $k \geq 0$ must have an upper bound. Hence, sequence $\{c_i^{(k)}(x_i)|k \geq 0\}$ must be bounded above. Thus it has a least upper bound, denoted as $c_i^{(\infty)}(x_i)$ here, and it converges to this bound.

Since the mapping of the assignment constraints c defined by (15) is continuous, $(c_i^{(\infty)}(x_i))$ must be a solution to the difference equations. §

Proof of Theorem 10

Proof: From (15),

$$
\begin{aligned}
c_i^{(k+1)}(x_i) &= \min_{x_j, j \neq i} (\lambda(w_i, c^{(k)}) + (1-\lambda)E_i) \\
&= \min_{x_j, j \neq i} \left((\lambda(w_i, c^{(\infty)}) + (1-\lambda)E_i) + \lambda(w_i, c^{(k)} - c^{(\infty)})\right) ,
\end{aligned}
$$

then

$$
c_i^{(k+1)}(x_i) \leq \min_{x_j, j \neq i} (\lambda(w_i, c^{(\infty)}) + (1-\lambda)E_i) + \lambda(w_i, (\|c^{(k)} - c^{(\infty)}\|_\infty)_{n \times 1}) ,
$$

and

$$
c_i^{(k+1)}(x_i) \geq \min_{x_j, j \neq i} (\lambda(w_i, c^{(\infty)}) + (1-\lambda)E_i) - \lambda(w_i, (\|c^{(k)} - c^{(\infty)}\|_\infty)_{n \times 1}) .
$$

According to Theorem 9,

$$
c_i^{(\infty)}(x_i) = \min_{x_j, j \neq i} (\lambda(w_i, c^{(\infty)}) + (1-\lambda)E_i), \quad \text{for any } x_i \in D_i \text{ and } 1 \leq i \leq n .
$$

Since $w_{ij} = w_{ji}$ and $\sum_i w_{ij} = 1$,

$$
(w_i, (\|c^{(k)} - c^{(\infty)}\|_\infty)_{n \times 1}) = \|c^{(k)} - c^{(\infty)}\|_\infty .
$$

Then

$$
c_i^{(k+1)}(x_i) \leq c_i^{(\infty)}(x_i) + \lambda\|c^{(k)} - c^{(\infty)}\|_\infty ,
$$

and

$$
c_i^{(k+1)}(x_i) \geq c_i^{(\infty)}(x_i) - \lambda\|c^{(k)} - c^{(\infty)}\|_\infty .
$$

That is

$$|c_i^{(k+1)}(x_i) - c_i^{(\infty)}(x_i)| \leq \lambda \|c^{(k)} - c^{(\infty)}\|_\infty .$$

Since the above inequality holds for any $x_i \in D_i$ and i, we have

$$\|c_i^{(k+1)} - c_i^{(\infty)}\|_\infty \leq \lambda \|c^{(k)} - c^{(\infty)}\|_\infty .$$

This implies

$$\|c_i^{(k)} - c_i^{(\infty)}\|_\infty \leq \lambda^k \|c^{(0)} - c^{(\infty)}\|_\infty . \quad \S$$

We now complete the proof of Theorem 9, i.e. Difference Equations (15) have a unique solution.
Proof: We have proved that the difference equations has one solution $c^{(\infty)}$. Suppose, for contradiction, there is another solution, denoted as $\hat{c}^{(\infty)}$, which satisfies the difference equations.

According to Theorem 10, with the choice of $c^{(0)} = \hat{c}^{(\infty)}$, we have

$$\|\hat{c}^{(\infty)} - c^{(\infty)}\|_\infty \leq \lambda^k \|\hat{c}^{(\infty)} - c^{(\infty)}\|_\infty . \tag{41}$$

Since $0 \leq \lambda < 1$, then from (41) we have

$$\|\hat{c}^{(\infty)} - c^{(\infty)}\|_\infty = 0 .$$

This implies

$$\hat{c}^{(\infty)} = c^{(\infty)} .$$

This contradicts the assumption that $\hat{c}^{(\infty)}$ and $c^{(\infty)}$ are different. Hence, the difference equations have only one solution. §

Proof of Theorem 11

Proof: According to Theorem 8, if a consensus solution is found at some step k with the choice of $\lambda_k = 0$, then $E(\tilde{x}^{(k)})$ is equal to E^*. This implies that a consensus solution found by the algorithm under the condition $\lambda_k = 0$ is also a global optimum.

Proof of Theorem 12

Proof: The proof of this theorem needs the following lemma.

Lemma 18: *Choose a propagation matrix W, then the lower bounds on the optimal cost obtained at two consecutive steps satisfy the following inequality:*

$$E_-^{(k+1)} \geq \lambda_{k+1} E_-^{(k)} + (1 - \lambda_{k+1}) \sum_i \tilde{E}_i^{(k)} . \tag{42}$$

Proof: From (15),

$$\begin{aligned}
E_-^{(k+1)} &= \sum_i c_i^{(k+1)}(\tilde{x}_i^{(k+1)}) \\
&= \sum_i \min_{x_j, j \neq i} (\lambda_{k+1}(w_i, c^{(k)}) + (1-\lambda_{k+1})E_i) \\
&= \lambda_{k+1} \sum_i \sum_j w_{ij} c_j^{(k)}(\tilde{x}_j^{(k)}(i)) + (1-\lambda_{k+1}) \sum_i \tilde{E}_i^{(k)} \\
&\geq \lambda_{k+1} \sum_i \sum_j w_{ij} c_j^{(k)}(\tilde{x}_j^{(k)}(j)) + (1-\lambda_{k+1}) \sum_i \tilde{E}_i^{(k)} \\
&= \lambda_{k+1} \sum_i c_i^{(k)}(\tilde{x}_i^{(k)}) + (1-\lambda_{k+1}) \sum_i \tilde{E}_i^{(k)} \\
&= \lambda_{k+1} E_-^{(k)} + (1-\lambda_{k+1}) \sum_i \tilde{E}_i^{(k)}
\end{aligned}$$

This completes the proof. §

Proof of the theorem: Let the consensus solution found at step k is $\tilde{x}^{(k)}$.

According to Lemma 18,

$$E_-^{(k+1)} \geq \lambda_{k+1} E_-^{(k)} + (1-\lambda_{k+1}) E(\tilde{x}^{(k)})$$

that is

$$(1-\lambda_{k+1})(E_-^{(k+1)} - E(\tilde{x}^{(k)})) \geq \lambda_{k+1}(E_-^{(k)} - E_-^{(k+1)})$$

Based on the condition $E_-^{(k+1)} \leq E_-^{(k)}$, we have

$$E_-^{(k+1)} - E(\tilde{x}^{(k)}) \geq 0$$

that is

$$E_-^{(k+1)} \geq E(\tilde{x}^{(k)}) \tag{43}$$

According to Theorem 6, when we choose $c^{(0)} = 0$ or $\lambda_1 = 0$,

$$E^* \geq E_-^{(k+1)}$$

using (43),

$$E^* \geq E(\tilde{x}^{(k)})$$

This implies

$$E^* = E(\tilde{x}^{(k)})$$

since $E(\tilde{x}^{(k)}) \geq E^*$. Hence $\tilde{x}^{(k)}$ is a global optimum. §

Proof of Theorem 13

Proof: This theorem can be simplified proved by using Theorem 8 get an upper bound for $c^{(k)}(x_i)$ and using Theorem 14 to get the inequity.

Proof of Theorem 14

Proof: Assume that $(x_1^*, x_2^*, \ldots, x_n^*)$ is a global optimum.

According to Theorem 6,

$$\sum_i c_i^{(k)}(x_i^*) \leq E(x_1^*, x_2^*, \ldots, x_n^*) = E^*$$

Also,

$$\begin{aligned}\sum_i c_i^{(k)}(x_i^*) &\geq c_i^{(k)}(x_i^*) + \sum_{j \neq i} c_j^{(k)}(\tilde{x}_j^{(k)}) \\ &= c_i^{(k)}(x_i^*) + E_-^{(k)} - c_i^{(k)}(\tilde{x}_i^{(k)})\end{aligned}$$

This implies

$$c_i^{(k)}(x_i^*) \leq (E^* - E_-^{(k)}) + c_i^{(k)}(\tilde{x}_i^{(k)})$$

because

$$\sum_i c_i^{(k)}(x_i^*) \leq E^*,$$

from Theorem 6. §

Proof of Theorem 15

The proof of this theorem needs the following lemma.

Lemma 19: *Choose a symmetric propagation matrix W. If x and y are two vectors satisfying*

$$\sum_i x_i = 0, \qquad \text{and} \qquad y = Wx$$

then y satisfies

$$\|y\|_2^2 \leq \alpha_2^2 \|x\|_2^2 \leq \|x\|_2^2$$

where $\|x\|_2$ is the Euclidean norm of x,

$$\|x\|_2 = \sqrt{\Sigma_i x_i^2}$$

and α_2 is the eigenvalue of W having the second maximum size.

Specifically, when W is irreducible,

$$\|y\|_2 < \|x\|_2,$$

and when $W = \frac{1}{n}(1)_{n \times n}$,

$$\|y\|_2 = 0.$$

Proof of the lemma: According to Property 5.1 in the Appendix, W can be expressed as

$$W = UAU^T$$

where $A = diag[\alpha_1, \alpha_2, \ldots, \alpha_n]$. α_1, ..., α_n are n real eigenvalues of W satisfying

$$1 = \alpha_1 \geq |\alpha_2| \geq \cdots \geq |\alpha_n| \geq 0$$

$U = (u^{(1)}, \ldots, u^{(n)})$, $u^{(1)}$, ..., $u^{(n)}$ are n eigenvectors corresponding to the n eigenvalues. U is unitary and orthogonal.

From the condition $y = Wx$,

$$\|y\|_2^2 = \|Wx\|_2^2 = (Wx, Wx) = (x, W^T Wx)$$

Write x as

$$x = \sum_{j=1}^{n} a_j u^{(j)}, \qquad \text{where } a_j = (x, u^{(j)})$$

then

$$\begin{aligned} W^T Wx &= \sum_{j=1}^{n} a_j W^T W u^{(j)} \\ &= \sum_{j=1}^{n} \alpha_j^2 a_j u^{(j)} \end{aligned}$$

thus

$$\begin{aligned} \|Wx\|_2^2 &= (\sum_{i=1}^{n} a_i u^{(i)}, \sum_{j=1}^{n} \alpha_j^2 a_j u^{(j)}) \\ &= \sum_{j=1}^{n} \alpha_j^2 a_j^2 \end{aligned}$$

Since $u^{(1)} = \frac{1}{\sqrt{n}}(1)_{n\times 1}$ and $\sum_{i=1}^{n} x_i = 0$, we have

$$a_1 = (x, u^{(1)}) = 0$$

thus

$$\|Wx\|_2^2 \leq \alpha_2^2 \sum_{j=1}^{n} a_j^2 = \alpha_2^2 \|x\|_2^2 \leq \|x\|_2^2$$

since $|\alpha_2| \leq 1$.

When W is irreducible, from Property 5.2 in the Appendix, 1 a simple eigenvalue of W. This implies

$$|\alpha_2| < 1 \qquad \text{and} \qquad \|y\|_2 < \|x\|_2$$

When $W = \frac{1}{n}(1)_{n\times n}$,

$$\alpha_2 = 0 \qquad \text{and} \qquad \|y\|_2 = 0$$

This completes the proof. §

Proof of the theorem: Let $\hat{c}_i$ $(1 \le i \le n)$ be real value functions defined as

$$\begin{pmatrix} \hat{c}_1 \\ \vdots \\ \hat{c}_n \end{pmatrix} = W^{(k)} \begin{pmatrix} E_1 \\ \vdots \\ E_n \end{pmatrix} \tag{44}$$

where $W^{(k)}$ is a $n \times n$ square matrix defined by the following recursive function:

$$\begin{cases} W^{(1)} = \lambda_1 W + (1-\lambda_1)I \\ W^{(k)} = \lambda_k W W^{(k-1)} + (1-\lambda_k)I \end{cases}$$

here I is the identity matrix.

Clearly, $W^{(k)}$ is also a symmetric propagation matrix.

Let $(x_1^*, \ldots, x_n^*)$ be an optimal solution. Substitute it into (44),

$$\begin{pmatrix} \hat{c}_1^* \\ \vdots \\ \hat{c}_n^* \end{pmatrix} = W^{(k)} \begin{pmatrix} E_1^* \\ \vdots \\ E_n^* \end{pmatrix} \tag{45}$$

Let $\mu = \frac{1}{n}\sum_i E_i^* = \frac{1}{n}E^*$, then

$$\frac{1}{n}\sum_i \hat{c}_i^* = \mu$$

since $W^{(k)}$ is a propagation matrix.

From (45),

$$\begin{pmatrix} \hat{c}_1^* - \mu \\ \vdots \\ \hat{c}_n^* - \mu \end{pmatrix} = W^{(k)} \begin{pmatrix} E_1^* - \mu \\ \vdots \\ E_n^* - \mu \end{pmatrix}$$

According to Lemma 19,

$$\sum_{i=1}^{n}(\hat{c}_i^* - \mu)^2 \le (\alpha_2^{(k)})^2 \sum_{i=1}^{n}(E_i^* - \mu)^2 \le (\alpha_2^{(k)})^2 \frac{n-1}{n}(E^*)^2$$

where $\alpha_2^{(k)}$ the eigenvalue of $W^{(k)}$ having the second maximum size, which can be computed as follows,

$$\begin{cases} \alpha_2^{(1)} = \lambda_1 \alpha_2 + (1 - \lambda_1) \\ \alpha_2^{(k)} = \lambda_k \alpha_2 \alpha_2^{(k-1)} + (1 - \lambda_k) \end{cases} \tag{46}$$

Then

$$\hat{c}_i^* \leq \frac{E^*}{n} + |\alpha_2^{(k)}| \sqrt{\frac{n-1}{n}} E^*, \qquad \text{for } 1 \leq i \leq n \tag{47}$$

Next we will prove

$$c_i^{(k)}(x_i^*) \leq \hat{c}_i^*, \qquad \text{for } 1 \leq i \leq n$$

by the principle of mathematical induction.

Let $k = 0$. From (15),

$$\begin{pmatrix} c_1^{(1)}(x_1^*) \\ \vdots \\ c_n^{(1)}(x_n^*) \end{pmatrix} = \begin{pmatrix} \min_{x_j, j \neq 1} (\lambda_1 (w_1, c^{(0)}) + (1 - \lambda_1) E_1) \\ \vdots \\ \min_{x_j, j \neq n} (\lambda_1 (w_n, c^{(0)}) + (1 - \lambda_1) E_n) \end{pmatrix}$$

then

$$\begin{pmatrix} c_1^{(1)}(x_1^*) \\ \vdots \\ c_n^{(1)}(x_n^*) \end{pmatrix} \leq \lambda_1 W \begin{pmatrix} c_1^{(0)}(x_1^*) \\ \vdots \\ c_n^{(0)}(x_n^*) \end{pmatrix} + (1 - \lambda_1) \begin{pmatrix} E_1^* \\ \vdots \\ E_n^* \end{pmatrix}$$

Under the general initial condition $c^{(0)} = 0$ or $\lambda_1 = 0$,

$$\begin{pmatrix} c_1^{(1)}(x_1^*) \\ \vdots \\ c_n^{(1)}(x_n^*) \end{pmatrix} \leq (\lambda_1 W + (1 - \lambda_1) I) \begin{pmatrix} E_1^* \\ \vdots \\ E_n^* \end{pmatrix} = W^{(1)} \begin{pmatrix} E_1^* \\ \vdots \\ E_n^* \end{pmatrix}$$

since $E_i^* \geq 0$ by the assumption of nonnegative constraints.

Assume that, for some $k \geq 0$,

$$\begin{pmatrix} c_1^{(k)}(x_1^*) \\ \vdots \\ c_n^{(k)}(x_n^*) \end{pmatrix} \leq W^{(k)} \begin{pmatrix} E_1^* \\ \vdots \\ E_n^* \end{pmatrix} \tag{48}$$

From (15),

$$\begin{pmatrix} c_1^{(k+1)}(x_1^*) \\ \vdots \\ c_n^{(k+1)}(x_n^*) \end{pmatrix} \leq \lambda_{k+1} W \begin{pmatrix} c_1^{(k)}(x_1^*) \\ \vdots \\ c_n^{(k)}(x_n^*) \end{pmatrix} + (1 - \lambda_{k+1}) \begin{pmatrix} E_1^* \\ \vdots \\ E_n^* \end{pmatrix}.$$

Substitute (48) into the above inequality, we have

$$\begin{pmatrix} c_1^{(k+1)}(x_1^*) \\ \vdots \\ c_n^{(k+1)}(x_n^*) \end{pmatrix} \le (\lambda_{k+1} W W^{(k)} + (1-\lambda_{k+1})) \begin{pmatrix} E_1^* \\ \vdots \\ E_n^* \end{pmatrix}$$

that is

$$\begin{pmatrix} c_1^{(k+1)}(x_1^*) \\ \vdots \\ c_n^{(k+1)}(x_n^*) \end{pmatrix} \le W^{(k+1)} \begin{pmatrix} E_1^* \\ \vdots \\ E_n^* \end{pmatrix} = \begin{pmatrix} \hat{c}_1^* \\ \vdots \\ \hat{c}_n^* \end{pmatrix}$$

Hence, the following inequality is proved,

$$c_i^{(k)}(x_i^*) \le \hat{c}_i^*, \qquad \text{for } 1 \le i \le n$$

Using (47),

$$c_i^{(k)}(x_i^*) \le \frac{E^*}{n} + \sqrt{\frac{n-1}{n}} |\alpha_2^{(k)}| E^*, \qquad \text{for } 1 \le i \le n \tag{49}$$

where $\alpha_2^{(k)}$ is computed by the recursive function (46).

If $W = \frac{1}{n}(1)_{n\times n}$, then $\alpha_2 = 0$ and

$$\alpha_2^{(k)} = (1-\lambda_k), \qquad \text{for } k \ge 1$$

using the recursive function (46).

From (49),

$$c_i^{(k)}(x_i^*) \le \frac{E^*}{n} + \sqrt{\frac{n-1}{n}} (1-\lambda_k) E^*, \qquad \text{for } 1 \le i \le n$$

This completes the proof. §

Proof of Theorem 16

Proof. The proof of the theorem is divided into two parts for two different cases. Let $c_i^{(k)}(x_i)$ be the assignment constraint for variable x_i at time k. Let $\tilde{x}_i^{(k)}$ be the choice of agent i at the time,

$$\tilde{x}_i^{(k)} = \arg\min_{x_i} c_i^{(k)}(x_i), \qquad \text{for any } i.$$

Case 1: $\lambda \to 1$ and $\mu = 0$.

Since $\lambda \to 1$, (16) is reduced to

$$\min_{x_j \in X \setminus x_i} \sum_j w_{ij} c_j^{(k)}(x_j) . \tag{50}$$

Therefore, the suggested choice for agent j from agent i is the same as the choice of agent j in this case. That is

$$\tilde{x}_j^{(k)}(\tilde{E}_i) = \tilde{x}_j^{(k-1)} \quad \text{for any } i \text{ and } j (j \neq i) \,. \tag{51}$$

Since $\mu = 0$, (17) is reduced to

$$c_i^{(k)}(x_i) = E_i(\tilde{x}(\tilde{E}_i(x_i))). \tag{52}$$

where $\tilde{x}(\tilde{E}_i(x_i))$ is the solution for minimizing $\tilde{E}_i$ with a given x_i.

Substitute (51) into (52), we have

$$c_i^{(k)}(x_i) = E_i(\tilde{x}_1^{(k-1)}, \ldots, \tilde{x}_{i-1}^{(k-1)}, x_i, \tilde{x}_{i+1}^{(k-1)}, \ldots, \tilde{x}_n^{(k-1)}) \,. \tag{53}$$

From (53), we have the choice for agent i at time k as

$$\begin{aligned} \tilde{x}_i^{(k)} &= \arg\min_{x_i} c_i^{(k)}(x_i) \\ &= \arg\min_{x_i} E_i(\tilde{x}_1^{(k-1)}, \ldots, \tilde{x}_{i-1}^{(k-1)}, x_i, \tilde{x}_{i+1}^{(k-1)}, \ldots, \tilde{x}_n^{(k-1)}) \,. \end{aligned}$$

Because $\{E_i\}$ is a good decomposition, then we have

$$\tilde{x}_i^{(k)} = \arg\min_{x_i} E(\tilde{x}_1^{(k-1)}, \ldots, \tilde{x}_{i-1}^{(k-1)}, x_i, \tilde{x}_{i+1}^{(k-1)}, \ldots, \tilde{x}_n^{(k-1)}) \,.$$

Therefore, in this case, the cooperative system with the cooperative scheme defined by (16) and (17) is equivalent to a local search algorithm. That is, it minimizes E with respect to different variables asynchronously.

Case 2: $\lambda \to 1$ and $0 < \mu < 1$.

In this case, since $\lambda \to 1$, (50) and (51) remain unchanged.

Substitute (50) into (17), we have

$$c_i^{(k)}(x_i) = \mu w_{ii} c_i^{(k-1)}(x_i) + \mu \sum_{j, j \neq i} w_{ij} c_j^{(k-1)}(\tilde{x}_j^{(k-1)}) + (1-\mu) E_i(\tilde{x}(\tilde{E}_i(x_i))). \tag{54}$$

Thus, according to (51) and (54), at time k, the choice for agent i is

$$\begin{aligned} \tilde{x}_i^{(k)} &= \arg\min_{x_i} c_i^{(k)}(x_i) \\ &= \arg\min_{x_i} \mu w_{ii} c_i^{(k-1)}(x_i) + \mu \sum_{j, j \neq i} w_{ij} c_j^{(k-1)}(\tilde{x}_j^{(k-1)}) + (1-\mu) E_i(\tilde{x}(\tilde{E}_i(x_i)) \\ &= \arg\min_{x_i} \mu w_{ii} c_i^{(k-1)}(x_i) + (1-\mu) E_i(\tilde{x}(\tilde{E}_i(x_i)). \end{aligned} \tag{55}$$

From the right side of (55), we have

$$\begin{aligned} &\mu w_{ii} c_i^{(k-1)}(\tilde{x}_i^{(k)}) + (1-\mu) E_i(\tilde{x}(\tilde{E}_i(\tilde{x}_i^{(k)})) \leq \\ &\quad \mu w_{ii} c_i^{(k-1)}(\tilde{x}_i^{(k-1)}) + (1-\mu) E_i(\tilde{x}(\tilde{E}_i(\tilde{x}_i^{(k-1)})). \end{aligned} \tag{56}$$

By the definition of $\tilde{x}_i^{(k-1)}$, we have

$$\tilde{x}_i^{(k-1)} = \arg\min_{x_i} c_i^{(k-1)}(x_i);$$

From the above, we have

$$c_i^{(k-1)}(\tilde{x}_i^{(k-1)}) \le c_i^{(k-1)}(\tilde{x}_i^{(k)}). \tag{57}$$

From (56), we have

$$(1-\mu)(E_i(\tilde{x}(\tilde{E}_i(\tilde{x}_i^{(k)}))-E_i(\tilde{x}(\tilde{E}_i(\tilde{x}_i^{(k-1)}))) \le \mu w_{ii}(c_i^{(k-1)}(\tilde{x}_i^{(k-1)})-c_i^{(k-1)}(\tilde{x}_i^{(k)})). \tag{58}$$

Since $w_{ii} \ge 0$ and $0 < \mu < 1$, combine (58) and (57), we get

$$(1-\mu)(E_i(\tilde{x}(\tilde{E}_i(\tilde{x}_i^{(k)})) - E_i(\tilde{x}(\tilde{E}_i(\tilde{x}_i^{(k-1)}))) \le 0. \tag{59}$$

According to condition $0 < \mu < 1$, the inequality (59) can be rewritten as

$$E_i(\tilde{x}(\tilde{E}_i(\tilde{x}_i^{(k)})) \le E_i(\tilde{x}(\tilde{E}_i(\tilde{x}_i^{(k-1)})) \tag{60}$$

To make it clear, (60) can be rewritten as:

$$\begin{aligned} &E_i(\tilde{x}_1^{(k-1)}, \ldots, \tilde{x}_{i-1}^{(k-1)}, \tilde{x}_i^{(k)}, \tilde{x}_{i+1}^{(k-1)}, \ldots, \tilde{x}_n^{(k-1)}) \le \\ &\quad E_i(\tilde{x}_1^{(k-1)}, \ldots, \tilde{x}_{i-1}^{(k-1)}, \tilde{x}_i^{(k-1)}, \tilde{x}_{i+1}^{(k-1)}, \ldots, \tilde{x}_n^{(k-1)}). \end{aligned} \tag{61}$$

Because $\{E_i\}$ is a good decomposition, from (61) we have

$$\begin{aligned} &E(\tilde{x}_1^{(k-1)}, \ldots, \tilde{x}_{i-1}^{(k-1)}, \tilde{x}_i^{(k)}, \tilde{x}_{i+1}^{(k-1)}, \ldots, \tilde{x}_n^{(k-1)}) \le \\ &\quad E(\tilde{x}_1^{(k-1)}, \ldots, \tilde{x}_{i-1}^{(k-1)}, \tilde{x}_i^{(k-1)}, \tilde{x}_{i+1}^{(k-1)}, \ldots, \tilde{x}_n^{(k-1)}). \end{aligned}$$

Therefore, in this case, the cooperative system with the cooperative scheme defined by (16) and (17) is equivalent to a local search algorithm of a lazy style. That is, E is decreased, not necessary to the best (so called lazy), at each time by adjusting the choice of one agent.

CHAPTER 8

COUPLED DETECTION RATES: AN INTRODUCTION

David E. Jeffcoat
Air Force Research Laboratory, Munitions Directorate
Eglin AFB, FL
david.jeffcoat@eglin.af.mil

The case of two cooperative searchers is examined, and the effect of cueing on the probability of target detection is derived from first principles using a Markov chain analysis. There are two main results: first, that the effect of cueing can be quantified, and second, that there is an upper bound on the benefit of cueing. Both results are presented in closed form. The joint probability of detection for two independent searchers is derived from Koopman's formula for a single searcher, and is shown to be a special case of one of the results in this chapter. Extensions of the model are discussed.

Keywords:
Cooperative search, target detection, cueing, Markov chain

1. Introduction

In any system-of-systems analysis, consideration of dependencies between systems is imperative. In this chapter, we consider a particular type of system interaction, called cueing. The interaction could be between similar systems, such as two or more wide area search munitions, or between dissimilar systems, such as a reconnaissance asset and a munition. In this introductory chapter, we consider two identical search vehicles cooperatively interacting via cueing.

In Shakespeare's day, the word "cue" meant a signal (a word, phrase, or bit of stage business) to a performer to begin a specific speech or action [7]. The word is now used more generally for anything serving a comparable purpose. In this chapter, we mean any information that provides focus to a search; e.g., information that limits the search area or provides a search heading.

Search theory is one of the oldest areas of operations research [10], with a solid foundation in mathematics, probability and experimental physics. Yet, search theory is clearly of more than academic interest. At times, a search can become an international priority, as in the 1966 search for the hydrogen bomb lost in the Mediterranean near Palomares, Spain.

That search was an immense operation involving 34 ships, 2,200 sailors, 130 frogmen and four mini-subs. The search took 75 days, but might have concluded much earlier if cueing had been utilized from the start. A Spanish fisherman had come forward quickly to say he'd seen something fall that looked like a bomb, but experts ignored him. Instead, they focused on four possible trajectories calculated by a computer, but for weeks found only airplane pieces. Finally, the fisherman, Francisco Simo, was summoned back. He sent searchers in the right direction, and a two-man sub, the Alvin, located the 10-foot-long bomb under 2,162 feet of water [14].

Cueing is a current topic in vision research. For example, Arrington, et al. [2] study the role of objects in guiding spatial attention through a cluttered visual environment. Magnetic resonance imaging is used to measure brain activity during cued discrimination tasks requiring subjects to orient attention either to a region bounded by an object or to an unbounded region of space in anticipation of an upcoming target. Comparison between the two tasks revealed greater brain activity when an object cues the subject's attention.

Bernard Koopman pioneered the application of mathematical process to military search problems during World War II [10]. Koopman [4] discusses the case in which a searcher inadvertently provides information to the target, perhaps allowing the target to employ evasive action. The use of receivers on German U-boats to detect search radar signals in World War II is a classic example. Koopman referred to this type of cueing as "target alerting."

This chapter uses a detection rate approach to examine the effect of cueing on probability of target detection. Koopman [5] used a similar approach in his discussion of target detection. In Koopman's terminology, a quantity γ was called the "instantaneous probability of detection." From this starting point, Koopman derived the probability of detection as a function of time. It is very clear that Koopman's instantaneous probability of detection is precisely the individual searcher detection rate used here. The main difference is that Koopman considered a single searcher, while we consider the case of two interdependent searchers.

Washburn [13] examines the case of a single searcher attempting to detect a randomly moving target at a discrete time. Given an effort distribution, bounded at each discrete time t, Washburn establishes an upper bound on the probability of target detection. It is noteworthy that Washburn mentions that the detection rate approach to computation of detection probabilities has proved to be more robust than approaches relying on geometric models.

In this chapter, we use a Markov chain analysis to examine cueing as a coupling mechanism between two searchers. A Markov chain approach to target detection can be found in [10], which deals with the optimal allocation of effort to detect a target. A prior distribution of the target's location is assumed known to the searcher. Stone uses a Markov chain analysis to deal with the search for targets whose motion is Markovian. In Stone's formulation, the states correspond to cells that contain a target at a discrete time with a specified probability. In this research, the states correspond to detection states for individual search vehicles.

Alpern and Gal discuss the problem of searching for a submarine with a known initial location [1]. Thomas and Washburn [11] considered "dynamic search games" in which the hider starts moving at time zero from a location known to both a searcher and a hider, while the searcher starts with a time delay known to both players; for example, a helicopter attempts to detect a submarine that reveals its position by torpedoing a ship.

2. Problem Description

Consider two cooperative searchers, and assume that cueing increases an individual searcher's detection capability by a factor of k. That is, let the nominal detection rate for an individual searcher be given by θ detections per unit of time, with the cued detection rate given by $k\theta/\text{time}$. We assume that once an individual searcher detects a target, it immediately cues the other searcher. This cue could take the form of a target coordinate, a search heading, or any other information that improves the second searcher's detection rate. We wish to examine the impact of cueing on the overall probability of target detection, denoted P_d.

3. Analysis

We first define four detection states for the two searchers, as shown in Table 1, in which "D" denotes detection, and "ND" denotes no detection.

We will obtain the state probabilities using a Markov chain approach,

Table 1. Detection States.

State	Searcher 1	Searcher 2
1	ND	ND
2	D	ND
3	ND	D
4	D	D

and then derive the probability of target detection from the state probabilities. Professor Andrei A. Markov (1856 - 1922) is well known for his study of sequences of mutually dependent variables. Today, we use the term Markov process to denote a random process whose future state probabilities are determined only by its current state. A Markov process with a discrete state space is called a Markov chain [3]. In our analysis, we have a continuous time Markov chain, because transitions between the discrete states can occur at any time. Figure 1 illustrates our four state Markov chain, with the transition rates between states. For example, the transition rate from state one (no detection by either searcher) to state two (detection by searcher 1 only) is given by θ, the detection rate for searcher 1. Once searcher 1 detects a target, searcher 2 is immediately cued, so that the transition rate from state two to state four (detection by both searchers) is given by $k\theta$, the cued detection rate of searcher 2.

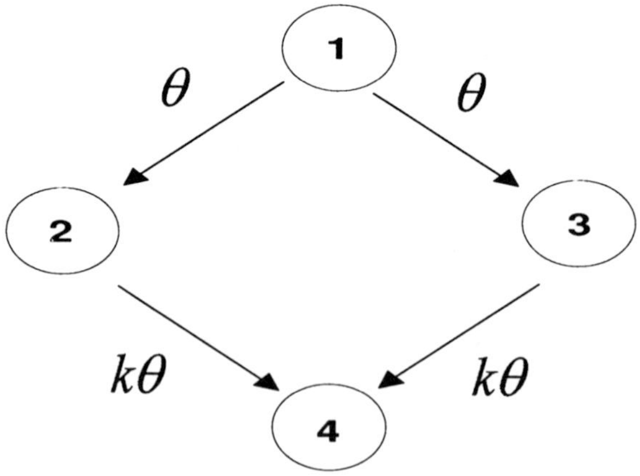

Fig. 1. Transition Rate Diagram.

We can use the transition rate diagram to write differential equations describing the change in states with respect to time. These equations are called Kolmogorov equations, after the Russian mathematician Andrei Kolmogorov (1903 - 1987), who was the first to derive these differential equations for continuous-time Markov chains.

$$\frac{d}{dt}P_1(t) = -2\theta \cdot P_1(t) \tag{1}$$

$$\frac{d}{dt}P_2(t) = +\theta \cdot P_1(t) - k\theta \cdot P_2(t) \tag{2}$$

$$\frac{d}{dt}P_3(t) = +\theta \cdot P_1(t) - k\theta \cdot P_3(t) \tag{3}$$

$$\frac{d}{dt}P_4(t) = +k\theta \cdot P_2(t) + k\theta \cdot P_3(t) \tag{4}$$

The initial conditions are defined by equations (5) through (8), based on the assumption that the process begins with no detections.

$$P_1(0) = 1 \tag{5}$$

$$P_2(0) = 0 \tag{6}$$

$$P_3(0) = 0 \tag{7}$$

$$P_4(0) = 0 \tag{8}$$

Given the four differential equations and the initial conditions defined by equations (1) through (8), we can find the state probability solutions using any technique familiar to the reader. To obtain the solutions below, we followed the approach of [6].

$$P_1(t) = e^{-2\theta t} \tag{9}$$

$$P_2(t) = [e^{-2\theta t} - e^{-k\theta t}]/(k-2) \tag{10}$$

$$P_3(t) = [e^{-2\theta t} - e^{-k\theta t}]/(k-2) \tag{11}$$

$$P_4(t) = [k - 2 - ke^{-2\theta t} + 2e^{-k\theta t}]/(k-2) \tag{12}$$

Note that all four functions are defined for any $t \geq 0$. Before moving to consideration of the probability of detection, we note with some concern that three of the state probabilities are not defined for $k = 2$. In particular, $P_2(t)$, $P_3(t)$, and $P_4(t)$ are indeterminate of the form $0/0$ when $k = 2$. We can address this issue using L'Hopital's rule [9], shown in Eq. (13) for the particular case of $k = 2$.

$$\lim_{k \to 2} \frac{f(k)}{g(k)} = \lim_{k \to 2} \frac{f'(k)}{g'(k)} \tag{13}$$

Taking the appropriate derivatives and then evaluating the limit, we find that

$$P_2(t) = \theta t e^{-2\theta t};\ k = 2 \tag{14}$$

So, $P_2(t)$ is defined for every nonnegative t and for every $k \geq 1$. Note that we have no interest here in values of k less than one, since that would imply a negative effect of cueing. As a quick check on Eq. (14), we can plot P_2 as a function of k, using Eq. (10) for all values of k except $k = 2$. Figure 2 shows such a plot for $\theta = 0.1$ and $t = 10$, with k ranging from 1.5 to 2.5. Although certainly not a proof, the plot gives us confidence that there is no problem at $k = 2$.

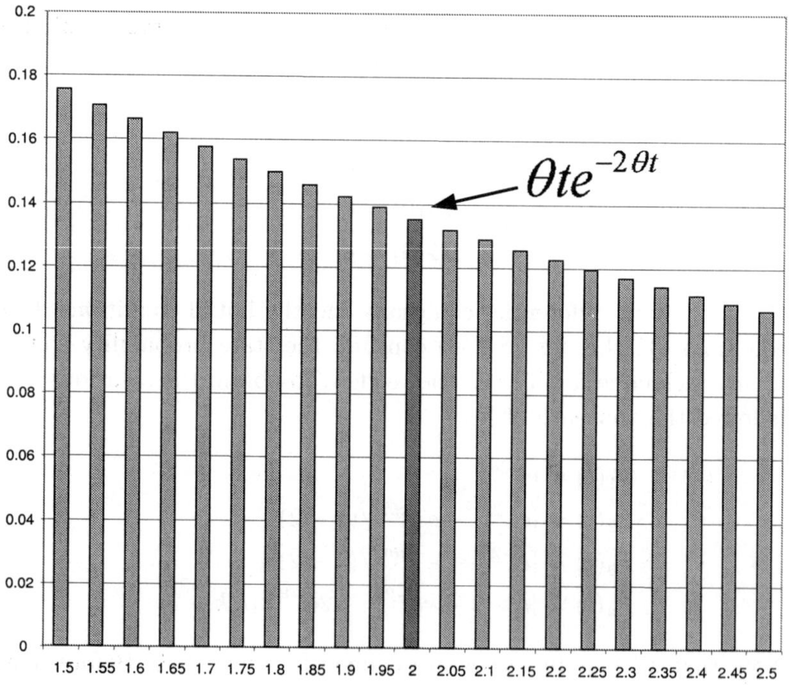

Fig. 2. $P_2(t, k)$ for $\theta = 0.1$.

For the remainder of this chapter, we will assume that $P_2(t)$ is continuous for all values of $k \geq 1$. The situation is similar for $P_4(t)$. Again using equation (13), we find that

$$P_4(t) = 1 - e^{-2\theta t}[1 + 2\theta t];\ k = 2 \tag{15}$$

so that $P_4(t)$ is defined for every nonnegative t and every $k \geq 1$.

Returning now to the state probabilities, Figure 3 provides all four state probability plots for the particular case $\theta = 0.1$ and $k = 1$; that is, with no cueing. Note the state probability plot for "exactly one searcher detects" represents two plots – one for each of two searchers.

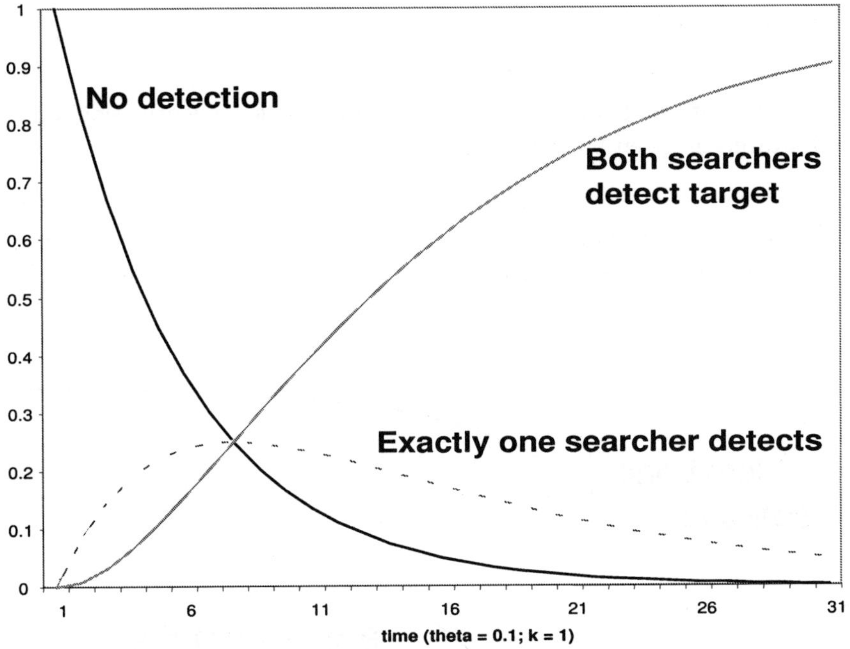

Fig. 3. State Probability Plots.

4. The Probability of Detection

With the state probabilities in hand, we can turn to the probability of detection. For example, the probability of detection by at least one searcher is given by

$$P_d(t;\ \text{at least one searcher}) = P_2(t) + P_3(t) + P_4(t) \tag{16}$$

The probability of detection by both searchers is given by

$$P_d(t;\ \text{both searchers}) = P_4(t) \tag{17}$$

As an aside, we find for the case $k = 1$ that

$$P_4(t) = 1 + e^{-2\theta t} - 2e^{-\theta t} \tag{18}$$

This is equivalent to the case of two independent searchers that derive no benefit from cueing. We can get the same result starting from Koopman's [5] single searcher formula for the probability of detection for a continuous search under unchanging conditions, where γ is the "instantaneous probability of detection."

$$p(t) = 1 - e^{-\gamma t} \tag{19}$$

The probability that two such searchers, working independently, would both find a target is given by $[p(t)]^2$, or

$$p(t) = 1 - 2e^{\gamma t} + e^{-2\gamma t} \tag{20}$$

which is precisely equation (18) with the substitution $\gamma = \theta$. Figure 4 shows two plots of probability of detection as a function of time, with $\theta = 0.1$, and $k = 1$; i.e., no cueing.

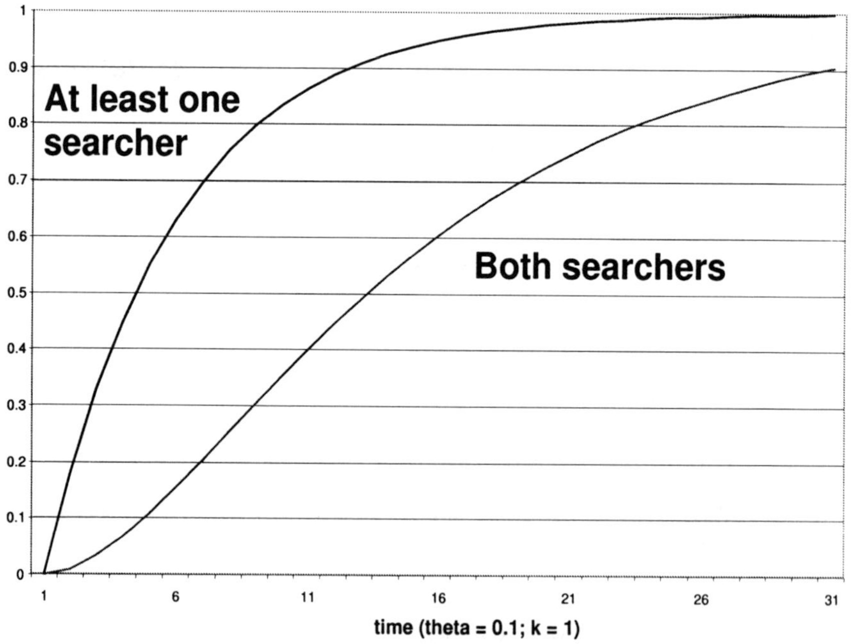

Fig. 4. Probability of Detection.

5. The Effect of Cueing

Figure 5 shows the effect of cueing on the probability of detection for $\theta = 1$. In this case, the P_d represents the probability of detection by both searchers.

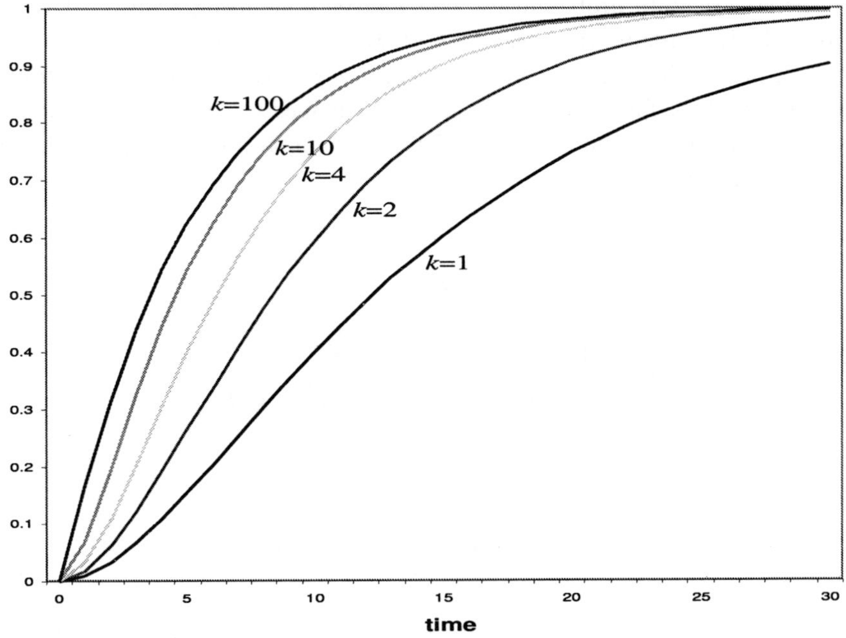

Fig. 5. Effect of Cueing on P_d.

Note that cueing can dramatically increase the aggregate probability of detection for two searchers. For example, at $t = 10$ and $k = 4$, we see that cueing essentially doubles the probability of detection (actual values are 0.4 for $k = 1$ and 0.75 for $k = 4$). Figure 5 also illustrates the diminishing return from cueing. The plots suggest that there is an upper bound to the benefit of cueing, at least for this problem. This can be verified by taking the limit of $P_4(t)$ in Eq. (12) as k approaches infinity. Again using L'Hopital's rule, we obtain the result shown in Equation (21).

$$\lim_{k\to\infty} P_4(t) = 1 - e^{-2\theta t} \tag{21}$$

6. Extending the Model

Although an obvious next step is to examine larger problems, it is clear that the approach outlined here is limited by the difficulty of solving large sets of coupled differential equations. There are at least two approaches that may prove fruitful. One is to ignore the transient effects and to solve only for the steady-state probabilities. This can be done using a linear algebra approach. To illustrate the basic method, we construct the transition matrix Q, such that each off-diagonal element q_{ij}, $i \neq j$, is the transition rate from state i to state j. The diagonal elements are defined to ensure that that the elements in each row sum to zero. For our example problem, Q would be as shown in Figure 6.

$$Q = \begin{bmatrix} -2\theta & \theta & \theta & 0 \\ 0 & -k\theta & 0 & k\theta \\ 0 & 0 & -k\theta & k\theta \\ 0 & 0 & 0 & 0 \end{bmatrix}$$

Fig. 6. The Transition Rate Matrix.

If we define $P = [p_1, p_2, p_3, p_4]$ as the steady-state probability vector, then we can solve the set of linear equations in Equations (22) and (23) to find P.

$$PQ = [0,\ 0,\ 0,\ 0] \tag{22}$$

$$\sum_i p_i = 1 \tag{23}$$

Solving these equations leads to the steady-state results $p_1 = p_2 = p_3 = 0$; with $p_4 = 1$, as expected since p_4 is clearly an absorbing state.

A second possible approach is matrix exponentiation, which has the potential to provide both transitional and steady state probabilities for large problems. Matrix exponentiation methods have been successfully applied to a broad class of problems in the theory of queues [8]. These methods exploit the structure of Markov chains to expedite numerical calculations.

7. Summary

We have shown that the effect of cueing on probability of detection can be quantified, and that cueing can dramatically affect the probability of detection over a fixed time interval. We have also shown that there is an upper bound on the steady-state benefit of cueing, at least for the problem defined. We have also introduced a line of inquiry into methods for addressing larger problems, which will be the subject of further research.

References

[1] S. Alpern and S. Gal, The Theory of Search Games and Rendezvous, Boston: Kluwer Academic Publishers, pages 161-162, 2003.

[2] C. Arrington, T. Carr, A. Mayer, and S. Rao, "Neural mechanisms of visual attention: object-based selection of a region in space," Journal of Cognitive Neuroscience, Vol. 12, Supplement 2, pages 106-117, 2000.

[3] L. Kleinrock, Queueing Systems, Volume I: Theory, New York: John Wiley & Sons, page 21, 1975.

[4] B. Koopman, Search and Screening: General Principles with Historical Applications. New York: Pergamon Press, pages 16-17, 1980.

[5] B. Koopman, "The Theory of Search II. Target Detection," Operations Research, Vol. 4, No. 5, October, pages 503-531, 1956.

[6] E. Lewis, Introduction to Reliability Engineering, 2nd Ed., New York: John Wiley & Sons, Inc., pages 326 - 334, 1994.

[7] Merriam-Webster's Collegiate Dictionary, 10th Ed., Springfield, MA: Merriam- Webster, Inc., 1999.

[8] M. Neuts, Matrix-Geometric Solutions in Stochastic Models: An Algorithmic Approach. New York: Dover Publications, Inc., 1981.

[9] R. Silverman, Modern Calculus and Analytic Geometry, New York: The Macmillan Company, pages 835 - 841, 1969.

[10] L. Stone, Theory of Optimal Search, 2nd Ed., Military Applications Section, Operations Research Society of America, pages 221-233, 1989.

[11] L. Thomas and A. Washburn, "Dynamic Search Games," Operations Research 39, No. 3, pages 415-422, 1991.

[12] A. Washburn, Search and Detection, 3rd Ed., Institute for Operations Research and the Management Sciences, 1996.

[13] A. Washburn, "Search for a Moving Target: Upper Bound on Detection Probability," in Search Theory and Applications, B. Haley and L. Stone, editors. New York: Plenum Press, pages 231-237, 1980.

[14] D. Woolls, "A Chronicle of Four Lost Nukes," Houston Chronicle, http://www.chron.com/cs/CDA/ssistory.mpl/world/1990826, July 12, 2003.

CHAPTER 9

DECENTRALIZED RECEDING HORIZON CONTROL FOR MULTIPLE UAVS

Yoshiaki Kuwata

Department of Aeronautics and Astronautics
Massachusetts Institute of Technology
kuwata@mit.edu

Jonathan P. How

Department of Aeronautics and Astronautics
Massachusetts Institute of Technology
jhow@mit.edu

This chapter presents recent work on the design and implementation of on-line trajectory optimization algorithms on our multi-vehicle testbed. This work extends the previous receding horizon control (RHC) for a single vehicle to handle scenarios with multiple vehicles by explicitly including collision avoidance constraints in a distributed planning system. The basic RHC trajectory design problem is encoded as a mixed-integer linear program (MILP), but this optimization is only solved for a detailed trajectory that extends part of the way towards the target waypoint. The rest of the trajectory is represented by an approximate cost-to-go function. This RHC approach enables us to exploit the power of the MILP formulation to encode the collision and obstacle avoidance constraints in a computationally tractable algorithm. However, even the solution times of the RHC scale poorly with the fleet size. To resolve this problem, we developed a technique for embedding the collision avoidance constraints in a distributed formulation of the RHC. In the new approach, vehicles plan their own trajectories using RHC while analyzing the published plans for conflicts. The reaction to a detected conflict is to solve a coupled MILP optimization that explicitly includes a cooperative maneuver. Real-time tests of this overall control system on the hardware testbed show that this approach could be scaled to much larger teams with a very small degradation in the performance.

1. Introduction

With the increasing number of UAVs that will be simultaneously involved in future UAV missions, the coordination of multiple vehicles is key technology for enhancing mission effectiveness [8, 7]. UAVs will be required to perform these tasks in complex environments in which threats or terrain restrict the flyable areas. Both obstacle and vehicle avoidance represent major difficulties that significantly complicate the vehicle guidance problem. Various papers [19, 11, 12, 5] using techniques such as potential functions, Voronoi diagrams, and probabilistic roadmap methods have presented different ways to solve this problem, but these typically do not optimize the control action for the vehicle guidance.

This chapter optimizes the trajectories for a fleet of vehicles using mixed-integer linear programming (MILP) and receding horizon control (RHC). MILP uses both integer and continuous variables to encode logical constraints and discrete decisions together with the continuous vehicle dynamics. Previous work has demonstrated the use of MILP in task allocation and trajectory design problems [3, 14, 6]. The RHC approach enables us to use the power of this MILP formulation in a computationally tractable algorithm. It solves a MILP for a detailed trajectory that only extends part of the way towards the goal. The remainder of the maneuver is represented by a cost-to-go function using path approximations.

However, problems arise for multi-UAV teams because the increased number of vehicles results in large optimizations that are computationally intractable for real-time applications. Distributed RHC planners, with each designing the trajectory for a vehicle in the team, can be used to solve these computational problems. The issue in this case is how to include collision avoidance to ensure that the vehicle maneuvers are feasible. The approach in this chapter is to use an algorithm that detects potential conflicts in the approximate trajectories beyond the planning horizon. The reaction to a detected conflict is to solve a coupled MILP optimization that explicitly includes a cooperative maneuver. These algorithms are tested on our ground-based truck testbed, and several results are presented in Section 4. The results also demonstrate two key features of RHC: replanning to account for uncertainty in the environment and real-time trajectory generation.

2. Path Planning System

This section discusses the distributed receding horizon controller with collision avoidance. Subsection 2.1 discusses how to distribute the compu-

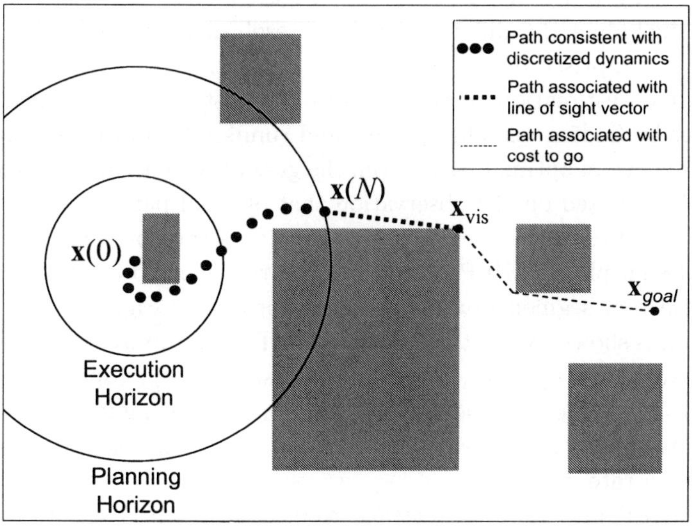

Fig. 1. Line-of-sight vector and cost-to-go [1].

tational load of the existing RHC. Subsection 2.2 then presents a collision avoidance planner in the receding horizon framework. Finally, Subsection 2.3 integrates the two approaches using an approximation to prune out future potential conflicts.

2.1. *Distributed Planners*

Initial results in [15, 9] used a single optimization to solve for all vehicle trajectories, but the results showed that the centralized approach is not well suited for real-time applications, even for a small team. If the vehicle collision avoidance constraints can be ignored, the trajectory generation problem for multiple vehicles naturally decomposes into several single vehicle trajectory generation problems. Since the vehicle avoidance constraints are typically not active for most of the mission, this is typically a reasonable approximation. However, there remains the issue of how to include the collision avoidance constraints when needed in the distributed maneuver optimization. The following sections present our distributed formulation of the receding horizon trajectory planner (RH-MILP) and discuss methods to include collision avoidance as needed.

The RH-MILP algorithm designs a minimum-time path to a fixed goal while avoiding a set of obstacles [17]. Figure 1 gives an overview of the method, including the different levels of resolution involved. The control strategy is comprised of two phases: cost estimation and trajectory design [2]. The cost estimation phase provides the cost-to-go from each obstacle corner by finding visibility graphs and running Dijkstra's algorithm. It produces a *tree* of optimal paths from the goal to each corner. This approximate cost is based on the observation that optimal paths (*i.e.*, minimum distance) tend to follow the edges and corners of the obstacles. In the trajectory design phase, MILP optimizations are solved to design a series of short trajectory segments over a planning horizon. In Figure 1, this section of the plan is shown by the thick dotted line. Each optimization finds a control sequence over n_p steps, but only the first $n_e(\leq n_p)$ control inputs are executed. The vehicle is modelled as a point mass moving in 2-D free-space with limited speed and acceleration to form a good approximate model for limited turn-rate vehicles.

The MILP also chooses a *visible point* $\vec{x}_{\text{vis}}$ which is visible from the *terminal point* $\vec{x}(n_p)$ from which the cost-to-go has been estimated in the previous phase. Note that the cost map is quite sparse: cost-to-go values are only known for the corners of the obstacles, but this still provides the trajectory optimization with the flexibility to choose between various routes around the obstacles.

The following avoidance constraints are applied at each point of the dynamic segment and at intermediate points between the terminal point and the visible point. Rectangular obstacles are used in this formulation to model *no-fly zones*, and are described by their lower left corner $[x_l, y_l]$ and upper right corner $[x_u, y_u]$. To avoid collisions, the following constraints must be satisfied at each point $[x, y]^T$ on the trajectory [18]

$$\begin{aligned} x &\leq x_l + M\ b_{\text{obst},1} \\ x &\geq x_u - M\ b_{\text{obst},2} \\ y &\leq y_l + M\ b_{\text{obst},3} \\ y &\geq y_u - M\ b_{\text{obst},4} \end{aligned} \tag{1}$$

$$\sum_{j=1}^{4} b_{\text{obst},j} \leq 3 \tag{2}$$

Note that this formulation is readily extended to polygonal and/or non-convex obstacles.

The trajectory cost involves two terms: the approximate straight-line cost from the terminal point to the visible point, and the cost from the

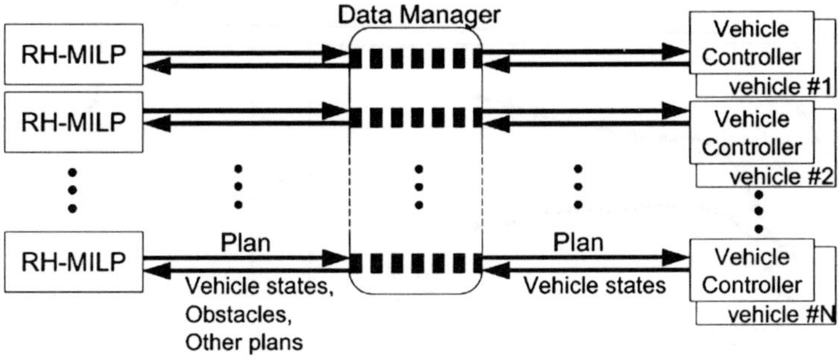

Fig. 2. Distributed system.

visible point to the goal. Referring to Figure 1, these represent the dotted line and the dashed line, respectively.

Figure 2 shows the distributed planner and vehicle system. Each vehicle has its own RH-MILP planner. The vehicle controller outputs the vehicle states, and the RH-MILP planner generates a list of waypoints for each vehicle based on the vehicle states and obstacle information. The central data manager stores each plan so that each planner has access to the plans of other vehicles. The on-line replanning procedure is as follows:

(1) Compute the cost map for the current environment.
(2) Solve MILP minimizing the distance to the target subject to dynamics and obstacle avoidance constraints, starting from the last waypoint uploaded (*or initial state if starting*).
(3) Upload the first n_e waypoints of the new plan to the vehicle.
(4) Monitor the world until the vehicle reaches the execution horizon of the previous plan, or until a change is detected in the environment.
(5) If a change is detected in the environment, go to (1), otherwise go to (2).

It is assumed in this work that the low-level controller can bring the vehicle to the execution horizon of the plan in step (4) . If the vehicle deviates from the nominal path, it is possible to use the propagated states as the next initial state in step (2) instead of the last waypoint uploaded to the vehicle.

2.2. *Collision Avoidance Planner*

Because the distributed RHC ignores the inter-vehicle couplings, another algorithm that focuses on avoiding collisions is required. This algorithm is applied only when the vehicle avoidance becomes dominant, and resolves

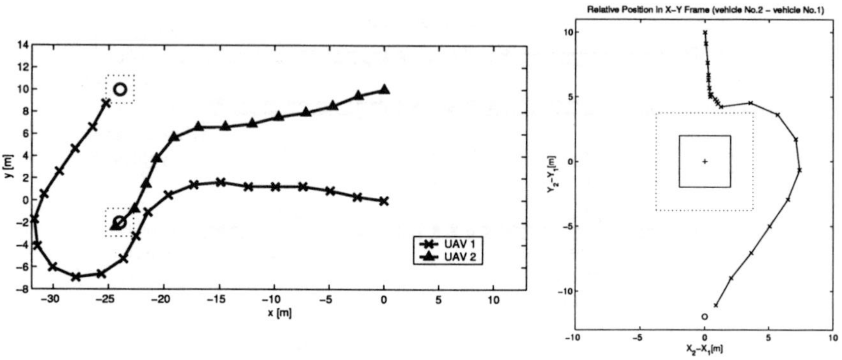

(a) Absolute position

(b) Relative position

Fig. 3. Collision avoidance maneuver with simple cost-to-go.

the conflict locally in a pairwise manner.

One approach is to just explicitly impose collision avoidance constraints during the detailed part of the trajectory, *i.e.*, at each time-step up to the planning horizon, while minimizing the distance to the targets. This is accomplished using the constraints:

$$\begin{aligned} x_1 &\leq x_2 + (d_1 + d_2) + M\, b_{\text{veh},1} \\ x_1 &\geq x_2 - (d_1 + d_2) - M\, b_{\text{veh},2} \\ y_1 &\leq y_2 + (d_1 + d_2) + M\, b_{\text{veh},3} \\ y_1 &\geq y_2 - (d_1 + d_2) - M\, b_{\text{veh},4} \end{aligned} \tag{3}$$

$$\sum_{j=1}^{4} b_{\text{veh},j} \leq 3 \tag{4}$$

where the d_i represents the vehicle size (including the safety distance) and M is a large number that is used when the binary variable relaxes the constraint.

Figure 3 shows that this approach can lead to a very poor set of maneuvers if collision avoidance is an important factor in determining the trajectory. In this example, the two vehicles start at the right heading towards the left using a planning horizon of four steps. Their goals are oriented such that the two vehicles must switch positions in the y direction. Figure 3(b) shows the relative positions beginning at the top of the figure and descending to the goal marked with $\circ$, where the relative frame for two vehicles 1 and 2 is defined as $\vec{x}_2 - \vec{x}_1$. The square in the center represents the vehicle avoidance constraints. Each vehicle tries to minimize the distance from its planning horizon to its goal in the absolute frame, while satisfying vehicle avoidance constraints over the planning horizon. In the relative frame, this is equivalent to moving straight to the goal, neglecting the vehicle avoidance box. The two vehicles do not start the collision avoidance maneuver until the vehicle avoidance constraints become active. As shown in Figure 3(a), when the goals become reachable within the horizon, one of the vehicles chooses to arrive at the goal in the planning horizon to reduce the terminal penalty. This decision causes the second vehicle to go around the first, resulting in a much longer trajectory, both in the absolute (Figure 3(a)) and the relative frames (Figure 3(b)).

The problem formulation can be greatly improved by including the relative vehicle positions in the cost function. As shown previously, the optimal trajectory for a vehicle flying in an environment with obstacles tends to touch the obstacle boundaries. In this case, the heuristic in [2] that uses the obstacle corners as cost points successfully finds the shortest path. In the vehicle avoidance case, a similar heuristic is that the optimal trajectory will tend to "follow" the vehicle avoidance box in the relative frame. Therefore, the modified formulation presented here uses the four corners of the vehicle avoidance box and a relative position of the two goals as the cost points $[\vec{x}_{\text{cp}}, \vec{y}_{\text{cp},}]^T$ in the relative frame. For any pair of vehicles j and k $(j < k)$, the following constraints are applied (in addition to the vehicle dynamics constraints):

I) Selection of visible point and the cost-to-go from there:

$$c_{\text{vis},jk} = \sum_{i=1}^{5} c_i \, b_{\text{cp},ijk} \tag{5}$$

$$\sum_{i=1}^{5} b_{\text{cp},ijk} = 1 \tag{6}$$

$$\begin{bmatrix} x_{\text{vis},jk} \\ y_{\text{vis},jk} \end{bmatrix} = \sum_{i=1}^{5} \begin{bmatrix} x_{\text{cp},ijk} \\ y_{\text{cp},ijk} \end{bmatrix} b_{\text{cp},ijk} \tag{7}$$

II) Visibility test in the relative frame:

$$\begin{bmatrix} x_{\text{LOS},jk} \\ y_{\text{LOS},jk} \end{bmatrix} = \begin{bmatrix} x_{\text{vis},jk} \\ y_{\text{vis},jk} \end{bmatrix} - \begin{bmatrix} x(n_p)_k - x(n_p)_j \\ y(n_p)_k - y(n_p)_j \end{bmatrix} \tag{8}$$

$$\begin{bmatrix} x_{\text{test},jkm} \\ y_{\text{test},jkm} \end{bmatrix} = \begin{bmatrix} x(n_p)_k - x(n_p)_j \\ y(n_p)_k - y(n_p)_j \end{bmatrix} + \frac{m}{n_t} \begin{bmatrix} x_{\text{LOS},jk} \\ y_{\text{LOS},jk} \end{bmatrix} \tag{9}$$

$$x_{\text{test},jkm} \leq -(d_j + d_k) + M\, b_{\text{vis},1jkm} \tag{10}$$

$$x_{\text{test},jkm} \geq (d_j + d_k) - M\, b_{\text{vis},2jkm} \tag{11}$$

$$y_{\text{test},jkm} \leq -(d_j + d_k) + M\, b_{\text{vis},3jkm} \tag{12}$$

$$y_{\text{test},jkm} \geq (d_j + d_k) - M\, b_{\text{vis},4jkm} \tag{13}$$

$$\sum_{n=1}^{4} b_{\text{vis},njkm} \leq 3 \tag{14}$$

$$j = 1, \ldots, n_v - 1, \qquad k = j+1, \ldots, n_v, \qquad m = 1, \ldots, n_t$$

where n_t represents a number of test points placed between the planning horizon and the selected cost point to ensure visibility in the relative frame [1]. Note that $\vec{x}_{\text{cp}}, \vec{x}_{\text{vis}}, \vec{x}_{\text{LOS}}, \vec{x}_{\text{test}}$ are in the relative frame whereas $\vec{x}(n_p)$ is measured in the absolute frame. The cost function includes the cost-to-go at the selected point, and the length of the line-of-sight vectors in the absolute frame (denoted by l_i for the ith vehicle) and the relative frame (denoted by $l_{\text{rel},jk}$ for a pair of vehicles j and k). Therefore, the problem statement is to minimize J subject to

$$J = \sum_{i=1}^{n_v} l_i + \sum_{j=1}^{n_v-1} \sum_{k=j+1}^{n_v} (\alpha\, l_{\text{rel},jk} + \beta\, c_{\text{vis},jk}) \tag{15}$$

$$l_i \geq \begin{bmatrix} x_{\text{goal},i} - (x_{n_p})_i \\ y_{\text{goal},i} - (y_{n_p})_i \end{bmatrix} \cdot \vec{i}_m \tag{16}$$

$$l_{\text{rel},jk} \geq \begin{bmatrix} x_{\text{LOS},jk} \\ y_{\text{LOS},jk} \end{bmatrix} \cdot \vec{i}_m \tag{17}$$

$$\vec{i}_m = \left[\cos\left(\frac{2\pi m}{n_l}\right), \sin\left(\frac{2\pi m}{n_l}\right) \right]^T \tag{18}$$

$$i = 1, \ldots, n_v, \quad j = 1, \ldots, n_v - 1, \quad k = j+1, \ldots, n_v, \quad m = 1, \ldots, n_l$$

where α is a weighting factor for the line-of-sight vector in the relative frame, as defined in (8), and β is a weighting factor for the cost-to-go at the cost point in the relative frame. If the goal is not visible from the initial position in the relative frame, (*i.e.*, the paths of the two vehicles intersect

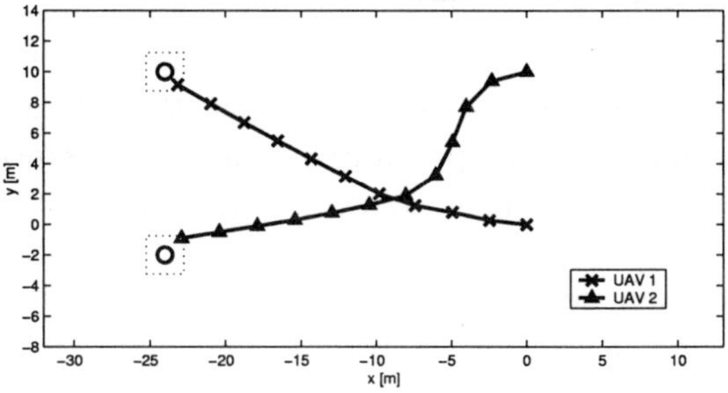

(a) Absolute position

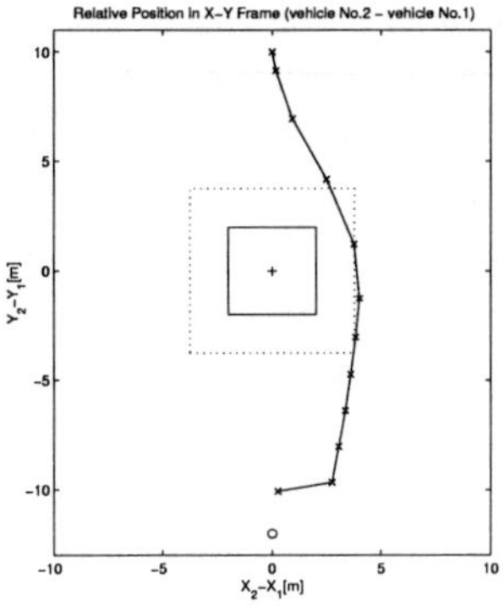

(b) Relative position

Fig. 4. Collision avoidance maneuver with improved cost-to-go.

in the absolute frame), the weights α and β navigate the vehicles along the vehicle avoidance box, initiating the collision avoidance action. Larger α and β result in faster avoidance maneuvers, but overly large weights can delay the completion of the mission because the distances of the vehicles from their goals have a smaller contribution to the objective function.

This approach easily extends to three or more vehicles by considering all pairs of interactions (*e.g.*, 2–1, 3–2, 1–3, for three vehicles). However, the multi-vehicle collision is much less likely to occur and it can lead to an exponential increase in the problem complexity. The approach presented in this chapter was designed to efficiently handle the most likely collision avoidance scenario (two vehicles).

Figure 4 shows this formulation applied to the same scenario presented in Figure 3. In contrast to Figure 3(a), vehicle 2 immediately begins a collision avoidance maneuver. Figure 4(b) shows that the relative trajectory successfully avoids the separation box, with some waypoints falling on the boundary.

2.3. *Integrated Planning System*

This subsection integrates the two planners presented in the previous subsections. The resulting controller guarantees the mission completion by the fleet of UAVs in finite time, while satisfying obstacle avoidance and vehicle avoidance constraints.

By default, the distributed planners in Subsection 2.1 continually generate trajectories for each vehicle ignoring the vehicle avoidance constraints.

However, the distributed RHC provides a detailed plan over the planning horizon, which enables us to predict when collision avoidance might be an issue. Each plan goes through a central station, which examines conflicts with other plans. When the central station detects a conflict, the collision avoidance (CA) controller in Subsection 2.2 generates a collision avoidance maneuver and overwrites the conflicting plans. When providing the initial states to the CA controller, the latest vehicle states are propagated forward along the nominal plans to account for different planning times. The distributed planners then solve for the next optimal trajectories starting from the execution horizon of the latest plans generated by the CA controller.

It is assumed here that not more than two vehicles have conflicts in their plans at the same location at the same time. This is a valid assumption in the typical UAV scenarios, but it can also be ensured by the pre-processing procedures discussed later. The algorithm of the integrated planner is as follows:

I) Solve distributed problems.
II) Perform pre-processing in a centralized manner.
III) Do the following until the vehicles reach their goals:
  (a) Solve distributed problems
  (b) Analyze detailed plans over the planning horizon
  (c) i. If there is no conflict, go to (a).
    ii. If there is a conflict, solve the pair-wise problems to obtain collision avoidance maneuver. Go to (a).

The pre-processing (step II) ensures that there always exists a CA maneuver around the nominal plans. First, it compares each pair of plans obtained in step I which consist of the detailed trajectories over the planning horizon and the approximate trajectories (*e.g.*, straight lines) beyond it. If $\|\vec{x}_i(t) - \vec{x}_j(t)\| \geq d,\ \forall t$, there will not be any conflict. If not, it tests if there exists a feasible maneuver around the conflicting trajectories. If feasible maneuver exists then the optimization by the CA planner will be successful (step (c)-ii). If not, the initial plans need to be revised to ensure that one exists. Figure 5 shows an approach that simply identifies the arc in the visibility graph that is causing the conflict, removes that connection from the visibility graph for one vehicle, and then re-solves the distributed problem for that vehicle. If vehicles are allowed to stop at the start positions, delaying the start time could also be used to ensure this feasibility. However, this approach requires complicated procedures and does not seem well suited to the problem of in-flight decision making.

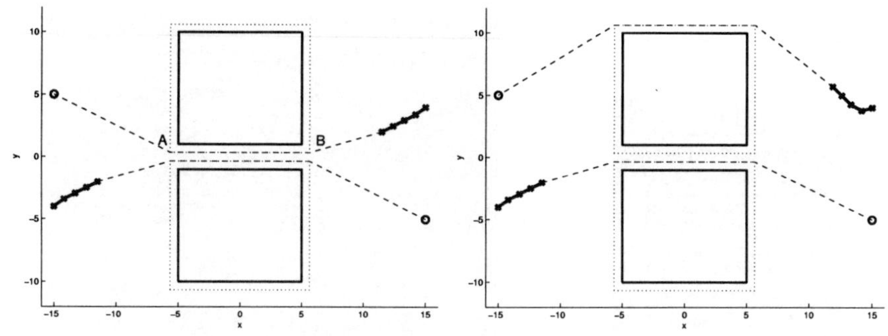

(a) Before pre-processing (b) After pre-processing

Fig. 5. Effect of the pre-processing. Before the pre-processing, two vehicles try to go through the same narrow passage. The pre-processing step detect a conflict in the two plans by analyzing the straight line trajectories. It removes the connection AB from the visibility graph for one vehicle, which prevents the potential collision.

3. Testbed Setup

In the experimental demonstration, the RHC is used as a high-level controller to compensate for uncertainty in the environment. It designs a series of waypoints for each vehicle to follow. A low-level vehicle controller then steers the vehicle to move along this path. The central data manager that monitors the vehicle positions and sends plan requests to the planner, receives planned waypoints and sends them to each vehicle controller.

Both the ground-based truck and autopilot testbeds have the same interface to the planner, and the planning algorithm can be demonstrated on both. All of the data is exchanged between the planners, data manager, and testbed vehicles via wireless TCP/IP local area network connections, which can flexibly accommodate additional vehicles or another module such as a mission level planner and GUI for a human operator. This wireless LAN communication has a bandwidth of 10Mbps, which is high enough to send vehicle states and planned waypoints. Figure 6 shows planner laptops that have Pentium 4, 2.4 GHz processors with 1 GB RAM.

The vehicles in the truck testbed have been modified to emulate the motions of UAVs, which would typically operate at a nominal speed, flying at a fixed altitude, and with the turning rate limited by the achievable

Fig. 6. Rack of planner CPUs.

Fig. 7. Four truck testbed showing the indoor GPS antennas, the Sony laptops, and the electronics package.

bank angle. The testbed described here consists of eight remote-controlled, wheel-steered miniature trucks, as shown in Figure 7. In order to capture the characteristics of UAVs, they are operated at constant speed. Due to the limited steering angles, the turn rate of the trucks is also restricted.

An indoor GPS sensing system produces position estimates accurate to about 2 cm. With an on-board laptop that performs the position estimation and low level control, the trucks can autonomously follow the waypoint commands. The more complex path planning is then performed off-board using the planner computer. This separation greatly simplifies the implementation (by eliminating the need to integrate the algorithms on one CPU and simplifying the debugging process) and is used for both testbeds.

The on-board laptop controls the cross-track error and the in-track error separately to follow the waypoint commands. The trucks are capable of steering with only the front wheels to change their heading. The heading controller drives the cross-track position error to zero using PD control. The speed control loop tracks the nominal speed while rejecting disturbances from the roughness of the ground and slope changes. In order to nullify any steady state error, a PI controller is implemented in this case.

This testbed has the following features: the trucks are physically moving vehicles and allow the tests to be conducted in a real environment; it is also able to stop, which makes debugging easier than with the flying vehicles; the test area does not need to be vast since they can move at a much slower speed; the hardware-in-the-loop tests done here are set up exactly the same as they will be when actual flight tests are conducted; it also enables numerous trials in a complex environment without the logistic work associated with aircraft experiments.

4. Results

4.1. *Truck Experiments*

The control laws for the low-level feedback loops account for the error in the cross-track direction and the error from the nominal reference speed. Although the PI speed controller does not have a steady state speed error, it cannot completely nullify the in-track position error, which translates into an error in the time-of-arrival at each waypoint. Failure to meet a timing constraint can cause a significant problem when coordinating multiple vehicles. This subsection demonstrates, using an example based on a collision avoidance maneuver, that the RH-MILP can re-optimize the trajectory on-line, accounting for in-track position errors.

The new formulation for the collision avoidance maneuver is experimentally tested using two trucks. In the previous work, a plan request is sent when the vehicle reaches the execution horizon [17], and the RHC re-optimizes the trajectory before the system reaches the end of the plan. In this two-truck case, a plan request is sent when either one of the vehicles reaches its horizon point. The speed controller in this experiment has a low bandwidth, and the RH-MILP controls the in-track position by adjusting the initial position of each plan, so that the vehicles reach waypoints at the right time. To see the effect of in-track adjustment by the RHC, three trials are conducted with different disturbances and control schemes:

Case–1: Small disturbance – no adjustment of in-track position.

Case–2: Small disturbance – adjustment of in-track position by RHC.

Case–3: Large disturbance – adjustment of in-track position by RHC.

The following parameters are used:

- $v\Delta t = 3.5$ [m], $v = 0.5$ [m/s], $r_{\min} = 5$ [m]

- $n_p = 4$, $n_e = 1$
- Safety box for each truck: 0.8 [m] × 0.8 [m]

Figure 8(a) shows the planned waypoints of the first scenario. The two vehicles start in the upper right of the figure and go to the lower left while switching their relative positions. In Figure 8(b), × marks represent the planned waypoints, and dots represent the position data reported from the trucks. The relative position starts in the lower right of Figure 8(b) and goes to the upper left. Although the vehicles avoided a collision, the relative position deviates from the planned trajectory by as much as 1.8 m. This error is mainly caused by the ground roughness in the test area, which acts as a disturbance to the speed control loop, resulting in in-track position errors for both vehicles.

One way to improve this situation is to introduce an in-track position control loop in the low-level feedback controller. This requires the use of the time stamp placed by the planner at each waypoint. Another approach presented here is to feed the in-track error back into the receding horizon control loop. Figure 9 illustrates this procedure. Let $d_{//}$ denote the in-track distance to the next waypoint. When $d_{//}$ of either one of the vehicles becomes smaller than a threshold, the vehicle sends a plan request to the planner. If vehicle 2 is slower than vehicle 1, as is the case in Figure 9, the position difference in the in-track direction $(d_{//})_2 - (d_{//})_1$ is propagated to the initial position of vehicle 2 in the next plan. This propagation is accomplished by moving the next initial position backward by $(d_{//})_2 - (d_{//})_1$. Note that the truck positions are reported at 2 Hz, and an in-track distance $d_{//}$ at a specific time is obtained through an interpolation.

Figure 10 shows the result of Case–2, where the in-track position is propagated and fed back to the next initial condition by the RHC. The outcome of the in-track position adjustment is apparent in Figure 10(b) as the discontinuous plans. The lower right of Figure 10(b) is magnified and shown in Figure 11 with further explanation. When the second plan request is sent, the difference between the planned relative position and the actual relative position is obtained (A), and is added as a correction term to the initial position of the next plan (A'). When the third plan request is sent, the difference at point B is fed back in the start position of the next plan (B'). This demonstrates that the replanning by the RHC can account for the relative position error of the two vehicles. Note that this feedback control by the RHC has a one-step delay, due to the computation time required by the RH-MILP. However, the computation time in this scenario is much smaller than the $\Delta t = 7$ [sec], and many more frequent updates are possible. Further research is being conducted to investigate this issue.

In Case–3, a larger disturbance was manually added to truck 2. As shown in Figure 12, vehicle 2 goes behind vehicle 1 as opposed to the results of Case–2 shown in Figure 10. This demonstrates the decision change by the RHC in an environment with strong disturbances. Further observations include: replanning by the RHC was able to correct the relative position errors; overly large disturbances can make the MILP problem infeasible; improvements of the current control scheme, which has a one step delay, will enable a further detailed timing control; similar performance could be achieved by updating the reference speed of the low-level PI speed controller. Future experiments will compare these two approaches.

4.2. *Integrated Planner*

This section shows a simulation result with the integrated planning system proposed in Subsection 2.3. The scenario considered has two obstacles and two vehicles. Figure 14 shows the resultant trajectories for both vehicles. The vehicles start in the bottom and go to the assigned targets marked with $\triangle$ while avoiding obstacles and the other vehicle. $\times$ marks show the planned waypoints for truck 1, and $\square$ marks show the planned waypoints for truck 2. Note that there are more waypoints when there is a conflict between the plans generated by the distributed planners. This is because the new plans solved by the CA planner overwrite the nominal plans.

Figure 13 shows the plans generated by the CA planner. The figures in the left column show the plans in the absolute frame, and the figures in the right column show the plans in the relative frame. The $\triangle$ marks show the short-term goals for vehicle. Each vehicle minimizes the distance to the visible point $\vec{x}_{\text{vis}}$, and hence, the visible point selected by the distributed planner is used as the short-term goal. From Figure 13(c) to Figure 13(e), because of the collision avoidance maneuver, the distributed planner made a different decision on the cost-to-go point ($\triangle$ marks). Note that the trajectory in the relative frame tends to follow the vehicle avoidance box, which is shown by the thick lines in the figures.

5. Conclusions and Future Work

This chapter presents a distributed trajectory planning system for a fleet of vehicles that combines two planners. The basic distributed RHC decouples the centralized problem by ignoring the vehicle interactions. The collision avoidance planner then efficiently handles conflicts in these plans by solving pairwise problems as they arise. The pre-processing of the combined planner ensures the existence of an initial feasible solution for the trajectory optimization, and the planning horizon of the RHC extended beyond the execution horizon maintains the feasibility over the mission. Several experiments and simulations are presented to show the successful integration of the planning system and demonstrate the use of MILP for on-line replanning to control vehicles in the presence of real-world disturbances. The results in this chapter focused on the most likely collision avoidance scenarios (*i.e.*, two vehicles), and additional results with larger teams will be presented in [10].

Acknowledgments

Research funded in part under Air Force grant # F49620-01-1-0453.

References

[1] Bellingham, J., Coordination and Control of UAV Fleets using Mixed-Integer Linear Programming. Master's thesis, Massachusetts Institute of Technology, 2002.

[2] Bellingham, J., Richards, A., and How, J., Receding Horizon Control of Autonomous Aerial Vehicles. In *Proceedings of the IEEE American Control Conference*, 2002.

[3] Bellingham, J., Tillerson, M., Richards, A., and How, J., Multi-Task Allocation and Path Planning for Cooperating UAVs. In *Second Annual Conference on Cooperative Control and Optimization*, 2001.

[4] Chandler, P., and Pachter, M., Hierarchical Control for Autonomous Teams. In *Proceedings of the AIAA Guidance, Navigation and Control Conference* AIAA, 2001.

[5] Dunbar, W. B., and Murray, R., Model predictive control of coordinated multi-vehicle formations. In *Proceedings of the IEEE Conference on Decision and Control*, 2002.

[6] Franz, R., Milam, M., , and Hauser, J., Applied Receding Horizon Control of the Caltech Ducted Fan. In *Proceedings of the IEEE American Control Conference*, 2002.

[7] Bay, J., DARPA, HURT: Heterogeneous Urban RSTA Team. available on-line at
http://dtsn.darpa.mil/ixo/solicitations/HURT/index.htm, 2003.

[8] Heise, S. A. DARPA Industry Day Briefing. available on-line at www.darpa.mil/ito/research/mica/MICA01mayagenda.html, 2001.

[9] Kuwata, Y., Real-time Trajectory Design for Unmanned Aerial Vehicles using Receding Horizon Control. Master's thesis, Massachusetts Institute of Technology, 2003.

[10] Bertuccelli, L., Alighanbari, M., and How, J., Robust Planning for Coupled and Cooperative UAV Missions, submitted to 43rd IEEE Conference on Decision and Control.

[11] Latombe, J. C., *Robot Motion Planning.* Kluwer Academic, 1991.

[12] Mao, Z. H., Feron, E., and Bilimoria, K., Stability and Performance of Intersecting Aircraft Flows Under Decentralized Conflict Avoidance Rules *IEEE Transactions on Intelligent Transportation Systems*, 2(2):101–109, 2001.

[13] Richards, A., How, J., Schouwenaars, T., and Feron, E., Plume Avoidance Maneuver Planning Using Mixed Integer. In *Proceedings of the AIAA Guidance, Navigation and Control Conference*, 2001.

[14] Richards, A., Schouwenaars, T., How, J., and Feron, E., Spacecraft Trajectory Planning With Collision and Plume Avoidance Using Mixed-Integer Linear Programming. *AIAA Journal of Guidance, Control and Dynamics*, 25(4):755–764, 2002.

[15] Richards, A., Trajectory Control Using Mixed Integer Linear Programming. Master's thesis, Massachusetts Institute of Technology, 2002.

[16] Richards, A. and How, J. P., Aircraft Trajectory Planning With Collision Avoidance Using Mixed Integer Linear Programming. In *Proceedings of the IEEE American Control Conference*, pages 1936–1941, Anchorage, AK, 2002.

[17] Richards, A., Kuwata, Y., and How, J., Experimental Demonstrations of Real-time MILP Control. In *Proceedings of the AIAA Guidance, Navigation and Control Conference*, Austin, TX, 2003.

[18] Schouwenaars, T., Moor, B. D., Feron, E., and How, J. Mixed Integer Programming for Multi-Vehicle Path Planning. In *Proceedings of the European Control Conference*, Porto, Portugal, 2001.

[19] Takahashi, O. and Schilling, R., Motion planning in a plane using generalized Voronoi diagrams. *IEEE Transactions on Robotics and Automation*, 5(2):143–150, 1989.

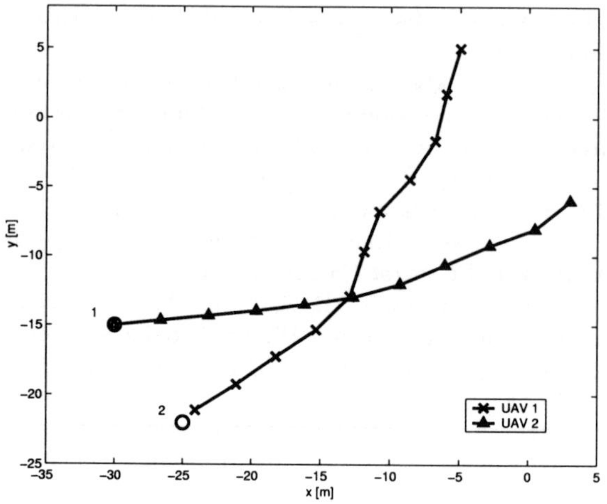

(a) Absolute position

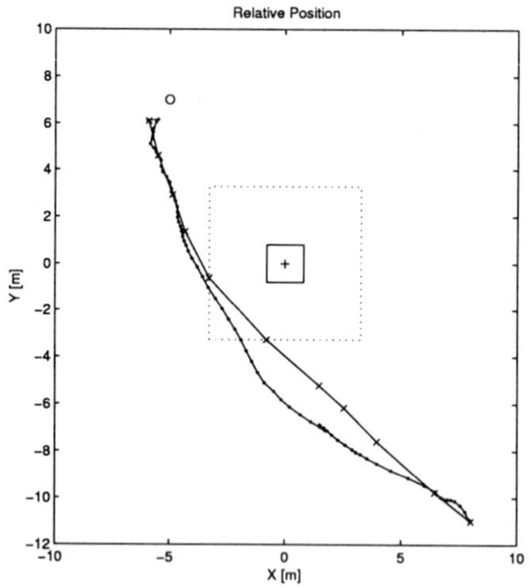

(b) Relative position

Fig. 8. Case–1. Cost-to-go in relative frame, no adjustment of in-track position.

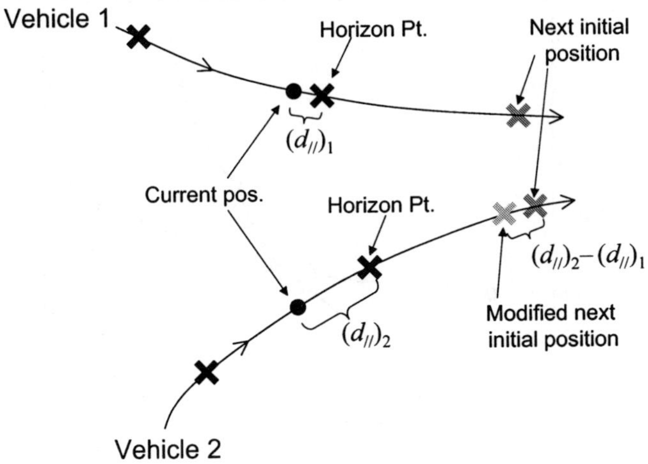

Fig. 9. Adjustment of in-track position for the next optimization.

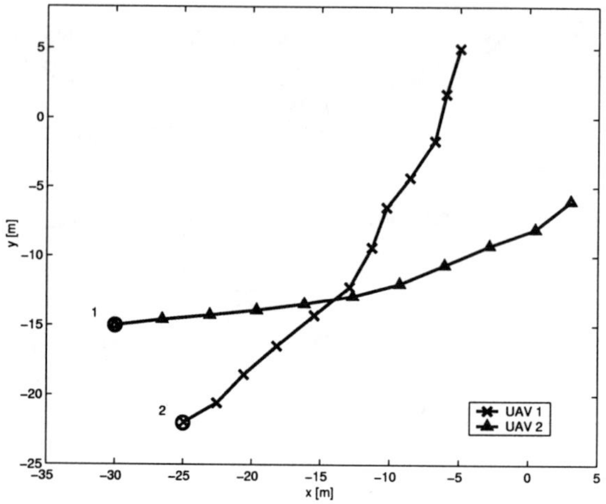

(a) Absolute position

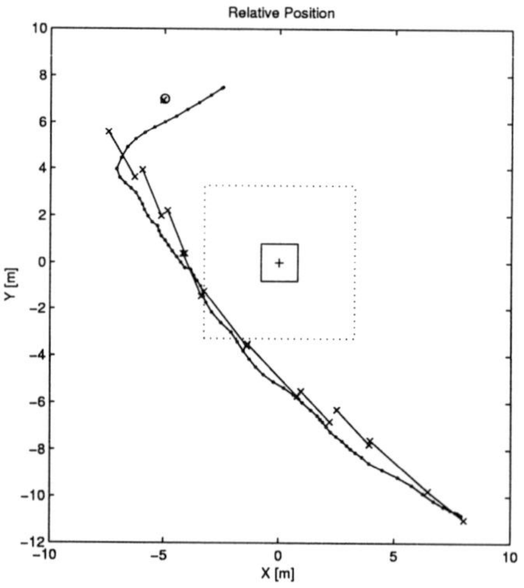

(b) Relative position

Fig. 10. Case–2. Cost-to-go in relative frame, with adjustment of in-track position.

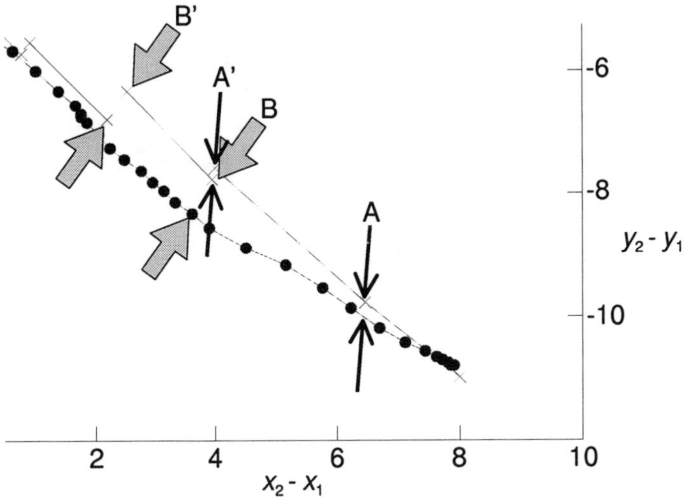

Fig. 11. Close-up of lower right corner of Figure 10(b). The position difference between the planned waypoint and the actual relative position (A) is fed back to the next initial position (A'). When the next initial position is reached, the position difference (B) is fed back such that the next-next start position is modified (B').

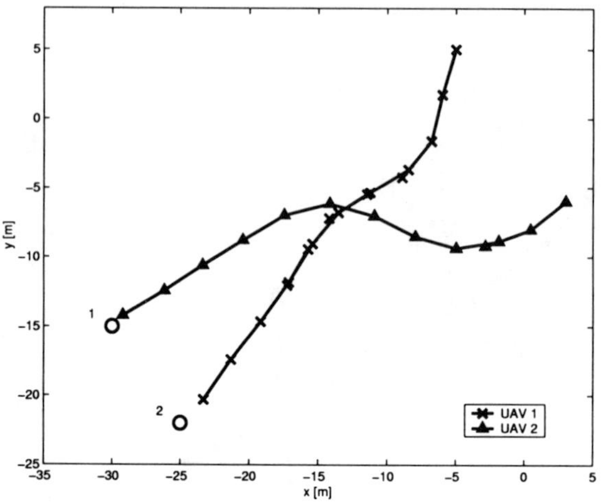

(a) Absolute position

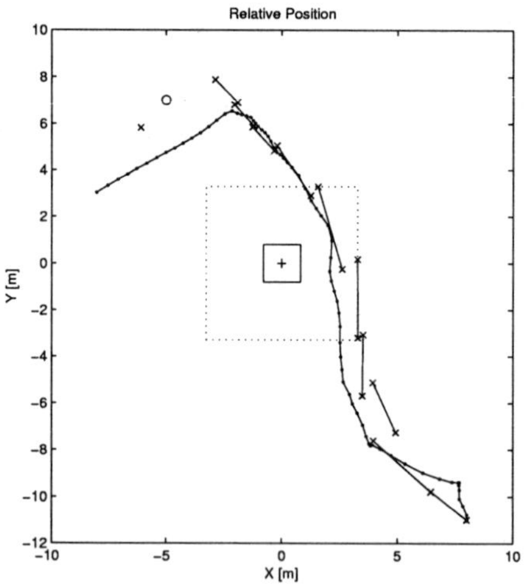

(b) Relative position

Fig. 12. Case–3. Cost-to-go in relative frame, with adjustment of in-track position. Large disturbance has been added. The square in solid lines is an actual vehicle avoidance box, and the square in dashed lines is an expanded safety box. The vehicle avoidance box is expanded to account for the time discretization.

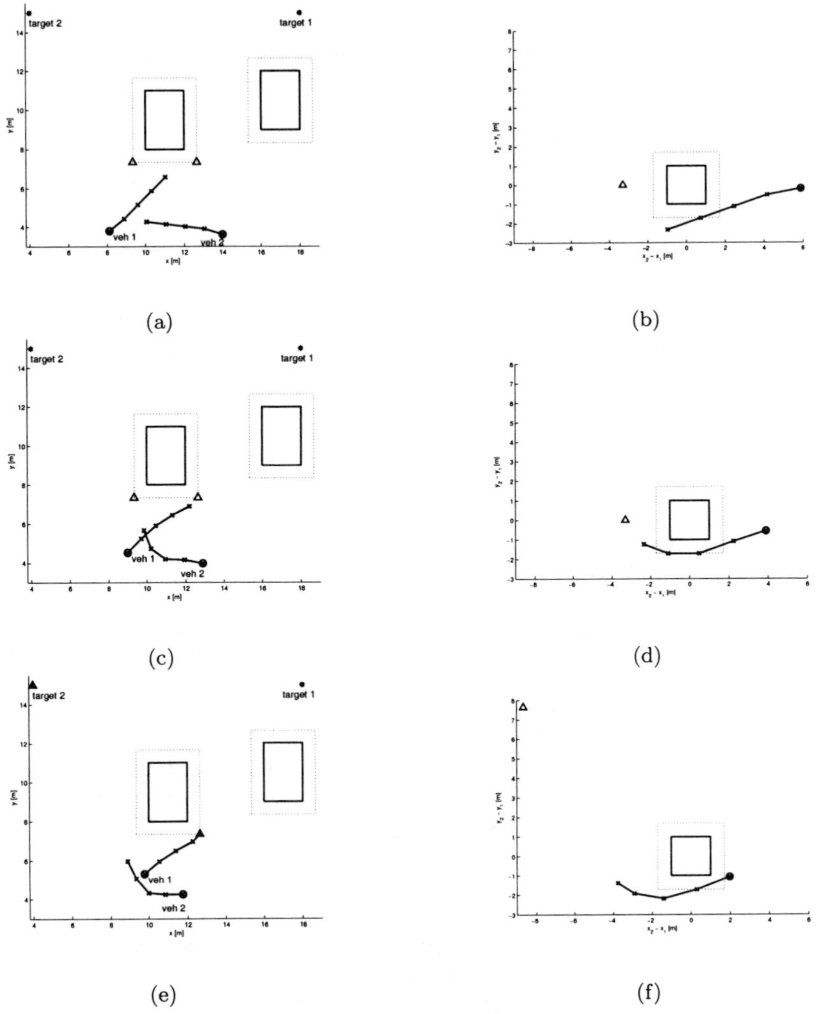

Fig. 13. Trajectories generated by the CA planner. Figures in the left column show the plans in the absolute frame. Figures in the right column show the plans in the relative frame. △ marks show the points that vehicles are aiming for.

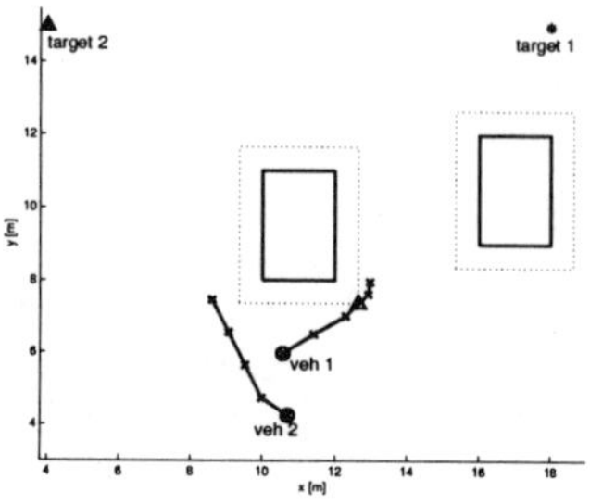

(g)

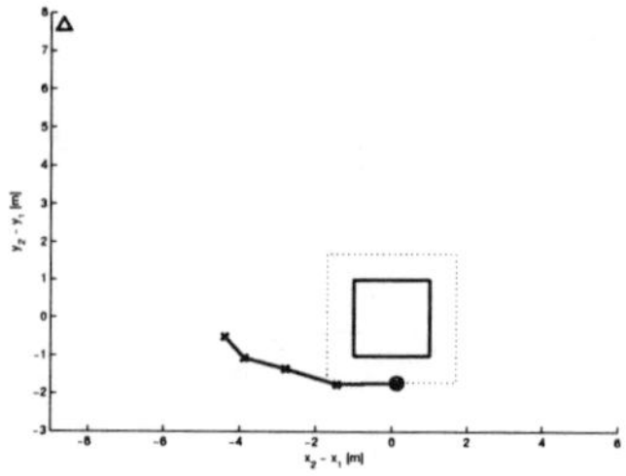

(h)

Fig. 13. (*Continued*)

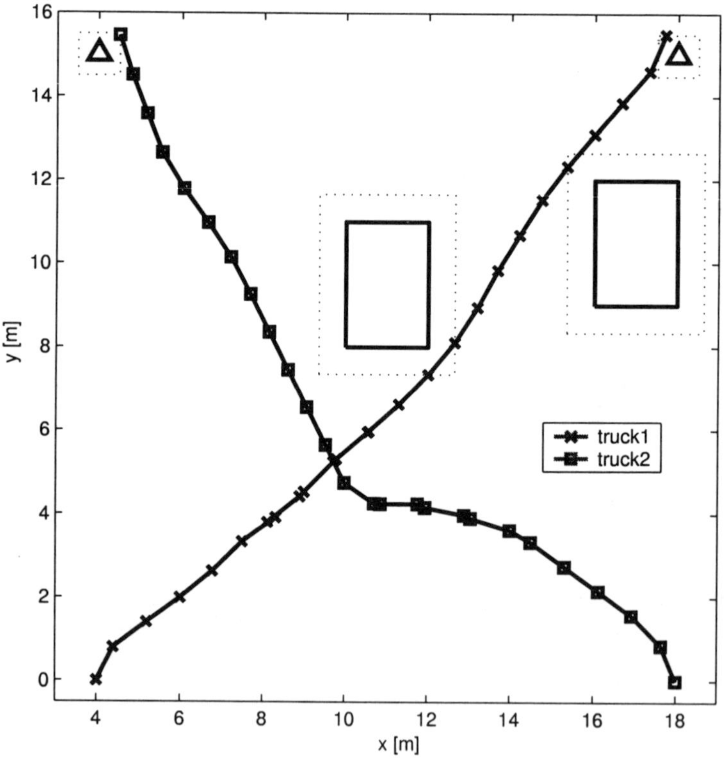

Fig. 14. Trajectories generated by the integrated planner.

CHAPTER 10

A STABLE AND EFFICIENT SCHEME FOR TASK ALLOCATION VIA AGENT COALITION FORMATION

Cuihong Li
Robotics Institute
Carnegie Mellon University
cuihong@cs.cmu.edu

Katia Sycara
Robotics Institute
Carnegie Mellon University
katia@cs.cmu.edu

Task execution in a multi-agent, multi-task environment often requires allocation of agents to different tasks and cooperation among agents. Agents usually have limited resources that cannot be regenerated, and are heterogeneous in capabilities and available resources. Agent coalition benefits the system because agents can complement each other by taking different functions and hence improve the performance of a task. Good task allocation decision in a dynamic and unpredictable environment must consider overall system optimization across tasks, and the sustainability of the agent society for the future tasks and usage of the resources. In this chapter we present an efficient scheme to solve the real time team/coalition formation problem. Our domain of applications is coalition formation of various Unmanned Aerial Vehicles (UAVs) for cooperative sensing and attack. In this scheme each agent bids the maximum affordable cost for each task. Based on the bidding information and the cost curves of the tasks, the agents are split into groups, one for each task, and cost division among the group members for each task is calculated. This cost sharing scheme provably guarantees the stability in cost division within each coalition in terms of the core in game theory, therefore achieves good sustainability of the agent society with balanced resource depletions across agents. Simulation results show that, under most conditions, our scheme greatly increases the total utility of the system compared with the traditional heuristics.

Keywords:
Coalition formation, task allocation, multi-agent coordination

1. Introduction

Task execution in a multi-agent, multi-task environment often requires allocation of agents to different tasks and coalition formation among agents. Good task allocation and coalition formation decisions must consider overall system optimization across tasks as well as agent heterogeneity in resources and capabilities. The cost division among coalition members is also important to sustain a well-functioning agent society in a dynamic and uncertain environment.

Consider a fleet of Unmanned Aerial Vehicles (UAVs) in missions over time to destroy targets that appear dynamically. The UAVs have different specializations in capabilities, although each can perform multiple functions subject to different costs (fuel). UAVs have limited resources (fuel) and cannot be refuelled during the process. The available resources of the UAVs are different because of the consumption of fuel on different levels. Coalitions of UAVs are desirable in executing the tasks because UAVs can complement each other by undertaking different functions that could be done more effectively with more participating UAVs. At each time there may be multiple targets that require different capabilities of UAVs to destroy. A UAV is capable of executing more than one task to destroy a target, and the UAVs have to be allocated to different coalitions/teams for different targets. The cost of, or the resource to be consumed by, a coalition to execute a task is deterministic. A coalition formation scheme decides the allocation of the UAVs into different coalitions, one for each target. A cost division scheme determines how much cost/effort a UAV should pay in participating in a coalition to destroy a target. We want the coalition configuration to be efficient so that the total performance of executing the multiple tasks is optimized. The cost sharing rule should be fair so that the resource depletions of the UAVs are balanced, and as many as possible UAVs can survive as long as possible through the usage horizon and complement each other in executing future tasks.

The task allocation and coalition formation problem can be characterized by the following important properties or requirements:

- Coalition formation: Agents can form a coalition and execute a task together. Agent coalitions benefit the system because agents can improve the task performance by taking different complementary functions. When there are more agents in a coalition, the average cost per agent for executing the task decreases. The relation between the total cost for executing a task and the number of agents[a] is characterized by a *cost curve*. Although the average cost decreases, the total cost for executing a task may increase with the number of agents in the coalition because more agents are involved. But the marginal cost imposed by an agent does not increase with the number of agents, in other words, the total cost is a non-decreasing concave function of the number of agents. Because of the contribution made by an agent to the task performance, it is always efficient to include an agent in a coalition if the agent can afford the marginal cost.

- Heterogeneity: The heterogeneity is in both the tasks and agents. The tasks are heterogeneous because the capability requirements and cost curves are different. An agent may be qualified in capabilities for participating in some tasks but not in others. The efficiency of an agent in executing a task is also different from executing other tasks. Agents are heterogeneous in capabilities and available resources. We use the *maximum affordable cost* of an agent as the measurement of the suitability of an agent to execute a task. The maximum affordable cost is a function of both the capability and the available resource. The maximum affordable cost is higher when an agent has more resources available, or the agent is more specialized in the capability desired for the task. In a quasi-linear form the maximum affordable cost can be expressed as the available resources plus a function of the capability that calculates the resource saving based on the capability. The index of the maximum affordable cost and the cost curve allow the comparison of the efficiency of allocating different agents with different capabilities and resources to different tasks.
- Sustainability: We want as many as agents to participate in the tasks to improve the efficiency, and also to minimize to the extent possible the depletion of resources across agents so as to retain

[a]Precisely the cost of a coalition depends on the functions of the agents and the coordination mechanism. Since we consider the task allocation on a high level and do not consider the specific function allocation or scheduling of the agents, the cost of a coalition is approximated as a function of the number of participants.

agents for future tasks. It is not desirable to have some agents consume their resources much faster than the others. Sustainability does not mean that all agents share the cost equally. The agents that can afford more cost are reasonably assumed to share more cost because they have a larger base.

The objectives of the coalition formation scheme include: (1) to optimize the total performance of the tasks, and (2) to divide the cost among agents in a fair way to achieve good sustainability. The first objective is important since it assures the efficiency in executing the current tasks by forming coalitions and matching agents with the tasks. The second objective considers the efficiency in executing future tasks by balancing the resource depletions across agents for current tasks.

If we consider the performance of a task as the value of the coalition for that task, the coalition formation problem can be translated into a weighted set packing problem, modelled as a set covering problem, which is well known as a NP-complete problem [1]. Task allocations often involve a large agent group in a scale of thousands or much higher. Additionally, time for calculating a solution is usually limited so the coalition formation must be performed in real time. Therefore an efficient algorithm is desired to ensure the real-time application for large scale problems. We present an efficient coalition formation scheme in polynomial time for the coalition formation problem. In this scheme each agent bids the maximum affordable cost for each task that it is capable of. Based on the bidding information and the cost curves of the tasks, the coordinator splits the agents into groups, one group for each task.

We use the *core*, a concept from cooperative game theory [4], to measure the fairness of cost division in a coalition. If the cost division is in the core, there are no agents that can get more total utility by deviating from the coalition and forming a coalition by themselves. Therefore a fair cost division scheme in the core achieves the *stability* of a coalition. In the task allocation situation the utility of an agent from a coalition is defined as the maximum affordable cost minus the cost to share. Agents in a coalition may pay different costs according to their maximum affordable costs.

As optimizing the total coalition values is, in general, computationally too complex as mentioned above, we take the following approach. When forming a coalition configuration, we try to maximize the value of the most valuable coalition, then maximize the value of the second valuable one, and continue recursively. Then we divide each coalition's cost within the coalition. We prove that our coalition formation scheme based on this approach guarantees the stability of cost division within each coalition in terms of the core in game theory. Simulation results show that, under most conditions, our scheme greatly increases the total utility of the system compared to the traditional heuristics.

This chapter is organized as follows. Section 2 describes prior work. In section 3 the problem is formulated. In section 4 we present the coalition formation scheme in detail. Section 5 analyzes the stability of the coalition formation scheme. Section 6 provides the experimental results. We conclude in section 7.

2. Prior Work

Works in game theory and microeconomics such as [4, 5] have provided concepts of coalition and its stability. A coalition is a set of agents which cooperate to achieve a common goal, and the stability requirement is that the outcome of a coalition be immune to deviations by individual agents or subsets of agents. Those concepts are important as criteria of coalition formation schemes, and we justify our scheme based on the core, one of the stability concepts in game theory. However, game theory does not provide efficient algorithms for coalition formation.

Finding the maximum total utility of coalitions can be translated into the weighted set packing problem [1]: Given a set B and collection of its subsets $Col = \{C_0, ..., C_n\}$ such that each C_i has its value $v(C_i)$, find a sub-collection $SubCol \subset Col$ of pairwise disjoint sets such that $\sum_{C_i \in SubCol} v(C_i)$ is the maximum among all sub-collections. We can interpret B as the set of agents, SubCol as a collection of coalitions, and v as a coalition's value. The weighted set packing problem is NP-complete, and several optimization algorithms have been proposed [1, 2]. However, these algorithms rely on the assumption that the maximum size of subsets in $SubCol$ is bounded by a relatively small number k. In the context of task allocation, bounding the coalition size by a small number is impractical.

Research on multi-agent systems also has investigated coalition formation of agents. [7] proved that, for a given set A, searching the best coalition configuration among $\{\{A\}\} \cup \{\{A_1, A_2\} \mid A_1 \cup A_2 = B, A_1 \cap A_2 = \emptyset\}$ guarantees that the largest coalition value found is within a bound from the optimal one by $|B|$, and that no other search algorithm can establish any bound while searching only $2^{|A|-1}$ coalition configurations or fewer. This result means, without some kind of heuristics or assumptions, bounding the group's total utility is virtually impossible because $|A|$ could be large.

[8, 10] have provided distributed coalition formation schemes for multi-agent systems mainly focusing on increasing the group's total utility. They also limit the highest coalition size by an integer k, which means the algorithms proposed cannot be applied to large coalitions. [9] aims both to increase the total utility and to reach the stable payoff division among agents. Yet, the algorithms restrict the size of each coalition to guarantee the practical computation time.

[3] has proposed a new model of coalition formation, and applied it to coalition formation among buyer agents in an e-marketplace. Their model treats agents as locally interacting entities; an agent may create a coalition when it encounters another agent, join an existing coalition, or leave a coalition. The model describes global behavior of a set of agents from the macroscopic view point by differential equations, and simulates well how buyer coalitions evolve and reach the steady state. However, the model does not assist individual agents to form a coalition nor to negotiate surplus distribution.

3. Problem Formulation

The terms and notations are defined and interpreted as follows.
Tasks and Cost Curves: $T = \{t_1, t_2, ..., t_m\}$ denotes the set of tasks. Let N and R be the set of natural numbers and real numbers respectively. A *cost curve* of t_i is represented as a descending function $p_i : N \rightarrow R$; $p_i(n)$ is the average cost per agent when n agents join the coalition for the task t_i.
Agents: Let $A = \{a_1, a_2, ..., a_n\}$ denote the group of agents to be allocated for the tasks. Agent a_k's maximum affordable cost for t_i is represented by $r_{ki} \geq 0$. The *maximum affordable cost* of an agent for a task comprises the agent's available resource, and the agent's capability for executing the task. An agent a_k's *utility* from participating in the task $t_i$at the cost p is defined as $u_{ki} = r_{ki} - p$.
Coalitions: Let $C_i \subset A$ denote a coalition for the task t_i. A coalition configuration is $Conf = \{C_1, ..., C_m\}$ such that $C_i \cap C_j = \emptyset$ for $i \neq j$. C_i can be empty. $Conf$ does not necessarily satisfy $\cup_{i=1,...,m} C_i = A$; some agents in A may not belong to any coalitions because their maximum affordable costs are too low.

The *value* $v_i(C)$ of a coalition C for the task t_i is defined as

$$v_i(C) \stackrel{\text{def}}{=} \sum_{a_k \in C} r_{ki} - cost_i(C)$$

where $cost_i(C)$ is the cost paid by the coalition C to execute the task t_i, i.e., $cost_i(C) = |C| \cdot p_i(|C|)$. ($|C|$ denotes the cardinality of C.) Since the cost of the coalition $cost_i(C)$ is shared by the agents in C, the value of a coalition is equal to the sum of the utilities of the agents in the coalition. A coalition C is formed for the task t_i only if it can afford to execute the task t_i, i.e., $v_i(C) \geq 0$. The higher the value $v_i(C)$, the more efficient the allocation of the coalition C to the task t_i. It is because a higher value $v_i(C)$ means that the task t_i requires lower cost from the coalition C, or the agents in the coalition C are more capable of executing the task t_i. To maximize the values of the coalitions is consistent with the objective to maximize the overall performance of the tasks.

A cost division scheme c_k, $a_k \in C$ for the coalition C_i is in the *core* if and only if there does not exist $S \subset C_i$ so that the value $v_i(S)$ of the coalition S is greater than the sum of the utilities of agents in S from the coalition C_i, i.e., $v_i(S) \leq \sum_{a_k \in S} u_{ki}$ for any $S \subset C_i$.

The problem of coalition formation can be formulated as

$$\max_{C_i \subset A, i=1,\ldots,m} \sum_{a_k \in C_i} r_{ki} - cost_i(C_i)$$

so that $C_i \cap C_j = \emptyset$ for $i \neq j$; and for each $i = 1, \ldots, m$, find c_k for each $a_k \in C_i$ so that for any $S \subset C_i$

$$\sum_{a_k \in S} (r_{ki} - c_k) \geq \sum_{a_k \in S} r_{ki} - cost_i(S).$$

4. Coalition Formation Scheme

We give a simple example to illustrate the model and the approach. Assume there are three tasks which have the same cost curve shown in Figure 1. The horizontal axis shows the number of agents in the coalition for a task, and the vertical axis indicates the average cost per agent. For instance, if there are three agents in the coalition, the average cost goes down to 90. Table 4 shows five agents to be allocated to these tasks. Each row shows an agent's affordable cost for each task that the agent is capable of performing. For instance, $a4$ is capable of participating in $task1$ or $task2$ and the maximum costs are 85 and 95 respectively.

Table 1. Sample agents' maximum affordable costs

agent	task0	task1	task2
a0	100		7
a1	80	95	95
a2		95	
a3		65	
a4		85	95

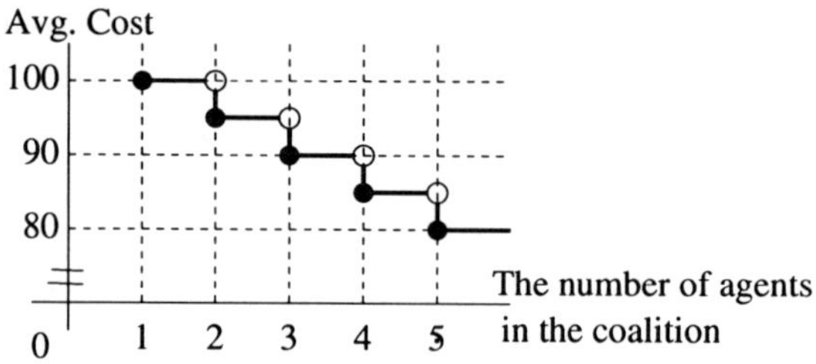

Fig. 1. A sample cost curve

The main issues we study are how to split the agents into coalitions, and how to distribute the cost of the group among agents. In this example, there are one possible coalition for task0 ($\{a0\}$), three for task1 ($\{a1, a2\}$, $\{a1, a2, a4\}$, $\{a1, a2, a3, a4\}$), and one for task2 ($\{a1, a4\}$). Our scheme derives the coalition configuration shown in Table 4; $\{a1, a2, a4\}$ as a 'task1' coalition has the largest surplus among all possible coalitions, and $\{a0\}$ as a 'task0' coalition is the only coalition which the rest of the agents can form. Each cell in the table contains the agents' cost to pay and the maximum affordable cost between parentheses. The costs to pay in a coalition differ depending on agents' maximum affordable costs. For example, $a1$ pays 92.5 ($a1$'s maximum affordable cost is 95), while $a4$ pays only 85 ($a4$'s maximum affordable cost is 85). If $a4$ did not join the coalition, $a1$ and $a2$ would have to pay 95 for executing task1. On the other hand, the coalition does not include $a3$ because $a3$ would bring no benefit to the coalition.

The rest of this section formally explains this coalition formation scheme.

Table 2. A sample coalition configuration

agent	task0	task1	task2
a0	100(100)		
a1		92.5 (95)	
a2		92.5 (95)	
a3			
a4		85.0 (85)	

4.1. *Coalition Configuration Algorithm*

As we have mentioned, it is not computationally feasible to obtain the optimal coalition configuration that maximizes the total coalition values. We design a computational heuristic to configure the coalitions that achieves fairly good efficiency in reasonable time. In the heuristic approach a coalition configuration $Conf = \{C_1, ..., C_m\}$ is formed so that the value of the most valuable coalition is maximized first, and then the utility of the second most one is maximized, etc. This algorithm is formalized as follows.

Algorithm 1: Coalition Configuration

(1) Set $Conf = \emptyset$, $RestOfTaskIDs = \{1, 2, ..., m\}$ and $RestOfAgents = B$.

(2) For every $i \in RestOfTaskIDs$, calculate a candidate coalition $C_i^* \subset RestOfAgents$, the set with the largest value as a t_i coalition, as follows.

$$AC_i \stackrel{\text{def}}{=} \{C \subset RestOfAgents \mid v_i(C) \geq 0\}$$
$$VC_i \stackrel{\text{def}}{=} \{C \in AC_i \mid v_i(C) \geq v_i(C') \, for \, \forall C' \in AC_i\}$$

(AC_i is the set of admissible coalitions, VCi the set of the most valuable coalitions.)

Select any one of $C_i^* \in VC_i$ if $VC_i \neq \emptyset$, $C_i^* = \emptyset$ otherwise. $Cand \stackrel{\text{def}}{=} \{C_i^* \mid i \in RestOfTaskIDs\}$ denotes the set of all candidates.

(3) If every $C_i^* \in Cand$ is empty, stop this procedure.

(4) If there exist non empty candidates in $Cand$, select one of them with the largest utility within $Cand$; that is, select C_i^* such that $v_k(C_k^*) \geq v_i(C_i^*)$ for $\forall C_i^* \in Cand$. Let $Conf = Conf \cup \{C_k^*\}$, $RestOfTaskIDs = RestOfTaskIDs \backslash \{k\}$, and $RestOfAgents = RestOfAgents \backslash C_k^*$.

(5) Go back to Step 2 if $RestOfTaskIDs \neq \emptyset$ and $RestOfAgents \neq \emptyset$. Otherwise, stop this procedure.

This algorithm can be considered as a variation of the greedy algorithm for the weighted set packing problem [2]. In general, finding a subset of A which has the largest value among all subsets could require $O(2^n)$ computations at worst.

However, we have an efficient algorithm to calculate our coalition configuration with order $O(n \cdot \log n)$, where n is the number of agents in B, and we assume the number of tasks can be bounded from above by a positive number K independently from n. This assumption makes sense even for very large coalitions. The complexity of searching C_i^* at each recursion is $O(n \cdot \log n)$ computations as explained below, each recursion includes at most K times of the search, and all coalitions are configured within K recursions. Thus, the entire complexity of the coalition configuration is $O(n \cdot \log n)$ computations.

To search C_i^* at each recursion, first arrange all agents in $RestOfAgents$ in the descending order in terms of the maximum affordable cost for t_i ($O(n \cdot \log n)$ computations). Then calculate the utility of subsets $C_{ij} \subset RestOfAgents$ for $j = 1, ..., t$ (t is at most n) which includes the top j agents in terms of the maximum affordable cost for t_i, and select C_i^* out of $\{C_{i1}, ..., C_{it}\}$. This requires $O(n)$ computations. (This algorithm is supported by Proposition 2 in the next section.)

4.2. *Cost Sharing in a Coalition*

Agents in a coalition share the cost within the coalition. Let the cost shared by the agent $a_k \in C_i$ be $c_k \geq 0$. The cost sharing rule is defined as follows.
Definition 1: Cost Sharing Rule When a coalition C_i has value $v_i(C_i) > 0$, the cost c_k shared by an agent $a_k \in C_i$ is

$$c_k \stackrel{\text{def}}{=} \begin{cases} h_{C_i} & (a_k \in \overline{C_i}) \\ r_{ki} & (a_k \notin \overline{C_i}) \end{cases}$$

where h_{C_i} and $\overline{C_i}$ satisfies the following conditions:

$$cost_i(C_i) = |\overline{C_i}| \cdot h_{C_i} + \textstyle\sum_{a_k \in C_i \backslash \overline{C_i}} r_{ki},$$
$$\overline{C_i} \stackrel{\text{def}}{=} \{a_k \in C_i \mid h_{C_i} \leq r_{ki}\}.$$

Figure 2 illustrates this definition. The graph shows each agent's maximum affordable cost, its share of cost, and its actual utility. Agents in $\overline{C_i}$ pay h_{C_i} which is equal to or lower than their maximum affordable costs. Others in $C_i \backslash \overline{C_i}$ pay just their maximum affordable costs.

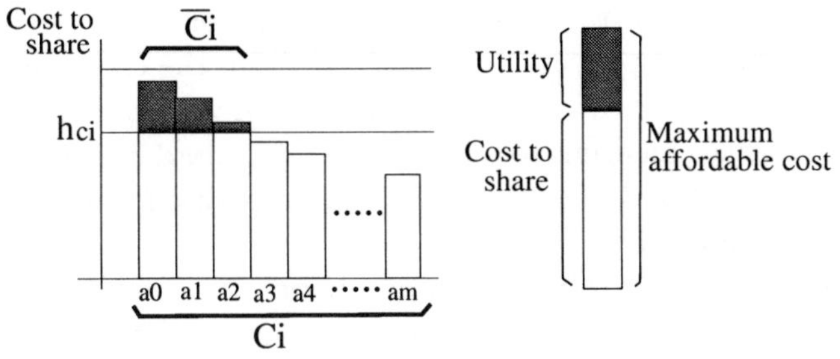

Fig. 2. The cost sharing rule

5. Stability of Coalition Configuration

As agents in a coalition pay different costs under our scheme, a fair share of the cost is essential to sustain the agent society, and guarantee the stability of the coalitions if agents are autonomous to choose the tasks driven by self-interest. If agents do not trust the fairness, they may not join a coalition, nor provide their maximum affordable costs truthfully, which could prevent successful coalition formation.

In this section, we discuss our scheme's stability in terms of the core in game theory [6, 4]. The core is defined as follows.

Definition 2: The Core [6]
A coalitional game with transferable payoff consists of (1) a finite set C of players, and (2) a utility function v which associates with every nonempty subset $S \subset C$ a real number $v(S)$. The core of the coalitional game with transferable payoff $< C, v >$ is

$$Core = \{(u_a)_{a \in C} \mid v(C) = \sum_{a \in C} u_a, v(S) \leq \sum_{a \in S} u_a \, for \, \forall S \subset C\}$$

In general, the core may contain multiple elements, or it may be empty. In our case each coalition C_i has a nonempty core. We prove that the cost distribution calculated by our cost sharing rule is within the core as the next proposition states.

Proposition 1 (Stability of a coalition)
For $\forall C_i \in Conf$, the cost distribution $(c_k)_{a_k \in C_i}$ calculated using the cost sharing rule (Definition 1) is in the core of the coalitional game with transferable payoff $< C_i, v_i >$. That is, $v_i(S) \leq \sum_{a_k \in S} u_k$ holds for $\forall S \subset C_i$.

The stability condition defined by the core is that no subset of agents in a coalition can obtain utility that exceeds the sum of the current utility of the members in the subset. Thus, even self-interested agents in a coalition would not be motivated to deviate from the coalition.

There can be multiple cost distributions within the core. Proposition 2 and 3 below characterize our cost distribution, and we expect these propositions will encourage an agent to tell its maximum affordable cost truthfully. (Note that Proposition 1 above is proved via Proposition 2 and 3. The proof is provided in Appendix.)

Proposition 2 (Members in a coalition)
At each recursion of coalition configuration in Algorithm 1, for $\forall a_k \in RestOfAgents$ and $\forall i \in RestOfTaskIDs$, if $\exists a_h \in C_i^*$ such that $r_{ki} > r_{hi}$, then $a_k \in C_i^*$.

Proposition 2 means that C_i^* consists of the top $|C_i^*|$ agents in terms of the maximum affordable cost. The higher an agent's maximum affordable cost is, the more likely it will be able to join a coalition.

Proposition 3 (Cost sharing)
At each recursion of coalition configuration in Algorithm 1, for $\forall i \in RestOfTaskIDs$ and $\forall C \in AC_i$, $h_{C_i^*} \leq h_C$.

The last proposition assures that, at each recursion, the highest cost anybody in C_i^* pays, $h_{C_i^*}$, is the lowest among all the costs afforded by any sets of agents.

6. Evaluation

We have conducted a series of simulations to evaluate the effectiveness of our coalition formation scheme in increasing the system's performance. We simulated agents' behaviors under three coalition formation schemes (our scheme, a traditional scheme and an optimal scheme) under particular conditions, and compared them by the groups' total utility.

6.1. *Assumptions*

We make the following assumptions.

Tasks and Cost Curves: The cost curve for each task is a predetermined non-increasing step function. The highest value of the function is called the *highest average cost*. There is no limit to how many agents can join a coalition.

Agents: An agent has several choices of tasks. We model the distribution of capabilities for multiple tasks by RAMT (the Ratio of Agents who are capable of Multiple Tasks). RAMT is an array $(ra_1, ..., ra_m)$, where m is the number of tasks and $ra_1 + ... + ra_m = 1$ holds. ra_i denotes the ratio of agents who can participate in i tasks out of m tasks. For instance, in the example shown by Table 4 in Section 4, RAMT is (0.4, 0.4, 0.2); out of five agents, two agents can only participate in one task, two agents can work for two tasks, and one agent can take part in three tasks. RAMT does not specify which particular tasks each agent is qualified for. An agent randomly selects the tasks that it is capable of performing.

Some agents' maximum affordable costs (MAC) for a given task may be greater or equal to the highest average cost. These agents are sure to be included in the candidate coalitions because they do not need the joining of other agents to form a coalition with a non-negative value. Let the ratio of the number of the agents with MACs no less than the highest average cost be called RMH (the Ratio of Maximum affordable costs which are the Highest average cost). Other MACs for the task are randomly distributed between its highest average cost and a certain lower value. We denote the lowest possible MAC by LAC. The environment (other agents' behaviors, cost curves, etc.) does not affect agents' capabilities or maximum affordable costs.

An Optimal Scheme: At every simulation, we calculate an optimal coalition configuration for comparison. The optimal scheme exhaustively searches all possible coalition configurations and selects one of the configurations which has the largest value[b]. Agents in a coalition share their cost within the coalition, but the optimal scheme does not care about how to share.

A Traditional Scheme: Under a traditional coalition formation scheme, each agent first selects one task, and then the agents that select the same task and can afford the cost are formed as a coalition. All agents in a coalition pay the same cost.

An agent can know the cost curve, current average cost and the number of agents in each coalition at any time. An agent a_k selects one task out of the tasks it is qualified for by following one of the selection rules listed below.

Random Rule: Randomly select a task.

Lowest Price Rule: Select a task whose current average cost is the lowest in proportion to the highest average cost.

Highest MAC Rule: Select a task with the highest maximum affordable cost in proportion to the highest average cost.

Highest Utility Rule: Select a task which currently brings the highest utility

[b]Exhaustive search is only computationally possible for a small problem size.

(maximum affordable cost - current cost share).

6.2. *Simulation and Parameters*

For every set of parameters, we simulate agents' behavior under our scheme, the optimal scheme and the traditional scheme 1000 times, and calculate the average data for the evaluation criteria. For the traditional scheme, we simulate four experimental conditions. At every condition, all agents follow the same selection rule out of four rules listed above.

Table 6.2 summarizes the simulation parameters in the evaluation. The range of the number of tasks is 1, 3 and 5. We assign the identical cost curve to all tasks such that the highest average cost is 100, the lowest is 80, and the average cost decreases by 5 in proportion to the number of agents. We only vary the average cost decreasing ratio (CDR), the ratio of 'the least number of agents which assures the lowest average cost' to 'the number of agents in a group.' CDR characterizes how steeply the average cost decreases. Figure 3 shows sample cost curves with CDR of 0.4 and 1.0, and 100 agents in a group. In the simulation, CDR varies among 0.2, 0.4, 0.6, 0.8 and 1.0.

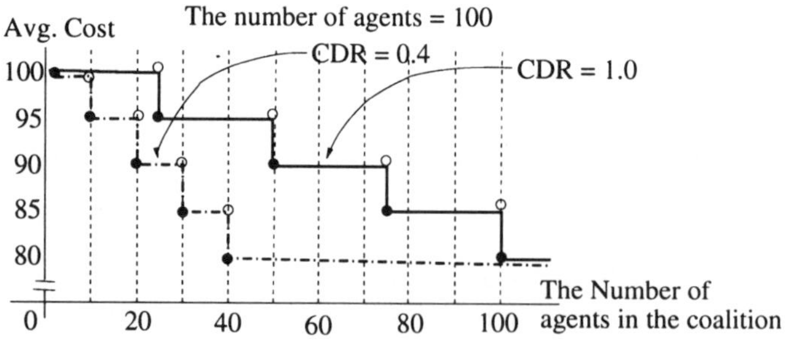

Fig. 3. Sample cost curves

The range of the number of agents is 50, 100, 200 and 400. We also vary RAMT, RMH and LAC as shown in Table 6.2 so that the effect of the agents' capability and resource distributions can be observed. Note that the optimal scheme can handle only the cases with 50 agents and RAMT of (1), (1,0,0) or (0.7, 0.2, 0.1) because of its high computational complexity.

Table 3. Simulation Parameters

	Parameter	Range
Tasks	The number of tasks	1, 3, 5
Cost Curve	CDR (price decreasing ratio)	0.2, 0.4, 0.6, 0.8, 1.0
Agents	The number of agents	100, 200, 400, 800
	RAMT (the ratio of agents capable of multiple tasks)	(1), (1, 0, 0), (.7, .2, .1), (.5, .3, .2), (1/3, 1/3, 1/3), (1, 0, 0, 0, 0), (.7, .2, .05, .03, .02), (.5, .3, .1, .05, .05), (.2, .2, .2, .2, .2)
	RMH (the ratio of MACs which are no less than the highest average cost)	0, 0.25
	LAC (the lowest MAC)	70, 80

6.3. *Results*

For a given number of agents and tasks, the three schemes showed common relations between agents' total utilities and the simulation parameters. The factors which affected the total coalition value favorably included smaller CDR, larger RMH and LAC, and more distributed RAMT (for instance, (1/3, 1/3, 1/3) brought a larger objective value than (1, 0, 0) did). Among them, CDR brought a clear contrast between the three schemes. Here, we analyze the simulation results focusing on CDR.

Out of the four experimental conditions for the traditional scheme, the one where all agents followed the highest utility rule produced the highest objective value in almost all simulations. Thus, in this section we refer only to this condition as the traditional scheme's output.

Optimality: First, we compare our scheme to the optimal one by examining the case that the number of tasks is 3, the number of agents is 50 and RAMT=(0.7, 0.2, 0.1). In summary, (1) our scheme came out more than 80 percent of the optimal utility under all conditions on average, and (2) as CDR became larger, the difference between our scheme and the optimal one became smaller; when CDR = 1.0, our scheme's outputs were nearly the same as the optimal ones.

Figure 4 shows the average objective value under the conditions where LRP = 70 and RRMP[c] = 0.25. The horizontal axis is CDR, and the vertical

[c]Similar results obtained for other combinations of LAC (70 or 80) and RMH (0 or 0.25).

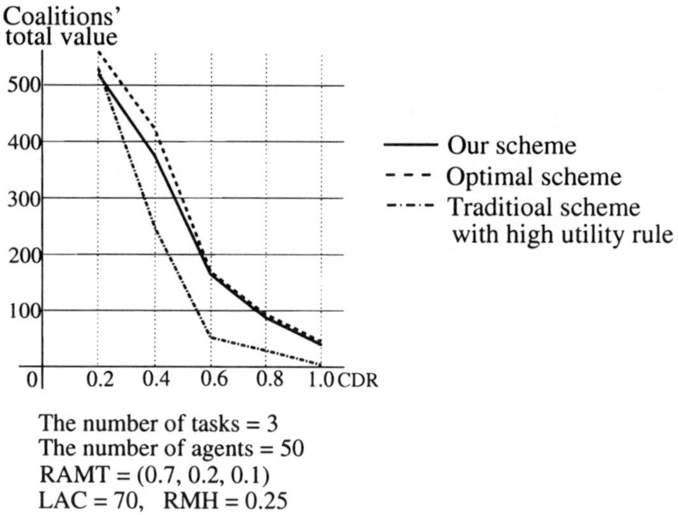

Fig. 4. Comparison between our scheme, the optimal one and the traditional one

axis is the total coalition value. When CDR is 0.2, the total utility gained by our scheme was slightly worse than the one by the optimal scheme and even the one by the traditional scheme. But, the average total utility under our scheme was still above 91 percent of the optimal one. As CDR became larger, our scheme performed better in the sense that the objective values became close to the optimal ones. When CDR ≥ 0.6, the objective value is within 96 percent of the optimal one. On the other hand, the traditional scheme became much worse when CDR was 0.4 or larger. When CDR = 1.0, the traditional scheme scarcely brought value to the system.

Cases with a large number of agents: Next, we examine the cases that 400 agents are involved in a group. (We compare only ours and the traditional scheme. Our implementation of optimal scheme could not handle such large number of agents.) Regardless of the number of agents, the comparison results showed the same tendency as the previous case of 50 agents: (1) when CDR=0.2, ours and the traditional scheme brought the best objective values, and the traditional scheme slightly outperformed ours under some conditions, and (2) as CDR became larger, our scheme performed better than the traditional one.

Figure 5 supports the above statements. The graph shows the ratio of the objective value by the traditional scheme to the one by our scheme. The horizontal axis of the graph is CDR. The vertical axis is the performance

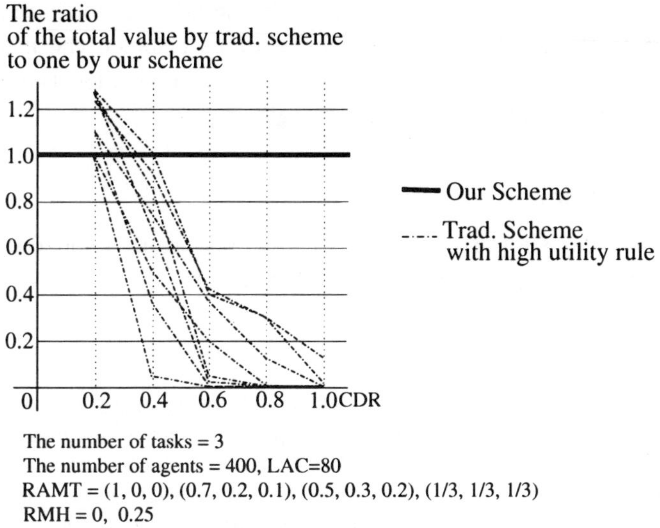

Fig. 5. Comparison between our scheme and the traditional scheme

ratio. The value 1.0 means two schemes have the same performance, the value under 1.0 indicates our scheme is better, and the value above 1.0 does the opposite. The graph includes the data under eight conditions; RAMT = (1,0,0), (0.7, 0.2, 0.1), (0.5, 0.3, 0.2) or (1/3, 1/3, 1/3), and RMH = 0 or 0.25. Other parameters are fixed (three tasks, 400 agents, and LAC = 80). In terms of the total coalition value, the traditional scheme outperformed ours only when CDR = 0.2. When CDR $\geq$ 0.4, our scheme was better under all conditions.

7. Conclusions and Future Work

In this chapter, a coalition formation scheme was proposed to allocate agents to different tasks and divide the task execution cost among coalition members, considering heterogeneity of agents and tasks. We showed that our scheme has enough scalability to handle a large number of agents, guarantees the stability in cost division within each coalition, and performs better in increasing the system's performance compared to a traditional coalition formation scheme.

Future work includes to investigate strategies of agents and the mechanism design. In the evaluation reported in this chapter, we simply assumed agents truthfully reveal their maximum affordable costs. Agents, however, may underreport the maximum affordable costs to share less cost in a coalition. We need to examine the relations between the mechanism design and agents' strategies to effectively solve the task allocation problem when agents are self-interested and strategic.

Acknowledgments

This research was supported in part by AFOSR grant F49620-01-1-0542 and by AFRL/MNK grant F08630-03-1-0005.

References

[1] Arkin, E. M. and Hassin, R., On local search for weighted k-set packing. In *Proceedings of the 7th Annual Europe Symposium on Algorithms*, 1999.

[2] Chandra, B. and Halldorsson, M. M., Greedy local improvement and weighted set packing approximation. In *Proceedings of the 10th Annual SIAM-ACM Symposium on Discrete Algorithms (SODA)*, 1999.

[3] Lerman, K. and Shehory, Coalition formation for large-scale electronic markets. In *Proceedings of the 4th International Conference on Multiagent Systems (ICMAS-2000).*

[4] Moulin, H., *Axioms of cooperative decision making.* Cambridge University Press, 1988.

[5] Moulin, H., *Cooperative Microeconomics: A Game-Theoretic Introduction.* Princeton University Press, 1995.

[6] Osborne, M. J. and Rubinstein, A., *A Course in Game Theory.* MIT Press, 1994.

[7] Sandholm, T., Larson, K., Andersson, M., Shehory, O., and Tohme, F., Coalition structure generation with worst case guarantees. *Artificial Intelligence*, 111(1-2):209–238, 1999.

[8] Shehory, O. and Kraus, S., Formation of overlapping coalitions for precedence-ordered task-execution among autonomous agents. In *Proceedings of the 2nd International Conference on Multiagent Systems (ICMAS-96)*, 1996.

[9] Shehory, O. and Kraus, S., Feasible formation of coalitions among autonomous agents in non-super-additive environments. *Computational Intelligence*, 15(3):218–251, 1999.

[10] Shehory, O., Sycara, K., and Jha, S., Multi-agent coordination through coalition formation. In Rao, A., Singh, M., and Wooldridge, M., editors, *Intelligent Agents IV (Lecture Notes in Artificial Intelligence no. 1365).* Springer-Verlag, 1997.

Appendix: Proof of Propositions

Proof of Prop.2. Suppose $\exists a_k \notin C_i^*, \exists a_h \in C_i^*$ such that $r_{ki} > r_{hi}$. From the definition of v_i, $v_i(C_i^* \cup \{a_k\} \backslash \{a_h\}) > v_i(C_i^*)$ holds, which contradicts the definition of C_i^* ($v_i(C_i^*)$ be the largest).

Lemma 1. For $\forall C \subset B$ and $\forall a_k \notin C$, if $h_C \leq r_{ki}$ then (1) $h_{C\cup\{a_k\}} \leq h_C$, and (2) $a_k \in \overline{C \cup \{a_k\}}$, where h_X and $\overline{X}$ for any X are calculated as a t_i coalition.

Proof of Lemma 1 (1). Suppose $h_{C\cup\{a_k\}} > h_C$, and we will show it leads to a contradiction, $cost_i(h_{C\cup\{a_k\}}) < cost_i(h_{C\cup\{a_k\}})$.

Let $D \stackrel{\text{def}}{=} C \cup \{a_k\}$. Then we have

$cost_i(D) \stackrel{\text{def}}{=} sum_{a_h \in C_i \backslash \overline{C_i}} r_{hi} + |\overline{D}| \cdot h_D$

$= \sum\{r_{hi} \mid a_h \in D, r_{hi} < h_D\} + |\overline{D}| \cdot h_D$

$> \sum\{r_{hi} \mid a_h \in D, r_{hi} < h_D\} + |\overline{D}| \cdot h_C$ (since $h_C < h_D$)

$= \sum\{r_{hi} \mid a_h \in D, r_{hi} < h_D\}$

$+ \sum\{r_{hi} \mid a_h \in D, h_C \leq r_{hi} < h_D\} + |\overline{D}| \cdot h_C$

$= \sum\{r_{hi} \mid a_h \in C, r_{hi} < h_C\} + \sum\{r_{hi} \mid a_h \in D, h_C \leq r_{hi} < h_D\}$

$+ |\{a_h \in D \mid \{h_D \leq r_{hi}\}| \cdot h_C$ (by $h_C \leq r_{ki}$ and Def. of $\overline{D}$)

$\geq \sum\{r_{hi} \mid a_h \in C, r_{hi} < h_C\} + |\{a_h \in D \mid h_C \leq r_{hi}\}| \cdot h_C$

$\geq \sum\{r_{hi} < h_C \mid a_h \in C\} + (|\overline{C}| + 1)h_C$

$= cost_i(C) + h_C$ (from Def. of $\overline{C}$)

$\geq |C| \cdot p_i(|C|) + p_i(|C|)$ (from Def. of $cost_i$ and $p_i(|C|) \leq h_C$)

$\geq |C| \cdot p_i(|D|) + p_i(|D|) = |D| \cdot p_i(|D|)$ (since $|D| = |C| + 1$)

$= cost_i(D)$.

Proof of Lemma 1 (2). From $h_C \leq r_{ki}$ and (1)$h_D \leq h_C$, we have $h_D \leq r_{ki}$, which means $a_k \in \overline{D} = \overline{C \cup \{a_k\}}$.

Lemma 2 (A general form of Lemma 1). For $\forall C \subset B$ and $\forall D \subset \{a_k \in B \mid h_C \leq r_{ki}\}$, (1)$h_{C\cup D} \leq h_C$, and (2)$D \subset \overline{C \cup D}$, where h_X and $\overline{X}$ for any X are calculated as a t_i coalition.

Proof of Lemma 2. The proof of Lemma 2 (1) is by induction on the cardinality of D. Begin with the first step. When $|D| = \{a_k\}$ and $a_k \in C$, (1) is trivial. If $a_k \notin C$, (1) is supported directly by Lemma 1. For the inductive step, suppose (1) holds for all D such that $|D| \leq n$, and we will show that (1) holds for $D \cup \{a_k\}$ where $a_k \notin D$ and $h_C \leq r_{ki}$. By the induction hypothesis, we have $h_{C\cup D} \leq h_C \leq r_{ki}$. In the case $a_k \notin C$, the above inequation and $a_k \notin C \cup D$ lead $h_{C\cup D\cup\{a_k\}} \leq h_{C\cup D}$ by using Lemma 1 (1). In the case $a_k \in C$, $h_{C\cup D\cup\{a_k\}} \leq h_{C\cup D}$ also holds since $C \cup D \cup \{a_k\} = C \cup D$. Using the induction hypothesis again, we have $h_{C\cup D\cup\{a_k\}} \leq h_C$. (2) follows trivial by (1).

Proof of Prop.3. Suppose $\exists C \in AC_i$ s.t. $h_C < h_{C_i^*}$ ... (1). By applying $\overline{C_i^*}$ to D in Lemma 2, we have $h_{C\cup\overline{C_i^*}} \leq h_C$ ... (2), and $\overline{C_i^*} \subset \overline{C \cup \overline{C_i^*}}$ ... (3). Using (1), (2) and (3), we see $v_i(C \cup \overline{C_i^*}) > C_i^*$ as follows, which contradicts that C_i^* be the largest by its definition.

$v_i(C \cup \overline{C_i^*}) = \sum_{a_k \in \overline{C\cup\overline{C_i^*}}} (r_{ki} - h_{C\cup\overline{C_i^*}})$

$> \sum_{a_k \in \overline{C\cup\overline{C_i^*}}} (r_{ki} - h_{C_i^*})$ (by combining (1) and (2))

$\geq \sum_{a_k \in \overline{C_i^*}}(r_{ki} - h_{C_i^*}) = v_i(C_i^*)$ (from (3)).

Lemma 3. For any coalition C_i and any subset $S \subset C_i$, $cost_i(S) \geq |S \cap \overline{C_i}| \cdot h_{C_i} + \sum_{a_k \in S \setminus \overline{C_i}} r_{ki}$.

Proof of Lemma 3. By Prop. 3, $h_{C_i} \leq h_S$...(1) holds. Then, the following two equations are straightforwardly proved using (1): $\overline{S} = \overline{S} \cap \overline{C_i}$, and $(S \setminus \overline{S}) \setminus \overline{C_i} = S \setminus \overline{C_i}$. Therefore,

$cost_i(S) = |\overline{S}| \cdot h_S + \sum_{a_k \in S \setminus \overline{S}} r_{ki}$

$= |\overline{S}| \cdot h_S + \sum_{a_k \in (S \setminus \overline{S}) \cap \overline{C_i}} r_{ki} + \sum_{a_k \in (S \setminus \overline{S}) \setminus \overline{C_i}} r_{ki}$

$\geq |\overline{S}| \cdot h_{C_i} + \sum_{a_k \in (S \setminus \overline{S}) \cap \overline{C_i}} h_{C_i} + \sum_{a_k \in (S \setminus \overline{S}) \setminus \overline{C_i}} r_{ki}$

$= |\overline{S} \cap \overline{C_i}| \cdot h_{C_i} + |(S \setminus \overline{S}) \cap \overline{C_i}| \cdot h_{C_i} + \sum_{a_k \in (S \setminus \overline{S}) \setminus \overline{C_i}} r_{ki}$

$= |S \cap \overline{C_i}| \cdot h_{C_i} + \sum_{a_k \in (S \setminus \overline{S}) \setminus \overline{C_i}} r_{ki}$

$= |S \cap \overline{C_i}| \cdot h_{C_i} + \sum_{a_k \in S \cap (C_i \setminus \overline{C_i})} r_{ki}$.

Proof of Prop.1. By Lemma 3 and the definition of group utility $v_i(S) \stackrel{\text{def}}{=} \sum_{a_k \in S} r_{ki} - cost_i(S)$, we have

$\sum_{a_k \in S} r_{ki} - v_i(S) \geq |S \cap \overline{C_i}| \cdot h_{C_i} + \sum_{a_k \in S \setminus \overline{C_i}} r_{ki}$.

Using Definition 2, this inequation yields

$v_i(S) \leq \sum_{a_k \in S} r_{ki} - \sum_{a_k \in S \setminus \overline{C_i}} r_{ki} - |S \cap \overline{C_i}| \cdot h_{C_i}$

$= \sum_{a_k \in S \cap \overline{C_i}} r_{ki} - |S \cap \overline{C_i}| \cdot h_{C_i}$

$= \sum_{a_k \in S \cap \overline{C_i}}(r_{ki} - h_{C_i}) = \sum_{a_k \in S} x_k$.

CHAPTER 11

COHESIVE BEHAVIORS OF MULTIPLE COOPERATIVE MOBILE DISCRETE-TIME AGENTS IN A NOISY ENVIRONMENT

Yanfei Liu
Department of Electrical Engineering
The Ohio State University
liu.336@osu.edu

Kevin M. Passino
Department of Electrical Engineering
The Ohio State University
passino@ee.eng.ohio-state.edu

Bacteria, bees and birds often work together in groups to find food. A group of robots can be designed to coordinate their activities. Networked cooperative UAVs are being developed for commercial and military applications. Suppose that we refer to all such groups of entities as "social foraging swarms." In order for such multi-agent systems to succeed it is often critical that they can both maintain cohesive behaviors and appropriately respond to environmental stimuli. In this chapter we focus on discrete-time case and use a Lyapunov approach to develop conditions under which local agent actions will lead to cohesive foraging even in the presence of sensing "noise." The results quantify earlier claims that social foraging is in a certain sense superior to individual foraging when noise is present, and provide clear connections between local agent-agent interactions and emergent group behavior.

Keywords:
Stability analysis, multiagent systems, discrete-time systems, biological systems, swarms, foraging

1. Introduction

Swarming has been studied extensively in biology [25, 4], and there is significant relevant literature in physics where collective behavior of "self-propelled particles" is studied. Swarms have also been studied in the context of engineering applications, particularly in collective robotics where there are teams of robots working together by communicating over a communication network [2, 27]. For example, the work in [26] on "social potential functions" is similar to how we view attraction-repulsion forces. Special types of swarms have been studied in "intelligent vehicle highway systems" [28] and in "formation control" for robots, aircraft, and cooperative control for uninhabited autonomous (air) vehicles. Early work on swarm stability is in [13, 3]. Also relevant is a study in [14] where the authors use virtual leaders and artificial potentials. Most work mentioned above is in continuous-time domain. Some work in discrete-time domain includes [15, 16, 17, 5, 19, 20], where the authors also consider asynchronism and time delays.

In this chapter, we continue some of our earlier work by studying stability properties of foraging swarms in [21, 22]. The main difference with our previous work is that we consider the effect of sensor errors ("noise") and errors in sensing the gradient of a "resource profile" (e.g., a nutrient profile) in the discrete-time case. We are able to show that even with noisy measurements the swarm can achieve cohesion and follow a nutrient profile in the proper direction. We illustrate that the agents can forage in noisy environments more efficiently as a group than individually, a principle that has been identified for some organisms [12, 18] and verified in [21, 22]. The work here builds on the work in (i) [6, 7] where the authors provide a class of attraction/repulsion functions and provide conditions for swarm stability (size of ultimate swarm size and ultimate behavior), and (ii) [8, 9, 10, 11] that represents progress in the direction of combining the study of aggregating swarms and how during this process decisions about foraging or threat avoidance can affect the collective/individual motion of the swarm/swarm members (i.e., typical characteristics influencing social foraging). Additional work on gradient climbing by swarms, including work on climbing noisy gradients, has been accomplished [1, 24]. There, similar to [8, 9], the authors study climbing gradients, but also consider noise effects and coordination strategies for climbing, something that we do not consider here.

The remainder of this chapter is organized as follows: In Section 2 we introduce a generic model for agents, interactions, and the foraging environment. Section 3 holds the main results on stability analysis of swarm cohesion. Section 4 holds the simulation results and some concluding remarks are provided in Section 5.

2. Basic Models

2.1. *Agents and Interactions*

Here, rather than focusing on the particular characteristics of one type of animal or autonomous vehicle we consider a swarm composed of an interconnection of N "agents," each of which has point mass dynamics given by

$$x^i((k+1)T) = x^i(kT) + v^i(kT)T$$
$$v^i((k+1)T) = v^i(kT) + \frac{1}{M_i}u^i(kT)T$$

where $x^i \in \Re^n$ is the position, $v^i \in \Re^n$ is the velocity, M_i is the mass, $u^i \in \Re^n$ is the control input for the ith agent, and T is the sampling time. To simplify notation, throughout the chapter we replace all "(kT)" with "(k)" whenever it does not lead to ambiguity. So we have

$$\begin{aligned} x^i(k+1) &= x^i(k) + v^i(k)T \\ v^i(k+1) &= v^i(k) + \frac{1}{M_i}u^i(k)T \end{aligned} \tag{1}$$

For some organisms like bacteria that move in highly viscous environments you can assume that $M_i = 0$ and if you use a velocity damping term in u^i for this you get the model studied in [6, 7, 8, 9, 10]. There, the authors view the choice of u^i as one that seeks to perform "energy minimization" which is consistent with other energy formalisms in mathematical biology. Here, we do not assume $M_i = 0$.

Agent to agent interactions considered here are of the "attract-repel" type where each agent seeks to be in a position that is "comfortable" relative to its neighbors (and for us all other agents are its neighbors). Attraction indicates that each agent wants to be close to every other agent and it provides the mechanism for achieving grouping and cohesion of the group of agents. Repulsion provides the mechanism where each agent does not want to be too close to any other agent (e.g., for animals to avoid collisions and excessive competition for resources). Attraction here will be represented in u^i in a form like $-k'\left(x^i - x^j\right)$ where $k' > 0$ is a scalar that represents the strength of attraction. For repulsion, we adopt 2-norm and use a repulsion term in u^i of the form

$$k_r \exp\left(\frac{-\frac{1}{2}\|x^i - x^j\|^2}{{r_s}^2}\right)\left(x^i - x^j\right) \tag{2}$$

where $k_r > 0$ and $r_s > 0$. Other types of attraction and repulsion terms are also possible.

2.2. *Environment Model*

Next, we will define the environment that the agents move in. While there are many possibilities, here we will simply consider the case where they move (forage) over a "resource profile" (e.g., nutrient profile) $J(x)$, where $x \in \Re^n$. Agents move in the direction of the negative gradient of $J(x)$ (i.e., in the direction of $-\nabla J(x) = -\frac{\partial J}{\partial x}$) in order to move away from "bad" areas and into "good" areas of the environment (e.g., to avoid noxious substances and find nutrients). So, a term that holds $\nabla J(x)$ will be used in u^i.

Clearly there are many possible shapes for $J(x)$, including ones with many peaks and valleys. Here, we list two simple forms for $J(x)$ as follows:

- *Plane:* In this case we have $J(x) = J_p(x)$ where

$$J_p(x) = R^\top x + r_p \tag{3}$$

  where $R \in \Re^n$ and r_p is a scalar. Here, $\nabla J_p(x) = R$.
- *Gaussian:* In this case we have $J(x) = J_g(x)$ where

$$J_g(x) = r_{m_1} \exp\left(-r_{m_2}\|x - R_c\|^2\right) + r_e$$

  where r_{m_1}, r_{m_2} and r_e are scalars, $r_{m_2} > 0$ and $R_c \in \Re^n$. Here, $\nabla J_g(x) = -2r_{m_1}r_{m_2} \exp\left(-r_{m_1}\|x - R_c\|^2\right)(x - R_c)$.

Below, we will study a family of profiles that is continuous with finite slope at all points.

3. Stability Analysis of Swarm Cohesion Properties

3.1. *Control and Error Dynamics*

Let $\bar{x}(k) = \frac{1}{N}\sum_{i=1}^N x^i(k)$ and $\bar{v}(k) = \frac{1}{N}\sum_{i=1}^N v^i(k)$ be the centroid position and velocity of the swarm at the kth time step, respectively. The objective of each agent is to move so as to end up at $\bar{x}$ and $\bar{v}$; in this way an emergent behavior of the group is produced where they aggregate dynamically and end up near each other and ultimately move in the same direction at nearly the same velocity (i.e., cohesion). Since all the agents are moving at the same time, $\bar{x}$ and $\bar{v}$ are time-varying. Hence, in order to study the stability of swarm cohesion, we study the dynamics of an error system with $e_p^i(k) = x^i(k) - \bar{x}(k)$ and $e_v^i(k) = v^i(k) - \bar{v}(k)$. Then the error dynamics are given by

$$
\begin{aligned}
e_p^i(k+1) &= e_p^i(k) + e_v^i(k)T \\
e_v^i(k+1) &= e_v^i(k) + \frac{1}{M_i}u^i(k)T - \frac{1}{N}\sum_{j=1}^{N}\frac{1}{M_j}u^j(k)T
\end{aligned}
\tag{4}
$$

We assume that each agent can sense its position and velocity relative to $\bar{x}$ and $\bar{v}$, but with some errors. In particular, let $d_p^i \in \Re^n$ and $d_v^i \in \Re^n$ be these sensing errors for agent i, respectively. We assume that $d_p^i(k)$ and $d_v^i(k)$ are any trajectories that are fixed a priori and and bounded by a known constant for all the time steps. We will refer to these terms somewhat colloquially as "noise" but clearly our framework is entirely deterministic. Thus, each agent actually senses

$$
\begin{aligned}
\hat{e}_p^i(k) &= e_p^i(k) - d_p^i(k) \\
\hat{e}_v^i(k) &= e_v^i(k) - d_v^i(k)
\end{aligned}
$$

and below we will also assume that it can sense its own velocity.

We assume the nutrient profile is continuous with finite slope at all points, i.e., $\|\nabla J(x(k))\| \leq \Lambda$, where Λ is a known constant. We assume the ith agent senses $\nabla J\left(x^i(k)\right)$, the gradient of the profile at its position, but with some bounded error $d_f^i(k)$ that is fixed a priori for all the time steps (as with d_p^i and d_v^i we will allow below d_f^i to be any in a certain class of trajectories) so each agent actually senses $\nabla J\left(x^i(k)\right) - d_f^i(k)$. For simplicity, we will write $\nabla J\left(x^i(k)\right)$ as ∇J^i from now on.

Suppose the general form of the control input for each agent at the kth step is

$$
\begin{aligned}
u^i(k) = &-M_i k_p \hat{e}_p^i(k) - M_i k_v \hat{e}_v^i(k) - M_i k_d v^i(k) \\
&+ M_i k_r \sum_{l=1, l\neq i}^{N} \exp\left(\frac{-\frac{1}{2}\left\|\hat{e}_p^i(k) - \hat{e}_p^l(k)\right\|^2}{{r_s}^2}\right)\left(\hat{e}_p^i(k) - \hat{e}_p^l(k)\right) \\
&- M_i k_f \left(\nabla J^i(k) - d_f^i(k)\right)
\end{aligned}
\tag{5}
$$

Here, we think of the scalars $k_p > 0$ and $k_v > 0$ as the "attraction gains" which indicate how aggressive each agent is in aggregating. The gain $k_d > 0$ works as a "velocity damping gain". The gain $k_r > 0$ is a "repulsion gain" which sets how much that agent wants to be away from others and r_s represents its repulsion range. The gain $k_f > 0$ in Equation (5) indicates that agent's desire to move along the negative gradient of the resource profile. Note that by writing the repulsion term as in Equation (5), we are assuming each agent can also sense, with some noise, its position relative to each other agent. Another option to construct this repulsion term is to replace $\hat{e}_p^i - \hat{e}_p^j$ with $\hat{x}^i - \hat{x}^j$, with $\hat{x}^i$ and $\hat{x}^j$ defined as the noise-contaminated positions of agent i and j, respectively, and $\hat{x}^i - \hat{x}^j = x^i - x^j + d_p^{ij}$, d_p^{ij} being the measurement noise. In physical sense, these two options are significantly different from each other since different variables are required to be measured. But note that

$$\begin{aligned}\hat{e}_p^i - \hat{e}_p^j &= \left((x^i - \bar{x}) - d_p^i\right) - \left((x^j - \bar{x}) - d_p^j\right) \\ &= \left(\hat{x}^i - \hat{x}^j\right) - \left[d_p^{ij} + \left(d_p^i - d_p^j\right)\right]\end{aligned}$$

It turns out that we will obtain the same stability properties with either option. A quick explanation is that the repulsion term is bounded by the same constants (in both directions), whether we adopt $\hat{e}_p^i - \hat{e}_p^j$ or $\hat{x}^i - \hat{x}^j$. This will become more clear by inspecting the proof in the later sections. From now on, we will use the one in Equation (5) throughout the chapter. Obviously if $d_p^i = 0$ for all i, there is no sensing error on repulsion, and a repulsion term of the form explained in Equation (2) is obtained.

To study stability properties, we will substitute the above choice for u^i into the error dynamics in Equation (4). To calculate $e_v^i(k+1)$, first notice that

$$\begin{aligned}\frac{1}{M_i}u^i(k) &= -k_p e_p^i(k) + k_p d_p^i(k) - k_v e_v^i(k) + k_v d_v^i(k) - k_d v^i(k) \\ &\quad + k_r \sum_{l=1, l\neq i}^{N} \exp\left(\frac{-\frac{1}{2}\left\|\hat{e}_p^i(k) - \hat{e}_p^l(k)\right\|^2}{{r_s}^2}\right)\left(\hat{e}_p^i(k) - \hat{e}_p^l(k)\right) \\ &\quad - k_f\left(\nabla J^i(k) - d_f^i(k)\right)\end{aligned} \tag{6}$$

Then we have

$$\begin{aligned}\frac{1}{N}\sum_{j=1}^{N}\frac{1}{M_i}u^j(k) &= \frac{1}{N}\sum_{j=1}^{N} k_p d_p^j(k) + \frac{1}{N}\sum_{j=1}^{N} k_v d_v^j(k) - k_d\bar{v}(k) \\ &\quad - \frac{1}{N}\sum_{j=1}^{N} k_f\left(\nabla J^j(k) - d_f^j(k)\right)\end{aligned} \tag{7}$$

where we used the facts of $\sum_{j=1}^{N} e_p^j = 0$, $\sum_{j=1}^{N} e_v^j = 0$ and

$$\frac{1}{N}\sum_{j=1}^{N} k_r \sum_{l=1, l\neq j}^{N} \exp\left(\frac{-\frac{1}{2}\|\hat{e}_p^j(k) - \hat{e}_p^l(k)\|^2}{{r_s}^2}\right)\left(\hat{e}_p^j(k) - \hat{e}_p^l(k)\right) = 0.$$

Define $E^i = [{e_p^i}^\top, {e_v^i}^\top]^\top$ and $E = [{E^1}^\top, {E^2}^\top, \ldots, {E^N}^\top]^\top$. From Equations (4), (6) and (7) we have

$$e_v^i(k+1) = -k_p T e_p^i(k) + (1 - k_v T - k_d T)\, e_v^i(k) + g^i(k) + \phi(k) + \delta^i(E(k))$$

where

$$g^i(k) = k_p T d_p^i(k) + k_v T d_v^i(k) + k_f T d_f^i(k)$$

$$\phi(k) = -\frac{1}{N}\sum_{j=1}^{N} k_p T d_p^j(k) - \frac{1}{N}\sum_{j=1}^{N} k_v T d_v^j(k) - \frac{1}{N}\sum_{j=1}^{N} k_f T d_f^j(k)$$

$$\delta^i(E(k)) = k_r T \sum_{j=1, j\neq i}^{N} \exp\left(\frac{-\frac{1}{2}\left\|\hat{e}_p^i - \hat{e}_p^j\right\|^2}{{r_s}^2}\right)\left(\hat{e}_p^i - \hat{e}_p^j\right)$$
$$- k_f T \left(\nabla J^i(k) - \frac{1}{N}\sum_{j=1}^{N} \nabla J^j(k)\right)$$

which is a nonlinear non-autonomous system.

With I an $n \times n$ identity matrix, the error dynamics of the ith agent may be written as

$$E^i(k+1) = \overbrace{\begin{bmatrix} I & TI \\ -k_p T I & (1 - k_v T - k_d T)\, I \end{bmatrix}}^{A} E^i(k)$$
$$+ \overbrace{\begin{bmatrix} 0 \\ I \end{bmatrix}}^{B} \overbrace{\left(g^i(k) + \phi(k) + \delta^i(E(k))\right)}^{C^i(k)} \tag{8}$$

If we regard the whole swarm as an interconnected system with each agent being a subsystem, then matrix A in Equation (8) specifies the internal system dynamics for each agent/subsystem in the error system, and $C^i(k)$ gives the external input for each agent i at time step k.

Lemma 1: *The matrix A in Equation (8) is convergent if*

$$T < \begin{cases} \frac{4}{(k_v + k_d) + \sqrt{(k_v + k_d)^2 - 4k_p}} & \text{if } (k_v + k_d)^2 - 4k_p \geq 0 \\ \frac{k_v + k_d}{k_p} & \text{if } (k_v + k_d)^2 - 4k_p < 0 \end{cases} \tag{9}$$

Proof: It can be proven that matrix A has n repeated values of the eigenvalues of matrix $\tilde{A} = \begin{bmatrix} 1 & T \\ -k_p T & 1 - k_v T - k_d T \end{bmatrix}$.

From $zI - \tilde{A} = \begin{bmatrix} z-1 & -T \\ k_pT & z-1+(k_v+k_d)T \end{bmatrix}$, we can write out the characteristic equation as

$$z^2 + [(k_v + k_d)T - 2]z + 1 - (k_v + k_d)T + k_pT^2 = 0$$

Solving the equation gives the eigenvalues

$$z_{1,2} = \frac{1}{2}\left[2 - (k_v + k_d)T \pm T\sqrt{(k_v + k_d)^2 - 4k_p}\right]$$

To have a convergent A,

- If $(k_v + k_d)^2 - 4k_p \geq 0$, then we need

$$-1 < \frac{1}{2}\left[2 - (k_v + k_d)T \pm T\sqrt{(k_v + k_d)^2 - 4k_p}\right] < 1$$

  Notice that $z_{1,2} < 1$ always holds with $k_p > 0$. To have $-1 < z_{1,2}$, we need

$$-2 < 2 - (k_v + k_d)T - T\sqrt{(k_v + k_d)^2 - 4k_p}$$

  that is

$$T < \frac{4}{(k_v + k_d) + \sqrt{(k_v + k_d)^2 - 4k_p}}$$

- If $(k_v + k_d)^2 - 4k_p < 0$, we need $\|z_{1,2}\| < 1$. So

$$(2 - (k_v + k_d)T)^2 + T^2\left[4k_p - (k_v + k_d)^2\right] < 4$$

$$4 - 4(k_v + k_d)T + 4k_pT^2 < 4$$

  That is

$$0 < T < \frac{k_v + k_d}{k_p}$$

This completes the proof. ∎

From now on, we assume T is sufficiently small such that the condition in Lemma 1 holds.

Fact 1: When matrix A is convergent, for any given matrix $\tilde{Q} = \tilde{Q}^\top > 0$, there exists a unique matrix $\tilde{P} = \tilde{P}^\top > 0$ which is the solution of the discrete Lyapunov equation $A^\top \tilde{P} A - \tilde{P} = -\tilde{Q}$.

Definition 1: Given $\tilde{P}$ and $\tilde{Q}$ that satisfy the discrete Lyapunov equation above, define β_M and β_m, respectively, as twice the values of the maximum and minimum eigenvalues of $\tilde{P}$ given $\tilde{Q} = I$, i.e., $\beta_M = 2\lambda_{max}\left(\tilde{P}|_{\tilde{Q}=I}\right)$ and $\beta_m = 2\lambda_{min}\left(\tilde{P}|_{\tilde{Q}=I}\right)$.

Fact 2: With $\tilde{P}$, $\tilde{Q}$ and β_M defined above, the minimum of function $f(\tilde{P},\tilde{Q}) = \frac{2\lambda_{max}(\tilde{P})}{\lambda_{min}(\tilde{Q})}$ is β_M, that is,

$$\beta_M = \left.\frac{2\lambda_{max}(\tilde{P})}{\lambda_{min}(\tilde{Q})}\right|_{\tilde{Q}=I} = \min_{\tilde{Q}} f(\tilde{P},\tilde{Q})$$

3.2. *Uniform Ultimate Boundedness of Cohesive Social Foraging with Noise*

Our analysis methodology involves viewing the error system in Equation (8) as generating $E^i(k)$ trajectories for a given $E^i(0)$ and the fixed sensing error trajectories $d_p^i(k)$, $d_v^i(k)$, an $d_f^i(k)$, $k \geq 0$. We do not consider, however, all possible sensing error trajectories. We only consider a class of ones that satisfy for all k

$$\begin{aligned} \|d_f^i(k)\| &\leq D_f \\ \|d_p^i(k)\| &\leq D_{p_1}\left\|E^i(k)\right\| + D_{p_2} \\ \|d_v^i(k)\| &\leq D_{v_1}\left\|E^i(k)\right\| + D_{v_2} \end{aligned} \tag{10}$$

for any i, where D_{p_1}, D_{p_2}, D_{v_1}, D_{v_2} and D_f are known non-negative constants. So we assume for position and velocity the sensing errors have linear relationship with the magnitude of the state of the error system. Basically the assumption means that when two agents are far away from each other, the sensing errors can increase. The noise d_f^i on the nutrient profile is unaffected by the position of an agent. By considering only this class of fixed sensing error trajectories we prune the set of possibilities for E^i trajectories and it is only that pruned set that our analysis holds for.

In this section, we will show some results which characterize the stability properties of the swarm system in the presence of noise. To do this, we use a Lyapunov approach to develop conditions under which local agent actions will lead to cohesive foraging. Before we present our main results, two Lemmas are shown first.

Lemma 2: *For the error dynamics model described in Equation (8), if the noise satisfies Equation (10), then it holds that*

$$\sum_{i=1}^{N} \left\|C^i(k)\right\|^2 \leq \gamma_1^2 \sum_{i=1}^{N} \left\|E^i(k)\right\|^2 + 4\gamma_1\gamma_3 \sum_{i=1}^{N} \left\|E^i(k)\right\| + 3\gamma_1\gamma_2 \sum_{i=1}^{N} \left\|E^i(k)\right\| \sum_{j=1}^{N} \left\|E^j(k)\right\| + N\gamma_3^2 \quad (11)$$

where $\gamma_1 = k_pTD_{p_1} + k_vTD_{v_1}$, $\gamma_2 = \frac{\gamma_1}{N}$, $\gamma_3 = \left(2k_pD_{p_2} + 2k_vD_{v_2} + 2k_fD_f + (N-1)k_rT\exp\left(-\frac{1}{2}\right) r_s + 2k_f\Lambda\right) T$ *are constants.*

Proof: Notice that any function $F(\psi) = \exp\left(\frac{-\frac{1}{2}\|\psi\|^2}{{r_s}^2}\right)\|\psi\|$, with ψ any real vector, has a unique maximum value of $\exp\left(-\frac{1}{2}\right) r_s$ which is achieved when $\|\psi\| = r_s$ [6]. So

$$\left\|\exp\left(\frac{-\frac{1}{2}\left\|\hat{e}_p^i - \hat{e}_p^j\right\|^2}{{r_s}^2}\right)\left(\hat{e}_p^i - \hat{e}_p^j\right)\right\| \leq \exp\left(-\frac{1}{2}\right) r_s$$

Recall that we assume that the resource profile has finite slope, i.e., $\|\nabla J(x(k))\| \leq \Lambda$, then

$$\left\|\delta^i(E(k))\right\| \leq (N-1)k_rT\exp\left(-\frac{1}{2}\right) r_s + 2k_fT\Lambda$$

Thus, from Equation (10) we have

$$\begin{aligned}
\left\|C^i(k)\right\| &\leq \left\|g^i(k)\right\| + \|\phi(k)\| + \left\|\delta^i(E(k))\right\| \\
&\leq k_pT\left(D_{p_1}\left\|E^i(k)\right\| + D_{p_2}\right) + k_vT\left(D_{v_1}\left\|E^i(k)\right\| + D_{v_2}\right) \\
&+ k_fTD_f + k_pT\frac{1}{N}\sum_{j=1}^{N}\left(D_{p_1}\left\|E^j(k)\right\| + D_{p_2}\right) \\
&+ k_vT\frac{1}{N}\sum_{j=1}^{N}\left(D_{v_1}\left\|E^j(k)\right\| + D_{v_2}\right) + k_fT\frac{1}{N}\sum_{j=1}^{N} D_f \\
&+ (N-1)k_rT\exp\left(-\frac{1}{2}\right) r_s + 2k_fT\Lambda \\
&= \gamma_1\left\|E^i(k)\right\| + \gamma_2\sum_{j=1}^{N}\left\|E^j(k)\right\| + \gamma_3 \quad (12)
\end{aligned}$$

with γ_1, γ_2 and γ_3 as given in the statement of the lemma. Then

$$\left\|C^i(k)\right\|^2 \leq \gamma_1^2 \left\|E^i(k)\right\|^2 + \gamma_2^2 \left(\sum_{j=1}^{N} \left\|E^j(k)\right\|\right)^2 + \gamma_3^2 + 2\gamma_3\gamma_1 \left\|E^i(k)\right\|$$
$$+ 2\gamma_1\gamma_2 \left\|E^i(k)\right\| \sum_{j=1}^{N} \left\|E^j(k)\right\| + 2\gamma_2\gamma_3 \sum_{j=1}^{N} \left\|E^j(k)\right\|$$

So we have

$$\sum_{i=1}^{N} \left\|C^i(k)\right\|^2 \leq \gamma_1^2 \sum_{i=1}^{N} \left\|E^i(k)\right\|^2 + N\gamma_2^2 \left(\sum_{i=1}^{N} \left\|E^i(k)\right\| \sum_{j=1}^{N} \left\|E^j(k)\right\|\right)$$
$$+ N\gamma_3^2 + 2\gamma_1\gamma_2 \sum_{i=1}^{N} \left\|E^i(k)\right\| \sum_{j=1}^{N} \left\|E^j(k)\right\|$$
$$+ 2N\gamma_2\gamma_3 \left(\sum_{i=1}^{N} \left\|E^i(k)\right\|\right) + 2\gamma_3\gamma_1 \sum_{i=1}^{N} \left\|E^i(k)\right\|$$
$$= \gamma_1^2 \sum_{i=1}^{N} \left\|E^i(k)\right\|^2 + 4\gamma_1\gamma_3 \sum_{i=1}^{N} \left\|E^i(k)\right\|$$
$$+ 3\gamma_1\gamma_2 \sum_{i=1}^{N} \left\|E^i(k)\right\| \sum_{j=1}^{N} \left\|E^j(k)\right\| + N\gamma_3^2$$

where we used the fact that $\gamma_1 = N\gamma_2$. ■

Lemma 3: *Given an $L \times L$ matrix $\tilde{S}$ specified by*

$$\tilde{s}_{jk} = \begin{cases} -(\tilde{\sigma} + \tilde{a}), & j = n \\ -\tilde{a}, & j \neq n \end{cases} \tag{13}$$

where $\tilde{\sigma} < 0$ and $\tilde{a} > 0$. Then $\tilde{S} > 0$ ($\tilde{S} > 0$ is positive definite) if and only if $L\tilde{a} < -\tilde{\sigma}$.

Proof: A necessary and sufficient condition for $\tilde{S} > 0$ is that its eigenvalues are all positive. We have $\left|\lambda I - \tilde{S}\right|$

$$= \begin{vmatrix} \lambda + (\tilde{\sigma} + \tilde{a}) & \tilde{a} & \tilde{a} & \dots & \tilde{a} \\ \tilde{a} & \lambda + (\tilde{\sigma} + \tilde{a}) & \tilde{a} & \dots & \tilde{a} \\ \tilde{a} & \tilde{a} & \lambda + (\tilde{\sigma} + \tilde{a}) & \dots & \tilde{a} \\ \vdots & \vdots & \vdots & \ddots & \vdots \\ \tilde{a} & \tilde{a} & \tilde{a} & \dots & \lambda + (\tilde{\sigma} + \tilde{a}) \end{vmatrix}$$

$$= \begin{vmatrix} \lambda + (\tilde{\sigma} + \tilde{a}) & \tilde{a} & \tilde{a} & \dots & \tilde{a} \\ -(\lambda + \tilde{\sigma}) & \lambda + \tilde{\sigma} & 0 & \dots & 0 \\ -(\lambda + \tilde{\sigma}) & 0 & \lambda + \tilde{\sigma} & \dots & 0 \\ \vdots & \vdots & \vdots & \ddots & \vdots \\ -(\lambda + \tilde{\sigma}) & 0 & 0 & \dots & \lambda + \tilde{\sigma} \end{vmatrix}$$

$$= \begin{vmatrix} \lambda + \tilde{\sigma} + L\tilde{a} & \tilde{a} & \tilde{a} & \dots & \tilde{a} \\ 0 & \lambda + \tilde{\sigma} & 0 & \dots & 0 \\ 0 & 0 & \lambda + \tilde{\sigma} & \dots & 0 \\ \vdots & \vdots & \vdots & \ddots & \vdots \\ 0 & 0 & 0 & \dots & \lambda + \tilde{\sigma} \end{vmatrix}$$

$$= (\lambda + \tilde{\sigma} + L\tilde{a})\,(\lambda + \tilde{\sigma})^{L-1}$$

Since $-\tilde{\sigma} > 0$, to have all the eigenvalues positive we need $\tilde{\sigma} + L\tilde{a} < 0$, that is, $L\tilde{a} < -\tilde{\sigma}$. ■

The results of Lemma 2 and 3 will be used in the proof of our main result, which we present next.

Theorem 4: Consider the error dynamics model described in Equation (8) and assume the noise satisfies Equation (10). Let β_M be defined in Fact 2. If we have

$$k_p D_{p_1} + k_v D_{v_1} < \frac{1}{T}\left(\sqrt{\|A\|^2 + \frac{2}{\beta_M}} - \|A\|\right) \tag{14}$$

and there exists some constant $0 < \theta < 1$ such that

$$k_p D_{p_1} + k_v D_{v_1} < \frac{1}{T}\left(\frac{\sqrt{(2-\theta)^2\|A\|^2 + \frac{2(4-\theta)(1-\theta)}{\beta_M}}}{4-\theta} - \frac{2-\theta}{4-\theta}\|A\|\right) \tag{15}$$

then the trajectories of the error system are uniformly ultimately bounded (UUB).

Proof: To study the stability of the error dynamics, it is convenient to choose a Lyapunov function for each agent as

$$V_i(k) = E^i(k)^\top P E^i(k) \tag{16}$$

with $P = P^\top > 0$ a $2n \times 2n$ positive definite matrix. Then we have

$$\begin{aligned} V_i(k+1) &= E^i(k+1)^\top P E^i(k+1) \\ &= E^i(k)^\top A^\top P A E^i(k) + 2C^i(k)^\top B^\top P A E^i(k) \\ &+ C^i(k)^\top B^\top P B C^i(k) \end{aligned}$$

So

$$\begin{aligned} \Delta V_i(k) &= V_i(k+1) - V_i(k) \\ &= E^i(k)^\top \underbrace{\left(A^\top P A - P\right)}_{-Q} E^i(k) + 2C^i(k)^\top B^\top P A E^i(k) \\ &+ C^i(k)^\top B^\top P B C^i(k) \end{aligned} \tag{17}$$

Note that given any $Q = Q^\top > 0$, the existence of a desired P is stated in Fact 1. Choose for the composite system

$$V(k) = \sum_{i=1}^{N} V_i(k)$$

where $V_i(k)$ is given in Equation (16). Since for any matrix $M = M^\top > 0$ and vector X

$$\lambda_{min}(M) X^\top X \leq X^\top M X \leq \lambda_{max}(M) X^\top X$$

where $\lambda_{min}(M)$ and $\lambda_{max}(M)$ denote the minimum and maximum eigenvalue of M, respectively, then we have

$$\sum_{i=1}^{N} \left(\lambda_{min}(P) \left\|E^i(k)\right\|^2\right) \leq V(k) \leq \sum_{i=1}^{N} \left(\lambda_{max}(P) \left\|E^i(k)\right\|^2\right) \tag{18}$$

Using Equations (11) and (12) from Lemma 2 and the fact that $\|B\| = 1$

we have

$$\begin{aligned}\Delta V(k) &= \sum_{i=1}^{N} \Delta V_i(k) \\ &\leq \sum_{i=1}^{N} \Big[-\lambda_{min}(Q) \left\|E^i(k)\right\|^2 + 2\lambda_{max}(P) \left\|C^i(k)\right\| \|A\| \left\|E^i(k)\right\| \\ &+ \lambda_{max}(P) \left\|C^i(k)\right\|^2 \Big] \\ &\leq \lambda_{min}(Q) \sum_{i=1}^{N} \left[-\left(1 - \beta'_M \gamma_1 \|A\| - \beta'_M \frac{\gamma_1^2}{2}\right) \left\|E^i(k)\right\|^2 \right. \\ &+ \beta'_M \gamma_3 \left(\|A\| + 2\gamma_1\right) \left\|E^i(k)\right\| + \frac{\beta'_M \gamma_3^2}{2} \\ &+ \left. \left\|E^i(k)\right\| \sum_{j=1}^{N} \beta'_M \gamma_2 \left(\|A\| + \frac{3}{2}\gamma_1\right) \left\|E^j(k)\right\| \right]\end{aligned}$$

where $\beta'_M = \frac{2\lambda_{max}(P)}{\lambda_{min}(Q)}$. By inspecting the above inequality, we can see that minimizing β'_M is desirable for achieving stability. Recall what is stated in Fact 2, we let $Q = I$ and thus, β'_M is minimized to β_M. Then with this choice of Q, we obtain

$$\begin{aligned}\Delta V(k) \leq \sum_{i=1}^{N} &\left[-c_1 \left\|E^i(k)\right\|^2 + c_2 \left\|E^i(k)\right\| \right. \\ &+ \left. \left\|E^i(k)\right\| \sum_{j=1}^{N} \left(a \left\|E^j(k)\right\|\right) \right] + c_3 \qquad (19)\end{aligned}$$

with c_1, c_2, c_3 and a constants and

$$\begin{aligned}c_1 &= 1 - \beta_M \gamma_1 \|A\| - \beta_M \frac{\gamma_1^2}{2} \\ c_2 &= \beta_M \gamma_3 \left(\|A\| + 2\gamma_1\right) \\ c_3 &= \frac{N \beta_M \gamma_3^2}{2} \\ a &= \beta_M \gamma_2 \left(\|A\| + \frac{3}{2}\gamma_1\right)\end{aligned}$$

Obviously $c_2 > 0$, $c_3 > 0$, and $a > 0$. To have $c_1 > 0$, we need

$$\frac{\beta_M}{2}\gamma_1^2 + \beta_M \|A\| \gamma_1 - 1 < 0. \qquad (20)$$

Solving this equation gives Equation (14).

Now, return to (19) and note that for any θ, $0 < \theta < 1$,

$$-c_1 \left\|E^i(k)\right\|^2 + c_2 \left\|E^i(k)\right\| \leq -(1-\theta)c_1 \left\|E^i(k)\right\|^2, \ \forall \ \left\|E^i(k)\right\| \geq r$$
$$= \sigma \left\|E^i(k)\right\|^2 \qquad (21)$$

where $r = \frac{c_2}{\theta c_1}$ and $\sigma = -(1-\theta)c_1 < 0$. This implies that as long as $\|E^i(k)\| \geq r$, the first two terms in Equation (19) combined will give a negative contribution to $\Delta V(k)$.

Next, we seek conditions under which $\Delta V(k) < 0$. To do this, we consider the third term in the brackets of Equation (19) and combine it with the above results. Note the general situation where some of the $E^i(k)$ are such that $\|E^i(k)\| < r$ and others are not. Accordingly, define sets

$$\Pi_O(k) = \left\{i : \left\|E^i(k)\right\| \geq r, \ i \in 1, \ldots, N\right\} = \left\{i_O^1, \ i_O^2, \ldots, \ i_O^{N_O(k)}\right\}$$

and

$$\Pi_I(k) = \left\{i : \left\|E^i(k)\right\| < r, \ i \in 1, \ldots, N\right\} = \left\{i_I^1, \ i_I^2, \ldots, \ i_I^{N_I(k)}\right\}$$

where $N_O(k)$ and $N_I(k)$ are the size of $\Pi_O(k)$ and $\Pi_I(k)$ at time step k, respectively. Also, $\Pi_O(k) \bigcup \Pi_I(k) = \{1, \ldots, N\}$ and $\Pi_O(k) \bigcap \Pi_I(k) = \phi$. Of course, we do not know the explicit sets $\Pi_O(k)$ and $\Pi_I(k)$; all we know is that they exist. For now, we assume $N_O(k) > 0$, that is, the set $\Pi_O(k)$ is non-empty. We will later discuss the $N_O(k) = 0$ case. Then using analysis ideas from the theory of stability of interconnected systems [23] and using Equations (19) and (21), we have

$$\begin{aligned}
\Delta V(k) \leq & \sum_{i \in \Pi_O(k)} \sigma \left\|E^i(k)\right\|^2 + \sum_{i \in \Pi_O(k)} \left(\left\|E^i(k)\right\| \sum_{j \in \Pi_O(k)} a \left\|E^j(k)\right\| \right) \\
& + \sum_{i \in \Pi_O(k)} \left(\left\|E^i(k)\right\| \sum_{j \in \Pi_I(k)} a \left\|E^j(k)\right\| \right) \\
& + \sum_{i \in \Pi_I(k)} \left(-c_1 \left\|E^i(k)\right\|^2 + c_2 \left\|E^i(k)\right\| \right) \\
& + \sum_{i \in \Pi_I(k)} \left(\left\|E^i(k)\right\| \sum_{j \in \Pi_O(k)} a \left\|E^j(k)\right\| \right) \\
& + \sum_{i \in \Pi_I(k)} \left(\left\|E^i(k)\right\| \sum_{j \in \Pi_I(k)} a \left\|E^j(k)\right\| \right) + c_3.
\end{aligned}$$

Note for each $N_O(k)$, with the corresponding $N_I(k) = N - N_O(k)$ there exist positive constants $K_1(N_I(k))$, $K_2(N_I(k))$ and $K_3(N_I(k))$ such that,

$$\begin{aligned} K_1(N_I(k)) &\geq \sum_{j\in\Pi_I(k)} a\left\|E^j(k)\right\| = \sum_{i\in\Pi_I(k)} \left\|E^i(k)\right\| a \\ K_2(N_I(k)) &\geq \sum_{i\in\Pi_I(k)} \left(-c_1\left\|E^i(k)\right\|^2 + c_2\left\|E^i(k)\right\|\right) \\ K_3(N_I(k)) &\geq \sum_{i\in\Pi_I(k)} \left(\left\|E^i(k)\right\| \sum_{j\in\Pi_I(k)} a\left\|E^j(k)\right\|\right) \end{aligned} \tag{22}$$

Then, we have

$$\begin{aligned} \Delta V(k) \leq & \sum_{i\in\Pi_O(k)} \sigma\left\|E^i(k)\right\|^2 + \sum_{i\in\Pi_O(k)} \left(\left\|E^i(k)\right\| \sum_{j\in\Pi_O(k)} a\left\|E^j(k)\right\|\right) \\ & + K_1 \sum_{i\in\Pi_O(k)} \left\|E^i(k)\right\| + K_2 + K_1 \sum_{j\in\Pi_O(k)} \left\|E^j(k)\right\| + K_3 + c_3 \\ = & \sum_{i\in\Pi_O(k)} \sigma\left\|E^i(k)\right\|^2 + \sum_{i\in\Pi_O(k)} \left(\left\|E^i(k)\right\| \sum_{j\in\Pi_O(k)} a\left\|E^j(k)\right\|\right) \\ & + \sum_{i\in\Pi_O(k)} 2K_1\left\|E^i(k)\right\| + K_2 + K_3 + c_3 \end{aligned}$$

Let $w(k)^\top = \left[\|E^{i_O^1}(k)\|, \|E^{i_O^2}(k)\|, \ldots, \|E^{i_O^{N_O(k)}}(k)\|\right]$ and the $N_O(k) \times N_O(k)$ matrix $S(k) = [s_{jk}]$ be specified by

$$s_{jk} = \begin{cases} -(\sigma + a), & j = n \\ -a, & j \neq n \end{cases}$$

so we have

$$\Delta V(k) \leq -w(k)^\top S(k) w(k) + \sum_{i\in\Pi_O(k)} 2K_1\left\|E^i(k)\right\| + K_2 + K_3 + c_3$$

From Lemma 3 we know that $S(k) > 0$ as long as $N_O(k)a < -\sigma$, while this holds if we have $Na < -\sigma$ since $N_O(k) \leq N$. In fact it can be proven that when Equation (15) holds, we have $Na < -\sigma$. This becomes clear when we write out $Na < -\sigma$ explicitly

$$N\beta_M\gamma_2\left(\|A\| + \frac{3}{2}\gamma_1\right) < (1-\theta)\left(1 - \beta_M\gamma_1\|A\| - \beta_M\frac{\gamma_1^2}{2}\right)$$

and solve the equation after manipulation

$$\frac{\beta_M(4-\theta)}{2}\gamma_1^2 + \beta_M\|A\|(2-\theta)\gamma_1 - (1-\theta) < 0$$

So when Equation (15) holds, we have $S(k) > 0$ and thus, $\lambda_{min}(S(k)) > 0$. Therefore

$$\Delta V(k) \leq -\lambda_{min}(S(k)) \sum_{i\in\Pi_O(k)} \left\|E^i(k)\right\|^2 + \sum_{i\in\Pi_O(k)} 2K_1 \left\|E^i(k)\right\| + K_2 + K_3 + c_3. \tag{23}$$

When the $\left\|E^i(k)\right\|$ for $i \in \Pi_O(k)$ are sufficiently large, the sign of $\Delta V(k)$ is determined by the term of $-\lambda_{min}(S(k))\sum_{i\in\Pi_O(k)}\left\|E^i(k)\right\|^2$ and $\Delta V(k) < 0$. This analysis is valid for any value of $N_O(k)$, $1 \leq N_O(k) \leq N$; hence for any $N_O(k) \neq 0$ the system is uniformly ultimately bounded.

To complete the proof, we need to consider the case when $N_O(k) = 0$. Note that when $N_O(k) = 0$, $\|E^i(k)\| < r$ for all i. If we have $N_O(k) = 0$ persistently, then we could simply take r as the uniform ultimate bound. If otherwise, at certain moment the system changes such that some $\|E^i(k)\| \geq r$, then we have $N_O(k) \geq 1$ immediately, then all the analysis above, which holds for any $1 \leq N_O(k) \leq N$, applies. Thus, in either case we obtain the uniform ultimate boundedness. This concludes the proof. ■

Remark 1: From Equations (14) and (15) we can see that it is the attraction gains k_p and k_v and damping gain k_d that determine if boundedness can be achieved for given parameters that quantify the size of the noise. Other parameters (k_r, r_s and k_f, etc.) do not affect the boundedness but only the bound.

Remark 2: On the noise side, it is the D_{p_1} and D_{v_1} that affect the uniform ultimate boundedness of the error system, while D_{p_2} and D_{v_2} do not. Note that when $D_{p_1} = D_{v_1} = 0$, Equations (14) and (15) are always satisfied, meaning when noise is constant or with constant bound, the trajectories of the error system are always UUB.

3.3. *Special Case: Constant-Bound Noise and Plane Profile*

In this section we assume the resource profile for each agent is a plane profile defined by $\nabla J^i(k) = R^i$, as seen in Equation (3). Also we assume that $d_p^i(k)$ and $d_v^i(k)$ are bounded by some constants for all i,

$$\|d_p^i\| \leq D_p$$
$$\|d_v^i\| \leq D_v \tag{24}$$

where $D_p \geq 0$ and $D_v \geq 0$ are known constants. The sensing error on the gradient of the nutrient profile is assumed to be bounded by known constant $D_f \geq 0$ such that for all i,

$$\|d_f^i\| \leq D_f \tag{25}$$

Theorem 5: Consider the error system described by the model in Equation (4). Assume the noise satisfies Equations (24) and (25). Let $\Lambda' = max_{1\leq i\leq N} \|R^i\|$. Then, the trajectories of the swarm error system are uniformly ultimately bounded, and E^i for all i will converge to the set Ω_b, where

$$\Omega_b = \left\{E: \ \|E^i\| \leq \frac{\gamma_3'}{2}\left(\beta_M + \sqrt{\|A\|^2\beta_M^2 + 2\beta_M}\right), \ i = 1, 2, \ldots, N\right\} \tag{26}$$

is attractive and compact, with β_M defined in Fact 2 and

$$\gamma_3' = \left(2k_pD_p + 2k_vD_v + 2k_fD_f + (N-1)k_rT\exp\left(-\frac{1}{2}\right)r_s + 2k_f\Lambda'\right)T \tag{27}$$

Moreover, the centroid velocity of the swarm $\bar{v}$ is uniformly ultimately bounded if we have

$$T < \frac{2}{k_d} \tag{28}$$

and $\bar{v}$ will converge to the set $\Omega_v = \{\bar{v}: \ \|v\| \leq v_M\}$, where

$$v_M = \begin{cases} \frac{\tau}{k_d} & \text{if } T \leq \frac{1}{k_d} \\ \frac{\tau T}{2-k_dT} & \text{if } \frac{1}{k_d} < T < \frac{2}{k_d} \end{cases} \tag{29}$$

with $\tau = k_pD_p + k_vD_v + k_fD_f + k_f\Lambda'$.

Proof: To find the set Ω_b, we use the same idea as in the last section. Since now the noise has a constant bound, we have $\gamma_1 = 0$ and $\gamma_2 = 0$. So Equation (12) is changed into $\|C^i(k)\| \leq \gamma_3'$ with γ_3' given by Equation (27). From Equations (19), we have

$$\Delta V_i(k) \leq -\left\|E^i(k)\right\|^2 + \beta_M\gamma_3'\|A\|\left\|E^i(k)\right\| + \frac{\beta_M{\gamma_3'}^2}{2} \tag{30}$$

where we let $Q = I$ by following the idea in Theorem 4 to obtain the above equation. Solving the equation gives that $\Delta V_i(k) < 0$ when

$$\|E^i(k)\| > \frac{\gamma_3'}{2}\left(\beta_M + \sqrt{\|A\|^2\beta_M^2 + 2\beta_M}\right)$$

So the set

$$\Omega_b = \left\{E: \ \|E^i\| \leq \frac{\gamma_3'}{2}\left(\beta_M + \sqrt{\|A\|^2\beta_M^2 + 2\beta_M}\right), \ i = 1, 2, \ldots, N\right\}$$

is attractive and compact.

To study the boundedness of $\bar{v}(k)$, choose a Lyapunov function $V_{\bar{v}}(k) = \bar{v}(k)^\top \bar{v}(k)$. Since

$$\begin{aligned} \bar{v}(k+1) &= \bar{v}(k) + \frac{1}{N}\sum_{j=1}^{N}\frac{1}{M_i}u^j(k)T \\ &= (1 - k_dT)\bar{v}(k) + \underbrace{\left(k_p\bar{d}_p + k_v\bar{d}_v + k_f\bar{d}_f - k_f\bar{R}\right)T}_{d(k)} \end{aligned} \tag{31}$$

where $\bar{d}_p(k) = \frac{1}{N}\sum_{i=1}^{N} d_p^i$. Similarly we define $\bar{d}_v(k)$ and $\bar{d}_f(k)$. Also $\bar{R} = \frac{1}{N}\sum_{i=1}^{N} R^i$. Obviously $\|d(k)\| \leq \tau$. Then we have

$$\begin{aligned} \Delta V_{\bar{v}}(k) &= \bar{v}(k+1)^\top\bar{v}(k+1) - \bar{v}(k)^\top\bar{v}(k) \\ &\leq \underbrace{k_dT(k_dT - 2)\|v(k)\|^2 + 2\tau T|1 - k_dT|\|v(k)\| + \tau^2T^2}_{F(v)} \end{aligned}$$

Obviously we need $k_dT - 2 < 0$, that is, $T < \frac{2}{k_d}$. Furthermore, it can be solved that the maximum root of $F(v) = 0$ are

$$\begin{aligned} v_M &= \frac{-2\tau T|1 - k_dT| - \sqrt{4\tau^2T^2(1 - k_dT)^2 - 4k_dT(k_dT - 2)\tau^2T^2}}{2k_dT(k_dT - 2)} \\ &= \frac{\tau|1 - k_dT| + \tau}{k_d(2 - k_dT)} \end{aligned}$$

If $1 - k_dT \geq 0$, then $v_M = \frac{\tau}{k_d}$; if otherwise, $v_M = \frac{\tau T}{2 - k_dT}$. Since $\Delta V_{\bar{v}}(k) < 0$ when $\|v(k)\| > v_M$, we have the attractive and compact set $\Omega_v = \{\bar{v}: \ \|v\| \leq v_M\}$. ∎

Remark 3: The size of Ω_b in Equation (26), which we denote by $|\Omega_b|$, is a function of several known parameters. If there are no sensing errors, i.e., $D_p = D_v = D_f = 0$, then Ω_b reduces to the set representing the no-noise case. If we increase r_s or k_r while keeping all other parameters unchanged, then each agent has a stronger repulsion effect to its neighbors so $|\Omega_b|$ is larger. If we let $N \to \infty$, then $|\Omega_b| \to \infty$ as expected.

Remark 4: Comparing Equation (9) with (28) we can see that the boundedness of $\bar{v}$ and the convergency of the system matrix A are independent. That is, it is possible for an error system to have infinitely increasing norm $\|E^i\|$ while having $\bar{v}$ bounded, and vice versa. Also, from Equation (29) we can see that the bound of $\|\bar{v}\|$ is affected by the noise bounds and the gradient of the plane profiles. Larger profile gradients or noise bounds will lead to larger v_M. Also when T is smaller than certain value ($\frac{2}{T}$ in this case), then the ultimate bound of $\|v(k)\|$ does not change with T any more; otherwise, it is a function of T.

Remark 5: If there is no noise, from Equation (31) we can see that it is the "averaged" profile gradient $\bar{R}$ that changes the moving directions of all the agents. That is, due to the desire to stay together, they each sacrifice following their own profile and compromise to follow the averaged profile. Furthermore, note that in this case Equation (31) changes into

$$\bar{v}(k+1) = (1 - k_d T)\bar{v}(k) - k_f \bar{R} T$$

when Equation (28) is satisfied, using this equation recursively, we have $\bar{v}(k) = (1 - k_d T)^k \bar{v}(0) - \frac{k_f \bar{R}}{k_d}$. This means as k goes to infinity, $\bar{v}(k)$ converges to a constant $-\frac{k_f \bar{R}}{k_d}$.

Remark 6: In reality noise always exists, but in some cases when the swarm is large (N is big) it can be that $\bar{d}_p \approx \bar{d}_v \approx \bar{d}_f \approx 0$ and thus, the group will still be able to follow the proper direction (i.e., the averaged profile). In the case when $N = 1$ (i.e., single agent), there is no opportunity for a cancellation of the sensor errors; hence an individual may not be able to climb a noisy gradient as easily as a group. This may be a reason why large group size is favorable for some organisms and this characteristic has been found in biological swarms [12, 18].

4. Simulations

In this section, we will show some simulation results for both the no-noise and noise cases. Unless otherwise stated, in all the following simulations the parameters are: $N = 50$, $k_p = 1$, $k_v = 1$, $k_d = 0.1$, $k_f = 0.1$, $k_r = 1$, $r_s = 1$, and the three dimensional nutrient plane profile $\nabla J_p^i(x) = R^i = [2,\ 4,\ 6]^\top$ for all i.

4.1. *No-Noise Case*

All the simulations in this case are run for 20 seconds. The position and velocity trajectories of the swarm agents are shown in Figure 1. All the agents are assigned initial velocities and positions randomly. At the beginning of the simulation, they appear to move around erratically. But soon, they swarm together and continuously reorient themselves as a group to slide down the plane profile. Note how these agents gradually catch up with each other while still keeping mutual spacing. Recall from the previous section that for this case $\bar{v}(k) \to -\frac{k_f}{k_d}R$ as $k \to \infty$, and this can be seen from Figure 1(b) since the final velocity of each swarm agent is indeed $-[2,\ 4,\ 6]^\top$.

4.2. *Noise Case*

In this case, we run the simulations for 80 seconds. All the parameters used in the no-noise case are kept unchanged except the number of agents in the swarm in certain simulations, which is specified in the relevant figures. Figures 2 and 3 illustrate the case with linear noise bounds for a typical simulation run. The noise bounds are $D_{p_1} = D_{v_1} = 0.1$, $D_{p_2} = D_{v_2} = 3$, and $D_f = 30$, respectively. According to the "Grunbaum principle" [12, 18], forming a swarm may help the agents go down the gradient of the nutrient profile without being significantly distracted by noise. Figure 2 shows that the existence of noise does affect the swarm's ability to follow the profile, which is indicated by the oscillation of the position trajectories. But with all the agents working together, especially when the agents number N is large, they are able to move in the right direction and thus, minimize the negative effects of noise. In comparison, Figure 3 shows the case when there is only one agent. Since the single agent cannot benefit from the averaging effects possible when there are many agents, the noise more adversely affects its performance in terms of accurately following the nutrient profile.

5. Concluding Remarks

In this chapter we focused on a discrete-time formulation and derived stability conditions under which social foraging swarms maintain cohesiveness and follow a resource profile even in the presence of sensor errors and noise on the profile. Our simulations illustrated advantages of social foraging in large groups relative to foraging alone since they show that a noisy resource profile can be more accurately tracked by a swarm than an individual.

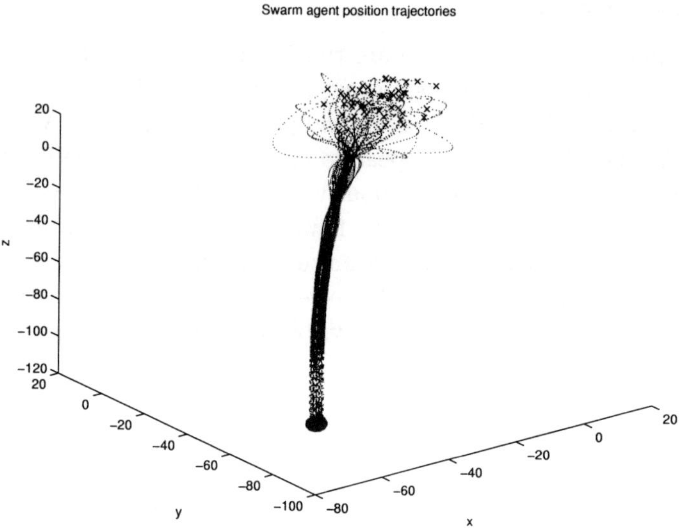

(a) Agent position trajectories.

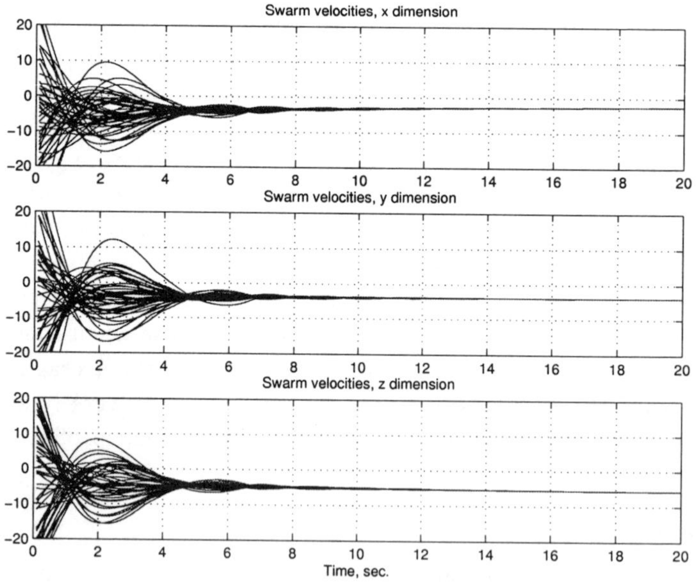

(b) Agent velocity trajectories.

Fig. 1. No noise case.

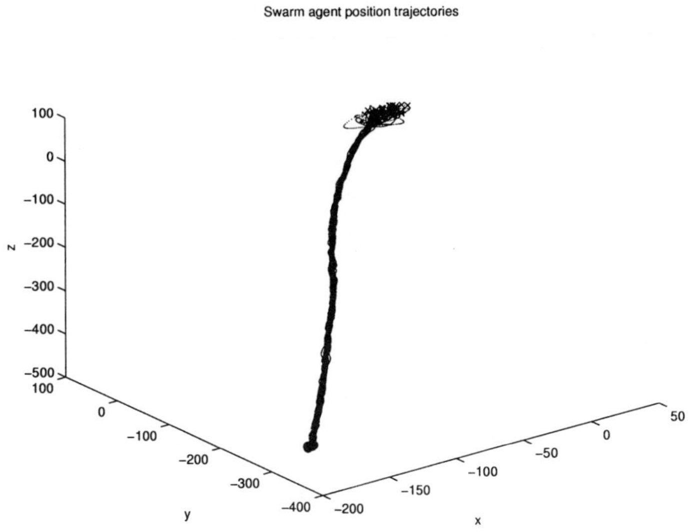

(a) Agent position trajectories.

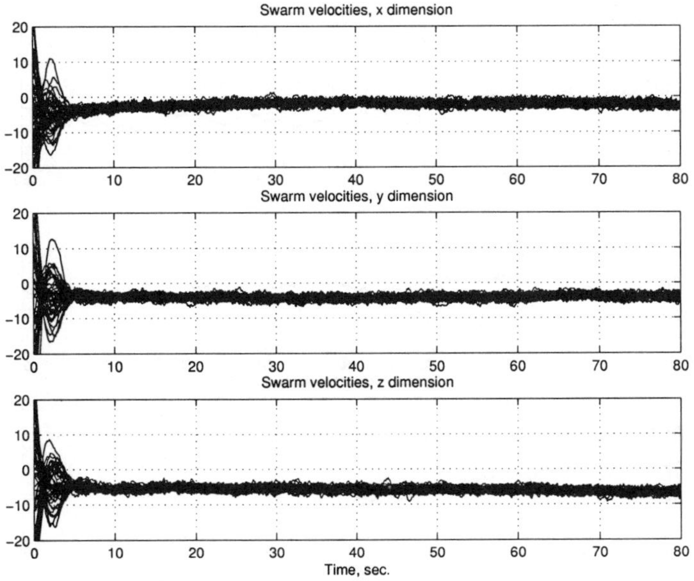

(b) Agent velocity trajectories.

Fig. 2. Linear noise bounds case ($N = 50$).

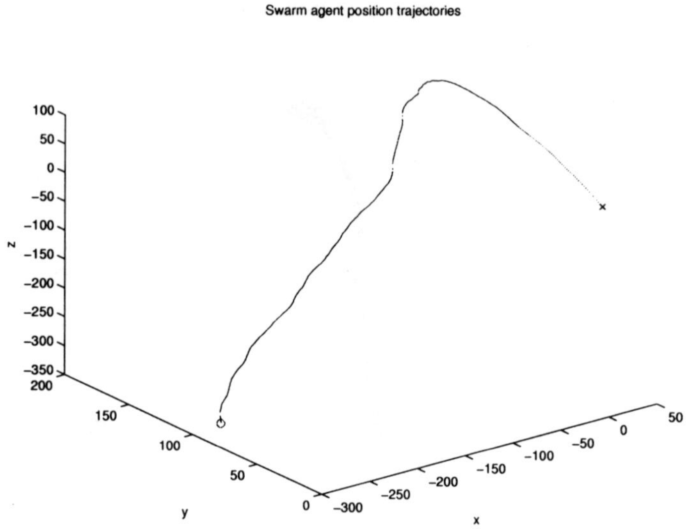

(a) Agent position trajectories.

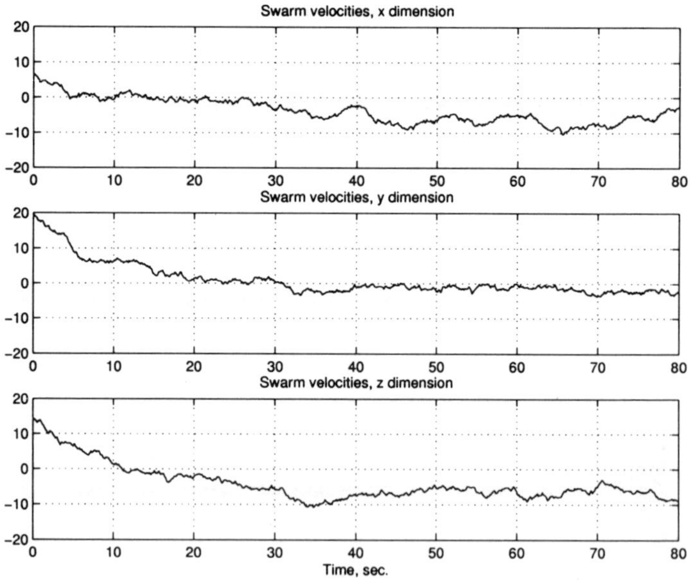

(b) Agent velocity trajectories.

Fig. 3. Linear noise bounds case ($N = 1$).

Acknowledgements

This work was supported by the DARPA MICA Program, via the Air Force Research Laboratory under Contract No. F33615-01-C-3151. This work was also supported in part by the AFRL/VA and AFOSR Collaborative Center of Control Science (Grant F33615-01-2-3154). Please address all correspondence to K. Passino, (614)-292-5716.

References

[1] R. Bachmayer and N. E. Leonard, "Vehicle networks for gradient descent in a sampled environment," in *Proc. of Conf. Decision Control*, (Las Vegas, Nevada), pp. 113–117, December 2002.

[2] T. Balch and R. C. Arkin, "Behavior-based formation control for multirobot teams," *IEEE Trans. on Robotics and Automation*, vol. 14, pp. 926–939, December 1998.

[3] G. Beni and P. Liang, "Pattern reconfiguration in swarms—convergence of a distributed asynchronous and bounded iterative algorithm," *IEEE Trans. on Robotics and Automation*, vol. 12, pp. 485–490, June 1996.

[4] L. Edelstein-Keshet, *Mathematical Models in Biology*. Brikhäuser Mathematics Series, New York: The Random House, 1989.

[5] V. Gazi and K. M. Passino, "Stability of a one-dimensional discrete-time asynchronous swarm," in *Proc. of the joint IEEE Int. Symp. on Intelligent Control/IEEE Conf. on Control Applications*, (Mexico City, Mexico), pp. 19–24, September 2001.

[6] V. Gazi and K. M. Passino, "Stability analysis of swarms," in *Proc. American Control Conf.*, (Anchorage, Alaska), pp. 1813–1818, May 2002.

[7] V. Gazi and K. M. Passino, "A class of attraction/repulsion functions for stable swarm aggregations," in *Proc. of Conf. Decision Control*, (Las Vegas, Nevada), pp. 2842–2847, December 2002.

[8] V. Gazi and K. M. Passino, "Stability analysis of swarms in an environment with an attractant/repellent profile," in *Proc. American Control Conf.*, (Anchorage, Alaska), pp. 1819–1824, May 2002.

[9] V. Gazi and K. M. Passino, "Stability analysis of social foraging swarms: Combined effects of attractant/repellent profiles," in *Proc. of Conf. Decision Control*, (Las Vegas, Nevada), pp. 2848–2853, December 2002.

[10] V. Gazi and K. M. Passino, "Modeling and analysis of the aggregation and cohesiveness of honey bee clusters and in-transit swarms," *Submitted for publication*, 2002.

[11] V. Gazi and K. M. Passino, "Stability analysis of social foraging swarms," *To appear, IEEE Trans. on Systems, Man, and Cybernetics*, 2004.

[12] D. Grunbaum, "Schooling as a strategy for taxis in a noisy environment," *Evolutionary Ecology*, vol. 12, pp. 503–522, 1998.

[13] K. Jin, P. Liang, and G. Beni, "Stability of synchronized distributed control of discrete swarm structures," in *Proc. of IEEE International IEEE Confer-*

ence on Robotics and Automation, (San Diego, California), pp. 1033–1038, May 1994.

[14] N. E. Leonard and E. Fiorelli, "Virtual leaders, artificial potentials and coordinated control of groups," in *Proc. of Conf. Decision Control*, (Orlando, FL), pp. 2968–2973, December 2001.

[15] Y. Liu, K. M. Passino, and M. Polycarpou, "Stability analysis of one-dimensional asynchronous swarms," in *Proc. American Control Conf.*, (Arlington, VA), pp. 716–721, June 2001.

[16] Y. Liu, K. M. Passino, and M. Polycarpou, "Stability analysis of one-dimensional asynchronous mobile swarms," in *Proc. of Conf. Decision Control*, (Orlando, FL), pp. 1077–1082, December 2001.

[17] Y. Liu, K. M. Passino, and M. M. Polycarpou, "Stability analysis of m-dimensional asynchronous swarms with a fixed communication topology," in *Proc. American Control Conf.*, (Anchorage, Alaska), pp. 1278–1283, May 2002.

[18] Y. Liu and K. M. Passino, "Biomimicry of social foraging behavior for distributed optimization: Models, principles, and emergent behaviors," *Journal of Optimization Theory and Applications*, vol. 115, pp. 603–628, Dec. 2002.

[19] Y. Liu, K. M. Passino, and M. M. Polycarpou, "Stability analysis of one-dimensional asynchronous swarms," *IEEE Transactions on Automatic Control*, vol. 48, no. 10, pp. 1848–1854, 2003.

[20] Y. Liu, K. M. Passino, and M. M. Polycarpou, "Stability analysis of M-dimensional asynchronous swarms with a fixed communication topology," *IEEE Transactions on Automatic Control*, vol. 48, no. 1, pp. 76–95, 2003.

[21] Y. Liu and K. M. Passino, "Stable social foraging swarms in a noisy environment," *IEEE Transactions on Automatic Control*, vol. 49, no. 1, 2004.

[22] Y. Liu and K. M. Passino, "Stability analysis of swarms in a noisy environment," in *Proc. of Conf. Decision Control*, (Maui, Hawaii), pp. 3573–3578, December 2003.

[23] A. N. Michel and R. K. Miller, *Qualitative Analysis of Large Scale Dynamical Systems*. New York: Academic Press, 1977.

[24] P. Ögren, E. Fiorelli, and N. E. Leonard, "Formations with a mission: Stable coordination of vehicle group maneuvers," *Proc. Symposium on Mathematical Theory of Networks and Systems*, August 2002.

[25] J. Parrish and W. Hamner, eds., *Animal Groups in Three Dimensions*. Cambridge, England: Cambridge Univ. Press, 1997.

[26] J. H. Reif and H. Wang, "Social potential fields: A distributed behavioral control for autonomous robots," *Robotics and Autonomous Systems*, vol. 27, pp. 171–194, 1999.

[27] I. Suzuki and M. Yamashita, "Distributed anonymous mobile robots: Formation of geometric patterns," *SIAM Journal on Computing*, vol. 28, no. 4, pp. 1347–1363, 1999.

[28] D. Swaroop, *String Stability of Interconnected systems: An Application to Platooning in Automated Highway Systems*. PhD thesis, Departnent of Mechanical Engineering, University of California, Berkeley 1995.

CHAPTER 12

MULTITARGET SENSOR MANAGEMENT OF DISPERSED MOBILE SENSORS

Ronald Mahler
Lockheed Martin Tactical Systems

The work described in this chapter is directed at a theoretically foundational but potentially practical control-theoretic basis for multisensor-multitarget sensor management using a comprehensive, intuitive, system-level Bayesian paradigm. Our approach is based on the following steps: (1) use point process theory to formulate all sensors and targets as a single joint dynamically evolving stochastic system; (2) propagate the state of this system using a multisensor-multitarget Bayes filter; (3) apply suitable objective functions that express global probabilistic goals; (4) apply suitable optimization strategies that hedge against the unknowability of future observation-collections; and (5) devise principled approximations of this general (but usually intractable) formulation. This chapter employs a new objective function and optimization-hedging strategy to generalize our previous results. Our refined approach now permits: preferential observation of targets of interest (ToIs); multistep look-ahead; non-ideal sensor dynamics; and modeling of communication drop-outs. It also addresses the dilemma of choosing among an infinitude of plausible objective functions, by focusing on "probabilistically natural" goals of sensor management.

1. Introduction

Sensor management is inherently an *optimal control problem*, albeit a very large, complex, and nonlinear one. On the one hand, data collected by various sources must be fused and interpreted to provide tactically useful information about targets of interest (ToIs). On the other hand, re-allocatable sources must be directed to optimize collection of useful information, both current and anticipated. These two processes—data collection and interpretation versus sensor coordination and control—should be tightly connected by a control-theoretic feedback loop that allows existing collections and anticipated future sensing and target conditions to influence the choice of future collections. Sensor management differs from standard control applications in that it is also inherently a *stochastic multi-object problem.* It involves *randomly varying sets of targets, randomly varying sets of sensors/sources, randomly varying sets of collected data*, and *randomly varying sets of sensor-carrying platforms.*

Like its predecessor at last year's International Conference on Cooperative Control and Optimization [14], this chapter describes current progress under a three-year basic research effort directed at a theoretically foundational but potentially practical control-theoretic basis for multisensor-multitarget sensor management using a comprehensive, intuitive, system-level Bayesian paradigm. It is based on the following steps:

- use point process theory / random set theory [2], [21] to formulate all sensors and targets as a single joint dynamically evolving stochastic system;
- propagate the state of this system using a joint multisensor-multitarget Bayes filter;
- apply suitable objective functions that express global probabilistic goals for sensor management;
- apply suitable optimization strategies that hedge against the inherent unknowability of future observation-collections; and
- devise principled approximations of this general (but usually intractable) formulation.

In particular, the last step means that we must devise principled, potentially tractable: multisensor-multitarget filters (MMFs); global objective functions (GOFs); and optimization-hedging strategies (OHSs).

1.1. *Summary of Previous Work*

In last year's work [14] we studied the following mix of approximations: MMFs = multi-hypothesis correlator (MHC) data fusion algorithms; GOFs = Csiszár information-theoretic functionals and their generalizations; and OHS = a "maxi-null" strategy [14], [10], [17]. The maxi-null strategy turned out to produce a too conservative prediction of the state of the future multitarget system, with the result that optimization based on it often did not perform well.

Our analysis uncovered a second, more subtle problem: *the impossibility of meaningfully deciding between an infinitude of plausible objective functions.* There are an infinite number of Csiszár and related objective functions. One could arbitrarily select a few candidates and compare them in a necessarily limited number of experiments. But how would we know that some unselected candidate would not be better still? Moreover, what reason is there to believe that optimizing *any* opaquely abstract information-concept will result in the fundamental goal of sensor management—sufficiently optimal collection of mission-relevant information?

We concluded that the only viable path out of this cul-de-sac is a *rigorous but intuitively sensible statistical formulation of "natural" sensor management goals.* Though there may be subsidiary goals, one minimal core "natural" objective should be to *maximize the number of well-resolved targets of interest.* But how does one precisely formulate objectives of this type in a statistically precise manner? The theory of finite-set statistics (FISST) [3], [15], [9], [16], [5] is key to answering this question. In our previous work we demonstrated the centrality of *probability generating functionals* (p.g.fl.'s) $G[h]$ and *multisensor probabilities of detection* $\breve{p}_D$ to the process of tractably integrating approximate multitarget filters with approximate optimization-hedging strategies and objective functions.

Towards this end, in [13] we studied a different mix of approximations: MMF = a *probability hypothesis density* (PHD) filter that propagates a first-order multitarget moment; GOF = posterior root-mean-square (RMS) expected number of targets; and OHS = maxi-mean. We derived a relatively simple approximate formula for the hedged objective function. Even so, the real-time computational tractability of this formula is doubtful because maxi-mean hedging requires the numerical evaluation of multidimensional integrals.

1.2. *Summary of Current Results*

The work described in this chapter corrects such deficiencies and generalizes our previous results. It is based on the following mix of approximations: MMFs = MHC or PHD filters (see sections 3.5 and 3.4); GOF = posterior expected number of targets (PENT, see section 4.4); and OHS = a new "maxi-PIMS" strategy (see section 4). Suppose that we want to determine the control-vector $\mathbf{u}_k$ at the current time-step k that will best position the field of view (FoV) of a single sensor at the next time-step $k+1$. The definition of PENT in this case has a precise statistical definition:

$$N_{k+1|k+1}(Z_{k+1}, \mathbf{u}_k) = \int |X| \cdot f_{k+1|k+1}(X|Z^{(k+1)})\delta X \tag{1}$$

where Z_{k+1} is the (unknowable) future observation-set; and where $f_{k+1|k+1}(X|Z^{(k)})$ is the multitarget posterior probability distribution at time-step $k+1$.

The new and potentially tractable "maxi-PIMS" optimization strategy hedges against the unknowability of future observation-sets such as Z_{k+1}. Intuitively speaking, we solve for those future placements of the sensor FoVs which will have the best chance of collecting the *predicted ideal measurement-set* (PIMS), denoted $\mathring{Z}_{k+1}$. This is the future observation-set that (1) contains no false alarms or clutter observations; and (2) contains a return from each target that is in the sensor FoVs, with each such return being uncontaminated by sensor noise. We select the control vector $\mathbf{u}_k$ as follows:

$$\mathring{\mathbf{u}}_k = \arg\sup_{\mathbf{u}} \mathring{N}_{k+1|k+1}(\mathbf{u}),\ \mathring{N}_{k+1|k+1}(\mathbf{u}_k) \stackrel{\text{def.}}{=} N_{k+1|k+1}(\mathring{Z}_{k+1}, \mathbf{u}_k) \qquad (2)$$

The maxi-PIMS strategy is not conservative in its modeling of the future multitarget system—if anything, it may be too optimistic. However, it allows us to greatly generalize our previous results; and preliminary simulations indicate that our refined approach results in good sensor management behavior. Our approach now encompasses:

- targets of current or potential tactical interest;
- multistep look-ahead (control of sensor resources throughout a future time-window);
- sensors with non-ideal dynamics, including sensors residing on moving platforms such as UAVs;
- sensors whose states are observed indirectly by internal actuator sensors; and
- possible communication drop-outs.

Our approach also:

- addresses the impossibility of deciding between an infinitude of plausible objective functions by concentrating on "probabilistically natural" core goals of sensor management, such as maximizing $\mathring{N}_{k+1|k+1}$.

Despite this progress, our work still has significant limitations (see section 9).

To illustrate our results, assume for the sake of clarity that there are no false alarms and that PENT is to be used with an MHC filter. Let $\hat{\mathbf{x}}_1, ..., \hat{\mathbf{x}}_N$ be the predicted target state-estimates produced by the MHC filter at time-step $k+1$; let $f_1(\mathbf{x}), ..., f_N(\mathbf{x})$ be their respective Gaussian track distributions; let $q_1, ..., q_N$ be the respective probabilities that these tracks exist; and let $f_j[h] \stackrel{\text{def.}}{=} \int h(\mathbf{x}) f_j(\mathbf{x}) d\mathbf{x}$. Then PENT has the following relatively simple formulas:

$$\mathring{N}_{k+1|k+1}(\overset{*}{\mathbf{x}}_{k+1}) = \sum_{j=1}^{N} (q_j f_j[1 - p_D] + p_D(\hat{\mathbf{x}}_j)) \tag{3}$$

$$\mathring{N}_{k+1|k+1}(\overset{*1}{\mathbf{x}}_{k+1}, \overset{*2}{\mathbf{x}}_{k+1}) = \sum_{j=1}^{N} (q_j f_j[1 - \breve{p}_D] + \breve{p}_D(\hat{\mathbf{x}}_i)) \tag{4}$$

$$\mathring{N}_{k+1|k+1}(\overset{*}{\mathbf{x}}_{k+1}, \overset{*}{\mathbf{x}}_{k+2}) = \sum_{j=1}^{N} (q_j f_j[1 - \mathring{p}_D] + \mathring{p}_D(\hat{\mathbf{x}}_j)) \tag{5}$$

$$\mathring{N}_{k+1|k+1}(\overset{*1}{\mathbf{x}}_{k+1}, \overset{*2}{\mathbf{x}}_{k+1}, \overset{*1}{\mathbf{x}}_{k+2}, \overset{*2}{\mathbf{x}}_{k+2}) = \sum_{j=1}^{N} (q_j f_j[1 - \mathring{p}_D] + \mathring{p}_D(\mathbf{x})) \tag{6}$$

The first equation (see Eq. (128)) addresses *single-sensor, single-step look-ahead*: p_D is the sensor field of view (FoV, Eq. (56)) and $\overset{*}{\mathbf{x}}_{k+1}$ is the next sensor state. The second equation (see Eq. (130)) addresses *two-sensor, single-step look-ahead*: $\breve{p}_D$ is the joint multisensor FoV (see Eq. (106)) and $\overset{*1}{\mathbf{x}}_{k+1}, \overset{*2}{\mathbf{x}}_{k+1}$ are the next sensor states. The third equation (see Eq. (141)) addresses *single-sensor, two-step look-ahead*: $\mathring{p}_D$ is the joint FoV in the two-step time window (see Eq. (138)), and $\overset{*}{\mathbf{x}}_{k+1}, \overset{*}{\mathbf{x}}_{k+2}$ are the sensor states in that window. The fourth equation (see Eq. (148)) addresses *two-sensor, two-step look-ahead*: $\mathring{p}_D$ is the joint FoV for both sensors in the two-step time window, and $\overset{*1}{\mathbf{x}}_{k+1}, \overset{*2}{\mathbf{x}}_{k+1}, \overset{*1}{\mathbf{x}}_{k+2}, \overset{*2}{\mathbf{x}}_{k+2}$ are the states for both sensors in that window.

Sensor management algorithms should be capable of directing sensing resources preferentially to targets of interest (ToIs)—i.e., to targets that have greater tactical importance than others. We extend the PENT objective function to include targets of interest as follows (see section 6). Rather than resorting to *ad hoc* techniques with inherent limitations, one should *integrally incorporate target preference into the fundamental statistical representation of multisensor-multitarget systems.* If a target has state $\mathbf{x}$, its relative tactical interest is expressed as the value of a function $0 \leq \rho(\mathbf{x}) \leq 1$. We show how such functions can be incorporated into the posterior p.g.f.l. using the formula $G^{\mathrm{ToI}}_{k+1|k+1}[h] = G_{k+1|k+1}[1 - \rho + h\rho]$ and, from there, into the PENT objective function. The resulting new objective function, the *posterior expected number of ToIs* (PENTI), preferentially directs sensor resources towards targets of current or potential tactical interest. For example, the PENTI analog of Eq. (3) is Eq. (154):

$$\mathring{N}^{\text{ToI}}_{k+1|k+1}(\overset{*}{\mathbf{x}}_{k+1})$$
$$= \sum_{j=1}^{N} \left(q_j f_j[\rho(1-p_D)] + p_D(\hat{\mathbf{x}}_j) \cdot \frac{\sum_{e=1}^{N} q_e f_e[\rho p_D L_{\eta(\hat{\mathbf{x}}_j)}]}{\sum_{e=1}^{N} q_e f_e[p_D L_{\eta(\hat{\mathbf{x}}_j)}]} \right) \tag{7}$$

We similarly extend the PENT objective function to the case of sensors that have non-ideal dynamics, whose states are observed by internal actuator sensors, and which can be affected by communication drop-out problems (see section 7). For example, in this case the analog of Eq. (4) is Eq. (195):

$$\mathring{N}_{k+1|k+1}(\mathbf{u}_k) = \sum_{j=1}^{N} \left(q_j (f_j \times \overset{*}{s})[1 - \overset{*}{p}_D p_D] + \overset{*}{p}_D(\overset{*}{\mathbf{x}}_0) \cdot p_D(\hat{\mathbf{x}}_j, \overset{*}{\mathbf{x}}_0) \right) \tag{8}$$

Here $\overset{*}{s}(\overset{*}{\mathbf{x}})$ is the distribution of the sensor state $\overset{*}{\mathbf{x}}$; $\overset{*}{\mathbf{x}}_0$ is the predicted sensor state; $\overset{*}{p}_D(\overset{*}{\mathbf{x}})$ is the communications FoV for the sensor; and $(h \times \overset{*}{h})[\eta] \overset{\text{def.}}{=} \int \eta(\mathbf{x}, \overset{*}{\mathbf{x}}) h(\mathbf{x}) \cdot \overset{*}{h}(\overset{*}{\mathbf{x}}) d\mathbf{x} d\overset{*}{\mathbf{x}}$.

1.3. *Organization of the Chapter*

We begin, in section 2, by specifying the mathematical foundations required to model sensor management problems. Section 3 describes our original core approach to sensor management, and our current refinement of it. The new maxi-PIMS optimization-hedging strategy is described in section 4. In the remaining sections we turn to the main results of the chapter. In section 5 we derive specific formulas for the PENT objective function for the following increasingly more complex stages: single-sensor with single-step look-ahead; multisensor with single-step look-ahead; single-sensor with multistep look-ahead; and multisensor with multistep look-ahead. In section 6 we show how to incorporate targets of interest (ToIs), resulting in another objective function, the posterior expected number of targets of interest (PENTI). In section 7, we show how to further extend our analysis to include actuator sensors, communication drop-outs, and non-ideal sensor dynamics. The more complicated mathematical proofs have been relegated to section 8, and conclusions may be found in section 9.

2. Modeling the Sensor Management Problem

The purpose of this section is to precisely specify the mathematical foundations of multisensor-multitarget sensor management. The section is organized as follows. We define joint multisensor-multitarget state space in section 2.1 and joint multisensor-multitarget measurement space in section 2.2. Integration on such spaces is summarized in section 2.3. Probability generating functionals (p.g.fl.'s) and their functional derivatives are described in sections 2.4 and 2.5, respectively. The first-order multitarget moment, the probability hypothesis density (PHD), is introduced in section 2.6. Section 2.7 describes the process of defining motion models and Markov transition densities for the joint multisensor-multitarget system. Section 2.8 repeats this discussion for measurement models and likelihood functions, including a detailed description of the multisensor-multitarget measurement model we will be assuming in the remainder of the chapter, particularly in section 7. The basic theoretical foundation for our sensor management approach, the joint multisensor-multitarget Bayes recursive filter, is described in section 2.9.

2.1. *State Spaces*

• **Single- and multi-target states:** The state space for single targets will be denoted $\mathfrak{X}$, with individual sensor states denoted as $\mathbf{x} \in \mathfrak{X}$. In general $\mathbf{x} = (\mathbf{x}_{\text{kin}}, c)$ where $\mathbf{x}_{\text{kin}}$ includes the kinematic state variables and c bundles together discrete state variables such as target class, target label, etc. We will assume that at least one kinematic state variable is continuous. The state of a multitarget system is modeled as a finite subset $X = \{\mathbf{x}_1, ..., \mathbf{x}_n\}$ of single-target states, with $n = 0, 1, ...$ The multitarget state space is the class of all such finite subsets, endowed with the Mathéron "hit-or-miss" topology (see p. 94 of [3] or p. 3 of [19]) and the induced Borel measure space, and is denoted by $\mathfrak{X}^\infty$.

• **Single- and multi-sensor states:** We assume that each sensor has associated with it a unique identifying sensor tag $j = 1, ..., s$. Once this tag is included as a state variable, the jth sensor will have its own unique state space $\overset{*j}{\mathfrak{X}}$, with individual sensor states denoted as $\overset{*j}{\mathbf{x}} \in \overset{*j}{\mathfrak{X}}$. (For example, assume a two-dimensional problem in which the sensor is on a platform that executes coordinated turns—i.e., the body frame axis of the platform is always tangent to the platform trajectory. Then we could have $\overset{*j}{\mathbf{x}} = (\overset{*}{x}, \overset{*}{y}, \overset{*}{v}_x, \overset{*}{v}_y, \overset{*}{\omega}, \overset{*}{\ell}, \mu, \chi, j)$ where $\overset{*}{x}, \overset{*}{y}$ are position parameters, $\overset{*}{v}_x, \overset{*}{v}_y$ are velocity parameters, $\overset{*}{\omega}$ is turn radius, $\overset{*}{\ell}$ is fuel level, μ is the sensor mode, χ is the datalink transmission channel currently used by the sensor, and j is the sensor tag.) If there are no more than s different sensors with respective state spaces $\overset{*1}{\mathfrak{X}}, ..., \overset{*s}{\mathfrak{X}}$, the joint state space for all sensors will be the topological sum (i.e., topologically disconnected disjoint union)

$$\overset{*}{\mathfrak{X}} = \overset{*1}{\mathfrak{X}} \uplus ... \uplus \overset{*s}{\mathfrak{X}} \tag{9}$$

We will write the state of a sensor with unidentified sensor tag as $\overset{*}{\mathbf{x}} \in \breve{\mathfrak{X}}$, so that a multisensor system will have state

$$\overset{*}{X} = \{\overset{*}{\mathbf{x}}_1, ..., \overset{*}{\mathbf{x}}_{\overset{*}{n}}\} \tag{10}$$

The space of all such multisensor states, endowed with the corresponding Mathéron topology, is denoted by $\overset{*}{\mathfrak{X}}^{\infty}$.

• **Joint multisensor-multitarget states:** The state of the joint multisensor-multitarget system is a finite subset of target and/or sensor states:

$$\breve{X} = \{\mathbf{x}_1, ..., \mathbf{x}_n, \overset{*}{\mathbf{x}}_1, ..., \overset{*}{\mathbf{x}}_{\overset{*}{n}}\} = X \cup \overset{*}{X} \tag{11}$$

This indicates that a particular multisensor-multitarget scene contains $n = 0, 1, ...$ targets and $\overset{*}{n} = 0, 1, ...$ sensors with their own respective types of states. In other words, a joint state is a finite subset of

$$\breve{\mathfrak{X}} = \mathfrak{X} \uplus \overset{*}{\mathfrak{X}} = \mathfrak{X} \uplus \overset{*1}{\mathfrak{X}} \uplus ... \uplus \overset{*s}{\mathfrak{X}} \tag{12}$$

The class of all such finite subsets, endowed with the induced Mathéron topology, is denoted as $\breve{\mathfrak{X}}^{\infty}$. We will denote the state of a target or a sensor as $\breve{\mathbf{x}} \in \breve{\mathfrak{X}}$, so that

$$\breve{X} = \{\breve{\mathbf{x}}_1, ..., \breve{\mathbf{x}}_{\breve{n}}\} \tag{13}$$

with $\breve{n} = n + \overset{*}{n}$.

2.2. *Measurement Spaces*

• **Single- and multi-sensor measurements of the targets:** We assume that any observation collected by a given sensor has that sensor's tag attached to it as an observation parameter. Consequently, the jth sensor will have its own unique measurement space $\overset{j}{\mathfrak{Z}}$, with individual measurements denoted as $\overset{j}{\mathbf{z}} \in \overset{j}{\mathfrak{Z}}$. So, the total single-sensor measurement space will be the topological sum

$$\mathfrak{Z} = \overset{1}{\mathfrak{Z}} \uplus ... \uplus \overset{s}{\mathfrak{Z}} \tag{14}$$

In general, the observation collected by whatever sensors might be present will be a finite subset of $\mathfrak{Z}$ of the form

$$Z = \{\overset{1}{\mathbf{z}}_{1,1}, ..., \overset{1}{\mathbf{z}}_{1,\overset{1}{m}}, ..., \overset{s}{\mathbf{z}}_{s,1}, ..., \overset{s}{\mathbf{z}}_{s,\overset{s}{m}}\} = \overset{1}{Z} \cup ... \cup \overset{s}{Z} \tag{15}$$

This indicates that the 1st sensor has collected $\overset{1}{m} = 0, 1, ...$ observations $\overset{1}{Z} = \{\overset{1}{\mathbf{z}}_{1,1}, ..., \overset{1}{\mathbf{z}}_{1,\overset{1}{m}}\}$, the 2nd sensor has collected $\overset{2}{m} = 0, 1, ...$ observations $\overset{2}{Z} = \{\overset{2}{\mathbf{z}}_{2,1}, ..., \overset{2}{\mathbf{z}}_{2,\overset{2}{m}}\}$, and so on. The set of all finite subsets of $\mathfrak{Z}$, endowed with the Mathéron topology, will be denoted by $\mathfrak{Z}^{\infty}$. We will denote a measurement with unidentified sensor tag as $\mathbf{z} \in \mathfrak{Z}$ and a finite subset of such observations as

$$Z = \{\mathbf{z}_1, ..., \mathbf{z}_m\} \tag{16}$$

where $m = \overset{1}{m} + ... + \overset{s}{m}$.

• **Single- and multi-sensor measurements of the sensors.** We assume that the states of the sensors cannot be known directly, but rather must be indirectly observed by internal actuator-sensors. We concatenate the observation-parameters for all actuator sensors for any given sensor into a single observation, along with the tag for the sensor. In this case the actuator sensor for the jth sensor will have its own unique measurement space $\overset{*j}{\mathfrak{Z}}$, with individual actuator-measurements denoted as $\overset{*j}{\mathbf{z}} \in \overset{*j}{\mathfrak{Z}}$. The total actuator-sensor measurement space is the topological sum

$$\overset{*}{\mathfrak{Z}} = \overset{*1}{\mathfrak{Z}} \uplus ... \uplus \overset{*s}{\mathfrak{Z}} \tag{17}$$

We will write a measurement collected by an actuator sensor with unidentified sensor tag as $\overset{*}{\mathbf{z}} \in \overset{*}{\mathfrak{Z}}$, and finite subsets of such observations as

$$\overset{*}{Z} = \{\overset{*}{\mathbf{z}}_1, ..., \overset{*}{\mathbf{z}}_m\} \tag{18}$$

The space of all such observation-sets is denoted as $\overset{*}{\mathfrak{Z}}^{\infty}$.

• **Joint multisensor-multitarget measurements:** Any measurement collected from the joint multisensor-multitarget system is a finite subset

$$\breve{Z} = \{\mathbf{z}_1, ..., \mathbf{z}_m, \overset{*}{\mathbf{z}}_1, ..., \overset{*}{\mathbf{z}}_{\overset{*}{m}}\} = Z \cup \overset{*}{Z} \tag{19}$$

This indicates that $m = 0, 1, ...$ measurements have been collected from the targets by the sensors; and $\overset{*}{m} = 0, 1, ...$ measurements from the sensors by the actuator sensors. In other words, a joint multisensor-multitarget observation is a finite subset of

$$\breve{\mathfrak{Z}} = \mathfrak{Z} \uplus \overset{*}{\mathfrak{Z}} = \overset{1}{\mathfrak{Z}} \uplus ... \uplus \overset{s}{\mathfrak{Z}} \uplus \overset{*1}{\mathfrak{Z}} \uplus ... \uplus \overset{*s}{\mathfrak{Z}} \tag{20}$$

We will write a measurement collected from a target or from a sensor as $\breve{\mathbf{z}} \in \breve{\mathfrak{Z}}$, so that

$$\breve{Z} = \{\breve{\mathbf{z}}_1, ..., \breve{\mathbf{z}}_{\breve{m}}\} \tag{21}$$

with $\breve{m} = m + \overset{*}{m}$. The space of all such joint observation-sets will be denoted $\breve{\mathfrak{Z}}^{\infty}$.

2.3. *Integrals*

• **Integration on joint single-object state space:** Functions defined on the joint single-target/sensor state space $\breve{\mathfrak{X}}$ have the form $\breve{h}(\breve{\mathbf{x}}) = \breve{h}(\mathbf{x})$ if $\breve{\mathbf{x}} = \mathbf{x}$; and $\breve{h}(\breve{\mathbf{x}}) = \breve{h}(\overset{*j}{\mathbf{x}})$ if $\breve{\mathbf{x}} = \overset{*j}{\mathbf{x}}$. In particular we will need the *joint Dirac delta function*

$$\breve{\delta}_{\breve{\mathbf{y}}}(\breve{\mathbf{x}}) \tag{22}$$

Note that $\breve{\delta}_{\mathbf{y}}(\overset{*i}{\mathbf{x}}) = 0$ and $\breve{\delta}_{\overset{*i}{\mathbf{y}}}(\mathbf{x}) = 0$ for all $i = 1, ..., s$ and $\breve{\delta}_{\overset{*i}{\mathbf{y}}}(\overset{*j}{\mathbf{x}}) = 0$ for all $i, j = 1, ..., s$ with $i \neq j$.

Integration on $\breve{\mathfrak{X}}$ on such functions has the form

$$\int \breve{h}(\breve{\mathbf{x}})d\breve{\mathbf{x}} \overset{\text{def.}}{=} \int \breve{h}(\mathbf{x})d\mathbf{x} + \int \breve{h}(\overset{*1}{\mathbf{x}})d\overset{*1}{\mathbf{x}} + ... + \int \breve{h}(\overset{*s}{\mathbf{x}})d\overset{*s}{\mathbf{x}} \tag{23}$$

• **Integration on joint single-object measurement space:** Functions defined on the joint single-sensor measurement space $\breve{\mathfrak{Z}}$ have the form $\breve{g}(\breve{\mathbf{x}}) = \breve{g}(\overset{j}{\mathbf{z}})$ if $\breve{\mathbf{z}} = \overset{j}{\mathbf{z}}$; and $\breve{g}(\breve{\mathbf{x}}) = \breve{g}(\overset{*j}{\mathbf{z}})$ if $\breve{\mathbf{z}} = \overset{*j}{\mathbf{z}}$. Integration on the joint single-object measurement space $\breve{\mathfrak{Z}}$ therefore has the form

$$\int \breve{g}(\breve{\mathbf{z}})d\breve{\mathbf{z}} = \int \breve{g}(\overset{1}{\mathbf{z}})d\overset{1}{\mathbf{z}} + ... + \int \breve{g}(\overset{s}{\mathbf{z}})d\overset{s}{\mathbf{z}} + \int \breve{g}(\overset{*1}{\mathbf{z}})d\overset{*1}{\mathbf{z}} + ... + \int \breve{g}(\overset{*s}{\mathbf{z}})d\overset{*s}{\mathbf{z}} \tag{24}$$

• **Integration on joint multi-object state space:** Let $\breve{f}(\breve{X})$ be a real-valued function of the finite-set variable $\breve{X}$. Then integration on the multi-object state space $\breve{\mathfrak{X}}^{\infty}$ is a "set integral" [3], [9]

$$\int_S \breve{f}(\breve{X})\delta\breve{X} \overset{\text{def.}}{=} \breve{f}(\emptyset) + \sum_{\breve{n}=1}^{\infty} \frac{1}{\breve{n}!} \int_{S^{\breve{n}}} \breve{f}(\{\breve{\mathbf{x}}_1, ..., \breve{\mathbf{x}}_{\breve{n}}\})d\breve{\mathbf{x}}_1 \cdots d\breve{\mathbf{x}}_{\breve{n}} \tag{25}$$

$$\int \breve{f}(\breve{X})\delta\breve{X} \overset{\text{def.}}{=} \breve{f}(\emptyset) + \sum_{\breve{n}=1}^{\infty} \frac{1}{\breve{n}!} \int \breve{f}(\{\breve{\mathbf{x}}_1, ..., \breve{\mathbf{x}}_{\breve{n}}\})d\breve{\mathbf{x}}_1 \cdots d\breve{\mathbf{x}}_{\breve{n}} \tag{26}$$

For purposes of integration, the quantity $\breve{f}(\{\breve{\mathbf{x}}_1, ..., \breve{\mathbf{x}}_{\breve{n}}\})$ must be treated as an ordinary function of $\breve{n}$ vector variables. By convention, we specify that

$$\breve{f}(\{\breve{\mathbf{x}}_1, ..., \breve{\mathbf{x}}_i, ..., \breve{\mathbf{x}}_j, ..., \breve{\mathbf{x}}_{\breve{n}}\}) = 0 \tag{27}$$

whenever $\mathbf{y}_i = \mathbf{y}_j$ for $i \neq j$; . (That is, no probability mass accrues to the finite set $\{\breve{\mathbf{x}}_1, ..., \breve{\mathbf{x}}_i, ..., \breve{\mathbf{x}}_j, ..., \breve{\mathbf{x}}_{\breve{n}}\}$ when $\mathbf{y}_i = \mathbf{y}_j$ for $i \neq j$, since any such mass should accrue to the $\breve{n} - 1$ term of the set integral.[a])

In what follows we will abbreviate

$$\breve{f}(\breve{X}) = \breve{f}(X \cup \mathring{X}) \overset{\text{abbr.}}{=} f(X, \mathring{X}) \tag{28}$$

In this case the set integral becomes

$$\begin{aligned}
&\int \breve{f}(\breve{X})\delta\breve{X} \\
&= \sum_{\breve{n}=0}^{\infty} \frac{1}{\breve{n}!} \int \breve{f}(\{\breve{\mathbf{x}}_1, ..., \breve{\mathbf{x}}_{\breve{n}}\}) d\breve{\mathbf{x}}_1 \cdots d\breve{\mathbf{x}}_{\breve{n}} \\
&= \sum_{\breve{n}=0}^{\infty} \sum_{n+\mathring{n}=\breve{n}} \frac{C_{n+\mathring{n},n}}{(n+\mathring{n})!} \int f(\{\mathbf{x}_1, ..., \mathbf{x}_n\}, \{\mathring{\mathbf{x}}_1, ..., \mathring{\mathbf{x}}_{\mathring{n}}\}) d\mathbf{x}_1 \cdots d\mathbf{x}_n d\mathring{\mathbf{x}}_1 \cdots d\mathring{\mathbf{x}}_{\mathring{n}} \\
&= \sum_{n,\mathring{n}=0}^{\infty} \frac{1}{n!\mathring{n}!} \int f(\{\mathbf{x}_1, ..., \mathbf{x}_n\}, \{\mathring{\mathbf{x}}_1, ..., \mathring{\mathbf{x}}_{\mathring{n}}\}) d\mathbf{x}_1 \cdots d\mathbf{x}_n d\mathring{\mathbf{x}}_1 \cdots d\mathring{\mathbf{x}}_{\mathring{n}} \\
&= \int f(X, \mathring{X}) \delta X \delta \mathring{X}
\end{aligned} \tag{29}$$

• **Integration on multi-object measurement space:** Let $\breve{g}(\breve{Z})$ be a real-valued function of the finite-set variable $\breve{Z}$. Then integration on the multi-object measurement space $\breve{\mathfrak{Z}}^\infty$ is also a set integral,

$$\int_S \breve{g}(\breve{Z})\delta\breve{Z} \overset{\text{def.}}{=} \breve{g}(\emptyset) + \sum_{\breve{m}=1}^{\infty} \frac{1}{\breve{m}!} \int_{\underbrace{S \times ... \times S}_{\breve{n}\ \text{times}}} \breve{g}(\{\breve{\mathbf{z}}_1, ..., \breve{\mathbf{z}}_{\breve{m}}\}) d\breve{\mathbf{z}}_1 \cdots d\breve{\mathbf{z}}_{\breve{m}} \tag{30}$$

As before, we will abbreviate

$$\breve{g}(\breve{Z}) = \breve{g}(Z \cup \mathring{Z}) \overset{\text{abbr.}}{=} g(Z, \mathring{Z}) \tag{31}$$

in which case, as before,

$$\int \breve{g}(\breve{Z})\delta\breve{Z} = \int g(Z, \mathring{Z})\delta Z \delta \mathring{Z} \tag{32}$$

[a] The functions $j_{\breve{n}}(\breve{\mathbf{x}}_1, ..., \breve{\mathbf{x}}_{\breve{n}}) = \breve{f}(\{\breve{\mathbf{x}}_1, ..., \breve{\mathbf{x}}_{\breve{n}}\})$ are known as the family of *Janossy densities*. Zero probability mass on the diagonals of the Janossy densities is a fundamental property of simple point processes. See Prop. 5.4.IV, p. 134 of [2], p. of [11], or [12].

2.4. *Probability Generating Functionals (p.g.fl.'s)*

Let $f(Y)$ be a multi-object probability density function defined on finite subsets Y of some space $\mathfrak{Y}$. If $\int f(Y)\delta Y = 1$ then $f(Y)$ can be interpreted as the probability distribution $f(Y) = f_\Psi(Y)$ of a random finite subset Ψ of $\mathfrak{Y}$. Given a measurable subset S of $\mathfrak{Y}$ let $\mathbf{1}_S(\mathbf{y})$ be the indicator function of S defined by $\mathbf{1}_S(\mathbf{y}) = 1$ if $\mathbf{y} \in S$ and $\mathbf{1}_S(\mathbf{y}) = 0$ otherwise. For any finite subset Y of $\mathfrak{Y}$ and any real-valued function $h(\mathbf{y})$ without units of measurement, define

$$h^Y = \begin{cases} 1 & \text{if} \quad Y = \emptyset \\ \prod_{\mathbf{y}\in Y} h(\mathbf{y}) & \text{if otherwise} \end{cases} \tag{33}$$

The *probability generating functional (p.g.fl.)* of Ψ or $f_\Psi(Y)$ (see pp. 141, 220 of [2]; [11]) is the expected value of the random real number h^Ψ:

$$G_\Psi[h] \stackrel{\text{def.}}{=} \mathrm{E}[h^\Psi] = \int h^Y \cdot f_\Psi(Y)\delta Y \tag{34}$$

The p.g.fl. is well-defined and finite-valued (see Eq. (202) section 8.1) if $h(\mathbf{y})$, called a "test function," has the form

$$h(\mathbf{y}) = h_0(\mathbf{y}) + w_1\delta_{\mathbf{w}_1}(\mathbf{y}) ... + w_n\delta_{\mathbf{w}_n}(\mathbf{y})$$

where: (1) $h_0(\mathbf{y})$ is some function without units of measurement such that $0 \leq h_0(\mathbf{y}) \leq 1$; (2) $\mathbf{w}_1, ..., \mathbf{w}_n$ are distinct elements of $\mathfrak{Y}$; and (3) $w_1, ..., w_n$ are nonnegative real numbers whose units of measurement are the same as $\mathbf{w}_1, ..., \mathbf{w}_n$.[b] The p.g.fl. is finite because, according to Eq. (27), $f(\{\mathbf{y}_1, ..., \mathbf{y}_i, ..., \mathbf{y}_j, ..., \mathbf{y}_m\}) = 0$ whenever $\mathbf{y}_i = \mathbf{y}_j$ for $i \neq j$. Therefore, undefined products of the form $\delta_{\mathbf{w}_i}(\mathbf{y})\delta_{\mathbf{w}_j}(\mathbf{y})$ cannot occur. Note that $G_\Psi[0] = f_\Psi(\emptyset)$, $G_\Psi[1] = 1$, and $0 \leq G_\Psi[h] \leq 1$ if $0 \leq h(\mathbf{y}) \leq 1$. If $h(\mathbf{y}) = \mathbf{1}_S(\mathbf{y})$ or $h(\mathbf{x}) = 1 - \mathbf{1}_T(\mathbf{x})$ where S is a closed subset and T an open subset of $\mathfrak{Y}$, then

$$G_\Psi[\mathbf{1}_S] = \beta_\Psi(S) = \Pr(\Psi \subseteq S) \tag{35}$$

$$1 - G_\Psi[1 - \mathbf{1}_T] = \pi_\Psi(T) = \Pr(\Psi \cap T \neq \emptyset) \tag{36}$$

[b] *ERRATUM:* In [14], the definition of the p.g.fl. was accidentally garbled by a typo which eliminated the coefficients $w_1, ..., w_n$. The definition given here is slightly more restrictive than that given in [14].

are the *belief-mass function* and *plausibility function* of Ψ, respectively.[c] One of the consequences of the *Choquet-Mathéron capacity theorem* (see p. 30 of [19], p. 96 of [3]) is that $\pi_\Psi(T)$, $\beta_\Psi(S)$, $G_\Psi[h]$, and $p_\Psi(O) = \Pr(\Psi \in O)$ are equivalent descriptions of the statistics of Ψ. Stated differently, $\pi_\Psi(T)$, $\beta_\Psi(S)$, and $G_\Psi[h]$ allow us to deal with finite random subsets Ψ using probability laws on conventional spaces with conventional topologies, rather than probability measures $p_\Psi(O)$ on an abstract probability space of subsets endowed with the Mathéron topology.

Eq. (35) provides an intuitive interpretation of $G_\Psi[h]$. Let $\mathfrak{Y} = \mathfrak{X}$ be single-target state space and $0 \leq h(\mathbf{y}) \leq 1$ for all $\mathbf{y}$, so that $h(\mathbf{y})$ is a fuzzy membership function on $\mathfrak{Y}$. Then $G_\Xi[h]$ can be regarded as an extension of $\beta_\Xi(S)$ from crisp sets S to fuzzy subsets h. In particular, let $\mathfrak{Y} = \mathfrak{X}$ be single-target state space, $\Psi = \Xi$ a random finite state-set, and $h = p_D$ the sensor probability of detection. Then $G_\Xi[p_D]$ can be interpreted as the probability that the random state-set Ξ is entirely contained within the sensor FoV p_D.

In what follows we will need the following results regarding p.g.fl.'s. First, from our discussion of integration it is clear that if $\breve{X} = X \cup \mathring{X}$ and $\breve{f}(\breve{X}) \overset{\text{def.}}{=} \breve{f}(X \cup \mathring{X}) = f(X, \mathring{X})$, then we can write

$$\breve{G}[\breve{h}] \overset{\text{def.}}{=} \int \breve{h}^{\breve{X}} \cdot \breve{f}(\breve{X}) = \int \breve{h}^{X} \cdot \breve{h}^{\mathring{X}} \cdot f(X, \mathring{X}) \delta X \delta \mathring{X} \tag{37}$$

2.5. *Functional Derivatives of p.g.fl.'s*

The *gradient derivatives* (Frechét derivatives) of a p.g.fl. $G[h]$ in the direction of the function g are

$$\begin{aligned} \frac{\partial G}{\partial g}[h] &\overset{\text{def.}}{=} \lim_{\varepsilon \to 0} \frac{G[h + \varepsilon g] - G[h]}{\varepsilon}, \\ \frac{\partial^n G}{\partial g_n \cdots \partial g_1}[h] &\overset{\text{def.}}{=} \frac{\partial}{\partial g_n} \frac{\partial^{n-1} G}{\partial g_{n-1} \cdots \partial g_1}[h] \end{aligned} \tag{38}$$

where the functional $g \longmapsto \frac{\partial G}{\partial g}[h]$ is assumed linear and continuous for each h. Gradient derivatives obey the usual "turn the crank" rules of undergraduate calculus, e.g. sum rule, product rule, etc. In physics, if

[c]This terminology arises from the Dempster-Shafer theory of evidence. If Ψ is a discrete random subset of $\mathfrak{Y}$ (i.e., $\Pr(\Psi = S) = 0$ for all but a finite number of S) and if $\Pr(\Psi = \emptyset) = 0$ then $m(S) = \Pr(\Psi = S)$ is a "basic mass assignment," $Bel_m(S) = \sum_{T \subseteq S} m(T) = \Pr(\Psi \subseteq S)$ is the "belief function" of m, and $Pl_m(S) = 1 - Bel_m(S^c) = \Pr(\Psi \cap S \neq \emptyset)$ is the "plausibility function" of m.

$g = \delta_{\mathbf{x}}$ then the gradient derivatives are known as *functional derivatives* (see pp. 173-174 of [20], pp. 140-141 of [11], or [12]). Using an abbreviated physics notation, write

$$\frac{\delta^n G_\Xi}{\delta \mathbf{x}_n \cdots \delta \mathbf{x}_1}[h] \stackrel{\text{def.}}{=} \frac{\partial^n G_\Xi}{\partial \delta_{\mathbf{x}_n} \cdots \partial \delta_{\mathbf{x}_1}}[h] \tag{39}$$

If $h = \mathbf{1}_S$ then the *set derivatives* of $\beta_\Xi(S)$ are [11], [13]:

$$\frac{\delta \beta_\Xi}{\delta \emptyset}(S) \stackrel{\text{def.}}{=} \beta_\Xi(S), \qquad \frac{\delta \beta_\Xi}{\delta \mathbf{x}}(S) \stackrel{\text{def.}}{=} \frac{\partial G_\Xi}{\partial \delta_{\mathbf{x}}}[\mathbf{1}_S] \tag{40}$$

$$\frac{\delta \beta_\Xi}{\delta X}(S) \stackrel{\text{def.}}{=} \frac{\delta^n \beta_\Xi}{\delta \mathbf{x}_n \cdots \delta \mathbf{x}_1}(S) = \frac{\partial^n G_\Xi}{\partial \delta_{\mathbf{x}_n} \cdots \partial \delta_{\mathbf{x}_1}}[\mathbf{1}_S] \tag{41}$$

for $X = \{\mathbf{x}_1, ..., \mathbf{x}_n\}$ with $\mathbf{x}_1, ..., \mathbf{x}_n$ distinct. The multitarget probability density function of Ξ is, therefore,

$$f_\Xi(X) = \frac{\delta \beta_\Xi}{\delta X}(\emptyset) = \frac{\delta^n G_\Xi}{\delta \mathbf{x}_n \cdots \delta \mathbf{x}_1}[0] \tag{42}$$

More generally, let G_Υ be the p.g.fl. of a random finite subset Υ of objects and $f_\Upsilon(Y)$ its multi-object probability density function. Let $r(\mathbf{y})$ be a unitless test function with $0 \leq r(\mathbf{y}) \leq 1$ for all $\mathbf{y}$. Then the following relationship between functional derivatives and set integrals is true (see section 8.1):[d]

$$\frac{\delta^n G_\Upsilon}{\delta \mathbf{y}_1 \cdots \delta \mathbf{y}_n}[r] = \int r^Y \cdot f_\Upsilon(Y \cup \{\mathbf{y}_1, ..., \mathbf{y}_n\}) \delta Y \tag{43}$$

In particular, note that if $r = 0$ then

$$\frac{\delta^n G_\Upsilon}{\delta \mathbf{y}_1 \cdots \delta \mathbf{y}_n}[0] = f_\Upsilon(\{\mathbf{y}_1, ..., \mathbf{y}_n\}) \tag{44}$$

and that if $r = 1$ and $Y = \{\mathbf{y}_1, ..., \mathbf{y}_n\}$,

$$m_\Upsilon(Y) \stackrel{\text{def.}}{=} \frac{\delta^n G_\Upsilon}{\delta \mathbf{y}_1 \cdots \delta \mathbf{y}_n}[1] = \int f_\Upsilon(W \cup \{\mathbf{y}_1, ..., \mathbf{y}_n\}) \delta W \tag{45}$$

The quantity $m_\Upsilon(Y)$ is called the *multitarget factorial moment density* of Υ [2, pp. 130, 122], [21, pp. 111,116], [11], [12].

---

[d] My thanks to Prof. Ba-Ngu Vo of the University of Melbourne, Australia, for sharing insights that led me to this formula [22], [4].

2.6. *Probability Hypothesis Densities*

In particular, if $n = 1$ and $r = 1$ then

$$D_\Upsilon(\mathbf{y}) \stackrel{\text{def.}}{=} D_\Upsilon(\{\mathbf{y}\}) = \frac{\delta G_\Upsilon}{\delta \mathbf{y}}[1] = \int f_\Upsilon(Y \cup \{\mathbf{y}\})\delta Y \tag{46}$$

is the *first-moment density* or *probability hypothesis density* (PHD) of Υ.

The PHD is characterized uniquely (almost everywhere) by the following property [12]: Its integral in any region S of state space is the expected number of objects in that region:

$$\int_S D_\Upsilon(\mathbf{y})d\mathbf{y} = \mathrm{E}[\ |S \cap \Upsilon|\ ] = \int |S \cap Y| \cdot f_\Upsilon(Y)\delta Y \tag{47}$$

Note that

$$\frac{\delta \log G_\Upsilon}{\delta \mathbf{y}}[1] = \left[\frac{\delta \log G_\Upsilon}{\delta \mathbf{y}}[h]\right]_{h=1} = \left[\frac{1}{G_\Upsilon[h]}\frac{\delta G_\Upsilon}{\delta \mathbf{y}}[h]\right]_{h=1} = D_\Upsilon(\mathbf{y}) \tag{48}$$

so that, in computing a PHD, one can use the often simpler functional $\log G_\Upsilon[h]$.

2.7. *Motion Models*

• **Single- and multi-target motion:** Individual target states are propagated between measurements using the Markov transition density $f_{k+1|k}(\mathbf{y}|\mathbf{x})$. The Markov transition density of the entire multitarget system has the form $f_{k+1|k}(Y|X)$. Generally speaking, it can be constructed from $f_{k+1|k}(\mathbf{y}|\mathbf{x})$ and from models for target appearance and/or disappearance using the techniques of finite-set statistics (see [9]). Since $\mathbf{y}, \mathbf{x}$ can contain information regarding target type or label, this means that different targets can have different motion models.

• **Single- and multi-sensor motion, with sensor controls:** Individual sensor states are propagated using the Markov densities

$$\overset{*1}{f}_{k+1|k}(\overset{*1}{\mathbf{y}}|\overset{*1}{\mathbf{x}}, \overset{1}{\mathbf{u}}_k) \quad ,..., \quad \overset{*s}{f}_{k+1|k}(\overset{*s}{\mathbf{y}}|\overset{*s}{\mathbf{x}}, \overset{s}{\mathbf{u}}_k)$$

where $\overset{j}{\mathbf{u}}_k$ is the *control vector* for the jth sensor at time-step k. In section 4 and thereafter in the chapter, we will assume that these Markov models have the additive form

$$\overset{*j}{f}_{k+1|k}(\overset{*j}{\mathbf{y}}|\overset{*j}{\mathbf{x}}, \overset{j}{\mathbf{u}}_k) = f_{\overset{*j}{\mathbf{V}}_k}(\overset{*j}{\mathbf{y}} - \overset{*j}{\varphi}_k(\overset{*j}{\mathbf{x}}, \overset{j}{\mathbf{u}}_k)) \tag{49}$$

Here $\overset{*j}{\mathbf{V}}_k$ is a zero-mean noise vector and the control $\overset{j}{\mathbf{u}}_k$ selects among a predetermined family $\overset{*j}{\mathbf{y}} = \overset{*j}{\varphi}_k(\overset{*j}{\mathbf{x}}, \overset{j}{\mathbf{u}})$ of deterministic sensor motion models, indexed by the control variable. These describe the possible range of motions of the sensor. For example, in linear control one typically presumes an affine family of models

$$\overset{*j}{\varphi}_k(\overset{*j}{\mathbf{x}}, \overset{j}{\mathbf{u}}) = \overset{*j}{F}_k \overset{*j}{\mathbf{x}} + \overset{*j}{E}_k \overset{j}{\mathbf{u}} \tag{50}$$

Remark 1: Strictly speaking, therefore, controls and sensor states always occur in pairs $(\overset{*j}{\mathbf{x}}, \overset{j}{\mathbf{u}})$. In what follows we will abuse notation by using pairs of the form $(\overset{*}{X}, U)$ to represent finite subsets of pairs $\{(\overset{*i_1}{\mathbf{x}}, \overset{i_1}{\mathbf{u}}_k), ..., (\overset{*i_e}{\mathbf{x}}, \overset{i_e}{\mathbf{u}}_k)\}$.

The Markov transition density for the entire multisensor system has the form $\overset{*}{f}_{k+1|k}(\overset{*}{Y}|\overset{*}{X}, U)$. Generally speaking, it can be constructed from the Markov transitions for the individual sensors, and from models for sensor appearance and disappearance, using the tools of finite-set statistics [9]. For example, suppose that the same sensors are always present in the scene (no appearance or disappearance of sensors), so that $\overset{*}{n} = e$ is constant. Then $\overset{*}{f}_{k+1|k}(\overset{*}{Y}|\overset{*}{X}, U) = 0$ unless it has the form

$$\begin{aligned} &\overset{*}{f}_{k+1|k}(\{\overset{*1}{\mathbf{y}}, ..., \overset{*e}{\mathbf{y}}\}|\{(\overset{*1}{\mathbf{x}}, \overset{1}{\mathbf{u}}_1), ..., (\overset{*e}{\mathbf{x}}, \overset{e}{\mathbf{u}}_e)\}) \\ &= \sum_\sigma \overset{*1}{f}_{k+1|k}(\overset{*1}{\mathbf{y}}|\overset{*\sigma 1}{\mathbf{x}}, \overset{\sigma 1}{\mathbf{u}}_{\sigma 1}) \cdots \overset{*e}{f}_{k+1|k}(\overset{*e}{\mathbf{y}}|\overset{*\sigma e}{\mathbf{x}}, \overset{\sigma e}{\mathbf{u}}_{\sigma e}) \end{aligned} \tag{51}$$

where the sum is over all permutations σ on the numbers $1, ..., e$.

- **Joint multisensor-multitarget motion:** We will assume that

$$\breve{f}_{k+1|k}(\breve{Y}|\breve{X}, U) = f_{k+1|k}(Y \cup \overset{*}{Y}|X \cup \overset{*}{X}, U) = f_{k+1|k}(Y|X) \cdot \overset{*}{f}_{k+1|k}(\overset{*}{Y}|\overset{*}{X}, U) \tag{52}$$

That is, the dynamic characteristics of the sensors are independent of the dynamic characteristics of the targets.

2.8. *Measurement Models*

In the sequel, and particularly in section 7, we will be constructing a likelihood function that implements the following generalization of the most commonly used multitarget observation model:[e]

[e]In actuality, the multisensor-multitarget likelihood function that results from these assumptions will be computationally intractable (see section 7.3.1), so we will be forced to produce a linearized approximation of it (see section 7.3.2).

(1) each platform carries one sensor;

(2) for each sensor, each target generates at most one observation and no observation is generated by more than one target;

(3) each observation collected from a target is contaminated by the sensor noise process;

(4) for each sensor, observations from different targets are conditionally independent upon target state;

(5) for each sensor, any multitarget observation is contaminated by a Poisson false alarm process that is independent of the target-generated observation process;

(6) for each sensor, the state of that sensor is observed by an internal actuator sensor;

(7) for each sensor, the observation collected by the actuator sensor may be contaminated by sensor noise and/or registration error;

(8) for each sensor, the actuator sensor observation may not be successfully collected because of obscuration, transmission channel drop-out, communication latency, and other effects;

(9) for each sensor, if transmission of the actuator observation does not occur, then neither target-generated observations nor clutter observations are transmitted either; and

(10) observations from different sensors are conditionally independent upon target state.

In more detail:

• **Sensor noise:** The noise characteristics of the sensors are modeled using likelihood functions:

$$\overset{1}{L}_{\overset{1}{\mathbf{z}},\overset{*1}{\mathbf{x}},k}(\mathbf{x}) = \overset{1}{f}_k(\overset{1}{\mathbf{z}}|\mathbf{x},\overset{*1}{\mathbf{x}}) \quad ,\ldots, \quad \overset{s}{L}_{\overset{s}{\mathbf{z}},\overset{*s}{\mathbf{x}},k}(\mathbf{x}) = \overset{s}{f}_k(\overset{s}{\mathbf{z}}|\mathbf{x},\overset{*s}{\mathbf{x}}) \tag{53}$$

We will abbreviate these as

$$\overset{j}{L}_{\overset{j}{\mathbf{z}},\overset{*j}{\mathbf{x}}}(\mathbf{x}) \overset{\text{abbr.}}{=} \overset{j}{f}_k(\overset{j}{\mathbf{z}}|\mathbf{x},\overset{*j}{\mathbf{x}}) \quad \text{or even} \quad \overset{j}{L}_{\overset{j}{\mathbf{z}}}(\mathbf{x}) \overset{\text{abbr.}}{=} \overset{j}{f}_k(\overset{j}{\mathbf{z}}|\mathbf{x},\overset{*j}{\mathbf{x}}) \tag{54}$$

In what follows we will assume that these likelihood functions have the additive form

$$\overset{j}{f}_k(\overset{j}{\mathbf{z}}|\mathbf{x},\overset{*j}{\mathbf{x}}) = f_{\overset{j}{\mathbf{W}}_k}(\overset{j}{\mathbf{z}} - \overset{j}{\eta}_k(\mathbf{x},\overset{*j}{\mathbf{x}})) \tag{55}$$

where $\overset{j}{\mathbf{W}}_k$ is a zero-mean noise vector and $\overset{j}{\eta}_k(\mathbf{x},\overset{*j}{\mathbf{x}})$ is a deterministic sensor model.

• **Sensor Fields of View (FoVs):** The FoVs of the sensors at time-step k will be modeled as state-dependent probabilities of detection:

$$\overset{1}{p}_D(\mathbf{x}, \overset{*1}{\mathbf{x}}_k) \qquad , \ldots, \qquad \overset{s}{p}_D(\mathbf{x}, \overset{*s}{\mathbf{x}}_k) \tag{56}$$

That is, the probability that the jth sensor will collect an observation from a target with state $\mathbf{x}$ at time-step k is $\overset{j}{p}_D(\mathbf{x}, \overset{*j}{\mathbf{x}}_k)$ if the state the sensor at that time-step is $\overset{*j}{\mathbf{x}}_k$. We will abbreviate

$$\overset{j}{p}_{D,k}(\mathbf{x}) \overset{\text{abbr.}}{=} \overset{j}{p}_D(\mathbf{x}, \overset{*j}{\mathbf{x}}_k) \quad \text{or even} \quad \overset{j}{p}_D(\mathbf{x}) \overset{\text{abbr.}}{=} \overset{j}{p}_D(\mathbf{x}, \overset{*j}{\mathbf{x}}_k) \tag{57}$$

• **Actuator-sensor errors:** The actuator sensors may have significant internal self-noise. Also, there may be biases in the observation of the sensor state caused by spatial and temporal registration error. Effects such as these can be modeled as likelihood functions

$$\overset{*1}{f}_k(\overset{*1}{\mathbf{z}}|\overset{*1}{\mathbf{x}}), \ldots, \overset{*s}{f}_k(\overset{*s}{\mathbf{z}}|\overset{*s}{\mathbf{x}}) \tag{58}$$

which will be abbreviated

$$\overset{*1}{L}_{\overset{*1}{\mathbf{z}},\overset{*1}{\mathbf{x}}}(\mathbf{x}) \overset{\text{abbr.}}{=} \overset{*1}{f}_k(\overset{*1}{\mathbf{z}}|\mathbf{x}, \overset{*1}{\mathbf{x}}) \quad \text{or even} \quad \overset{*j}{L}_{\overset{*j}{\mathbf{z}}}(\mathbf{x}) \overset{\text{abbr.}}{=} \overset{*j}{f}_k(\overset{*j}{\mathbf{z}}|\mathbf{x}, \overset{*j}{\mathbf{x}}) \tag{59}$$

• **Transmission errors:** Even though a sensor may have collected observations from targets that are located in its FoV, neither these observations nor the actuator-sensor observation may actually be available to the collection site. This may be because of transmission drop-outs such as atmospheric interference, terrain blockage, latency, etc. These effects can be modeled using actuator-sensor probabilities of detection,

$$\overset{*1}{p}_{D,k}(\overset{*1}{\mathbf{x}}), \ldots, \overset{*s}{p}_{D,k}(\overset{*s}{\mathbf{x}}) \tag{60}$$

• **Joint sensor, actuator-sensor multitarget measurement models:** In multitarget problems, the sensor likelihoods will take the more general multitarget form

$$\overset{1}{f}_k(\overset{1}{Z}|X, \overset{*1}{\mathbf{x}}) \qquad , \ldots, \qquad \overset{s}{f}_k(\overset{s}{Z}|X, \overset{*s}{\mathbf{x}}) \tag{61}$$

Generally speaking, these likelihoods can be constructed from the individual sensor likelihoods $\overset{j}{f}_k(\overset{j}{\mathbf{z}}|\mathbf{x}, \overset{*j}{\mathbf{x}})$ and FoVs $\overset{j}{p}_D(\mathbf{x}, \overset{*j}{\mathbf{x}})$, together with false alarm and/or clutter models, using the techniques of finite-set statistics [9].

In what follows we will assume the following observation model for target-generated observations for each sensor:

$$\overset{j}{f}_k(\overset{j}{Z},|\mathbf{x},\overset{*j}{\mathbf{x}}_k) = \begin{cases} 1-\overset{j}{p}_D(\mathbf{x},\overset{*j}{\mathbf{x}}) & \text{if } \overset{j}{Z}=\emptyset \\ \overset{j}{p}_D(\mathbf{x},\overset{*j}{\mathbf{x}})\cdot \overset{j}{L}_{\overset{j}{\mathbf{z}}}(\mathbf{x},\overset{*j}{\mathbf{x}}) & \text{if } \overset{j}{Z}=\{\overset{j}{\mathbf{z}}\} \\ 0 & \text{if otherwise}\end{cases} \tag{62}$$

The likelihoods of Eq. (61) for the jth sensor collecting from a multitarget state $X = \{\mathbf{x}_1, ..., \mathbf{x}_n\}$ are, assuming conditional independence upon target and sensor state,

$$\overset{j}{f}_k(\overset{j}{Z},|X,\overset{*j}{\mathbf{x}}) = \overset{j}{f}_k(\overset{j}{Z},|\mathbf{x}_1,\overset{*j}{\mathbf{x}}) \cdots \overset{j}{f}_k(\overset{j}{Z},|\mathbf{x}_n,\overset{*j}{\mathbf{x}}) \tag{63}$$

Finally, we will assume that the target observations from the jth sensor are contaminated by a Poisson false alarm process. That is, at time-step k the spatial distribution of the false alarms are governed by a probability distribution $\overset{j}{c}_k(\overset{j}{\mathbf{z}})$ and that the time-arrival of these false alarms are governed by a Poisson distribution with expected value $\overset{j}{\lambda}_k$. The likelihood function for the false alarms is

$$\overset{j}{c}_k(\overset{j}{Z}) = e^{-\overset{j}{\lambda}_k} \cdot \prod_{\overset{j}{\mathbf{z}}\in\overset{j}{Z}} \overset{j}{\lambda}_k \overset{j}{c}_k(\overset{j}{\mathbf{z}}) \tag{64}$$

So, multitarget observations from the jth sensor, contaminated by the false alarm process, are governed by the likelihood (see p. 35 of [9])

$$\overset{j}{f}_{\text{tot},k}(\overset{j}{Z}|X,\overset{*j}{\mathbf{x}}) = \sum_{\overset{j}{W}\subseteq\overset{j}{Z}} \overset{j}{f}_k(\overset{j}{W}|X,\overset{*j}{\mathbf{x}}) \cdot \overset{j}{c}_k(\overset{j}{Z}-\overset{j}{W}) \tag{65}$$

Now take the actuator sensors into account. The joint observation collected by the jth sensor and its actuator sensor is

$$\overset{j}{f}_k(\overset{j}{Z},\overset{*j}{Z}|X,\overset{*j}{\mathbf{x}}) = \begin{cases} 1-\overset{*j}{p}_D(\overset{*j}{\mathbf{x}}) & \text{if } \overset{*j}{Z}=\emptyset \\ \overset{*j}{p}_D(\overset{*j}{\mathbf{x}})\cdot \overset{*j}{L}_{\overset{*j}{\mathbf{z}}}(\overset{*j}{\mathbf{x}})\cdot \overset{j}{f}_{\text{tot},k}(\overset{j}{Z}|X,\overset{*j}{\mathbf{x}}) & \text{if } \overset{*j}{Z}=\{\overset{*j}{\mathbf{z}}\} \\ 0 & \text{if } |\overset{*j}{Z}|\geq 2\end{cases} \tag{66}$$

• **Joint multisensor-multitarget measurement models.** We must specify the general form of the joint multisensor-multitarget likelihood $\breve{f}_k(\breve{Z}|\breve{X})$. That is, we must specify the likelihood when multiple sensors and multiple targets are present. Suppose that the sensors present in the scene have states $\overset{*}{X} = \{\overset{*i_1}{\mathbf{x}}_1, ..., \overset{*i_e}{\mathbf{x}}_e\}$ and so the joint multisensor-multitarget state is $\breve{X} = X \cup \overset{*}{X} = X \cup \{\overset{*i_1}{\mathbf{x}}_1, ..., \overset{*i_e}{\mathbf{x}}_e\}$. Then $\breve{f}_k(\breve{Z}|\breve{X}) = 0$ unless $\breve{Z}$ has the form

$$\breve{Z} = \overset{i_1}{Z} \cup ... \cup \overset{i_e}{Z} \cup \overset{*i_1}{Z} \cup ... \cup \overset{*i_e}{Z} \tag{67}$$

In this case we specify cross-sensor conditional independence on the sensor states:

$$\begin{aligned}\breve{f}_k(\breve{Z}|\breve{X}) &= \breve{f}_k(\overset{i_1}{Z}_1 \cup ... \cup \overset{i_e}{Z}_e \cup \overset{*i_1}{Z}_1 \cup ... \cup \overset{*i_e}{Z}_e | X \cup \{\overset{*i_1}{\mathbf{x}}_1, ..., \overset{*i_e}{\mathbf{x}}_e\}) \\ &= \overset{i_1}{f}_k(\overset{i_1}{Z}_1, \overset{*i_1}{Z}_1 | X, \overset{*i_1}{\mathbf{x}}) \cdots \overset{i_e}{f}_k(\overset{i_e}{Z}_e, \overset{*i_e}{Z}_e | X, \overset{*i_e}{\mathbf{x}})\end{aligned} \tag{68}$$

2.9. *The Joint Multisensor-Multitarget Bayes Filter*

In Eq. (52) we noted that the general multisensor-multitarget Markov transition $\breve{f}_{k+1|k}(\breve{Y}|\breve{X}, U) = f_{k+1|k}(Y|X) \cdot \overset{*}{f}_{k+1|k}(\overset{*}{Y}|\overset{*}{X}, U)$ depends on a set U of control vectors. For the sake of notational clarity we suppress the control vectors in what follows. Given this simplification, the Bayes filter for the joint multisensor-multitarget system is given by the equations

$$\breve{f}_{k+1|k}(\breve{X}|\breve{Z}^{(k)}) = \int \breve{f}_{k+1|k}(\breve{X}|\breve{W}) \cdot \breve{f}_{k|k}(\breve{W}|\breve{Z}^{(k)}) \delta\breve{W} \tag{69}$$

$$\breve{f}_{k+1|k+1}(\breve{X}|\breve{Z}^{(k+1)}) = \frac{\breve{f}_{k+1}(\breve{Z}_{k+1}|\breve{X})\ \breve{f}_{k+1|k}(\breve{X}|\breve{Z}^{(k)})}{\breve{f}_{k+1}(\breve{Z}_{k+1}|\breve{Z}^{(k)})} \tag{70}$$

where

$$\breve{f}_{k+1}(\breve{Z}_{k+1}|\breve{Z}^{(k)}) = \int \breve{f}_{k+1}(\breve{Z}_{k+1}|\breve{X})\ \breve{f}_{k+1|k}(\breve{X}|\breve{Z}^{(k)}) \delta\breve{X} \tag{71}$$

Thus $\breve{f}_{k|k}(\breve{W}|\breve{Z}^{(k)})$ implicitly depends on the choices of the control vectors introduced by the Markov transition at each of the previous recursive steps. This filter describes the time-evolution of all sensors and all targets, when regarded as a single composite physical system.

Written in the alternative notation introduced in Eqs. (28) and (29), Eqs. (69), (70), and (71) become:

$$f_{k+1|k}(X, \overset{*}{X}|\breve{Z}^{(k)}) = \int f_{k+1|k}(X, \overset{*}{X}|W, \overset{*}{W}) \cdot f_{k|k}(W, \overset{*}{W}|\breve{Z}^{(k)}) \delta W \delta\overset{*}{W} \tag{72}$$

$$f_{k+1|k+1}(X, \overset{*}{X}|\breve{Z}^{(k+1)}) = \frac{f_{k+1}(Z_{k+1}, \overset{*}{Z}_{k+1}|X, \overset{*}{X}) \cdot f_{k+1|k}(X, \overset{*}{X}|\breve{Z}^{(k)})}{f_{k+1}(Z_{k+1}, \overset{*}{Z}_{k+1}|\breve{Z}^{(k)})} \tag{73}$$

where

$$f_{k+1}(Z_{k+1}, \mathring{Z}_{k+1}|\breve{Z}^{(k)}) = \int f_{k+1}(Z_{k+1}, \mathring{Z}_{k+1}|X, \mathring{X}) \cdot f_{k+1|k}(X, \mathring{X}|\breve{Z}^{(k)})\delta X \delta \mathring{X} \tag{74}$$

3. Sensor Management

In this section we describe our core approach to sensor management and our current reformulation and refinement of it. The section is organized as follows:

• Section 3.1: We summarize the core approach that we proposed in March 1996 [8] and, with a slight generalization, again in 1998 [7]. It is based on the use of Csiszár information-theoretic objective functions defined in terms of multitarget posterior and predicted probability distributions $f_{k+1|k+1}(X|Z^{(k+1)})$ resp. $f_{k+1|k}(X|Z^{(k)})$.

• Section 3.2 summarizes the revision of the core approach. The Bayes filter Eqs. (72) and (73) are reformulated in terms of probability generating functionals (section 2.4). The p.g.fl.'s $G_{k+1|k+1}$ and $G_{k+1|k}$ are used in place of $f_{k+1|k+1}$ and $f_{k+1|k}$, respectively. Since $G_{k+1|k}[h]$ can be expressed in terms of $G_{k|k}[h]$ and $G_{k+1|k+1}[h]$ can be expressed in terms of $G_{k+1|k}[h]$, we can devise predictor and corrector equations for approximate filters. If $G_{k+1|k}[h]$ is assumed to have a simplified form (e.g., Eqs. (93), (96), or (99)) then $G_{k+1|k+1}[h]$ and any objective function defined in terms of it will also have relatively simple forms.

• Section 3.3: Finding an effective but tractable optimization-hedging strategy has been the major stumbling block to practical realization of the core approach. We discuss "hedged" versions $\mathring{G}_{k+1|k+1}$ of $G_{k+1|k+1}$—i.e., ones which do not depend on future observation collections. Our proposed solution, maxi-PIMS optimization-hedging, will be introduced in section 4.

• Sections 3.4 and 3.5: We describe sensor management using objective functions in conjunction with two approximate multitarget filters: the probability hypothesis density (PHD) filter and the multi-hypothesis correlator (MHC) filter.

3.1. *The Core Sensor Management Approach*

The core approach we proposed in 1996 and 1998 was general enough to encompass sensors on independent platforms, as well as the dynamics of those sensors and platforms. The basic idea is as follows (see section 3.2

of [14]). Recall the joint multisensor-multitarget filter Eqs. (72) and (73). Recall the fact that these equations implicitly depend on to-be-determined control vectors. Assume that the sensors are constant in number, in which case we can replace the sensor state-set $\overset{*}{X}_k = \{\overset{*1}{\mathbf{x}}_k, ..., \overset{*s}{\mathbf{x}}_k\}$ by a total state-vector $\overset{*}{\mathbf{x}}_k = (\overset{*1}{\mathbf{x}}_k, ..., \overset{*s}{\mathbf{x}}_k)$ and replace the control-set $U_k = \{\overset{1}{\mathbf{u}}_k, ..., \overset{s}{\mathbf{u}}_k\}$ by a total control-vector $\mathbf{u}_k = (\overset{1}{\mathbf{u}}_k, ..., \overset{s}{\mathbf{u}}_k)$ (see Remark 1 of section 2.7).

First consider the simplest kind of sensor management, *single-step look-ahead*: we need only determine the best choice of the control-vector $\mathbf{u}_k$ introduced by the Markov transition $f_{k+1|k}(X, \overset{*}{X}|W, \overset{*}{W})$. Integrate the multisensor state $\overset{*}{\mathbf{x}}$ out of Eqs. (72) and (73) as a nuisance variable:

$$\hat{f}_{k+1|k}(X) \overset{\text{def.}}{=} \int f_{k+1|k}(X, \overset{*}{\mathbf{x}}|\breve{Z}^{(k)})d\overset{*}{\mathbf{x}} \tag{75}$$

$$\hat{f}_{k+1|k+1}(X) \overset{\text{def.}}{=} \int f_{k+1|k+1}(X, \overset{*}{\mathbf{x}}|\breve{Z}^{(k+1)})d\overset{*}{\mathbf{x}} \tag{76}$$

Note that if $f_{k+1|k}(X, \overset{*}{\mathbf{x}}|W, \overset{*}{\mathbf{w}}) = f_{k+1|k}(X|W) \cdot f_{k+1|k}(\overset{*}{\mathbf{x}}|\overset{*}{\mathbf{w}})$ then $\hat{f}_{k+1|k}(X)$ has no functional dependence on the unknown control vectors:

$$\begin{aligned} &\int f_{k+1|k}(X, \overset{*}{\mathbf{x}}|\breve{Z}^{(k)})d\overset{*}{\mathbf{x}} \\ &= \int\int f_{k+1|k}(X, \overset{*}{\mathbf{x}}|W, \overset{*}{\mathbf{w}}) \cdot f_{k|k}(W, \overset{*}{\mathbf{x}}|\breve{Z}^{(k)})\delta W d\overset{*}{\mathbf{w}} d\overset{*}{\mathbf{x}} \\ &= \int f_{k+1|k}(X|W) \cdot \hat{f}_{k|k}(W|\breve{Z}^{(k)})\delta W \end{aligned}$$

where $\hat{f}_{k|k}(W|\breve{Z}^{(k)}) = \int f_{k|k}(W, \overset{*}{\mathbf{w}}|\breve{Z}^{(k)})d\overset{*}{\mathbf{w}}$.

For the sake of clarity assume for the remainder of this subsection that:

- controls are sensor states and that sensor control consists of direct selection of the next sensor state—i.e., $\mathbf{u}_k = \overset{*}{\mathbf{x}}_{k+1}$; and
- the sensor has perfect response to control commands: $f_{k+1|k}(\overset{*}{\mathbf{y}}|\overset{*}{\mathbf{x}}, \mathbf{u}) = \delta_{\mathbf{u}}(\overset{*}{\mathbf{y}})$ where $\delta_{\mathbf{u}}(\overset{*}{\mathbf{y}})$ denotes the Dirac delta function concentrated at $\mathbf{u}$.

Then it can be shown (see section 2.2 of [14]) that the joint multisensor-multitarget Bayes filter reduces to the conventional multitarget Bayes filter

$$\hat{f}_{k+1|k}(X|Z^{(k)}) = \int f_{k+1|k}(\mathbf{x}|\mathbf{w}) \cdot \hat{f}_{k|k}(W|Z^{(k)})\delta W \tag{77}$$

$$\hat{f}_{k+1|k+1}(X|Z^{(k+1)}) = \frac{f_{k+1}(Z_{k+1}|\mathbf{x}, \overset{*}{\mathbf{x}}_{k+1}) \cdot \hat{f}_{k+1|k}(\mathbf{x}|Z^{(k)})}{\hat{f}_{k+1}(Z_{k+1}|Z^{(k)})} \tag{78}$$

where $\hat{f}_{k+1}(Z_{k+1}|Z^{(k)}) = \int f_{k+1}(Z_{k+1}|\mathbf{x}, \overset{*}{\mathbf{x}}_{k+1}) \cdot \hat{f}_{k+1|k}(\mathbf{x}|Z^{(k)})\delta X$.

By analogy with linear control, regard $\hat{f}_{k+1|k}(X)$ as a "reference" distribution and $\hat{f}_{k+1|k+1}(X)$ as a "controlled" distribution. To compare these two distributions, define a multisensor-multitarget Csiszár objective functional [14]

$$I_c(Z, \overset{*}{\mathbf{x}}_{k+1}) = \int c\left(\frac{\hat{f}_{k+1|k+1}(X|Z, \overset{*}{\mathbf{x}}_{k+1})}{\hat{f}_{k+1|k}(X)}\right) \cdot \hat{f}_{k+1|k}(X)\delta X \tag{79}$$

and then determine the value $\overset{*}{\mathbf{x}}_{k+1}$that maximizes this quantity. If $c(x) = 1-x+x\log x$ then $I(Z, \overset{*}{\mathbf{x}}_{k+1})$ is the multitarget Kullback-Leibler objective function proposed in [8]. If $c(x) = |x-1|$ it is the L^1 metric; if $c(x) = (\sqrt{x}-1)^2$ it is the Hellinger distance; and so on.

Since the future observation-set Z cannot be known ahead of time, we must hedge against this uncertainty. The two most familiar optimization-hedging strategies are, respectively, *maxi-mean* and *maxi-min*:

$$\overset{*}{\mathbf{x}}{}^{\text{mean}}_{k+1} = \arg_{\overset{*}{\mathbf{x}}} I_c(\overset{*}{\mathbf{x}}), \quad I_c(\overset{*}{\mathbf{x}}) \overset{\text{def.}}{=} \int I_c(Z, \overset{*}{\mathbf{x}}) f_{k+1}(Z)\delta Z \tag{80}$$

$$\overset{*}{\mathbf{x}}{}^{\text{min}}_{k+1} = \arg\sup_{\overset{*}{\mathbf{x}}} I_c(\overset{*}{\mathbf{x}}), \quad I_c(\overset{*}{\mathbf{x}}) \overset{\text{def.}}{=} \inf_Z I_c(Z, \overset{*}{\mathbf{x}}) \tag{81}$$

Eq. (80) hedges against the average future observation, whereas Eq. (81) hedges against the worst-case future observation (e.g., the data-set that highly non-cooperative targets might produce). Both strategies are computationally intractable in general. So, in [8], [7] we proposed a more tractable "maxi-null" optimization strategy,

$$\overset{*}{\mathbf{x}}{}^{\text{null}}_{k+1} = \arg\sup_{\overset{*}{\mathbf{x}}} I_c(\overset{*}{\mathbf{x}}), \qquad I_c(\overset{*}{\mathbf{x}}) \overset{\text{def.}}{=} I_c(\emptyset, \overset{*}{\mathbf{x}}) \tag{82}$$

This bears resemblance to maxi-min in that it hedges against the non-informative observation-set $Z = \emptyset$ instead of (as with maxi-min) the least-informative observation-set.

This reasoning is easily extended to *multistep look-ahead*: we are to determine the best sequence $\overset{*}{\mathbf{x}}_{k+1}, ..., \overset{*}{\mathbf{x}}_{k+M}$ of sensor states in a future time-window. To do this we iterate Eqs. (77) and (78) until we construct the multitarget posterior $f_{k+M|k+M}(X|Z^{(k+M)})$. We then form the objective function

$$\begin{aligned} &I_c(Z_{k+1}, ..., Z_{k+M}, \overset{*}{\mathbf{x}}_{k+1}, ..., \overset{*}{\mathbf{x}}_{k+M}) \\ &= \int c\left(\frac{f_{k+M|k+M}(X|Z^{(k+M)})}{f_{k+1|k}(X)}\right) \cdot f_{k+1|k}(X)\delta X \end{aligned} \tag{83}$$

After hedging against the unknowable future observation-sets $Z_{k+1}, ..., Z_{k+M}$, we select those $\mathring{\mathbf{x}}_{k+1}, ..., \mathring{\mathbf{x}}_{k+M}$ which jointly maximize the hedged objective function.

3.2. *p.g.fl. Representation of Multitarget Bayes Filter*

In this section we show how to transform the multitarget Bayes filter of Eqs. (72) and (73) into probability generating functionals (p.g.fl.'s). We do this first for the multitarget prediction integral and then for the multitarget Bayes' rule. We no longer make the special assumptions used in the previous section.

- *p.g.fl. representation of the multitarget prediction integral (Eq. (77))* From Eq. (72) the p.g.fl. of $f_{k+1|k}(X, \mathring{X}|\breve{Z}^{(k)})$ is

$$\begin{aligned}
&\breve{G}_{k+1|k}[\breve{h}] \\
&= \int \breve{h}^X \cdot \breve{h}^{\mathring{X}} \cdot f_{k+1|k}(X, \mathring{X}|\breve{Z}^{(k)}) \delta X \delta \mathring{X} \\
&= \int \breve{h}^X \cdot \breve{h}^{\mathring{X}} \cdot \left(\int f_{k+1|k}(X, \mathring{X}|W, \mathring{W}) \cdot f_{k|k}(W, \mathring{W}|\breve{Z}^{(k)}) \delta W \delta \mathring{W} \right) \delta X \delta \mathring{X} \\
&= \int \breve{G}_{k+1|k}[\breve{h}|W, \mathring{W}] \cdot f_{k|k}(W, \mathring{W}|\breve{Z}^{(k)}) \delta W \delta \mathring{W}
\end{aligned} \tag{84}$$

where

$$\breve{G}_{k+1|k}[\breve{h}|W, \mathring{W}] \overset{\text{def.}}{=} \int \breve{h}^X \cdot \breve{h}^{\mathring{X}} \cdot f_{k+1|k}(X, \mathring{X}|W, \mathring{W}) \delta X \delta \mathring{X} \tag{85}$$

Eq. (84) is the p.g.fl. representation of the multitarget prediction integral, Eq. (72).

- *p.g.fl. representation of the multitarget Bayes' rule (Eq. (78)).* From Eq. (73) the p.g.fl. of $f_{k+1|k+1}(X, \mathring{X}|\breve{Z}^{(k+1)})$ is

$$\breve{G}_{k+1|k+1}[\breve{h}] = \frac{\int \breve{h}^X \cdot \breve{h}^{\mathring{X}} \cdot f_{k+1}(\breve{Z}_{k+1}|X, \mathring{X}) \, f_{k+1|k}(X, \mathring{X}|\breve{Z}^{(k)}) \delta X \delta \mathring{X}}{\int f_{k+1}(\breve{Z}_{k+1}|X, \mathring{X}) \, f_{k+1|k}(X, \mathring{X}|\breve{Z}^{(k)}) \delta X \delta \mathring{X}} \tag{86}$$

Define the two-variable p.g.fl. $\breve{F}_{k+1}[\breve{g}, \breve{h}]$ by

$$\begin{aligned}
&\breve{F}_{k+1}[\breve{g}, \breve{h}] \\
&\overset{\text{def.}}{=} \int \breve{h}^X \cdot \breve{h}^{\mathring{X}} \cdot \breve{g}^Z \cdot \breve{g}^{\mathring{Z}} \cdot f_{k+1}(Z, \mathring{Z}|X, \mathring{X}) \, f_{k+1|k}(X, \mathring{X}|\breve{Z}^{(k)}) \delta X \delta \mathring{X} \delta Z \delta \mathring{Z}
\end{aligned} \tag{87}$$

and note that

$$\breve{F}_{k+1}[\breve{g}, \breve{h}] = \int \breve{h}^X \cdot \breve{h}^{\overset{*}{X}} \cdot \breve{G}_{k+1}[\breve{g}|X, \overset{*}{X}] \cdot f_{k+1|k}(X, \overset{*}{X}|\breve{Z}^{(k)}) \delta X \delta \overset{*}{X} \quad (88)$$

where

$$\breve{G}_{k+1}[\breve{g}|X, \overset{*}{X}] \overset{\text{def.}}{=} \int \breve{g}^Z \cdot \breve{g}^{\overset{*}{Z}} \cdot f_{k+1}(Z, \overset{*}{Z}|X, \overset{*}{X}) \delta Z \delta \overset{*}{Z}$$

is the p.g.fl. of the likelihood function $f_{k+1}(Z, \overset{*}{Z}|X, \overset{*}{X})$. From Eq. (42) we know that, taking functional derivatives of $\breve{F}_{k+1}$ with respect to its first variable $\breve{g}$,

$$\frac{\delta \breve{F}_{k+1}}{\delta Z \delta \overset{*}{Z}}[0, \breve{h}] = \int \breve{h}^X \cdot \breve{h}^{\overset{*}{X}} \cdot f_{k+1}(Z, \overset{*}{Z}|X, \overset{*}{X}) \, f_{k+1|k}(X, \overset{*}{X}|\breve{Z}^{(k)}) \delta X \delta \overset{*}{X}$$

Thus Eq. (86) becomes

$$\breve{G}_{k+1|k+1}[\breve{h}] = \frac{\frac{\delta \breve{F}_{k+1}}{\delta Z \delta \overset{*}{Z}}[0, \breve{h}]}{\frac{\delta \breve{F}_{k+1}}{\delta Z \delta \overset{*}{Z}}[0, 1]} \quad (89)$$

This is the p.g.fl. representation of the multitarget Bayes' rule, Eq. (73).

3.3. *Hedged posterior p.g.fl.'s*

This section summarizes the difficulties associated with devising a "hedged" version $\mathring{G}_{k+1|k+1}$ of $\breve{G}_{k+1|k+1}$—i.e., one that does not depend on Z_{k+1}. For the sake of clarity we once again make the simplifying assumptions employed in section 3.1. Abbreviate

$$f_{k+1|k+1}(X|Z, \overset{*}{\mathbf{x}}_{k+1}) \overset{\text{abbr.}}{=} f_{k+1|k+1}(X|Z^{(k+1)})$$

$$f_{k+1|k+1}(X) \overset{\text{abbr.}}{=} f_{k+1|k}(X|Z^{(k)})$$

$$f_{k+1}(Z) \overset{\text{abbr.}}{=} f_{k+1}(Z|Z^{(k)})$$

Let $G_{k+1|k+1}[h|Z, \overset{*}{\mathbf{x}}_{k+1}]$ denote the p.g.fl. of $f_{k+1|k+1}(X|Z, \overset{*}{\mathbf{x}}_{k+1})$. The most obvious optimization-hedging strategy would be maxi-mean:

$$\mathring{G}_{k+1|k+1}[h|\overset{*}{\mathbf{x}}_{k+1}] = \int G_{k+1|k+1}[h|Z, \overset{*}{\mathbf{x}}_{k+1}] \cdot f_{k+1}(Z) \delta Z \quad (90)$$

However, note that in this case

$$\mathring{G}_{k+1|k+1}[h|\overset{*}{\mathbf{x}}_{k+1}] = \int \left(\int h^X \cdot \frac{f_{k+1}(Z|X) \cdot f_{k+1|k}(X)}{f_{k+1}(Z)} \delta X \right) f_{k+1}(Z) \delta Z$$

$$= \int h^X \cdot f_{k+1|k}(X) \delta X = \breve{G}_{k+1|k}[h]$$

which no longer depends on $\overset{*}{\mathbf{x}}_{k+1}$ and thus cannot be used for purposes of sensor management. The same fact is true for any objective functions that is defined linearly in terms of $f_{k+1|k+1}(X)$. One can get around this by averaging various nonlinear transforms of $G_{k+1|k}[h|Z,\ \overset{*}{\mathbf{x}}_{k+1}]$ such as

$$\mathring{G}_{k+1|k+1}[h] = \int \breve{G}_{k+1|k+1}[h|Z, \overset{*}{\mathbf{x}}_{k+1}]^2 \cdot f_{k+1}(Z)\delta Z$$

but these are inherently computationally intractable. Maxi-min hedging $\mathring{G}_{k+1|k+1}[h|\overset{*}{\mathbf{x}}_{k+1}] = \inf_Z \mathring{G}_{k+1|k+1}[h|Z\overset{*}{\mathbf{x}}_{k+1}]$ will be equally intractable and maxi-null has proved to be too conservative. This leads us to the "maxi-PIMS" strategy proposed in section 4 below.

3.4. *The Probability Hypothesis Density (PHD) Filter*

The purpose of this section is to briefly describe an approximation of the general multitarget Bayes filter (equations (72) and (73)) by an approximate multitarget filter that propagates the first multitarget moments of the $f_{k|k}(X|Z^{(k)})$ rather than the $f_{k|k}(X|Z^{(k)})$ themselves. We also explain how this filter is used, in conjunction with objective functions defined from a hedged posterior p.g.fl. $\mathring{G}_{k+1|k+1}[h]$, for sensor management. (The same basic reasoning will be re-applied in section 7.)

The PHD of a multitarget posterior $f_{k|k}(X|Z^{(k)})$ is, according to Eq. (46),

$$D_{k|k}(\mathbf{x}|Z^{(k)}) \overset{\text{def.}}{=} \int f_{k|k}(W \cup \{\mathbf{x}\}|Z^{(k)})\delta W = \frac{\delta G_{k|k}}{\delta \mathbf{x}}[1] \tag{91}$$

where

$$G_{k|k}[h] \overset{\text{def.}}{=} \int h^X \cdot f_{k|k}(X|Z^{(k)})\delta X$$

is the p.g.fl. of $f_{k|k}(X|Z^{(k)})$.

• *PHD Predictor Equation*: Assuming multitarget motion models like those described in section 2.7, it can be shown that the predicted p.g.fl. $G_{k+1|k}[h]$ can be written in terms of $G_{k|k}[h]$. Eq. (91) can then be applied to derive a predictor equation, see [12]. For the purposes of this chapter, this is:

$$\begin{aligned} D_{k+1|k}&(\mathbf{y}|Z^{(k)}) \\ &= b_{k+1|k}(\mathbf{y}) \\ &+ \int \left(s_{k+1|k}(\mathbf{x}) \cdot f_{k+1|k}(\mathbf{y}|\mathbf{x}) + b_{k+1|k}(\mathbf{y}|\mathbf{x})\right) D_{k|k}(\mathbf{x}|Z^{(k)})d\mathbf{x} \end{aligned} \tag{92}$$

where $f_{k+1|k}(\mathbf{y}|\mathbf{x})$ is the single-target Markov transition; where $s_{k+1|k}(\mathbf{x})$ is the probability that a target will disappear at time-step $k+1$ if it had state $\mathbf{x}$ at time-step k; where $b_{k+1|k}(Y)$ is the probability that targets with state-set Y will appear at time-step $k+1$; where $b_{k+1|k}(Y|\mathbf{x})$ is the probability that a target with state $\mathbf{x}$ at time-step k will spawn targets with state-set Y at time-step $k+1$; and where $b_{k+1|k}(\mathbf{y}) = \int b_{k+1|k}(Y \cup \{\mathbf{y}\})\delta Y$ and $b_{k+1|k}(\mathbf{y}|\mathbf{x}) = \int b_{k+1|k}(Y \cup \{\mathbf{y}\}|\mathbf{x})\delta Y$ are the respective PHDs.

• *PHD Corrector Equation*: Assuming a single-sensor, multitarget measurement model of the kind described in section 2.8, it can be shown that the posterior p.g.fl. $G_{k+1|k+1}[h]$ can be written in terms of the predicted p.g.fl. $G_{k+1|k}[h]$ (see [12]). Even so, we must assume that $G_{k+1|k}[h]$ has a simple form in order to get a closed-form formula. Write $D_{k+1|k}[h] \stackrel{\text{def.}}{=} \int h(\mathbf{x}) \cdot D_{k+1|k}(\mathbf{x}|Z^{(k)})d\mathbf{x}$ and $N_{k+1|k} \stackrel{\text{def.}}{=} D_{k+1|k}[1]$. Then we assume that $G_{k+1|k}$ is Poisson:

$$G_{k+1|k}[h] \cong \exp\left(-N_{k+1|k} + D_{k+1|k}[h]\right) \tag{93}$$

Given this, it follows that the two-variable p.g.fl. of Eq. (87) is

$$F_{k+1}[g,h] \cong \exp\left(-\lambda - N_{k+1|k} + \lambda c_g + D_{k+1|k}[h(1 - p_D + p_D p_g)]\right) \tag{94}$$

where

$$\lambda \stackrel{\text{abbr.}}{=} \lambda_{k+1}, \qquad c_g \stackrel{\text{def.}}{=} \int g(\mathbf{z}) \cdot c_{k+1}(\mathbf{z})d\mathbf{z}\ , \quad p_g(\mathbf{x}) \stackrel{\text{def.}}{=} \int g(\mathbf{z}) \cdot f_{k+1}(\mathbf{z}|\mathbf{x})d\mathbf{z}$$

From Eq. (88) we get a closed-form formula for $G_{k+1|k}[h]$. From this, Eq. (91) allows us to derive the following corrector equation for the PHD [12]:

$$\begin{aligned} &D_{k+1|k+1}(\mathbf{x}|Z^{(k)}) \\ &\quad\cong \left(1 - p_D(\mathbf{x}) + \sum_{\mathbf{z} \in Z_{k+1}} \frac{p_D(\mathbf{x}) \cdot L_{\mathbf{z}}(\mathbf{x})}{\lambda c(\mathbf{z}) + D_{k+1|k}[p_D L_{\mathbf{z}}]}\right) \\ &\quad\cdot D_{k+1|k}(\mathbf{x}|Z^{(k)}) \end{aligned} \tag{95}$$

where as usual $L_{\mathbf{z}}(\mathbf{x}) \stackrel{\text{def.}}{=} f_{k+1}(\mathbf{z}|\mathbf{x})$ and $D_{k+1|k}[h] \stackrel{\text{def.}}{=} \int h(\mathbf{x})D_{k+1|k}(\mathbf{x}|Z^{(k)})d\mathbf{x}$.

The multisensor case can be dealt with using the same corrector equation. If the observation-sets from two sensors arrive at different time-steps, apply the corrector equations corresponding to those sensors at the appropriate times. If the two observation-sets arrive simultaneously, then apply the corrector equations corresponding to those sensors twice in a row, without any intervening predictor step.

The predictor and corrector equations (92) and (95) can be used to approximately implement a single-step look-ahead, control-theoretic sensor management scheme of the general type described in section 3.1. Consider single-step look-ahead first. Use the predictor equation to extrapolate $D_{k|k}(\mathbf{x}|Z^{(k)})$ to $D_{k+1|k}(\mathbf{x}|Z^{(k)})$. Use the hedged single-step objective function to determine the next sensor state (or, in the general case, the next sensor control). Using the field of view (FoV) corresponding to this choice, collect the next observation-set Z_{k+1}. Then use the PHD corrector equation to update $D_{k+1|k}(\mathbf{x}|Z^{(k)})$ to $D_{k+1|k+1}(\mathbf{x}|Z^{(k+1)})$. For multi-step look-ahead, use the hedged multistep objective function to determine the sensor states/controls in the future time-window. Then run the filter as usual during that window, collecting observations using the optimally chosen FoVs.

3.5. *The Multi-Hypothesis Correlator (MHC) Filter*

The purpose of this section is to briefly describe an approximation of the general multitarget Bayes filter (equations (72) and (73) [1]) by an approximate multi-hypothesis correlator (MHC) tracker [14], [10]. We also explain how this filter is used for purposes of sensor management in conjunction with objective functions defined from a hedged posterior p.g.fl. $\mathring{G}_{k+1|k+1}[h]$.

MHC algorithms have the same recursive form as the multitarget Bayes filter (i.e., prediction followed by correction followed by prediction, etc.). At each recursion step they produce a set of "hypotheses" as outputs, along with a probability that each of the hypotheses is a valid representation of ground truth. Each hypothesis is a subset of a "track table" consisting of N tracks for some N. Each track in the track table has a linear-Gaussian probability distribution $f_j(\mathbf{x}) = N_{P_j}(\mathbf{x} - \mathbf{x}_j)$ where $\mathbf{x}_j$ is the estimated state of the track and P_j is its error covariance matrix. The tracks in the track table are statistically independent. This is because of the measurement model specified in section 2.8. Measurements are assumed to be independent when conditioned on target states, and any measurement is assigned to at most one track. Consequently, the $f_1(\mathbf{x}), ..., f_N(\mathbf{x})$ are posterior densities that have been constructed from a partition of the time-accumulated measurements—they share no measurements in common.

Any given track has a "track probability" q_j, which is the sum of the hypothesis probabilities of all hypotheses that contain that track; and which can be interpreted as the probability that the jth track exists. Unlike the tracks, the track probabilities $q_1, ..., q_N$ are not necessarily independent because they do not arise from a unique partition of the accumulated measurements. Nevertheless, the following equation for the predicted p.g.fl. can be assumed to be approximately true:

$$G_{k+1|k}[h] \cong \prod_{j=1}^{N} (1 - q_j + q_j f_j[h]) \tag{96}$$

where $f_j[h] \overset{\text{def.}}{=} \int h(\mathbf{x}) f_j(\mathbf{x}) d\mathbf{x}$ and where q_j is the probability that the jth predicted track exists and $f_j(\mathbf{x})$ is the distribution of the jth predicted track. Thus in this case the two-variable p.g.fl. of Eq. (87) is

$$F_{k+1}[g, h] = e^{\lambda c[g] - \lambda} \cdot \prod_{j=1}^{N} (1 - q_j + q_j p_j[h] - q_j f_j[h p_D (1 - p_g)]) \tag{97}$$

This formula is too complicated to produce practical closed-form formulas. It can be further simplified by noting that the PHD of the p.g.fl. of Eq. (96) is, by Eq. (48),

$$\begin{aligned} D_{k+1|k}(\mathbf{x}) &= \frac{\delta \log G_{k+1|k}}{\delta \mathbf{x}}[1] \\ &= \left[\sum_{j=1}^{N} \frac{\frac{\delta}{\delta \mathbf{x}} q_j f_j[h]}{1 - q_j + q_j p_j[h]} \right]_{h=1} = \left[\sum_{j=1}^{N} \frac{q_j f_j(\mathbf{x})}{1 - q_j + q_j p_j[h]} \right]_{h=1} \\ &= \sum_{j=1}^{N} q_j f_j(\mathbf{x}) \end{aligned} \tag{98}$$

So, in the place of the approximation of Eq. (96), assume the simpler Poisson approximation

$$G_{k+1|k}[h] \cong \exp\left(-q + \sum_{j=1}^{N} q_j f_j[h]) \right) \tag{99}$$

where $q \overset{\text{def.}}{=} \sum_{j=1}^{N} q_j = N_{k+1|k}$ is the predicted expected number of targets. In this case, Eq. (94) becomes

$$F_{k+1}[g, h] = \exp\left(-\lambda - q + \lambda c_g + \sum_{j=1}^{N} q_j f_j[h(1 - p_D + p_D p_g)]) \right) \tag{100}$$

This can be used to derive a hedged p.g.fl. and any objective function definable in terms of the hedged p.g.fl., using the procedure outlined in section 4.

The MHC predictor and corrector steps can be used to approximately implement a single-step look-ahead, control-theoretic sensor management scheme of the general type described in section 3.1. That is, use the MHC predictor equation to extrapolate the current tracks and hypotheses to the time-step $k+1$ of the next data collection. Use the hedged objective function to determine the next sensor state (or, in the general case, the next sensor control). Using the field of view (FoV) corresponding to this choice, collect the next observation-set Z_{k+1}. Then use the MHC corrector step to update the track table and hypotheses. This can be further generalized to multistep look-ahead (see below).

4. "Maxi-PIMS" Optimization-Hedging

In section 3.3 we sketched the difficulties involved with obtaining useful objective functions and tractable, useful optimization-hedging strategies. The purpose of this section is to propose a new optimization-hedging strategy that is both potentially tractable and effective. In the single-sensor case the basic idea is this: *choose the FoV that will have the best chance of producing an "ideal" future observation-set.* By "ideal," we mean (in the single-sensor case) that no clutter observations are collected, that every target in the FoV generated an observation, and that target-generated observations are noise-free.

The section is organized as follows. We define the ideal observation-set in section 4.1. In section 4.2 we use it to construct the hedged posterior p.g.fl. $\mathring{G}_{k+1|k+1}[h]$. Since $\mathring{G}_{k+1|k+1}[h]$ functionally depends on the future sensor state $\overset{*}{\mathbf{x}}_{k+1}$, we can use it to define objective functions for sensor management. For example, the posterior expected number of targets (PENT) may be constructed by first finding the PHD $\mathring{D}_{k+1|k+1}(\mathbf{x})$ of $\mathring{G}_{k+1|k+1}[h]$ (section 4.3) and then constructing its integral (section 4.4). In section 4.5 we show how to extend the approach to the multisensor case. Since this extension will produce very complicated formulas in general, we propose a simplified version in section 4.6.

4.1. *Step 1: Predicted Ideal Measurement-Set (PIMS)*

Recall that in Eq. (55) of section 2.8 we assumed that single-sensor likelihood functions have the additive form $L_{\mathbf{z}}(\mathbf{x}) \overset{\text{abbr.}}{=} f_{k+1}(\mathbf{z}|\mathbf{x}, \overset{*}{\mathbf{x}}_{k+1}) = f_{\mathbf{W}_{k+1}}(\mathbf{z} - \eta_{k+1}(\mathbf{x}, \overset{*}{\mathbf{x}}_{k+1}))$ where $f_{\mathbf{W}_{k+1}}(\mathbf{z})$ is the probability density function of a measurement noise process $\mathbf{W}_{k+1}$. Abbreviate $\eta(\mathbf{x}) \overset{\text{abbr.}}{=} \eta_{k+1}(\mathbf{x}, \overset{*}{\mathbf{x}}_{k+1})$. Assume that we have some estimate of the number $\hat{n}$ and states $\hat{\mathbf{x}}_1, ..., \hat{\mathbf{x}}_{\hat{n}}$ of the predicted tracks. (An MHC filter inherently provides such estimates in the form of a track table (see section 3.5). In the case of a PHD filter, see Eq. (121).) Suppose that the sensor FoV is a cookie cutter: $p_D = \mathbf{1}_S$. Then an "ideal" noise- and clutter-free single-sensor observation-set at time-step $k+1$ would be

$$Z_{k+1} = \bigcup_{\hat{\mathbf{x}}_i \in S} \{\eta(\hat{\mathbf{x}}_i)\}$$

If the FoV is not a cookie cutter we must account for the fact that $p_D(\hat{\mathbf{x}}_i)$ may be neither zero nor one. We proceed as follows. For the sake of clarity, begin with a special case. Let S_0 be some subset of states and suppose that $p_D(\mathbf{x}) = 1$ when $\mathbf{x} \in S_0$ and, otherwise, $p_D(\mathbf{x}) = \varepsilon$ for some small positive number $\varepsilon < 1$. (In other words, we are modifying a cookie-cutter FoV S_0 to include the possibility of a small probability of detection outside of the FoV.) Next, let $S_a(p_D) \overset{\text{def.}}{=} \{\mathbf{x}|\ a \leq p_D(\mathbf{x})\}$. Then $S_a(p_D) = \mathfrak{X}$ (all of state space) when $a \leq \varepsilon$ and $S_a(p_D) = S_0$ otherwise. Let A be a uniformly distributed random number on the unit interval $[0,1]$. Then $\omega \longmapsto S_{A(\omega)}(p_D)$ defines a random subset of states with two instantiations: $S_A(p_D) = \mathfrak{X}$ with probability $\Pr(S_A(p_D) = \mathfrak{X}) = \Pr(A \leq \varepsilon) = \varepsilon$ and $S_A(p_D) = S_0$ with probability $\Pr(S_A(p_D) = S_0) = \Pr(A > \varepsilon) = 1 - \varepsilon$. In other words, $S_A(p_D)$ can be interpreted as a *random FoV* that is almost always equal to the cookie cutter FoV S_0; but that has some small probability of being infinite in extent. (Stated differently: There is a small probability that observations will be collected even if the target is not in S_0.)

If p_D is a general FoV then the random set $S_A(p_D)$ can be regarded as selecting among a range of possible alternative cookie-cutter FoVs, the shapes of which are specified by p_D. Moreover, this random set contains exactly the same information as p_D since p_D can be recovered from it: $\Pr(\mathbf{x} \in S_A(p_D)) = \Pr(A \leq p_D(\mathbf{x})) = p_D(\mathbf{x})$. Also note that, for fixed $\mathbf{x}$, the expected value of the random number $\mathbf{1}_{S_A(p_D)}(\mathbf{x})$ is

$$\begin{aligned} \mathrm{E}[\mathbf{1}_{S_A(p_D)}(\mathbf{x})] &= \int_0^1 \mathbf{1}_{S_a(p_D)}(\mathbf{x}) \cdot f_A(a)da \qquad (101) \\ &= \int_0^1 \mathbf{1}_{S_a(p_D)}(\mathbf{x})da = \int_0^{p_D(\mathbf{x})} 1 \cdot da = p_D(\mathbf{x}) \end{aligned}$$

In the next section, this equation will allow us to account for observations that are generated by sensors with arbitrary fields of view.

4.2. *Step 2: The Hedged Posterior p.g.fl.*

Assume that it is possible to approximate the posterior p.g.fl. in the form

$$G_{k+1|k+1}[h] \cong G[h] \cdot \prod_{\mathbf{z} \in Z_{k+1}} \gamma_{\mathbf{z}}[h] \qquad (102)$$

for some $G[h]$ such that $G[1] = 1$ and which has no dependence upon Z_{k+1}; and for some family of functionals $\gamma_{\mathbf{z}}[h]$ such that $\gamma_{\mathbf{z}}[h] = 1$ for all $\mathbf{z}$. (In what follows this will prove to be the case, for example, if Poisson-type approximations such as Eqs. (93) or (99) are made.) Taking the logarithm,

$$\log G_{k+1|k+1}[h] \cong \log G[h] + \sum_{\mathbf{z} \in Z_{k+1}} \log \gamma_{\mathbf{z}}[h]$$

Choose some fixed instantiation $S_a(p_D)$ of the random FoV $S_A(p_D)$. Then we will not be able to collect the ideal observation $\eta(\hat{\mathbf{x}}_i)$ unless $\hat{\mathbf{x}}_i \in S_a(p_D)$—i.e., unless $\mathbf{1}_{S_a(p_D)}(\hat{\mathbf{x}}_i) \neq 0$. So, the log-posterior p.g.fl. corresponding to the ideal observation-set must be

$$\log G_{k+1|k+1}[h] \cong \log G[h] + \sum_{i=1}^{\hat{n}} \mathbf{1}_{S_a(p_D)}(\hat{\mathbf{x}}_i) \cdot \log \gamma_{\eta(\hat{\mathbf{x}}_i)}[h]$$

This equation corresponds to only one of the possible FoVs defined by p_D. We must produce an equation that corresponds to an "average FoV." By Eq. (101) the expected log-posterior, averaged over all possible FoVs is

$$\begin{aligned} \mathrm{E}[\log G_{k+1|k+1}[h]] &\cong \log G[h] + \sum_{i=1}^{\hat{n}} \mathrm{E}[\mathbf{1}_{S_A(p_D)}(\hat{\mathbf{x}}_i)] \cdot \log \gamma_{\eta(\hat{\mathbf{x}}_i)}[h] \\ &= \log G[h] + \sum_{i=1}^{\hat{n}} p_D(\hat{\mathbf{x}}_i) \cdot \log \gamma_{\eta(\hat{\mathbf{x}}_i)}[h] \end{aligned}$$

Taking the exponential, we get what we will call the *hedged posterior p.g.fl.*:

$$\mathring{G}_{k+1|k+1}[h] = G[h] \cdot \prod_{i=1}^{\hat{n}} \gamma_{\eta(\hat{\mathbf{x}}_i)}[h]^{p_D(\hat{\mathbf{x}}_i)} \tag{103}$$

Note that $\mathring{G}_{k+1|k+1}[1] = 1$, as must be the case with any p.g.fl.

4.3. *Step 3: PHD of the Hedged p.g.fl.*

According to Eq. (48), the PHD of $\mathring{G}_{k+1|k+1}[h]$ may be computed as:

$$\begin{aligned}\mathring{D}_{k+1|k+1}(\mathbf{x}) &= \frac{\delta \log \mathring{G}_{k+1|k+1}}{\delta \mathbf{x}}[1] \qquad (104)\\ &= \left[\frac{1}{G[h]}\frac{\delta G}{\delta \mathbf{x}}[h] + \sum_{i=1}^{\hat{n}} p_D(\hat{\mathbf{x}}_i) \cdot \frac{1}{\gamma_{\eta(\hat{\mathbf{x}}_i)}[h]} \frac{\delta \gamma_{\eta(\hat{\mathbf{x}}_i)}}{\delta \mathbf{x}}[h]\right]_{h=1}\\ &= \frac{\delta G}{\delta \mathbf{x}}[1] + \sum_{i=1}^{\hat{n}} p_D(\hat{\mathbf{x}}_i) \cdot \frac{\delta \gamma_{\eta(\hat{\mathbf{x}}_i)}}{\delta \mathbf{x}}[1]\end{aligned}$$

4.4. *Step 4: Posterior Expected Number of Tracks (PENT)*

According to Eq. (47), the integral of this PHD yields the expected number of tracks (given that we succeeded in selecting the future FoV so that the ideal observation-set is collected):

$$\mathring{N}_{k+1|k+1} \cong \int \mathring{D}_{k+1|k+1}(\mathbf{x}) d\mathbf{x} \tag{105}$$

4.5. *Extending PENT to the Multisensor Case*

The concepts presented in the previous sections can be extended to the multisensor case as follows. Using the notation introduced in section 2.8, assume for convenience that two sensors are present. (The multisensor case will follow immediately by analogy.) As in section 2.8 assume that

$$\overset{j}{f}_{k+1}(\overset{j}{\mathbf{z}}|\mathbf{x}, \overset{*j}{\mathbf{x}}_{k+1}) = f_{\overset{j}{\mathbf{W}}_{k+1}}(\overset{j}{\mathbf{z}} - \overset{j}{\eta}_{k+1}(\mathbf{x}, \overset{*j}{\mathbf{x}}_{k+1}))$$

where in what follows we abbreviate $\overset{j}{\eta}(\mathbf{x}) \overset{\text{abbr.}}{=} \overset{j}{\eta}_{k+1}(\mathbf{x}, \overset{*j}{\mathbf{x}}_{k+1})$. We assume that the sensors have respective Poisson false alarm processes $\overset{1}{\lambda}_{k+1}, \overset{1}{c}_{k+1}(\overset{1}{\mathbf{z}})$ and $\overset{2}{\lambda}_{k+1}, \overset{2}{c}_{k+1}(\overset{2}{\mathbf{z}})$ where we abbreviate $\overset{1}{\lambda} \overset{\text{abbr.}}{=} \overset{1}{\lambda}_{k+1}$, $\overset{1}{c} \overset{\text{abbr.}}{=} \overset{1}{c}_{k+1}$, $\overset{2}{\lambda} \overset{\text{abbr.}}{=} \overset{2}{\lambda}_{k+1}$, $\overset{2}{c} \overset{\text{abbr.}}{=} \overset{2}{c}_{k+1}$. The fields of view are

$$\overset{1}{p}_D(\mathbf{x}) \overset{\text{abbr.}}{=} \overset{1}{p}_D(\mathbf{x}, \overset{*1}{\mathbf{x}}_{k+1}), \qquad \overset{2}{p}_D(\mathbf{x}) \overset{\text{abbr.}}{=} \overset{2}{p}_D(\mathbf{x}, \overset{*2}{\mathbf{x}}_{k+1})$$

Unfortunately, we cannot proceed as before because a strictly rigorous development of the two-sensor case leads to intractable formulas. Instead we approximate by modeling the two sensors as a single imaginary "pseudo-sensor." This will allow us to apply the reasoning used for the single-sensor case. Since an ideal-observation set contains ideal observations collected from each target by at least one (but not necessarily both) sensors, we can take the sensor field of view to be

$$\breve{p}_D(\mathbf{x}) \overset{\text{abbr.}}{=} \breve{p}_D(\mathbf{x}, \overset{*1}{\mathbf{x}}, \overset{*2}{\mathbf{x}}) = 1 - \left(1 - \overset{1}{p}_D(\mathbf{x}, \overset{*1}{\mathbf{x}})\right)\left(1 - \overset{2}{p}_D(\mathbf{x}, \overset{*2}{\mathbf{x}})\right) \tag{106}$$

In the multisensor case this will be

$$\breve{p}_D(\mathbf{x}) \overset{\text{abbr.}}{=} \breve{p}_D(\mathbf{x}, \overset{*1}{\mathbf{x}}, ..., \overset{*s}{\mathbf{x}}) = 1 - \left(1 - \overset{1}{p}_D(\mathbf{x}, \overset{*1}{\mathbf{x}})\right) \cdots \left(1 - \overset{2}{p}_D(\mathbf{x}, \overset{*s}{\mathbf{x}})\right) \tag{107}$$

In either case $\breve{p}_D(\mathbf{x})$ is the probability that at least one of the sensors will collect an observation if a target with state $\mathbf{x}$ is present.

We assume that the imaginary sensor collects observations of the form $\tilde{\mathbf{z}} = \overset{1}{\mathbf{z}}$, $\tilde{\mathbf{z}} = \overset{2}{\mathbf{z}}$, or $\tilde{\mathbf{z}} = (\overset{1}{\mathbf{z}}, \overset{2}{\mathbf{z}})$ and has the likelihood function

$$\tilde{L}_{\overset{1}{\mathbf{z}}}(\mathbf{x}, \overset{*1}{\mathbf{x}}, \overset{*2}{\mathbf{x}}) \overset{\text{abbr.}}{=} \frac{\overset{1}{p}_D(\mathbf{x}, \overset{*1}{\mathbf{x}}) \cdot (1 - \overset{2}{p}_D(\mathbf{x}, \overset{*2}{\mathbf{x}}))}{\breve{p}_D(\mathbf{x}, \overset{*1}{\mathbf{x}}, \overset{*2}{\mathbf{x}})} \cdot f_{k+1}(\overset{1}{\mathbf{z}}|\mathbf{x}, \overset{*1}{\mathbf{x}}) \tag{108}$$

$$\tilde{L}_{\overset{2}{\mathbf{z}}}(\mathbf{x}, \overset{*1}{\mathbf{x}}, \overset{*2}{\mathbf{x}}) \overset{\text{abbr.}}{=} \frac{(1 - \overset{1}{p}_D(\mathbf{x}, \overset{*1}{\mathbf{x}})) \cdot \overset{2}{p}_D(\mathbf{x}, \overset{*2}{\mathbf{x}})}{\breve{p}_D(\mathbf{x}, \overset{*1}{\mathbf{x}}, \overset{*2}{\mathbf{x}})} \cdot f_{k+1}(\overset{2}{\mathbf{z}}|\mathbf{x}, \overset{*2}{\mathbf{x}}) \tag{109}$$

$$\tilde{L}_{(\overset{1}{\mathbf{z}}, \overset{2}{\mathbf{z}})}(\mathbf{x}, \overset{*1}{\mathbf{x}}, \overset{*2}{\mathbf{x}}) \overset{\text{abbr.}}{=} \frac{\overset{1}{p}_D(\mathbf{x}, \overset{*1}{\mathbf{x}}) \cdot \overset{2}{p}_D(\mathbf{x}, \overset{*2}{\mathbf{x}})}{\breve{p}_D(\mathbf{x}, \overset{*1}{\mathbf{x}}, \overset{*2}{\mathbf{x}})} \cdot f_{k+1}(\overset{1}{\mathbf{z}}|\mathbf{x}, \overset{*1}{\mathbf{x}}) \cdot f_{k+1}(\overset{2}{\mathbf{z}}|\mathbf{x}, \overset{*2}{\mathbf{x}}) \tag{110}$$

Note that this likelihood is well-defined since

$$\int \tilde{L}_{\overset{1}{\mathbf{z}}}(\mathbf{x}, \overset{*1}{\mathbf{x}}, \overset{*2}{\mathbf{x}}) d\overset{1}{\mathbf{z}} + \int \tilde{L}_{\overset{2}{\mathbf{z}}}(\mathbf{x}, \overset{*1}{\mathbf{x}}, \overset{*2}{\mathbf{x}}) d\overset{2}{\mathbf{z}} + \int \tilde{L}_{(\overset{1}{\mathbf{z}}, \overset{2}{\mathbf{z}})}(\mathbf{x}, \overset{*1}{\mathbf{x}}, \overset{*2}{\mathbf{x}}) d\overset{1}{\mathbf{z}} d\overset{2}{\mathbf{z}} = 1$$

We must also specify a Poisson false alarm process for the pseudo-sensor—specifically, an expected number $\tilde{\lambda}_{k+1}$ of false alarms and a spatial distribution $\tilde{c}_{k+1}(\tilde{\mathbf{z}})$ such that $\int \tilde{c}_{k+1}(\overset{1}{\mathbf{z}}) d\overset{1}{\mathbf{z}} + \int \tilde{c}_{k+1}(\overset{2}{\mathbf{z}}) d\overset{2}{\mathbf{z}} + \int \tilde{c}_{k+1}(\overset{1}{\mathbf{z}}, \overset{2}{\mathbf{z}}) d\overset{1}{\mathbf{z}} d\overset{2}{\mathbf{z}} = 1$. Abbreviate $\tilde{\lambda} \overset{\text{abbr.}}{=} \tilde{\lambda}_{k+1}$ and $\tilde{c} \overset{\text{abbr.}}{=} \tilde{c}_{k+1}$. The actual joint false alarm process for the two sensors is also Poisson and is given by

$$\overset{12}{c}(\{\overset{1}{\mathbf{z}}_1, ..., \overset{1}{\mathbf{z}}_{\overset{1}{m}}, \overset{2}{\mathbf{z}}_1, ..., \overset{2}{\mathbf{z}}_{\overset{2}{m}}\}) = e^{-\overset{1}{\lambda} - \overset{2}{\lambda}} \cdot (\overset{1}{\lambda} + \overset{2}{\lambda})^{\overset{1}{m} + \overset{2}{m}} \cdot \tilde{c}(\overset{1}{\mathbf{z}}_1) \cdots \tilde{c}(\overset{1}{\mathbf{z}}_{\overset{1}{m}}) \cdot \tilde{c}(\overset{2}{\mathbf{z}}_1) \cdots \tilde{c}(\overset{2}{\mathbf{z}}_{\overset{2}{m}})$$

So we should set $\tilde{\lambda} = \overset{1}{\lambda} + \overset{2}{\lambda}$. The probability that the first sensor will collect false alarms but the second will not is $e^{-\overset{2}{\lambda}}$; and the probability that the second sensor will collect false alarms but the first will not is $e^{-\overset{2}{\lambda}}$. So set

$$\tilde{\lambda}\cdot\tilde{c}(\overset{1}{\mathbf{z}}) = e^{-\overset{2}{\lambda}}\cdot\overset{1}{\lambda}\overset{1}{c}(\overset{1}{\mathbf{z}}), \qquad \tilde{\lambda}\cdot\tilde{c}(\overset{2}{\mathbf{z}}) = e^{-\overset{1}{\lambda}}\cdot\overset{2}{\lambda}\overset{2}{c}(\overset{2}{\mathbf{z}}) \tag{111}$$

$$\tilde{\lambda}\cdot\tilde{c}(\overset{1}{\mathbf{z}},\overset{2}{\mathbf{z}}) = \left((1-e^{-\overset{2}{\lambda}})\overset{1}{\lambda} + (1-e^{-\overset{1}{\lambda}})\overset{2}{\lambda}\right)\cdot\overset{1}{c}(\overset{1}{\mathbf{z}})\cdot\overset{2}{c}(\overset{2}{\mathbf{z}})$$

Note that the original single-sensor false alarm models are limiting cases of this model. For example, if $\overset{2}{\lambda}=0$ then $\tilde{\lambda}\cdot\tilde{c}(\overset{2}{\mathbf{z}})=0$, $\tilde{\lambda}\cdot\tilde{c}(\overset{1}{\mathbf{z}},\overset{2}{\mathbf{z}})=0$, and $\tilde{\lambda}\cdot\tilde{c}(\overset{1}{\mathbf{z}})=\overset{1}{\lambda}\overset{1}{c}(\overset{1}{\mathbf{z}})$.

Now let the observation-set for the imaginary sensor at time-step $k+1$ be $\tilde{Z}_{k+1} = \overset{1}{Z}_{k+1}\cup\overset{2}{Z}_{k+1}\cup\overset{12}{Z}_{k+1}$where $\overset{1}{Z}_{k+1}$ consists of the observations in $\tilde{Z}_{k+1}$ of the form $\overset{1}{\mathbf{z}}$; where $\overset{2}{Z}_{k+1}$ consists of the observations in $\tilde{Z}_{k+1}$ of the form $\overset{2}{\mathbf{z}}$; and where $\overset{12}{Z}_{k+1}$ consists of the observations in $\tilde{Z}_{k+1}$ of the form $(\overset{1}{\mathbf{z}},\overset{2}{\mathbf{z}})$. Then we assume that, for this pseudo-sensor, it is possible to make the same type of approximation as in Eq. (102).

$$\begin{aligned} G_{k+1|k+1}[h] &\cong G[h]\cdot\prod_{\tilde{\mathbf{z}}\in\tilde{Z}_{k+1}}\tilde{\gamma}_{\tilde{\mathbf{z}}}[h] \\ &= G[h]\cdot\left(\prod_{\overset{1}{\mathbf{z}}\in\overset{1}{Z}_{k+1}}\tilde{\gamma}_{\overset{1}{\mathbf{z}}}[h]\right)\cdot\left(\prod_{\overset{2}{\mathbf{z}}\in\overset{2}{Z}_{k+1}}\tilde{\gamma}_{\overset{2}{\mathbf{z}}}[h]\right)\cdot\left(\prod_{(\overset{1}{\mathbf{z}},\overset{2}{\mathbf{z}})\in\overset{12}{Z}_{k+1}}\tilde{\gamma}_{(\overset{1}{\mathbf{z}},\overset{2}{\mathbf{z}})}[h]\right)\end{aligned}$$

in which case the log-posterior is

$$\begin{aligned}&\log G_{k+1|k+1}[h] \\ &\quad\cong \log G[h] + \sum_{\overset{1}{\mathbf{z}}\in\overset{1}{Z}_{k+1}}\log\tilde{\gamma}_{\overset{1}{\mathbf{z}}}[h] + \sum_{\overset{2}{\mathbf{z}}\in\overset{2}{Z}_{k+1}}\log\tilde{\gamma}_{\overset{2}{\mathbf{z}}}[h] + \sum_{(\overset{1}{\mathbf{z}},\overset{2}{\mathbf{z}})\in\overset{12}{Z}_{k+1}}\log\tilde{\gamma}_{(\overset{1}{\mathbf{z}},\overset{2}{\mathbf{z}})}[h]\end{aligned}$$

Assume that the sensor FoVs are cookie cutters: $\overset{1}{p}_D(\mathbf{x}) = \mathbf{1}_{S_1}(\mathbf{x})$ and $\overset{2}{p}_D(\mathbf{x}) = \mathbf{1}_{S_2}(\mathbf{x})$. Then ideal pseudo-sensor observations cannot be collected unless the following relationships hold:

$$\begin{array}{rl} \overset{1}{\eta}(\hat{\mathbf{x}}_i) & \text{collected if } \hat{\mathbf{x}}_i\in S_1-(S_1\cap S_2) \\ \overset{2}{\eta}(\hat{\mathbf{x}}_i) & \text{collected if } \hat{\mathbf{x}}_i\in S_2-(S_1\cap S_2) \\ (\overset{1}{\eta}(\hat{\mathbf{x}}_i),\ \overset{2}{\eta}(\hat{\mathbf{x}}_i)) & \text{collected if } \quad \hat{\mathbf{x}}_i\in S_1\cap S_2 \end{array}$$

So, the log-posterior corresponding to the ideal observation-set is

$$
\begin{aligned}
&\log G_{k+1|k+1}[h] \\
&\quad \cong \log G[h] + \sum_{i=1}^{\hat{n}} \mathbf{1}_{S_1-(S_1\cap S_2)}(\hat{\mathbf{x}}_i) \cdot \log \tilde{\gamma}_{\overset{1}{\eta}(\hat{\mathbf{x}}_i)}[h] \\
&\quad + \sum_{i=1}^{\hat{n}} \mathbf{1}_{S_2-(S_1\cap S_2)}(\hat{\mathbf{x}}_i) \cdot \log \tilde{\gamma}_{\overset{2}{\eta}(\hat{\mathbf{x}}_i)}[h] \\
&\quad + \sum_{i=1}^{\hat{n}} \mathbf{1}_{S_1\cap S_2}(\hat{\mathbf{x}}_i) \cdot \log \tilde{\gamma}_{(\overset{1}{\eta}(\hat{\mathbf{x}}_i),\overset{2}{\eta}(\hat{\mathbf{x}}_i))}[h]
\end{aligned}
\tag{112}
$$

where

$$
\begin{aligned}
\mathbf{1}_{S_1\cap S_2}(\hat{\mathbf{x}}_i) &= \mathbf{1}_{S_1}(\hat{\mathbf{x}}_i) \cdot \mathbf{1}_{S_2}(\hat{\mathbf{x}}_i) \\
\mathbf{1}_{S_1-(S_1\cap S_2)}(\hat{\mathbf{x}}_i) &= \mathbf{1}_{S_1}(\hat{\mathbf{x}}_i) - \mathbf{1}_{S_1}(\hat{\mathbf{x}}_i) \cdot \mathbf{1}_{S_2}(\hat{\mathbf{x}}_i) \\
\mathbf{1}_{S_2-(S_1\cap S_2)}(\hat{\mathbf{x}}_i) &= \mathbf{1}_{S_2}(\hat{\mathbf{x}}_i) - \mathbf{1}_{S_1}(\hat{\mathbf{x}}_i) \cdot \mathbf{1}_{S_2}(\hat{\mathbf{x}}_i)
\end{aligned}
$$

Assume next that S_1 and S_2 are instantiations of random sets $S_{A_1}(\overset{1}{p}_D)$ and $S_2 = S_{A_2}(\overset{2}{p}_D)$ defined by the FoVs: $S_1 = S_{a_1}(\overset{1}{p}_D)$ and $S_2 = S_{a_2}(\overset{2}{p}_D)$. Here A_1 and A_2 are uniformly distributed random numbers on $[0,1]$. We assume that A_1 and A_2 independent (i.e., the detection processes for the two sensors are independent), so that $f_{A_1,A_2}(a_1,a_2) = f_{A_1}(a_1) \cdot f_{A_2}(a_2)$. Then the expected log-posterior p.g.fl. is

$$
\begin{aligned}
\mathrm{E}[\log D_{k+1|k+1}[h]] &\simeq \int_0^1 \int_0^1 \log G_{k+1|k+1}[h] \cdot f_{A_1}(a_1) f_{A_2}(a_2) da_1 da_2 \qquad (113) \\
&= \log G[h] + \sum_{i=1}^{\hat{n}} \overset{1}{p}_D(\hat{\mathbf{x}}_i) \cdot (1 - \overset{2}{p}_D(\hat{\mathbf{x}}_i)) \cdot \log \gamma_{\overset{1}{\eta}(\hat{\mathbf{x}}_i)}[h] \\
&\quad + \sum_{i=1}^{\hat{n}} \overset{2}{p}_D(\hat{\mathbf{x}}_i) \cdot (1 - \overset{1}{p}_D(\hat{\mathbf{x}}_i)) \cdot \log \gamma_{\overset{2}{\eta}(\hat{\mathbf{x}}_i)}[h] \\
&\quad + \sum_{i=1}^{\hat{n}} \overset{1}{p}_D(\hat{\mathbf{x}}_i) \cdot \overset{2}{p}_D(\hat{\mathbf{x}}_i) \cdot \log \gamma_{(\overset{1}{\eta}(\hat{\mathbf{x}}_i),\overset{2}{\eta}(\hat{\mathbf{x}}_i))}[h]
\end{aligned}
$$

Consequently, the hedged posterior is

$$\mathring{G}_{k+1|k+1}[h] = G[h] \cdot \prod_{i=1}^{\hat{n}} \tilde{\gamma}_{\overset{1}{\eta}(\hat{\mathbf{x}}_i)}[h]^{\overset{1}{p}_D(\hat{\mathbf{x}}_i)\cdot(1-\overset{2}{p}_D(\hat{\mathbf{x}}_i))} \tag{114}$$
$$\cdot \prod_{i=1}^{\hat{n}} \tilde{\gamma}_{\overset{2}{\eta}(\hat{\mathbf{x}}_i)}[h]^{\overset{2}{p}_D(\hat{\mathbf{x}}_i)\cdot(1-\overset{1}{p}_D(\hat{\mathbf{x}}_i))}$$
$$\cdot \prod_{i=1}^{\hat{n}} \tilde{\gamma}_{\overset{1}{\eta}(\hat{\mathbf{x}}_i)}[h]^{\overset{1}{p}_D(\hat{\mathbf{x}}_i)\cdot(1-\overset{2}{p}_D(\hat{\mathbf{x}}_i))}$$

In this case the PHD of the hedged posterior has the form

$$\mathring{D}_{k+1|k+1}(\mathbf{x}) \cong \frac{\delta G}{\delta \mathbf{x}}[1] + \sum_{i=1}^{\hat{n}} \overset{1}{p}_D(\hat{\mathbf{x}}_i) \cdot (1 - \overset{2}{p}_D(\hat{\mathbf{x}}_i)) \cdot \frac{\delta \tilde{\gamma}_{\overset{1}{\eta}(\hat{\mathbf{x}}_i)}}{\delta \mathbf{x}}[1]$$
$$+ \sum_{i=1}^{\hat{n}} \overset{2}{p}_D(\hat{\mathbf{x}}_i) \cdot (1 - \overset{2}{p}_D(\hat{\mathbf{x}}_i)) \cdot \frac{\delta \tilde{\gamma}_{\overset{2}{\eta}(\hat{\mathbf{x}}_i)}}{\delta \mathbf{x}}[1]$$
$$+ \sum_{i=1}^{\hat{n}} \overset{1}{p}_D(\hat{\mathbf{x}}_i) \cdot \overset{2}{p}_D(\hat{\mathbf{x}}_i) \cdot \frac{\delta \tilde{\gamma}_{(\overset{1}{\eta}(\hat{\mathbf{x}}_i), \overset{2}{\eta}(\hat{\mathbf{x}}_i))}}{\delta \mathbf{x}}[1] \tag{115}$$

4.6. *Multisensor PENT: Simplified Version*

Unfortunately, in general the "pseudo-sensor" approximation just outlined will produce excessively complicated formulas for the posterior expected number of targets. We can simplify things using a stronger approximation. Assume that both sensors collect a relatively large number of observations, so that $\tilde{\lambda} \cdot \tilde{c}(\overset{1}{\mathbf{z}}) \cong 0$ and $\tilde{\lambda} \cdot \tilde{c}(\overset{2}{\mathbf{z}}) \cong 0$. Also assume that both sensors collect ideal observations from all targets, regardless of the placement of the FoVs. This means that all observations will be pairs $(\overset{1}{\mathbf{z}}, \overset{2}{\mathbf{z}})$. (Note that this approximation will produce more optimism than is justified.) Then we can replace the pseudo-sensor likelihood of equations (108-110) by the pseudo-sensor likelihood

$$\tilde{L}_{(\overset{1}{\mathbf{z}}, \overset{2}{\mathbf{z}})}(\mathbf{x}, \overset{*1}{\mathbf{x}}, \overset{*2}{\mathbf{x}}) \overset{\text{abbr.}}{=} f_{k+1}(\overset{1}{\mathbf{z}}|\mathbf{x}, \overset{*1}{\mathbf{x}}) \cdot f_{k+1}(\overset{2}{\mathbf{z}}|\mathbf{x}, \overset{*2}{\mathbf{x}}) \tag{116}$$

Likewise the false alarm model of equations (111) can be replaced by the model $\tilde{\lambda} = \overset{1}{\lambda} + \overset{2}{\lambda}$ and

$$\tilde{\lambda} \cdot \tilde{c}(\overset{1}{\mathbf{z}}) = 0, \quad \tilde{\lambda} \cdot \tilde{c}(\overset{2}{\mathbf{z}}) = 0, \quad \tilde{\lambda} \cdot \tilde{c}(\overset{1}{\mathbf{z}}, \overset{2}{\mathbf{z}}) = \tilde{\lambda} \cdot \overset{1}{c}(\overset{1}{\mathbf{z}}) \cdot \overset{2}{c}(\overset{2}{\mathbf{z}}) \tag{117}$$

Thus Eq. (113) becomes

$$
\begin{aligned}
&\mathrm{E}[\,\log G_{k+1|k+1}[h]] \\
&\qquad \cong \log G[h] + \sum_{i=1}^{\hat{n}} \overset{1}{p}_D(\hat{\mathbf{x}}_i) \cdot (1 - \overset{2}{p}_D(\hat{\mathbf{x}}_i)) \cdot \log \tilde{\gamma}_{(\overset{1}{\eta}(\hat{\mathbf{x}}_i), \overset{2}{\eta}(\hat{\mathbf{x}}_i))}[h] \\
&\qquad + \sum_{i=1}^{\hat{n}} \overset{2}{p}_D(\hat{\mathbf{x}}_i) \cdot (1 - \overset{1}{p}_D(\hat{\mathbf{x}}_i)) \cdot \log \tilde{\gamma}_{(\overset{1}{\eta}(\hat{\mathbf{x}}_i), \overset{2}{\eta}(\hat{\mathbf{x}}_i))}[h] \\
&\qquad + \sum_{i=1}^{\hat{n}} \overset{1}{p}_D(\hat{\mathbf{x}}_i) \cdot \overset{2}{p}_D(\hat{\mathbf{x}}_i) \cdot \log \tilde{\gamma}_{(\overset{1}{\eta}(\hat{\mathbf{x}}_i), \overset{2}{\eta}(\hat{\mathbf{x}}_i))}[h] \\
&\qquad = \log G[h] + \sum_{i=1}^{\hat{n}} \breve{p}_D(\hat{\mathbf{x}}_i) \cdot \log \tilde{\gamma}_{(\overset{1}{\eta}(\hat{\mathbf{x}}_i), \overset{2}{\eta}(\hat{\mathbf{x}}_i))}[h]
\end{aligned}
\tag{118}
$$

where the last equation follows from Eq. (106). In this case the hedged posterior has the simplified form

$$
\mathring{G}_{k+1|k+1}[h] = G[h] \cdot \prod_{i=1}^{\hat{n}} \tilde{\gamma}_{(\overset{1}{\eta}(\hat{\mathbf{x}}_i), \overset{2}{\eta}(\hat{\mathbf{x}}_i))}[h]^{\breve{p}_D(\hat{\mathbf{x}}_i)}
\tag{119}
$$

5. Posterior Expected Number of Targets (PENT)

In sections 3.4 and 3.5 we described the process of integrating a maxi-PIMS hedged objective function with the PHD and MHC multitarget filters. In this section we derive concrete formulas for PENT that can be used with these two filters.

The section is organized as follows. In section 5.1 we begin with the single-sensor, multitarget case and single-step look-ahead. The behavior of PENT in this case is illustrated with a simple example in section 5.2. The analysis is extended to the multisensor-multitarget case with single-step look-ahead in section 5.3. Sections 5.4 and 5.5 address the multistep look-ahead case, for a single sensor and multiple sensors, respectively.

5.1. *PENT: Single-Sensor, One-Step Look-Ahead*

We follow the procedure outlined in sections 3 and 4. We begin by deriving a formula for PENT that can be used with the PHD filter of section 3.4. From Eq. (94) we know that the two-variable p.g.fl. for the Bayes update at time-step $k+1$ is:

$$
F_{k+1}[g,h] \cong \exp\left(-\lambda - N_{k+1|k} + \lambda c_g + D_{k+1|k}[h(1 - p_D + p_D p_g)]\right)
$$

Taking iterated functional derivatives with respect to the measurements in $Z_{k+1} = \{\mathbf{z}_1, ..., \mathbf{z}_m\}$ yields

$$\frac{\delta^m F_{k+1}}{\delta \mathbf{z}_1 \cdots \delta \mathbf{z}_m}[g,h] = F_{k+1}[g,h] \cdot \prod_{i=1}^{m} \left(\lambda c(\mathbf{z}_i) + D_{k+1|k}[hp_D L_{\mathbf{z}_i}]\right) \tag{120}$$

So, the posterior p.g.fl. conditioned on any given observation-set $Z_{k+1} = \{\mathbf{z}_1, ..., \mathbf{z}_m\}$ is given by Eq. (89):

$$\begin{aligned} G_{k+1|k+1}[h] &= \frac{\frac{\delta^m F_{k+1}}{\delta \mathbf{z}_1 \cdots \delta \mathbf{z}_m}[0,h]}{\frac{\delta^m F_{k+1}}{\delta \mathbf{z}_1 \cdots \delta \mathbf{z}_m}[0,1]} \\ &= e^{D_{k+1|k}[(h-1)(1-p_D)]} \cdot \prod_{i=1}^{m} \frac{\lambda c(\mathbf{z}_i) + D_{k+1|k}[hp_D L_{\mathbf{z}_i}]}{\lambda c(\mathbf{z}_i) + D_{k+1|k}[p_D L_{\mathbf{z}_i}]} \end{aligned}$$

Because this has the form that was assumed in Eq. (102), we can construct a hedged posterior p.g.fl. Use some technique such as the expectation-maximization (EM) algorithm to approximate the predicted PHD as a linear combination of Gaussian distributions:

$$D_{k+1|k}(\mathbf{x}) \cong q_1 f_1(\mathbf{x}) + ... + q_N f_N(\mathbf{x}) \tag{121}$$

where $q_1 + ... + q_N \cong N_{k+1|k}$ is the expected number of targets and where $f_j(\mathbf{x}) = N_{P_j}(\mathbf{x} - \hat{\mathbf{x}}_j)$. Then according to Eq. (103) the hedged posterior p.g.fl. is:

$$\begin{aligned} \mathring{G}_{k+1|k+1}[h] &= e^{D_{k+1|k}[(h-1)(1-p_D)]} \\ &\cdot \prod_{j=1}^{N} \left(\frac{\lambda c(\eta(\hat{\mathbf{x}}_j)) + D_{k+1|k}[hp_D L_{\eta(\hat{\mathbf{x}}_j)}]}{\lambda c(\eta(\hat{\mathbf{x}}_j)) + D_{k+1|k}[p_D L_{\eta(\hat{\mathbf{x}}_j)}]} \right)^{p_D(\hat{\mathbf{x}}_j)} \end{aligned} \tag{122}$$

Likewise we can construct its PHD according to Eq. (104):

$$\begin{aligned} &\mathring{D}_{k+1|k+1}(\mathbf{x}) \\ &= \left(1 - p_D(\mathbf{x}) + \sum_{j=1}^{N} p_D(\hat{\mathbf{x}}_j) \cdot \frac{p_D(\mathbf{x}) L_{\eta(\hat{\mathbf{x}}_j)}(\mathbf{x})}{\lambda c(\eta(\hat{\mathbf{x}}_j)) + D_{k+1|k}[p_D L_{\eta(\hat{\mathbf{x}}_j)}]} \right) \\ &\cdot D_{k+1|k}(\mathbf{x}) \end{aligned} \tag{123}$$

From section 4.4 we know that the integral of this yields the posterior expected number of targets:

$$\mathring{N}_{k+1|k+1}(\mathring{\mathbf{x}}_{k+1}) = D_{k+1|k}[1-p_D] + \sum_{j=1}^{N} p_D(\hat{\mathbf{x}}_j) \cdot \frac{D_{k+1|k}[p_D L_{\eta(\hat{\mathbf{x}}_j)}]}{\lambda c(\eta(\hat{\mathbf{x}}_j)) + D_{k+1|k}[p_D L_{\eta(\hat{\mathbf{x}}_j)}]}$$

or, using partial fractions,

$$\mathring{N}_{k+1|k+1}(\overset{*}{\mathbf{x}}_{k+1}) = D_{k+1|k}[1-p_D] + \sum_{j=1}^{N} p_D(\hat{\mathbf{x}}_j) \cdot \left(1 - \frac{\lambda c(\eta(\hat{\mathbf{x}}_j))}{\lambda c(\eta(\hat{\mathbf{x}}_j)) + D_{k+1|k}[p_D L_{\eta(\hat{\mathbf{x}}_j)}]}\right) \tag{124}$$

This formula for the PENT is intended for use with the PHD filter of section 3.4.

In the case of the MHC filter of section 3.5, we know from Eq. (98) that the predicted PHD already has the form of Eq. (121). Therefore, the formula for PENT suitable for use with an MHC filter is

$$\mathring{N}_{k+1|k+1}(\overset{*}{\mathbf{x}}_{k+1}) = \sum_{j=1}^{N} q_j f_j[1-p_D] + p_D(\hat{\mathbf{x}}_j) + \sum_{j=1}^{N} p_D(\hat{\mathbf{x}}_j) \cdot \left(1 - \frac{\lambda c(\eta(\hat{\mathbf{x}}_j))}{\lambda c(\eta(\hat{\mathbf{x}}_j)) + \sum_{e=1}^{N} q_e f_e[p_D L_{\eta(\hat{\mathbf{x}}_j)}]}\right) \tag{125}$$

Equations (124) and (125) can both be written in terms of closed-form formulas if one assumes that the probability of detection has linear-Gaussian form:

$$p_D(\mathbf{x}) = \exp\left(-\frac{1}{2}(A_{k+1}\mathbf{x} - \overset{*}{A}_{k+1}\overset{*}{\mathbf{x}}_{k+1})^T L_{k+1}^{-1}(A_{k+1}\mathbf{x} - \overset{*}{A}_{k+1}\overset{*}{\mathbf{x}}_{k+1})\right) \tag{126}$$

If there are no false alarms, then equations (124) and (125) respectively reduce to the simple form

$$\mathring{N}_{k+1|k+1}(\overset{*}{\mathbf{x}}_{k+1}) = D_{k+1|k}[1-p_D] + \sum_{j=1}^{N} p_D(\hat{\mathbf{x}}_j) \tag{127}$$

and

$$\mathring{N}_{k+1|k+1}(\overset{*}{\mathbf{x}}_{k+1}) = \sum_{j=1}^{N} (q_j f_j[1-p_D] + p_D(\hat{\mathbf{x}}_j)) \tag{128}$$

5.2. *PENT: Simple Example*

In this section we use a simple example to illustrate the behavior of PENT in the single-sensor, single-step look-ahead case. Suppose that the sensor FoV is a cookie cutter: $p_D = \mathbf{1}_T$ where T is a subset of state space that has fixed shape but can be translated to any location. Then $f_j[p_D] = p_j(T)$

where $p_j(T)$ is the amount of probability mass in the jth predicted track that is contained in T. (So, $p_j(T)$ is a measure of the degree to which the track has been localized.) Abbreviate $q = \sum_{j=1}^{N} q_j$. If T is in free space (not over any track) then $\breve{p}_D(\hat{\mathbf{x}}_j) = 0$ and $p_j(T) = 0$ for all $j = 1, ..., N$ and so $N_{k+1|k+1} = q$. But if the FoV is over the eth track then $p_D(\hat{\mathbf{x}}_j) = 0$ and $p_j(T) = 0$ for all $j \neq e$, whereas $p_D(\hat{\mathbf{x}}_e) = 1$. So

$$N_{k+1|k+1} = q + 1 - q_e p_e(T) \geq q$$

and so $N_{k+1|k+1}$ is globally maximized by placing T over some track rather than in free space. Given this, $N_{k+1|k+1}$ is maximized by choosing that track such that the product $q_e p_e(T)$ is minimized. In other words, more firm and more well-localized tracks (tracks with larger q_e and larger $p_e(T)$) will be ignored in favor of less firm and less localized tracks. This is what one would hope to see, since over-all information is not increased by placing the FoV over tracks for which one already possesses sufficient information. But also note that the choice depends on a balance between firmness q_e and degree of localization $p_e(T)$. The FoV may be placed over a relatively firm track ($q_e \cong 1$) if it is sufficiently poorly localized ($p_e(T) \cong 0$); and vice-versa. Yet, broadly speaking, the FoV will be placed over the track that is simultaneously least firm and least localized.

Similar simple examples can be constructed for the two-sensor case. These examples show that the PENT exhibits other desirable behavior. For example, if in the previous example one has two sensors and one sensor has an FoV large enough to encompass two of the tracks, the PENT objective function tends to direct the larger FoV to encompass the two tracks and the second FoV to encompass a third one.

5.3. *PENT: Multisensor, One-Step Look-Ahead*

We use the procedure outlined in sections 3 and 4. We consider only the two-sensor case since the multisensor case will be a self-evident extension. We apply the same analysis as in section 5.1, but with the simplified pseudo-sensor of section 4.6 substituted for the single sensor of section 5.1. According to Eq. (119), Eq. (122) must be replaced by

$$\mathring{G}_{k+1|k+1}[h] = e^{D_{k+1|k}[(h-1)(1-\breve{p}_D)]}$$
$$\cdot \prod_{j=1}^{N} \left(\frac{\tilde{\lambda}\tilde{c}(\overset{1}{\eta}(\hat{\mathbf{x}}_j), \overset{2}{\eta}(\hat{\mathbf{x}}_j)) + D_{k+1|k}[h\breve{p}_D \tilde{L}_{(\overset{1}{\eta}(\hat{\mathbf{x}}_j), \overset{2}{\eta}(\hat{\mathbf{x}}_j))}]}{\tilde{\lambda}\tilde{c}(\overset{1}{\eta}(\hat{\mathbf{x}}_j), \overset{2}{\eta}(\hat{\mathbf{x}}_j)) + D_{k+1|k}[\breve{p}_D \tilde{L}_{(\overset{1}{\eta}(\hat{\mathbf{x}}_j), \overset{2}{\eta}(\hat{\mathbf{x}}_j))}]} \right)^{\breve{p}_D(\hat{\mathbf{x}}_i)}$$

where the simplified pseudo-sensor is defined by Eq. (116). Since $\tilde{L}_{(\overset{1}{\eta}(\hat{\mathbf{x}}_j),\overset{2}{\eta}(\hat{\mathbf{x}}_j))} = L_{\overset{1}{\eta}(\hat{\mathbf{x}}_j)} \cdot L_{\overset{2}{\eta}(\hat{\mathbf{x}}_j)}$ the formula for the PENT becomes

$$\mathring{N}_{k+1|k+1}(\overset{*1}{\mathbf{x}}_{k+1}, \overset{*2}{\mathbf{x}}_{k+1}) = D_{k+1|k}[1 - \breve{p}_D] + \sum_{j=1}^{N} \breve{p}_D(\hat{\mathbf{x}}_i) \cdot \frac{D_{k+1|k}[\breve{p}_D L_{\overset{1}{\eta}(\hat{\mathbf{x}}_j)} \cdot L_{\overset{2}{\eta}(\hat{\mathbf{x}}_j)}]}{\tilde{\lambda}\tilde{c}(\overset{1}{\eta}(\hat{\mathbf{x}}_j), \overset{2}{\eta}(\hat{\mathbf{x}}_j)) + D_{k+1|k}[\breve{p}_D L_{\overset{1}{\eta}(\hat{\mathbf{x}}_j)} \cdot L_{\overset{2}{\eta}(\hat{\mathbf{x}}_j)}]}$$

or, using partial fractions,

$$\begin{aligned} &\mathring{N}_{k+1|k+1}(\overset{*1}{\mathbf{x}}_{k+1}, \overset{*2}{\mathbf{x}}_{k+1}) \\ &= D_{k+1|k}[1 - \breve{p}_D] \\ &\quad + \sum_{j=1}^{N} \breve{p}_D(\hat{\mathbf{x}}_i) \cdot \left(1 - \frac{\tilde{\lambda}\tilde{c}(\overset{1}{\eta}(\hat{\mathbf{x}}_j), \overset{2}{\eta}(\hat{\mathbf{x}}_j))}{\tilde{\lambda}\tilde{c}(\overset{1}{\eta}(\hat{\mathbf{x}}_j), \overset{2}{\eta}(\hat{\mathbf{x}}_j)) + D_{k+1|k}[\breve{p}_D L_{\overset{1}{\eta}(\hat{\mathbf{x}}_j)} \cdot L_{\overset{2}{\eta}(\hat{\mathbf{x}}_j)}]}\right) \end{aligned} \tag{129}$$

This formula is to be used in conjunction with the PHD filter of section 3.4. In the case of the MHC filter of section 3.5, it becomes

$$\begin{aligned} &\mathring{N}_{k+1|k+1}(\overset{*1}{\mathbf{x}}_{k+1}, \overset{*2}{\mathbf{x}}_{k+1}) \\ &= \sum_{j=1}^{N} q_j f_j[1 - \breve{p}_D] \\ &\quad + \sum_{j=1}^{N} \breve{p}_D(\hat{\mathbf{x}}_i) \cdot \left(1 - \frac{\tilde{\lambda}\tilde{c}(\overset{1}{\eta}(\hat{\mathbf{x}}_j), \overset{2}{\eta}(\hat{\mathbf{x}}_j))}{\tilde{\lambda}\tilde{c}(\overset{1}{\eta}(\hat{\mathbf{x}}_j), \overset{2}{\eta}(\hat{\mathbf{x}}_j)) + \sum_{e=1}^{N} q_e f_e[\breve{p}_D L_{\overset{1}{\eta}(\hat{\mathbf{x}}_j)} \cdot L_{\overset{2}{\eta}(\hat{\mathbf{x}}_j)}]}\right) \end{aligned} \tag{130}$$

When there are no false alarms these formulas reduce to the simple form

$$\mathring{N}_{k+1|k+1}(\overset{*1}{\mathbf{x}}_{k+1}, \overset{*2}{\mathbf{x}}_{k+1}) = D_{k+1|k}[1 - \breve{p}_D] + \sum_{j=1}^{N} \breve{p}_D(\hat{\mathbf{x}}_j) \tag{131}$$

$$\mathring{N}_{k+1|k+1}(\overset{*1}{\mathbf{x}}_{k+1}, \overset{*2}{\mathbf{x}}_{k+1}) = \sum_{j=1}^{N} (q_j f_j[1 - \breve{p}_D] + \breve{p}_D(\hat{\mathbf{x}}_i)) \tag{132}$$

Note that these are the same equations that one would get if one used the unsimplified pseudo-sensor approximation (section 4.5) and assumed no false alarms.

5.4. *PENT: Single-Sensor, Multistep Look-Ahead*

We address only the case of single-sensor, two-step look-ahead. The multistep case is an obvious extension. We first show how to reformulate the single-sensor, two-step look-ahead problem as a two-sensor, single-step look-ahead problem. Having done this, we will turn to the problem of deriving formulas for the PENT.

Abbreviate the sensor FoVs at time-steps $k+1$ and $k+2$ by

$$p_D(\mathbf{x}) \stackrel{\text{abbr.}}{=} p_D(\mathbf{x},\mathbf{\mathring{x}}_{k+1}), \qquad \stackrel{+}{p}_D(\mathbf{x}) \stackrel{\text{abbr.}}{=} p_D(\mathbf{x},\mathbf{\mathring{x}}_{k+2}) \tag{133}$$

Assume that the multitarget Markov transition at time-step $k+2$ does not model target appearances or disappearances:

$$f_{k+2|k+1}(X|W) = \sum_{\sigma} f_{k+2|k+1}(\mathbf{x}_1|\mathbf{w}_{\sigma 1}) \cdots f_{k+2|k+1}(\mathbf{x}_e|\mathbf{w}_{\sigma e}) \tag{134}$$

where

$$f_{k+1|k}(\mathbf{x}|\mathbf{w}) = f_{\mathbf{V}_k}(\mathbf{x} - \varphi_k(\mathbf{w})), \quad f_{k+2|k+1}(\mathbf{x}|\mathbf{w}) = f_{\mathbf{V}_{k+1}}(\mathbf{x} - \varphi_{k+1}(\mathbf{w})) \tag{135}$$

are the single-target Markov transitions at time-steps k and $k+1$ and where we abbreviate $\varphi \stackrel{\text{abbr.}}{=} \varphi_k$ and $\stackrel{+}{\varphi} \stackrel{\text{abbr.}}{=} \varphi_{k+1}$.

Define the *retrodiction* of the FoV $\stackrel{+}{p}_D$ at time-step $k+2$, back to an FoV $\stackrel{+\downarrow}{p}_D$ at time-step $k+1$, as

$$\stackrel{+\downarrow}{p}_D(\mathbf{w}) \stackrel{\text{abbr.}}{=} \stackrel{+\downarrow}{p}_D(\mathbf{w},\mathbf{\mathring{x}}_{k+2}) = \int \stackrel{+}{p}_D(\mathbf{x}) \cdot f_{k+2|k+1}(\mathbf{x}|\mathbf{w})d\mathbf{x} \tag{136}$$

Note that it is a well-defined FoV: $0 \leq \stackrel{+\downarrow}{p}_D(\mathbf{w}) \leq 1$. The quantity $G_{k+2|k}[\stackrel{+}{p}_D]$ is the probability that the *twice-predicted* tracks (from time-step k to time-step $k+2$) will all be contained in the sensor FoV at time-step $k+2$. In section 8.2 we prove that this is identical to $G_{k+1|k}[\stackrel{+\downarrow}{p}_D]$, which is the probability that the singly-predicted tracks will all be contained in the *retrodicted* FoV at time-step $k+1$:

$$G_{k+2|k}[\stackrel{+}{p}_D] = G_{k+1|k}[\stackrel{+\downarrow}{p}_D] \tag{137}$$

Since this equation is true for arbitrary $G_{k+1|k}$, it shows that the FoV $\stackrel{+}{p}_D$ at the future time-step $k+2$ is equivalent, in a probabilistic sense, to the retrodicted FoV $\stackrel{+\downarrow}{p}_D$ at time-step $k+1$. So, $\stackrel{+\downarrow}{p}_D$ and the original FoV p_D can be treated as though they were the FoVs of two different sensors at time-step $k+1$. This allows us to transform the single-sensor, two-step look-ahead problem to a two-sensor, single-step look-ahead problem and then apply the reasoning of sections 5.3 and 4.5.

First, from Eq. (106) the probability that at least one of these sensors will detect a target in step $k+1$ or in step $k+2$ is

$$\mathring{p}_D(\mathbf{x}) \stackrel{\text{abbr.}}{=} \mathring{p}_D(\mathbf{x}, \mathbf{\mathring{x}}_{k+2}, \mathbf{\mathring{x}}_{k+1}) = 1 - (1 - \overset{+\downarrow}{p}_D(\mathbf{x}, \mathbf{\mathring{x}}_{k+2}))(1 - p_D(\mathbf{x}, \mathbf{\mathring{x}}_{k+1})) \tag{138}$$

This is the analog of the multisensor FoV $\breve{p}_D(\mathbf{x})$ used previously. In the case of multistep look-ahead it will have the form

$$\begin{aligned} \mathring{p}_D(\mathbf{x}) &\stackrel{\text{abbr.}}{=} \mathring{p}_D(\mathbf{x}, \mathbf{\mathring{x}}_{k+2}, \mathbf{\mathring{x}}_{k+1}) \\ &= 1 - (1 - \overset{+\downarrow M}{p}_D(\mathbf{x}, \mathbf{\mathring{x}}_{k+M})) \cdots (1 - \overset{+\downarrow 2}{p}_D(\mathbf{x}, \mathbf{\mathring{x}}_{k+2}))(1 - p_D(\mathbf{x}, \mathbf{\mathring{x}}_{k+1})) \end{aligned} \tag{139}$$

where

$$\overset{+\downarrow a}{p}_D(\mathbf{w}) \stackrel{\text{abbr.}}{=} \int p_D(\mathbf{x}, \mathbf{\mathring{x}}_{k+a}) \cdot f_{k+a|k+1}(\mathbf{x}|\mathbf{w}) d\mathbf{x}$$

and where $f_{k+a|k+1}(\mathbf{x}|\mathbf{w})$ is the a-step Markov transition.

As in the previous section we use the simpler pseudo-sensor model of Eq. (116):

$$\tilde{L}_{(\mathbf{z}_{k+1}, \mathbf{z}_{k+2})}(\mathbf{x}) \stackrel{\text{abbr.}}{=} f_{k+1}(\mathbf{z}_{k+1}|\mathbf{x}, \mathbf{\mathring{x}}_{k+1}) \cdot f_{k+2}(\mathbf{z}_{k+2}|\mathbf{x}, \mathbf{\mathring{x}}_{k+2})$$

along with the false alarm model $\tilde{\lambda} = \lambda_{k+1} + \lambda_{k+2}$ and $\tilde{c}(\mathbf{z}_{k+1}, \mathbf{z}_{k+2}) = c_{k+1}(\mathbf{z}_{k+1}) \cdot c_{k+2}(\mathbf{z}_{k+2})$.

Also, an ideal observation at time-step $k+2$ has the form $\eta_{k+2}(\varphi_{k+2|k+1}((\hat{\mathbf{x}}_j))$. So, the formula for the PENT is, for the MHC filter case, obtained from Eq. (130) is:

$$\begin{aligned} &\mathring{N}_{k+1|k+1}(\mathbf{\mathring{x}}_{k+1}, \mathbf{\mathring{x}}_{k+2}) \\ &= \sum_{j=1}^{N} q_j f_j [1 - \mathring{p}_D] \\ &\quad + \sum_{j=1}^{N} \mathring{p}_D(\hat{\mathbf{x}}_j) \\ &\quad \cdot \left(1 - \frac{\tilde{\lambda}\tilde{c}(\eta_{k+1}(\hat{\mathbf{x}}_j), \eta_{k+2}(\varphi_{k+2|k+1}(\hat{\mathbf{x}}_j)))}{\begin{array}{c}\tilde{\lambda}\tilde{c}(\eta_{k+1}(\hat{\mathbf{x}}_j), \eta_{k+2}(\varphi_{k+2|k+1}(\hat{\mathbf{x}}_j))) \\ + \sum_{e=1}^{N} q_e f_e [\mathring{p}_D L_{\eta_{k+1}(\hat{\mathbf{x}}_j)} L_{\eta_{k+2}(\varphi_{k+2|k+1}(\hat{\mathbf{x}}_j))}]\end{array}} \right) \end{aligned} \tag{140}$$

When there are no false alarms this reduces to the simple form

$$\mathring{N}_{k+1|k+1}(\mathbf{\mathring{x}}_{k+1}, \mathbf{\mathring{x}}_{k+2}) = \sum_{j=1}^{N} (q_j f_j [1 - \mathring{p}_D] + \mathring{p}_D(\hat{\mathbf{x}}_j)) \tag{141}$$

5.5. PENT: Multisensor, Multistep Look-Ahead

Multistep look-ahead for the multisensor case follows the same reasoning used in the previous section. We consider the two-sensor, two-step case. The more general situation follows directly. Abbreviate the sensor FoVs at time-steps $k+1$ and $k+2$ by

$$\overset{1}{p}_D(\mathbf{x}) \overset{\text{abbr.}}{=} \overset{1}{p}_D(\mathbf{x},\overset{*1}{\mathbf{x}}_{k+1}), \qquad \overset{2}{p}_D(\mathbf{x}) \overset{\text{abbr.}}{=} \overset{2}{p}_D(\mathbf{x},\overset{*2}{\mathbf{x}}_{k+1}) \tag{142}$$

$$\overset{+1}{p}_D(\mathbf{x}) \overset{\text{abbr.}}{=} \overset{1}{p}_D(\mathbf{x},\overset{*1}{\mathbf{x}}_{k+2}), \qquad \overset{+2}{p}_D(\mathbf{x}) \overset{\text{abbr.}}{=} \overset{2}{p}_D(\mathbf{x},\overset{*2}{\mathbf{x}}_{k+2}) \tag{143}$$

Let

$$\overset{+1\downarrow}{p}_D(\mathbf{w}) = \int \overset{+1}{p}_D(\mathbf{x}) \cdot f_{k+2|k+1}(\mathbf{x}|\mathbf{w})d\mathbf{x} \tag{144}$$

$$\overset{+2\downarrow}{p}_D(\mathbf{w}) = \int \overset{+2}{p}_D(\mathbf{x}) \cdot f_{k+2|k+1}(\mathbf{x}|\mathbf{w})d\mathbf{x} \tag{145}$$

denote the respective retrodictions as defined in Eq. (136). Then the probability that at least one of the sensors will detect a target in step $k+1$ or in step $k+2$ is

$$\begin{aligned} \mathring{p}_D(\mathbf{x}) &\overset{\text{abbr.}}{=} \mathring{p}_D(\mathbf{x}, \overset{*1}{\mathbf{x}}_{k+2}, \overset{*2}{\mathbf{x}}_{k+2}, \overset{*1}{\mathbf{x}}_{k+1}, \overset{*2}{\mathbf{x}}_{k+1}) \\ &= 1 - (1 - \overset{+1\downarrow}{p}_D(\mathbf{x}, \overset{*1}{\mathbf{x}}_{k+2}))(1 - \overset{+2\downarrow}{p}_D(\mathbf{x}, \overset{*2}{\mathbf{x}}_{k+2})) \\ &\quad \cdot(1 - \overset{1}{p}_D(\mathbf{x}, \overset{*1}{\mathbf{x}}_{k+1}))(1 - \overset{2}{p}_D(\mathbf{x}, \overset{*2}{\mathbf{x}}_{k+1})) \end{aligned} \tag{146}$$

As in the previous section we use the simplified joint likelihood of section 4.6:

$$\begin{aligned} \tilde{L}_{(\overset{1}{\mathbf{z}}_{k+1},\overset{1}{\mathbf{z}}_{k+2},\overset{2}{\mathbf{z}}_{k+1},\overset{2}{\mathbf{z}}_{k+2})}(\mathbf{x}) &= \overset{1}{f}_{k+1}(\overset{1}{\mathbf{z}}_{k+1}|\mathbf{x}, \overset{*1}{\mathbf{x}}_{k+1}) \cdot \overset{1}{f}_{k+2}(\overset{1}{\mathbf{z}}_{k+2}|\mathbf{x}, \overset{*1}{\mathbf{x}}_{k+2}) \\ &\quad \cdot \overset{2}{f}_{k+1}(\overset{2}{\mathbf{z}}_{k+1}|\mathbf{x}, \overset{*2}{\mathbf{x}}_{k+1}) \cdot \overset{2}{f}_{k+2}(\overset{2}{\mathbf{z}}_{k+2}|\mathbf{x}, \overset{*2}{\mathbf{x}}_{k+2}) \end{aligned}$$

and proceed as before. When there are no false alarms we get

$$\mathring{N}_{k+1|k+1}(\overset{*1}{\mathbf{x}}_{k+1}, \overset{*2}{\mathbf{x}}_{k+1}, \overset{*1}{\mathbf{x}}_{k+2}, \overset{*2}{\mathbf{x}}_{k+2}) = \sum_{j=1}^{N} \left(D_{k+1|k}[1 - \mathring{p}_D] + \mathring{p}_D(\mathbf{x})\right) \tag{147}$$

$$\mathring{N}_{k+1|k+1}(\overset{*1}{\mathbf{x}}_{k+1}, \overset{*2}{\mathbf{x}}_{k+1}, \overset{*1}{\mathbf{x}}_{k+2}, \overset{*2}{\mathbf{x}}_{k+2}) = \sum_{j=1}^{N} (q_j f_j[1 - \mathring{p}_D] + \mathring{p}_D(\mathbf{x})) \tag{148}$$

6. Posterior Expected Number of Targets of Interest (PENTI)

Targets of Interest (ToIs) are targets that have greater tactical importance than others. This may be because they have high immediacy (e.g., are threateningly near friendly installations or forces), high intrinsic value (e.g., tanks and missile launchers), and so on. Sensor management algorithms must be capable of directing sensing resources preferentially to known or potential ToIs. The obvious approach would be to wait until accumulated information strongly suggests that particular targets are probable ToIs and then bias sensor management towards these targets. However, *ad hoc* techniques of this sort have inherent limitations:

- Information about target type accumulates incrementally, not suddenly. Preferential biasing of sensors towards targets should likewise be accomplished incrementally, only to the degree supported by accumulated evidence.
- Information about target type may be erroneous, and may be reversed by later, better data. So, it may not be possible to recover from an erroneous hard-and-fast decision to ignore a target, since the target has been lost because of that decision.
- Target preference may not be an either-or choice, since ToIs themselves may be ranked in order of tactical importance. For example, missile launchers and tanks both have high tactical value, but the former even more so than the latter.

Rather than resorting to *ad hoc* techniques with inherent limitations, one should *integrally incorporate target preference into the fundamental statistical representation of multisensor-multitarget systems.* The purpose of this section is to describe how this can be done.

Begin with single-step look-ahead. The Bayes multitarget posterior $f_{k+1|k+1}(X|Z^{(k)})$ is the probability distribution of a randomly varying state-set $\Xi_{k+1|k+1}$. The corresponding probability generating functional $G_{k+1|k+1}[h]$ contains the same information as $f_{k+1|k+1}(X|Z^{(k)})$ and has the following intuitive interpretation (see section 2.4). If h is a fuzzy membership function, then $G_{k+1|k+1}[h]$ is the probability that $\Xi_{k+1|k+1}$ is completely contained in the corresponding fuzzy set.

First consider the simplest possible situation: any given target is either a ToI or a non-ToI. In this case, there is a specific set S of all possible ToIs. The random finite subset $\Xi_{k+1|k+1} \cap S$ is the set of all targets at time-step $k+1$ that are of current interest. It can be shown (see proposition 23, p.

164 of [3]) that the p.g.fl. of $\Xi_{k+1|k+1} \cap S$ is

$$G^{\text{ToI}}_{k+1|k+1}[h] \stackrel{\text{def.}}{=} G_{k+1|k+1}[h \vee \mathbf{1}^c_S] \tag{149}$$

where in general $h^c \stackrel{\text{def.}}{=} 1 - h$ and $h_1 \vee h_2 \stackrel{\text{def.}}{=} h_1 + h_2 - h_1 h_2$.

More generally, suppose that if a target has state $\mathbf{x}$ then there is a relative ranking $0 \leq \rho(\mathbf{x}) \leq 1$ regarding tactical importance. If $\rho(\mathbf{x}) = 0$ then a target with state $\mathbf{x}$ has no tactical importance, whereas if $\rho(\mathbf{x}) = 1$ then the target has the highest possible tactical importance. And if $\rho(\mathbf{x})$ is neither zero nor one, it has some intermediate degree of tactical importance. Then $\rho(\mathbf{x})$ can be regarded as a fuzzy membership function defined on target states. By analogy, the p.g.fl. corresponding to all targets of tactical interest is

$$G^{\text{ToI}}_{k+1|k+1}[h] \stackrel{\text{def.}}{=} G_{k+1|k+1}[h \vee \rho^c] = G_{k+1|k+1}[1 - \rho + h\rho] \tag{150}$$

Given this, the procedure outlined in section 4 provides us with a means of deriving formulas for the *posterior expected number of targets of interest* (PENTI).

6.1. *PENTI: Single-Sensor, One-Step Look-Ahead*

From Eq. (103) of section 4.2 we know that the hedged p.g.fl. is

$$\mathring{G}_{k+1|k+1}[h] = e^{D_{k+1|k}[(h-1)(1-p_D)]} \cdot \prod_{j=1}^{N} \left(\frac{\lambda c(\eta(\hat{\mathbf{x}}_j)) + D_{k+1|k}[h p_D L_{\eta(\hat{\mathbf{x}}_j)}]}{\lambda c(\eta(\hat{\mathbf{x}}_j)) + D_{k+1|k}[p_D L_{\eta(\hat{\mathbf{x}}_j)}]} \right)^{p_D(\hat{\mathbf{x}}_j)}$$

Consequently, the restriction of the hedged p.g.fl. to ToIs is

$$\begin{aligned} \mathring{G}^{\text{ToI}}_{k+1|k+1}[h] &= \mathring{G}_{k+1|k+1}[1 - \rho + h\rho] \\ &= e^{D_{k+1|k}[\rho(h-1)(1-p_D)]} \\ &\quad \cdot \prod_{j=1}^{N} \left(\frac{\lambda c(\eta(\hat{\mathbf{x}}_j)) + D_{k+1|k}[(1 - \rho + h\rho) p_D L_{\eta(\hat{\mathbf{x}}_j)}]}{\lambda c(\eta(\hat{\mathbf{x}}_j)) + D_{k+1|k}[p_D L_{\eta(\hat{\mathbf{x}}_j)}]} \right)^{p_D(\hat{\mathbf{x}}_j)} \end{aligned}$$

The PHD of the restricted p.g.fl. is

$$\begin{aligned} \mathring{D}^{\text{ToI}}_{k+1|k+1}(\mathbf{x}) &= \frac{\delta \log \mathring{G}^{\text{ToI}}_{k+1|k+1}}{\delta \mathbf{x}}[1] \\ &= \left(\rho(\mathbf{x}) \cdot (1 - p_D(\mathbf{x})) + \sum_{j=1}^{N} p_D(\hat{\mathbf{x}}_j) \cdot \frac{\rho(\mathbf{x}) p_D(\mathbf{x}) L_{\eta(\hat{\mathbf{x}}_j)}(\mathbf{x})}{\lambda c(\eta(\hat{\mathbf{x}}_j)) + D_{k+1|k}[p_D L_{\eta(\hat{\mathbf{x}}_j)}]} \right) \\ &\quad \cdot D_{k+1|k}(\mathbf{x}) \end{aligned}$$

Therefore, the PENTI is

$$\mathring{N}^{\text{ToI}}_{k+1|k+1}(\mathring{\mathbf{x}}_{k+1}) = \int \mathring{D}^{\text{ToI}}_{k+1|k+1}(\mathbf{x})d\mathbf{x} \tag{151}$$
$$= D_{k+1|k}[\rho(1-p_D)]$$
$$+ \sum_{j=1}^{N} p_D(\hat{\mathbf{x}}_j) \cdot \frac{D_{k+1|k}[\rho p_D L_{\eta(\hat{\mathbf{x}}_j)}]}{\lambda c(\eta(\hat{\mathbf{x}}_j)) + D_{k+1|k}[p_D L_{\eta(\hat{\mathbf{x}}_j)}]}$$

This is the formula that is appropriate for use with the PHD filter of section 3.4. The corresponding formula for the MHC filter of section 3.5 is, from Eq. (98),

$$\mathring{N}^{\text{ToI}}_{k+1|k+1}(\mathring{\mathbf{x}}_{k+1}) = \sum_{j=1}^{N} q_j f_j[\rho(1-p_D)] \tag{152}$$
$$+ \sum_{j=1}^{N} p_D(\hat{\mathbf{x}}_j) \cdot \frac{\sum_{e=1}^{N} q_e f_e[\rho p_D L_{\eta(\hat{\mathbf{x}}_j)}]}{\lambda c(\eta(\hat{\mathbf{x}}_j)) + \sum_{e=1}^{N} q_e f_e[p_D L_{\eta(\hat{\mathbf{x}}_j)}]}$$

When there are no false alarms ($\lambda = 0$), these two equations respectively reduce to the somewhat simpler form

$$\mathring{N}^{\text{ToI}}_{k+1|k+1}(\mathring{\mathbf{x}}_{k+1}) = D_{k+1|k}[\rho(1-p_D)] + \sum_{j=1}^{N} p_D(\hat{\mathbf{x}}_j) \cdot \frac{D_{k+1|k}[\rho p_D L_{\eta(\hat{\mathbf{x}}_j)}]}{D_{k+1|k}[p_D L_{\eta(\hat{\mathbf{x}}_j)}]} \tag{153}$$

$$\mathring{N}^{\text{ToI}}_{k+1|k+1}(\mathring{\mathbf{x}}_{k+1}) = \sum_{j=1}^{N} q_j f_j[\rho(1-p_D)] + \sum_{j=1}^{N} p_D(\hat{\mathbf{x}}_j) \cdot \frac{\sum_{i=1}^{N} q_i f_i[\rho p_D L_{\eta(\hat{\mathbf{x}}_j)}]}{\sum_{e=1}^{N} q_e f_e[p_D L_{\eta(\hat{\mathbf{x}}_j)}]} \tag{154}$$

Note that these two equations can be evaluated in closed form if we assume that p_D and ρ have linear-Gaussian form:

$$p_D(\mathbf{x}) = \exp\left(-\frac{1}{2}(A_{k+1}\mathbf{x} - \mathring{A}_{k+1}\mathring{\mathbf{x}}_{k+1})^T L_{k+1}^{-1}(A_{k+1}\mathbf{x} - \mathring{A}_{k+1}\mathring{\mathbf{x}}_{k+1})\right) \tag{155}$$

$$\rho(\mathbf{x}) = \exp\left(-\frac{1}{2}(\mathbf{x} - \mathbf{x}_{\text{ToI}})^T B_{\text{ToI}}^{-1}(\mathbf{x} - \mathbf{x}_{\text{ToI}})\right) \tag{156}$$

Here, $\mathbf{x}_{\text{ToI}}$ denotes the most ToI-like target state, and the positive-definite matrix B_{ToI} models the uncertainty in the definition of a ToI.

6.2. *PENTI: Multisensor, One-Step Look-Ahead*

The analysis follows that of sections 4.5, 6.2 and 6.1. We will not go through the full derivation here since it is so similar to that of section 6.2. Eq. (129), the simplified formula for the multisensor PENT for use with the PHD filter of section 3.4, becomes the following formula for PENTI:

$$\mathring{N}_{k+1|k+1}(\overset{*1}{\mathbf{x}}_{k+1}, \overset{*2}{\mathbf{x}}_{k+1}) = D_{k+1|k}[\rho(1-\breve{p}_D)] + \sum_{j=1}^{N} \breve{p}_D(\hat{\mathbf{x}}_i) \cdot \frac{D_{k+1|k}[\rho \breve{p}_D L_{\overset{1}{\eta}(\hat{\mathbf{x}}_j)} L_{\overset{2}{\eta}(\hat{\mathbf{x}}_j)}]}{\tilde{\lambda}\tilde{c}(\overset{1}{\eta}(\hat{\mathbf{x}}_j), \overset{2}{\eta}(\hat{\mathbf{x}}_j)) + D_{k+1|k}[\breve{p}_D L_{\overset{1}{\eta}(\hat{\mathbf{x}}_j)} L_{\overset{2}{\eta}(\hat{\mathbf{x}}_j)}]} \tag{157}$$

Eq. (130), the simplified formula for the multisensor PENT for use with the MHC filter of section 3.5, becomes the following formula for PENTI:

$$\mathring{N}_{k+1|k+1}(\overset{*1}{\mathbf{x}}_{k+1}, \overset{*2}{\mathbf{x}}_{k+1}) = \sum_{j=1}^{N} q_j f_j[\rho(1-\breve{p}_D)] + \sum_{j=1}^{N} \breve{p}_D(\hat{\mathbf{x}}_i) \cdot \frac{\sum_{i=1}^{N} q_i f_i[\rho \breve{p}_D L_{\overset{1}{\eta}(\hat{\mathbf{x}}_j)} L_{\overset{2}{\eta}(\hat{\mathbf{x}}_j)}]}{\tilde{\lambda}\tilde{c}(\overset{1}{\eta}(\hat{\mathbf{x}}_j), \overset{2}{\eta}(\hat{\mathbf{x}}_j)) + \sum_{e=1}^{N} q_e f_e[\breve{p}_D L_{\overset{1}{\eta}(\hat{\mathbf{x}}_j)} L_{\overset{2}{\eta}(\hat{\mathbf{x}}_j)}]} \tag{158}$$

6.3. *PENTI: Single-Sensor, Multistep Look-Ahead*

This case is dealt with in the same manner as in section 5.4. For example, the single-sensor, two-step look-ahead case is the following analog of the Eq. (158) in the previous section:

$$\begin{aligned} &\mathring{N}_{k+1|k+1}(\overset{*}{\mathbf{x}}_{k+1}, \overset{*}{\mathbf{x}}_{k+2}) \\ &= \sum_{j=1}^{N} q_j f_j[\rho(1-\mathring{p}_D)] \\ &\quad + \sum_{j=1}^{N} \mathring{p}_D(\hat{\mathbf{x}}_i) \\ &\quad \cdot \frac{\sum_{i=1}^{N} q_i f_i[\rho \mathring{p}_D L_{\eta_{k+1}(\hat{\mathbf{x}}_j)} L_{\eta_{k+2}(\varphi_{k+2|k+1}(\hat{\mathbf{x}}_j))}]}{\tilde{\lambda}\tilde{c}(\eta_{k+1}(\hat{\mathbf{x}}_j), \eta_{k+2}(\varphi_{k+2|k+1}(\hat{\mathbf{x}}_j))) + \sum_{e=1}^{N} q_e f_e[\mathring{p}_D L_{\eta_{k+1}(\hat{\mathbf{x}}_j)} L_{\eta_{k+2}(\varphi_{k+2|k+1}(\hat{\mathbf{x}}_j))}]} \end{aligned} \tag{159}$$

We will not go into further detail regarding this matter here.

6.4. *PENTI: Multisensor, Multistep Look-Ahead*

This case is dealt with in the same manner as in section 5.5. We will not go into further detail regarding this matter here.

7. Dispersed Mobile Sensors

In this section we extend our previous results to deal with sensors carried by autonomous and other platforms. First, we no longer assume that sensor states are perfectly observed. The state of a sensor is observed by an actuator sensor whose observations may be corrupted by various noise sources. Second, the transmission of both sensor observations and actuator-sensor observations may be blocked for various reasons. Third, we no longer assume that sensor dynamics are ideal. Sensor motion is limited by physical or other constraints, and these motions influenced indirectly via control vectors rather than by directly choosing future sensor states. We must assume that sensors are known in number. In this section our goal is to derive three things: the predictor equation and corrector equation for a dispersed-sensor version of the PHD filter of section 3.4; and a PENT formula for use with this filter. Once this has been accomplished we can also derive PENT formulas for use with MHC filters of section 3.5.

The section is organized as follows. We begin in section 7.1 by specifying how the mathematical foundations of section 2 are affected by the assumption that the sensors are known and, in particular, of known number. The PHD predictor equation is derived in section 7.2 and the PHD corrector equation in section 7.3 in the single-sensor. A formula for PENT in the single-sensor, single-step look-ahead case is derived in section 7.4. The results are extended to multisensor, single-step look-ahead in section 7.5.

7.1. *Restriction to the $\mathring{n}$-Sensor Case*

Assume that we know, on an *a priori* basis, that there are $\mathring{n}$ sensors present, with no sensor appearances or disappearances. In this case all multi-object states have the form

$$\breve{X} = \{\mathbf{x}_1, ..., \mathbf{x}_n, \mathring{\mathbf{x}}_1, ..., \mathring{\mathbf{x}}_{\mathring{n}}\} \qquad \text{or} \qquad \breve{X} = \{\mathring{\mathbf{x}}_1, ..., \mathring{\mathbf{x}}_{\mathring{n}}\} \tag{160}$$

where $n = 0, 1, ...$ is variable and $\mathring{n}$ is fixed. Multi-object measurements have the form

$$\breve{Z} = \{\mathbf{z}_1, ..., \mathbf{z}_m, \mathring{\mathbf{z}}_1, ..., , \mathring{\mathbf{z}}_{\mathring{m}}\} \tag{161}$$

where both $m = 0, 1, ...$ and $\mathring{m} = 0, 1, ...$ are variable.

The set integral (Eq. (26)) now has the form

$$\int \breve{f}(\breve{X})\delta\breve{X} = \int\int_{|\mathring{X}|=\mathring{n}} f(X, \mathring{X})\delta X \delta\mathring{X} \tag{162}$$

Likewise, the p.g.fl. has the form

$$\breve{G}[\breve{h}] = \int\int_{|\mathring{X}|=\mathring{n}} \breve{h}^X \cdot \breve{h}^{\mathring{X}} \cdot f(X, \mathring{X})\delta X \delta\mathring{X} \tag{163}$$

According to Eq. (43) the first functional derivative is

$$\frac{\delta\breve{G}}{\delta\breve{\mathbf{x}}}[\breve{h}] = \int \breve{h}^{\breve{X}} \cdot \breve{f}(\breve{X} \cup \{\breve{\mathbf{x}}\})\delta\breve{X}$$

The first functional derivative with respect to a target state vector $\breve{\mathbf{y}} = \mathbf{y}$ is, therefore

$$\begin{aligned}\frac{\delta\breve{G}}{\delta\mathbf{y}}[\breve{h}] &= \int \breve{h}^{\breve{X}} \cdot \breve{f}(\breve{X} \cup \{\mathbf{y}\})\delta\breve{X} \\ &= \int\int_{|\mathring{X}|=\mathring{n}} \breve{h}^X \cdot \breve{h}^{\mathring{X}} \cdot \breve{f}(X \cup \{\mathbf{y}\}, \mathring{X})\delta X \delta\mathring{X}\end{aligned}$$

and so the value of the joint PHD at $\breve{\mathbf{y}} = \mathbf{y}$ is

$$\breve{D}(\mathbf{y}) = \int f(X \cup \{\mathbf{y}\}, \mathring{X})\delta X \delta\mathring{X} \tag{164}$$

When there is only one sensor,

$$\breve{D}(\mathbf{y}) = \int f(X \cup \{\mathbf{y}\}, \mathring{\mathbf{x}})\delta X d\mathring{\mathbf{x}} \tag{165}$$

Likewise, the first functional derivative with respect to a sensor state $\breve{\mathbf{y}} = \mathring{\mathbf{y}}$ is

$$\begin{aligned}\frac{\delta\breve{G}}{\delta\mathring{\mathbf{y}}}[\breve{h}] &= \int \breve{h}^{\breve{W}} \cdot \breve{f}(\breve{W} \cup \{\mathring{\mathbf{y}}\})\delta\breve{W} \\ &= \int\int_{|\mathring{X}|=\mathring{n}-1} \breve{h}^X \cdot \breve{h}^{\mathring{X}} \cdot f(X, \mathring{X} \cup \{\mathring{\mathbf{y}}\})\delta X \delta\mathring{X}\end{aligned}$$

and the value of the joint PHD at $\breve{\mathbf{y}} = \mathring{\mathbf{y}}$ is

$$\breve{D}(\mathring{\mathbf{y}}) = \int\int_{|\mathring{X}|=\mathring{n}-1} f(X, \mathring{X} \cup \{\mathring{\mathbf{y}}\})\delta X \delta\mathring{X} \tag{166}$$

In particular, when there is only one sensor,

$$\breve{D}(\overset{*}{\mathbf{y}}) = \int f(X, \overset{*}{\mathbf{y}})\delta X \tag{167}$$

7.2. *PHD Filter Predictor Step*

We are to predict the joint PHD $\breve{D}_{k|k}(\breve{\mathbf{y}})$ at time-step k to the joint PHD $\breve{D}_{k+1|k}(\breve{\mathbf{y}})$ at time-step $k+1$. Let $f_{k+1|k}(\mathbf{y}|\mathbf{x})$ be the Markov transition density for a single target. Let $1 - s_{k+1|k}(\mathbf{x})$ be the probability that a target will disappear at time-step $k+1$ if it had state $\mathbf{x}$ at time-step k. Let $b_{k+1|k}(Y|\mathbf{x})$ be the probability that a target will spawn a set of new targets Y at time-step $k+1$ if it had state $\mathbf{x}$ at time-step k; and let $b_{k+1|k}(\mathbf{y}|\mathbf{x}) = \int b_{k+1|k}(W \cup \{\mathbf{y}\}|\mathbf{x})\delta W$ be its PHD. Let $b_{k+1|k}(Y)$ be the probability that a set Y of targets will appear spontaneously at time-step $k+1$ and let $b_{k+1|k}(\mathbf{y}|\mathbf{x}) = \int b_{k+1|k}(W \cup \{\mathbf{y}\}|\mathbf{x})\delta W$ be its PHD. Finally, let $\overset{*i}{f}_{k+1|k}(\overset{*i}{\mathbf{y}}|\overset{*i}{\mathbf{x}}, \mathbf{u}_i)$ be the Markov transition for the i'th sensor. Then the predictor equations for the joint PHD for the cases $\breve{\mathbf{y}} = \mathbf{y}$ (targets) and $\breve{\mathbf{y}} = \overset{*i}{\mathbf{y}}$ (i^{th} sensor) are:

$$\breve{D}_{k+1|k}(\mathbf{y}) = b_{k+1|k}(\mathbf{y}) \tag{168}$$
$$+ \int \left(s_{k+1|k}(\mathbf{x}) \cdot f_{k+1|k}(\mathbf{y}|\mathbf{x}) + b_{k+1|k}(\mathbf{y}|\mathbf{x})\right) \cdot \breve{D}_{k+1|k}(\mathbf{x})d\mathbf{x}$$

$$\breve{D}_{k+1|k}(\overset{*i}{\mathbf{y}}) = \int \overset{*i}{f}_{k+1|k}(\overset{*i}{\mathbf{y}}|\overset{*i}{\mathbf{x}}, \mathbf{u}_i) \cdot \breve{D}_{k|k}(\overset{*i}{\mathbf{x}})d\overset{*i}{\mathbf{x}} \tag{169}$$

The proof may be found in section 8.3.

In other words, the predictor Eq. (168) for the target part of the joint PHD is just the usual PHD predictor of Eq. (92). The predictor Eq. (169) for the sensor part of the joint PHD states that the individual sensors are predicted using a conventional prediction integral. If it is the case that $\int \breve{D}_{k|k}(\overset{*i}{\mathbf{x}})d\overset{*i}{\mathbf{x}} = 1$ (i.e., $\breve{D}_{k|k}(\overset{*i}{\mathbf{x}})$ is a single-object posterior distribution) then Eq. (169) is a conventional single-object Bayes filter prediction step.

7.3. *PHD Filter Corrector Step: Single-Sensor Case*

Assume that multiple targets are interrogated by a single moving sensor. We are to update the predicted joint PHD $\breve{D}_{k+1|k}(\breve{\mathbf{y}})$ at time-step k to the joint PHD $\breve{D}_{k+1|k+1}(\breve{\mathbf{y}})$ at time-step $k+1$. Recall that in section 2.8 we specified the following observation model: $\overset{*}{p}_D(\overset{*}{\mathbf{x}})$ is the actuator-sensor probability of detection; $\overset{*}{L}_{\overset{*}{\mathbf{z}}}(\overset{*}{\mathbf{x}}) = \overset{*}{f}_{k+1}(\overset{*}{\mathbf{z}}|\overset{*}{\mathbf{x}})$ is the actuator-sensor likelihood; $p_D(\mathbf{x}, \mathring{\mathbf{x}})$ is the sensor probability of detection; $L_{\mathbf{z},\mathring{\mathbf{x}}}(\mathbf{x}) = f_{k+1}(\mathbf{z}|\mathbf{x}, \mathring{\mathbf{x}})$ is the sensor likelihood function; and $c_{k+1}(\mathbf{z})$ is the spatial distribution of a Poisson false alarm process and λ_{k+1} is the expected number of false alarms. At time-step $k+1$ let $Z_{k+1} = \{\mathbf{z}_1, ..., \mathbf{z}_m\}$ and $\mathring{\mathbf{z}}_{k+1} = \mathring{\mathbf{z}}$ be collected. Then the PHD filter corrector equations for the cases $\breve{\mathbf{y}} = \mathbf{y}$ (targets) and $\breve{\mathbf{y}} = \mathring{\mathbf{y}}^i$ (i^{th} sensor) are:

$$\breve{D}_{k+1|k+1}(\mathbf{x}) = \mathring{D}_{k+1|k}[1 - \mathring{p}_D p_{D,\mathbf{x}}] \cdot \breve{D}_{k+1|k}(\mathbf{x}) + \sum_{i=1}^{m} \frac{\mathring{D}_{k+1|k}[p_{D,\mathbf{x}}\mathring{p}_D L_{\mathbf{z}_i,\mathbf{x}}] \cdot \breve{D}_{k+1|k}(\mathbf{x})}{\mathring{D}_{k|k}[\mathring{p}_D] \cdot \lambda_{k+1} c_{k+1}(\mathbf{z}_i) + (D_{k+1|k} \times \mathring{D}_{k+1|k})[p_D \mathring{p}_D L_{\mathbf{z}_i}]} \tag{170}$$

$$\breve{D}_{k+1|k+1}(\mathring{\mathbf{x}}) = \left(1 - \mathring{p}_D(\mathring{\mathbf{x}}) + \frac{\mathring{p}_D(\mathring{\mathbf{x}})\mathring{L}_{\mathring{\mathbf{z}}}(\mathring{\mathbf{x}})]}{\mathring{D}_{k+1|k}[\mathring{p}_D \mathring{L}_{\mathring{\mathbf{z}}}]}\right) \breve{D}_{k+1|k}(\mathring{\mathbf{x}}) \tag{171}$$

Here we have used the notation: $L_{\mathbf{z}_i,\mathbf{x}}(\mathring{\mathbf{x}}) \overset{\text{def.}}{=} f_{k+1}(\mathbf{z}|\mathbf{x}, \mathring{\mathbf{x}})$; $p_{D,\mathbf{x}}(\mathring{\mathbf{x}}) \overset{\text{def.}}{=} p_D(\mathbf{x}, \mathring{\mathbf{x}})$; $(h \times \mathring{h})(\mathbf{x}, \mathring{\mathbf{x}}) \overset{\text{def.}}{=} h(\mathbf{x}) \cdot \mathring{h}(\mathring{\mathbf{x}})$; and $\mathring{D}_{k+1|k}(\mathring{\mathbf{x}}) \overset{\text{def.}}{=} \mathring{N}_{k+1|k}^{-1} \cdot \breve{D}_{k+1|k}(\mathring{\mathbf{x}})$ where $\mathring{N}_{k+1|k} \overset{\text{def.}}{=} \int \breve{D}_{k+1|k}(\mathring{\mathbf{x}}) d\mathring{\mathbf{x}}$.

Note that in general form, Eq. (171) is what one would get if one took the usual PHD corrector Eq. (95) and applied it to the sensor update by assuming that there are no false alarms and there is at most one target present (and hence at most one observation to collect). If we assume that the sensor truly does exist then we can substitute $\mathring{p}_D = 1$ in Eq. (171), resulting in a standard Bayes' rule update of $\breve{D}_{k+1|k}(\mathring{\mathbf{x}})$.

As a check, let $\mathring{p}_D = 1$ and $\mathring{D}_{k+1|k}(\mathring{\mathbf{y}}) = \delta_{\mathring{\mathbf{x}}_{k+1}}(\mathring{\mathbf{y}})$. Then the target corrector Eq. (170) reduces to

$$\breve{D}_{k+1|k+1}(\mathbf{y}) = (1 - p_D(\mathbf{y})) \cdot D_{k+1|k}(\mathbf{y}) + \sum_{i=1}^{m} \frac{p_D(\mathbf{y}) \cdot L_{\mathbf{z}_i}(\mathbf{y}) \cdot D_{k+1|k}(\mathbf{y})}{\lambda c(\mathbf{z}_i) + D_{k+1|k}[p_D L_{\mathbf{z}_i}]}$$

where $p_D(\mathbf{y}) \overset{\text{abbr.}}{=} p_D(\mathbf{y}, \mathring{\mathbf{x}}_{k+1})$ and $L_{\mathbf{z}}(\mathbf{y}) \overset{\text{abbr.}}{=} L_{\mathbf{z}}(\mathbf{y}, \mathring{\mathbf{x}}_{k+1})$. That is, the posterior PHD reduces to the usual PHD corrector Eq. (95).

The derivation of Eqs. (170) and (171) proceeds as follows. We can construct the joint posterior PHD $\breve{D}_{k+1|k+1}(\breve{\mathbf{y}})$ from the joint posterior p.g.fl. $\breve{G}_{k+1|k+1}[\breve{h}]$ using Eq. (48). Derivation of both $\breve{G}_{k+1|k+1}[\breve{h}]$ and the hedged p.g.fl. $\mathring{G}_{k+1|k+1}[\breve{h}]$ requires that we first construct the joint two-variable functional $\breve{F}_{k+1}[\breve{g}, \breve{h}]$ of Eq. (87) from the sensor model. This sensor model was specified in detail in section 2.8.

In section 7.3.1 we derive $\breve{F}_{k+1}[\breve{g}, \breve{h}]$ from this sensor model and then approximate it by a generalized Poisson p.g.fl. Even with this approximation, the formula for $\log \breve{F}_{k+1}[\breve{g}, \breve{h}]$ is so nonlinear that it cannot be used to derive useful formulas. So, in section 7.3.2, we show how to linearize $\log \breve{F}_{k+1}[\breve{g}, \breve{h}]$ while preserving the major features of the sensor model of section 2.8.

Given this, the actual derivation of the corrector equations can be found in section 8.4.

7.3.1. *Derivation of the Joint p.g.fl. of the Full Sensor Model*

Recall that the joint two-variable functional $\breve{F}_{k+1}[\breve{g}, \breve{h}]$ of Eq. (87) is central to the construction of the joint posterior p.g.fl. $\breve{G}_{k+1|k+1}[\breve{h}]$ and the hedged joint posterior p.g.fl. $\mathring{G}_{k+1|k+1}[\breve{h}]$. In this section we derive this p.g.fl., assuming the detailed observation model we specified in section 2.8. We begin by transforming the joint likelihood function $f_{k+1}(Z, \overset{*}{Z}|X, \overset{*}{\mathbf{x}})$ of Eq. (66) into its corresponding p.g.fl.

$$\breve{G}_{k+1}[\breve{g}|X, \overset{*}{\mathbf{x}}] = \int \breve{g}^Z \cdot \breve{g}^{\overset{*}{Z}} \cdot f_{k+1}(Z, \overset{*}{Z}|X, \overset{*}{\mathbf{x}}) \delta Z \delta \overset{*}{Z} \tag{172}$$

From Eq. (66) we find that

$$\breve{G}_{k+1}[\breve{g}|X, \overset{*}{\mathbf{x}}] = 1 - \overset{*}{p}_D(\overset{*}{\mathbf{x}}) + \overset{*}{p}_D(\overset{*}{\mathbf{x}}) \cdot \overset{*}{p}_{\breve{g}}(\overset{*}{\mathbf{x}}) \cdot \breve{G}_{\text{tot},k+1}[\breve{g}|X, \overset{*}{\mathbf{x}}] \tag{173}$$

where $\breve{G}_{\text{tot},k+1}[\breve{g}|X, \overset{*j}{\mathbf{x}}]$ is the p.g.fl. of $f_{\text{tot},k}(Z|X, \overset{*}{\mathbf{x}})$ and where

$$\overset{*}{p}_{\breve{g}}(\overset{*}{\mathbf{x}}) \overset{\text{def.}}{=} \int \breve{g}(\overset{*}{\mathbf{z}}) \cdot \overset{*}{f}_{k+1}(\overset{*}{\mathbf{z}}|\overset{*}{\mathbf{x}}) d\overset{*}{\mathbf{z}} \tag{174}$$

Because of the conditional independence expressed by Eq. (65), the corresponding p.g.fl. is

$$\breve{G}_{\text{tot},k+1}[\breve{g}|X, \overset{*}{\mathbf{x}}] = \breve{G}_{\text{targ},k+1}[\breve{g}|X, \overset{*}{\mathbf{x}}] \cdot \breve{G}_{\text{clutt},k+1}[\breve{g}] \tag{175}$$

where $\breve{G}_{\text{targ},k+1}[\breve{g}|X, \overset{*}{\mathbf{x}}]$ is the p.g.fl. of $f_{k+1}(Z|X, \overset{*}{\mathbf{x}})$ and $\breve{G}_{\text{clutt},k+1}[\breve{g}]$ is the p.g.fl. of $c_{k+1}(Z)$. By Eq. (64),

$$\breve{G}_{\text{clutt},k+1}[\breve{g}] = e^{-\lambda + \lambda c_{\breve{g}}} \tag{176}$$

where

$$\lambda \overset{\text{abbr.}}{=} \lambda_{k+1}\,, \qquad c_{\breve{g}} \overset{\text{abbr.}}{=} \int \breve{g}(\mathbf{z}) \cdot c_{k+1}(\mathbf{z}) d\mathbf{z}.$$

and by Eq. (63)

$$\breve{G}_{\text{targ},k+1}[\breve{g}|X, \overset{*}{\mathbf{x}}] = \breve{G}_{k+1}[\breve{g}|\mathbf{x}_1, \overset{*}{\mathbf{x}}] \cdots \breve{G}_{k+1}[\breve{g}|\mathbf{x}_n, \overset{*}{\mathbf{x}}] \tag{177}$$

Thus so far Eq. (173) can be rewritten as

$$\breve{G}_{k+1}[\breve{g}|X,\mathring{\mathbf{x}}] = 1 - \mathring{p}_D(\mathring{\mathbf{x}}) + \mathring{p}_D(\mathring{\mathbf{x}}) \cdot \mathring{p}_{\breve{g}}(\mathring{\mathbf{x}}) \cdot e^{-\lambda+\lambda c_{\breve{g}}} \cdot \breve{G}_{k+1}[\breve{g}|\mathbf{x}_1,\mathring{\mathbf{x}}] \cdots \breve{G}_{k+1}[\breve{g}|\mathbf{x}_n,\mathring{\mathbf{x}}] \tag{178}$$

Next, by Eq. (62),

$$\breve{G}_{k+1}[\breve{g}|\mathbf{x}_i,\mathring{\mathbf{x}}] = 1 - p_D(\mathbf{x}_i,\mathring{\mathbf{x}}) + p_D(\mathbf{x}_i,\mathring{\mathbf{x}}) \cdot p_{\breve{g}}(\mathbf{x}_i,\mathring{\mathbf{x}})$$

where

$$p_{\breve{g}}(\mathbf{x}_i,\mathring{\mathbf{x}}) \overset{\text{def.}}{=} \int \breve{g}(\mathbf{z}) \cdot f_{k+1}(\mathbf{z}|\mathbf{x}_i,\mathring{\mathbf{x}}) d\mathbf{z}$$

So Eq. (178) becomes

$$\begin{aligned}\breve{G}_{k+1}[\breve{g}|X,\mathring{\mathbf{x}}] = 1 &- \mathring{p}_D(\mathring{\mathbf{x}}) + \mathring{p}_D(\mathring{\mathbf{x}}) \cdot \mathring{p}_{\breve{g}}(\mathring{\mathbf{x}}) \cdot e^{-\lambda+\lambda c_{\breve{g}}} \\ &\cdot (1 - p_D(\mathbf{x}_1,\mathring{\mathbf{x}}) + p_D(\mathbf{x}_1,\mathring{\mathbf{x}}) \cdot p_{\breve{g}}(\mathbf{x}_1,\mathring{\mathbf{x}})) \\ &\cdots (1 - p_D(\mathbf{x}_n,\mathring{\mathbf{x}}) + p_D(\mathbf{x}_n,\mathring{\mathbf{x}}) \cdot p_{\breve{g}}(\mathbf{x}_n,\mathring{\mathbf{x}}))\end{aligned} \tag{179}$$

We must transform this equation into a more useful form. We extend p.g.fl.'s $\breve{G}[\breve{h}]$ defined on functions $\breve{h}$ to p.g.fl.'s defined on more general test functions of the form $\eta(\mathbf{x},\mathring{\mathbf{x}})$:

$$\breve{G}[\eta] \overset{\text{def.}}{=} \int \eta^{X\times\mathring{X}} \cdot f(X,\mathring{X}) \delta X \delta \mathring{X} \tag{180}$$

where $X \times \mathring{X}$ denotes Cartesian product and where

$$\eta^{X\times\mathring{X}} \overset{\text{def.}}{=} \begin{cases} 1 & \text{if } X \times \mathring{X} = \emptyset \\ \prod_{(\mathbf{x},\mathring{\mathbf{x}})\in X\times\mathring{X}} \eta(\mathbf{x},\mathring{\mathbf{x}}) & \text{if otherwise} \end{cases}$$

Define

$$(\breve{h}_1 \times \breve{h}_2)(\mathbf{x},\mathring{\mathbf{x}}) \overset{\text{def.}}{=} \breve{h}_1(\mathbf{x}) \cdot \breve{h}_2(\mathring{\mathbf{x}}) \tag{181}$$

If $\mathring{X} \neq \emptyset$ then $\breve{h}^X \cdot \breve{h}^{\mathring{X}} = (\breve{h} \times \breve{h})^{X\times\mathring{X}}$. Given this, we can rewrite Eq. (179) in the form

$$\begin{aligned}&\breve{G}_{k+1}[\breve{g}|X,\mathring{\mathbf{x}}] \\ &= \left(1 - (\mathring{p}_D \times 1) + (\mathring{p}_D \times 1) \cdot (\mathring{p}_{\breve{g}} \times 1) \cdot e^{-\lambda+\lambda c_{\breve{g}}} \cdot (1 - p_D + p_D \cdot p_{\breve{g}})\right)^{X\times\{\mathring{\mathbf{x}}\}}\end{aligned} \tag{182}$$

Hereafter, we abuse notation and implicitly understand that $\mathring{p}_D$ means the same thing as $\mathring{p}_D \times 1$. Then Eq. (182) simplifies as

$$\breve{G}_{k+1}[\breve{g}|X,\mathring{\mathbf{x}}] = \left(1 - \mathring{p}_D + \mathring{p}_D \cdot \mathring{p}_{\breve{g}} \cdot e^{-\lambda+\lambda c_{\breve{g}}} \cdot (1 - p_D + p_D \cdot p_{\breve{g}})\right)^{X\times\{\mathring{\mathbf{x}}\}}$$

Next, from Eq. (88) we have

$$\begin{aligned}
&\breve{F}_{k+1}[\breve{g},\breve{h}] \\
&= \int \breve{h}^X \cdot \breve{h}(\mathring{\mathbf{x}}) \cdot \breve{G}_{k+1}[\breve{g}|X,\mathring{\mathbf{x}}] \cdot f_{k+1|k}(X,\mathring{\mathbf{x}}|\breve{Z}^{(k)})\delta X d\mathring{\mathbf{x}} \\
&= \int (\breve{h}\times\breve{h})^{X\times\{\mathring{\mathbf{x}}\}} \cdot \breve{G}_{k+1}[\breve{g}|X,\mathring{\mathbf{x}}] \cdot f_{k+1|k}(X,\mathring{\mathbf{x}}|\breve{Z}^{(k)})\delta X d\mathring{\mathbf{x}} \\
&= G_{k+1|k}[(\breve{h}\times\breve{h})(1-\mathring{p}_D + e^{-\lambda+\lambda c_{\breve{g}}} \cdot \mathring{p}_D \cdot \mathring{p}_{\breve{g}} \cdot (1-p_D+p_D\cdot p_{\breve{g}}))]
\end{aligned} \tag{183}$$

Using a Poisson approximation analogous to that of Eq. (93) we can assume that

$$\breve{G}_{k+1|k}[\eta] \cong \exp\left(-\mu\mathring{\mu} + \mu\mathring{\mu}\cdot(s\times\mathring{s})[\eta]\right) \tag{184}$$

where $\int s(\mathbf{x})d\mathbf{x} = 1$ and $\int \mathring{s}(\mathring{\mathbf{x}})d\mathring{\mathbf{x}} = 1$ and where

$$\mu s(\mathbf{x}) \overset{\text{abbr.}}{=} \breve{D}_{k+1|k}(\mathbf{x}), \qquad \mathring{\mu}\mathring{s}(\mathring{\mathbf{x}}) \overset{\text{abbr.}}{=} \breve{D}_{k+1|k}(\mathring{\mathbf{x}}) \tag{185}$$

$$(s\times\mathring{s})[\eta] \overset{\text{def.}}{=} \int \eta(\mathbf{x},\mathring{\mathbf{x}})\cdot s(\mathbf{x})\cdot\mathring{s}(\mathring{\mathbf{x}})d\mathbf{x}d\mathring{\mathbf{x}} \tag{186}$$

In this case we get

$$\breve{F}_{k+1}[\breve{g},\breve{h}] \cong e^{-\mu\mathring{\mu}+\mu\mathring{\mu}\cdot(s\times\mathring{s})[(\breve{h}\times\breve{h})(1-\mathring{p}_D+e^{-\lambda+\lambda c_{\breve{g}}}\cdot\mathring{p}_D\cdot\mathring{p}_{\breve{g}}\cdot(1-p_D+p_D\cdot p_{\breve{g}}))]} \tag{187}$$

Unfortunately, this formula is far too complex to produce a usable closed-form formula because of the presence of the highly nonlinear term $e^{-\lambda+\lambda c_{\breve{g}}} \cdot \mathring{p}_{\breve{g}} \cdot p_{\breve{g}}$.

7.3.2. *Linearization of the Joint p.g.fl. of the Full Sensor Model*

Consequently, we must further simplify by devising *a linearization of Eq. (187) that preserves most of the features of the observation model described in section 2.8.* Inspection of Eq. (187) leads to the following linearization:

- *Linearized Actuator Sensor Joint p.g.fl.* The two-variable p.g.fl. of the actuator sensor is of Poisson form in observations:

$$\breve{F}^{\text{actu}}_{k+1}[\breve{g},\breve{h}] \cong \exp\left(-\mathring{\mu} + \mathring{\mu}\mathring{s}[\breve{h}(1-\mathring{p}_D+\mathring{p}_D\mathring{p}_{\breve{g}})]\right) \tag{188}$$

In other words, the probability that an actuator-sensor observation will be collected and transmitted is $\mathring{p}_D$ and, in that case, its statistics are governed by $\mathring{L}_{\mathring{\mathbf{z}}}(\mathring{\mathbf{x}})$.

- *Linearized Sensor Joint p.g.fl.*: The two-variable p.g.fl. for the sensor itself, excluding clutter, is of Poisson form in observations:

$$\breve{F}^{\text{sens}}_{k+1}[\breve{g},\breve{h}] \cong \exp\left(-\mu + \mu(s\times\mathring{s})[(\breve{h}\times 1)(1-\mathring{p}_D p_D + p_D\mathring{p}_D p_{\breve{g}})]\right) \tag{189}$$

In other words, the probability that any sensor observation will be collected and transmitted is $\mathring{p}_D p_D$ and, in that case, its statistics are governed by $L_{\mathbf{z}}(\mathbf{x})$.

• *Linearized Sensor False Alarm Joint p.g.fl.*: Sensor observations are corrupted by a Poisson false alarm process of the form:

$$\breve{F}^{\text{clutt}}_{k+1}[\breve{g}] = \exp\left(-\lambda + \lambda\mathring{s}[1 - \mathring{p}_D + \mathring{p}_D c_{\breve{g}}]\right) \tag{190}$$

In other words, the probability that clutter observations will be collected and transmitted is $\mathring{p}_D$, and, in that case, their statistics are governed by $c_{k+1}(\mathbf{z})$ and λ_{k+1}.

• *Linearized Joint p.g.fl.*: Multiplying these three p.g.fl.'s, we get an approximate two-variable p.g.fl. for Bayes' rule:

$$\breve{F}_{k+1}[\breve{g}, \breve{h}] \cong \exp\begin{pmatrix} -\mathring{\mu} - \lambda - \mu \\ +\mathring{\mu}\mathring{s}[\breve{h}(1 - \mathring{p}_D + \mathring{p}_D \mathring{p}_{\breve{g}})] + \lambda\mathring{s}[1 - \mathring{p}_D + \mathring{p}_D c_{\breve{g}}] \\ +\mu(s \times \mathring{s})[(\breve{h} \times 1)(1 - \mathring{p}_D p_D + p_D \mathring{p}_D p_{\breve{g}})] \end{pmatrix} \tag{191}$$

Note that $\log \breve{F}_{k+1}[\breve{g}, \breve{h}]$ is now linear in $\breve{g}$ and linear in $\breve{h}$. As a check, note that if $\mathring{p}_D = 0$ then this reduces to

$$\breve{F}_{k+1}[\breve{g}, \breve{h}] \cong \exp\left(-\mathring{\mu} - \mu + \mathring{\mu}\mathring{s}[\breve{h}] + \mu s[\breve{h}]\right) \tag{192}$$

Since only null observations can be collected when $\mathring{p}_D = 0$, it follows that the posterior p.g.fl. is always

$$\breve{G}_{k+1|k}[\breve{h}] = \breve{G}_{k+1|k+1}[\breve{h}] = \frac{\breve{F}_{k+1}[0, \breve{h}]}{\breve{F}_{k+1}[0, 1]} = \exp\left(-\mathring{\mu} - \mu + \mathring{\mu}\mathring{s}[\breve{h}] + \mu s[\breve{h}]\right) \tag{193}$$

In other words and as one would expect, $\breve{G}_{k+1|k}[\breve{h}]$ is Poisson with an average number of $\mu + \mathring{\mu}$ objects.

7.4. *Joint PENT (Single-Sensor, Single-Step Look-Ahead)*

The derivation of the formula for PENT is nearly identical to the derivation of the joint PHD corrector equation in section 8.4. The primary difference is that the hedged p.g.fl. is used in place of the posterior p.g.fl. and the PIMS is used in place of the arbitrary observation-set. That is, we employ the optimization-hedging strategy described in section 4. From the predicted joint PHD $\breve{D}_{k+1|k}(\mathbf{x})$, extract an estimate $\hat{\mathbf{x}}_1, ..., \hat{\mathbf{x}}_{\hat{n}}$ of the number and states of the predicted tracks as in Eq. (121). Likewise, from $\breve{D}_{k+1|k}(\mathring{\mathbf{x}})$

extract an estimate $\mathring{\mathbf{x}}_0$ of the predicted track of the sensor. Then from equations (103) and (209) the hedged p.g.fl. is

$$\begin{aligned}
&\mathring{G}_{k+1|k+1}[\breve{h}] \\
&\quad = \exp\left(\mathring{\mu}\mathring{s}[(\breve{h}-1)(1-\mathring{p}_D)] + \mu(s\times\mathring{s})[(\breve{h}\times 1 - 1)(1-\mathring{p}_D p_D)]\right) \\
&\quad \cdot \prod_{i=1}^{m}\left(\frac{\mathring{s}[\breve{h}\mathring{p}_D]\cdot\lambda c(\eta(\hat{\mathbf{x}}_i)) + \mu(s\times\mathring{s})[(\breve{h}\times 1)p_D\mathring{p}_D L_{\eta(\hat{\mathbf{x}}_i)}]}{\mathring{s}[\mathring{p}_D]\cdot\lambda c(\eta(\hat{\mathbf{x}}_i)) + \mu(s\times\mathring{s})[p_D\mathring{p}_D L_{\eta(\hat{\mathbf{x}}_i)}]}\right)^{\mathring{p}_D(\mathring{\mathbf{x}}_0)\cdot p_D(\hat{\mathbf{x}}_i,\mathring{\mathbf{x}}_0)} \\
&\quad \cdot \left(\frac{\mathring{s}[\breve{h}\mathring{p}_D\mathring{L}_{\mathring{\eta}(\mathring{\mathbf{x}}_0)}]}{\mathring{s}[\mathring{p}_D\mathring{L}_{\mathring{\eta}(\mathring{\mathbf{x}}_0)}]}\right)^{\mathring{p}_D(\mathring{\mathbf{x}}_0)}
\end{aligned} \tag{194}$$

The derivation then proceeds exactly as in the previous section. We derive formulas for $\mathring{D}_{k+1|k+1}(\mathbf{x})$ and $\mathring{D}_{k+1|k+1}(\mathring{\mathbf{x}})$, respectively. By Eq. (105) the formula for PENT is

$$\begin{aligned}
\mathring{N}_{k+1|k+1}(\mathbf{u}_k) &= \int \breve{D}_{k+1|k+1}(\mathbf{y})d\mathbf{y} \\
&= \mu(s\times\mathring{s})[1-\mathring{p}_D p_D] \\
&+ \sum_{i=1}^{\hat{n}} \mathring{p}_D(\mathring{\mathbf{x}}_0)\cdot p_D(\hat{\mathbf{x}}_i,\mathring{\mathbf{x}}_0)\cdot \frac{\mu(s\times\mathring{s})[p_D\mathring{p}_D L_{\eta(\hat{\mathbf{x}}_i)}]}{\mathring{s}[\mathring{p}_D]\cdot\lambda c(\eta(\hat{\mathbf{x}}_i)) + \mu(s\times\mathring{s})[p_D\mathring{p}_D L_{\eta(\hat{\mathbf{x}}_i)}]}
\end{aligned}$$

or

$$\begin{aligned}
\mathring{N}_{k+1|k+1}(\mathbf{u}_k) = \mu(s\times\mathring{s})[1-\mathring{p}_D p_D] & \\
+ \sum_{i=1}^{\hat{n}} \mathring{p}_D(\mathring{\mathbf{x}}_0)\cdot p_D(\hat{\mathbf{x}}_i,\mathring{\mathbf{x}}_0) & \\
\cdot\left(1 - \frac{\mathring{s}[\mathring{p}_D]\cdot\lambda c(\eta(\hat{\mathbf{x}}_i))}{\mathring{s}[\mathring{p}_D]\cdot\lambda c(\eta(\hat{\mathbf{x}}_i)) + \mu(s\times\mathring{s})[p_D\mathring{p}_D L_{\eta(\hat{\mathbf{x}}_i)}]}\right) &
\end{aligned} \tag{195}$$

This formula applies to the PHD filter of section 3.4. It can also be used in conjunction with the MHC filter of section 3.5 by using Eq. (98):

$\mu s(\mathbf{x}) = D_{k+1|k}(\mathbf{x}) = \sum_{j=1}^{N} q_j f_j(\mathbf{x})$. Then Eq. (195) becomes

$$\mathring{N}_{k+1|k+1}(\mathbf{u}_k) = \sum_{j=1}^{N} q_j (f_j \times \overset{*}{s})[1 - \overset{*}{p}_D p_D] + \sum_{i=1}^{N} \overset{*}{p}_D(\overset{*}{\mathbf{x}}_0) \cdot p_D(\hat{\mathbf{x}}_i, \overset{*}{\mathbf{x}}_0) \cdot \left(1 - \frac{\overset{*}{s}[\overset{*}{p}_D] \cdot \lambda c(\eta(\hat{\mathbf{x}}_i))}{\overset{*}{s}[\overset{*}{p}_D] \cdot \lambda c(\eta(\hat{\mathbf{x}}_i)) + \sum_{e=1}^{N} q_e (f_e \times \overset{*}{s})[p_D \overset{*}{p}_D L_{\eta(\hat{\mathbf{x}}_i)}]}\right) \tag{196}$$

If there are no false alarms it simplifies to

$$\mathring{N}_{k+1|k+1}(\mathbf{u}_k) = \sum_{j=1}^{N} \left(q_j (f_j \times \overset{*}{s})[1 - \overset{*}{p}_D p_D] + \overset{*}{p}_D(\overset{*}{\mathbf{x}}_0) \cdot p_D(\hat{\mathbf{x}}_j, \overset{*}{\mathbf{x}}_0)\right) \tag{197}$$

7.5. *Joint PENT (Multisensor, Single-Step Look-Ahead)*

In the multisensor case we can derive a formula for PENT using the simplified pseudo-sensor approximation of section 4.6. Assume the notation and assumptions for the multisensor, single-step look-ahead case used in section 5.3. Under our current assumptions, the joint actuator-sensor probability of detection is

$$\overset{*}{\breve{p}}_D(\mathbf{x}, \overset{*1}{\mathbf{x}}_{k+1}, \overset{*2}{\mathbf{x}}_{k+1}) \overset{\text{def.}}{=} 1 - (1 - \overset{*1}{p}_D(\overset{*1}{\mathbf{x}}_{k+1})) \cdot (1 - \overset{*2}{p}_D(\overset{*2}{\mathbf{x}}_{k+1}))$$

and the joint multisensor probability of detection is

$$\breve{p}_D(\mathbf{x}, \overset{*1}{\mathbf{x}}_{k+1}, \overset{*2}{\mathbf{x}}_{k+1}) \overset{\text{def.}}{=} 1 - (1 - \overset{*1}{p}_D(\overset{*1}{\mathbf{x}}_{k+1}) \overset{1}{p}_D(\mathbf{x}, \overset{*1}{\mathbf{x}}_{k+1})) \cdot (1 - \overset{*1}{p}_D(\overset{*2}{\mathbf{x}}_{k+1}) \overset{2}{p}_D(\mathbf{x}, \overset{*2}{\mathbf{x}}_{k+1})) \tag{198}$$

Let $\breve{h}$ be a joint function on target states and pseudo-sensor states. Then the formula for the hedged p.g.fl. in the single-sensor case (Eq. (194)) becomes:

$$\begin{aligned}&\mathring{G}_{k+1|k+1}[\breve{h}] \\ &= e^{\overset{*1}{\mu}\overset{*2}{\mu}(\overset{*1}{s}\times\overset{*2}{s})[(\breve{h}\times\breve{h}-1)(1-\overset{*}{\breve{p}}_D)]+\mu(s\times\overset{*1}{s}\times\overset{*2}{s})[(\breve{h}\times 1\times 1-1)(1-\breve{p}_D)]} \\ &\cdot\prod_{i=1}^{\hat{n}}\left(\frac{(\overset{*1}{s}\times\overset{*2}{s})[\overset{*}{\breve{p}}_D]\cdot\tilde{\lambda}\tilde{c}(\overset{1}{\eta}(\hat{\mathbf{x}}_i),\overset{2}{\eta}(\hat{\mathbf{x}}_i)) + \mu(s\times\overset{*1}{s}\times\overset{*2}{s})[(\breve{h}\times 1\times 1)\breve{p}_D\overset{1}{L}_{\overset{1}{\eta}(\hat{\mathbf{x}}_i)}\overset{2}{L}_{\overset{2}{\eta}(\hat{\mathbf{x}}_i)}]}{(\overset{*1}{s}\times\overset{*2}{s})[\overset{*}{\breve{p}}_D]\cdot\tilde{\lambda}\tilde{c}(\overset{1}{\eta}(\hat{\mathbf{x}}_i),\overset{2}{\eta}(\hat{\mathbf{x}}_i)) + \mu(s\times\overset{*1}{s}\times\overset{*2}{s})[\breve{p}_D\overset{1}{L}_{\overset{1}{\eta}(\hat{\mathbf{x}}_i)}\overset{2}{L}_{\overset{2}{\eta}(\hat{\mathbf{x}}_i)}]}\right)^{\breve{p}_D(\mathbf{x},\overset{*1}{\mathbf{x}}_0,\overset{*2}{\mathbf{x}}_0)} \\ &\cdot\left(\frac{(\overset{*1}{s}\times\overset{*2}{s})[(\breve{h}\times\breve{h})\overset{*}{\breve{p}}_D\overset{*1}{L}_{\overset{*1}{\eta}(\overset{*1}{\mathbf{x}}_0)}\overset{*2}{L}_{\overset{*2}{\eta}(\overset{*2}{\mathbf{x}}_0)}]}{(\overset{*1}{s}\times\overset{*2}{s})[\overset{*}{\breve{p}}_D\overset{*1}{L}_{\overset{*1}{\eta}(\overset{*1}{\mathbf{x}}_0)}\overset{*2}{L}_{\overset{*2}{\eta}(\overset{*2}{\mathbf{x}}_0)}]}\right)^{\overset{*}{\breve{p}}_D(\mathbf{x},\overset{*1}{\mathbf{x}}_0,\overset{*2}{\mathbf{x}}_0)}\end{aligned} \tag{199}$$

The usual procedure outlined in section 4 leads to

$$\begin{aligned}&\mathring{N}_{k+1|k+1}(\overset{*1}{\mathbf{x}}_{k+1},\overset{*2}{\mathbf{x}}_{k+1}) \\ &= (\breve{D}_{k+1|k}\times\overset{*1}{D}_{k+1|k}\times\overset{*2}{D}_{k+1|k})[1-\breve{p}_D] \\ &\quad+\sum_{i=1}^{\hat{n}}\breve{p}_D(\hat{\mathbf{x}}_i,\overset{*1}{\mathbf{x}}_0,\overset{*2}{\mathbf{x}}_0) \\ &\quad\cdot\left(1-\frac{(\overset{*1}{D}_{k+1}\times\overset{*2}{D}_{k+1})[\overset{*}{p}_D]\cdot\tilde{\lambda}\tilde{c}(\overset{1}{\eta}(\hat{\mathbf{x}}_i),\overset{2}{\eta}(\hat{\mathbf{x}}_i))}{(\overset{*1}{D}_{k+1}\times\overset{*2}{D}_{k+1})[\overset{*}{\breve{p}}_D]\cdot\tilde{\lambda}\tilde{c}(\overset{1}{\eta}(\hat{\mathbf{x}}_i),\overset{2}{\eta}(\hat{\mathbf{x}}_i)) + (\breve{D}_{k+1|k}\times\overset{*1}{D}_{k+1}\times\overset{*2}{D}_{k+1})[\breve{p}_D\overset{1}{L}_{\overset{1}{\eta}(\hat{\mathbf{x}}_i)}\overset{2}{L}_{\overset{2}{\eta}(\hat{\mathbf{x}}_i)}]}\right)\end{aligned} \tag{200}$$

where $\overset{*1}{D}_{k+1}$ and $\overset{*2}{D}_{k+1}$ are defined as in Eqs. (170) and (171).

When there are no false alarms and this equation is suitably modified for use with the MHC filter of section 3.5 using Eq. (98), this equation becomes

$$\mathring{N}_{k+1|k+1}(\overset{*1}{\mathbf{x}}_{k+1},\overset{*2}{\mathbf{x}}_{k+1}) = \sum_{j=1}^{N}\Big(q_j(f_j\times\overset{*1}{s}\times\overset{*2}{s})[1-\breve{p}_D]+\breve{p}_D(\hat{\mathbf{x}}_j,\overset{*1}{\mathbf{x}}_0,\overset{*2}{\mathbf{x}}_0)\Big) \tag{201}$$

8. Mathematical Proofs

This section contains the proofs of some of the more complicated mathematical derivations.

8.1. *Proof of Eq. (45)*

Let G_Υ be the p.g.fl. of a random finite subset Υ of objects and $f_\Upsilon(Y)$ its multi-object probability density function. Let $r(\mathbf{y})$ be a unitless test function with $0 \leq r(\mathbf{y}) \leq 1$ for all $\mathbf{y}$. We are to show that

$$\frac{\delta^n G_\Upsilon}{\delta \mathbf{y}_1 \cdots \delta \mathbf{y}_n}[r] = \int r^Y \cdot f_\Upsilon(Y \cup \{\mathbf{y}_1, ..., \mathbf{y}_n\}) \delta Y$$

Begin with $n = 1$. By definition

$$\frac{\delta G_\Upsilon}{\delta \mathbf{y}}[r] = \lim_{\varepsilon \to 0} \frac{G_\Upsilon[r + \varepsilon \delta_\mathbf{y}] - G_\Upsilon[r]}{\varepsilon}$$

Hold r and $\mathbf{y}$ fixed and expand $G_\Upsilon[r + \varepsilon \delta_\mathbf{y}]$ in a Taylor's series around $\varepsilon = 0$:

$$G_\Upsilon[r + \varepsilon \delta_\mathbf{y}] = G_\Upsilon[r] + \sum_{i=1}^{\infty} \frac{1}{i!} \left[\frac{\partial^i}{\partial x^i} G_\Upsilon[r + x \delta_\mathbf{y}] \right]_{x=0} \varepsilon^i$$

Note that

$$\left[\frac{\partial}{\partial x} G_\Upsilon[r + x \delta_\mathbf{y}] \right]_{x=0} = \int \left[\frac{\partial}{\partial x} (r + x \delta_\mathbf{y})^Y \right]_{x=0} \cdot f_\Upsilon(Y) \delta Y$$

where

$$\begin{aligned} &\left[\frac{\partial}{\partial x} (r + x \delta_\mathbf{y})^{\{\mathbf{y}_1, \ldots, \mathbf{y}_n\}} \right]_{x=0} \\ &= \left[\sum_{i=1}^{n} (r(\mathbf{x}_1) + x \delta_\mathbf{y}(\mathbf{y}_1)) \cdots \delta_\mathbf{y}(\mathbf{y}_i) \cdots (r(\mathbf{y}_n) + x \delta_\mathbf{y}(\mathbf{y}_n)) \right]_{x=0} \\ &= \sum_{i=1}^{n} r(\mathbf{y}_1) \cdots \delta_\mathbf{y}(\mathbf{y}_i) \cdots r(\mathbf{y}_n) \end{aligned}$$

and so

$$\left[\frac{\partial}{\partial x} G_\Upsilon[r + x \delta_\mathbf{y}] \right]_{x=0} = \int r^W \cdot f_\Upsilon(W \cup \{\mathbf{y}\}) \delta \breve{W}$$

On the other hand

$$\begin{aligned} &\left[\frac{\partial^2}{\partial x^2} (r + x \delta_\mathbf{y})^{\{\mathbf{y}_1, \ldots, \mathbf{y}_n\}} \right]_{x=0} \\ &= \left[\sum_{1 \leq i \neq j \leq n} (r(\mathbf{y}_1) + x \delta_\mathbf{y}(\mathbf{y}_1)) \cdots \delta_\mathbf{y}(\mathbf{y}_i) \cdots \delta_\mathbf{y}(\mathbf{y}_j) \cdots (r(\mathbf{y}_n) + x \delta_\mathbf{y}(\mathbf{y}_n)) \right]_{x=0} \\ &= \sum_{i=1}^{\breve{n}} r(\mathbf{y}_1) \cdots \delta_\mathbf{y}(\mathbf{y}_i) \cdots \delta_\mathbf{y}(\mathbf{y}_j) \cdots r(\mathbf{y}_{\breve{n}}) \end{aligned}$$

Since $f_\Upsilon(\{\mathbf{y}_1, ..., \mathbf{y}_i, ..., \mathbf{y}_j, ..., \mathbf{y}_n\})$ vanishes whenever $\mathbf{y}_i = \mathbf{y}_j$ for $i \neq j$ (see Eq. (27)), it follows that

$$\delta_{\mathbf{y}}(\mathbf{y}_i) \cdot \delta_{\mathbf{y}}(\mathbf{y}_j) \cdot f_\Upsilon(\{\mathbf{y}_1, ..., \mathbf{y}_i, ..., \mathbf{y}_j, ..., \mathbf{y}_n\}) = 0$$

whenever $\mathbf{y}_i = \mathbf{y}_j$ for $i \neq j$. So

$$\left[\frac{\partial^i}{\partial x^i}(r + x\delta_{\mathbf{y}})^{\{\mathbf{y}_1,...,\mathbf{y}_n\}}\right]_{x=0} = 0$$

for $i \geq 2$ and thus

$$G_\Upsilon[r + \varepsilon\delta_{\mathbf{y}}] = G_\Upsilon[r] + \varepsilon \cdot \int r^W \cdot f_\Upsilon(W \cup \{\mathbf{y}\})\delta W \tag{202}$$

Consequently

$$\frac{\delta G_\Upsilon}{\delta \mathbf{y}}[r] = \lim_{\varepsilon \to 0} \frac{G_\Upsilon[r + \varepsilon\delta_{\mathbf{y}}] - G_\Upsilon[r]}{\varepsilon} = \int r^W \cdot f_\Upsilon(W \cup \{\mathbf{y}\})\delta W$$

By iteration we get the desired result.

8.2. *Proof of Eq. (137)*

We are to show that under the assumptions described in section 5.4,

$$G_{k+2|k}[\overset{+}{p}_D] = G_{k+1|k}[\overset{+\downarrow}{p}_D]$$

First, note that the probability that single-predicted tracks (i.e., from time-step $k+1$ to time-step $k+2$) will be in the FoV $\overset{+}{p}_D$ at time-step $k+2$, given that they had state-set W at time-step $k+1$, is

$$\begin{aligned}
&G_{k+2|k+1}[\overset{+}{p}_D|W] \\
&\quad = \int \overset{+}{p}_D^X \cdot f_{k+2|k+1}(X|W)\delta X \\
&\quad = \frac{1}{e!}\int \overset{+}{p}_D(\mathbf{x}_1)\cdots \overset{+}{p}_D(\mathbf{x}_e) \\
&\quad \cdot \left(\sum_\sigma f_{k+2|k+1}(\mathbf{x}_1|\mathbf{w}_{\sigma 1})\cdots f_{k+2|k+1}(\mathbf{x}_e|\mathbf{w}_{\sigma e})\right) d\mathbf{x}_1\cdots d\mathbf{x}_e \\
&\quad = \frac{1}{e!}\sum_\sigma \left(\int \overset{+}{p}_D(\mathbf{x}_1) f_{k+1|k}(\mathbf{x}_1|\mathbf{w}_{\sigma 1})d\mathbf{x}_1\right) \\
&\quad \cdots \left(\int \overset{+}{p}_D(\mathbf{x}_e) f_{k+1|k}(\mathbf{x}_e|\mathbf{w}_{\sigma e})d\mathbf{x}_e\right) \\
&\quad = \left(\int \overset{+}{p}_D(\mathbf{x}) f_{k+1|k}(\mathbf{x}_1|\mathbf{w}_1)d\mathbf{x}\right)\cdots\left(\int \overset{+}{p}_D(\mathbf{x}) f_{k+1|k}(\mathbf{x}_e|\mathbf{w}_e)d\mathbf{x}\right) \\
&\quad = \overset{+\downarrow}{p}_D(\mathbf{w}_1)\cdots \overset{+\downarrow}{p}_D(\mathbf{w}_e) \\
&\quad = \overset{+\downarrow}{p}_D^W
\end{aligned}$$

where the second equation follows from Eq. (134). Therefore, the probability that *twice-predicted* tracks (from time-step k to time-step $k+2$) will all be contained in the FoV at time-step $k+2$ is:

$$\begin{aligned}
G_{k+2|k}[\overset{+}{p}_D] &= \int \overset{+}{p}_D^X \cdot f_{k+2|k}(X)\delta X \\
&= \int \overset{+}{p}_D^X \cdot \int f_{k+2|k+1}(X|W)\cdot f_{k+1|k}(W)\delta W\delta X \\
&= \int \left(\int \overset{+}{p}_D^X \cdot f_{k+2|k+1}(X|W)\delta X\right)\cdot f_{k+1|k}(W)\delta W \\
&= \int G_{k+2|k+1}[\overset{+}{p}_D|W]\cdot f_{k+1|k}(W)\delta W \\
&= \int \overset{+\downarrow}{p}_D^W \cdot f_{k+1|k}(W)\delta W \\
&= G_{k+1|k}[\overset{+\downarrow}{p}_D]
\end{aligned}$$

as claimed.

8.3. *Proof of Eqs. (168) and (169)*

We are to derive the following predictor equations:

$$\breve{D}_{k+1|k}(\mathbf{y}) = b_{k+1|k}(\mathbf{y}) + \int \left(s_{k+1|k}(\mathbf{x}) \cdot f_{k+1|k}(\mathbf{y}|\mathbf{x}) + b_{k+1|k}(\mathbf{y}|\mathbf{x})\right) \cdot \breve{D}_{k+1|k}(\mathbf{x})d\mathbf{x}$$

$$\breve{D}_{k+1|k}(\overset{*i}{\mathbf{y}}) = \int \overset{*i}{f}_{k+1|k}(\overset{*i}{\mathbf{y}}|\overset{*i}{\mathbf{x}}, \mathbf{u}_i) \cdot \breve{D}_{k|k}(\overset{*i}{\mathbf{x}})d\overset{*i}{\mathbf{x}}$$

Let the set of sensor states and sensor controls at time-step k be (see Remark 1 of section 2.7)

$$(\overset{*}{X}, U) = \{(\overset{*1}{\mathbf{x}}, \overset{1}{\mathbf{u}}_1), ..., (\overset{*e}{\mathbf{x}}, \overset{e}{\mathbf{u}}_e)\}$$

Then the predicted joint multisensor-multitarget posterior is

$$f_{k+1|k}(Y, \overset{*}{Y}) = \int\int_{|\overset{*}{X}|=\overset{*}{n}} f_{k+1|k}(Y, \overset{*}{Y}|X, \overset{*}{X}, U) \cdot f_{k|k}(X, \overset{*}{X})\delta X \delta \overset{*}{X}$$

and its p.g.fl. is

$$\breve{G}_{k+1|k}[\breve{h}] = \int\int_{|\overset{*}{X}|=\overset{*}{n}} \breve{G}_{k+1|k}[\breve{h}|X, \overset{*}{X}, U] \cdot f_{k|k}(X, \overset{*}{X})\delta X \delta \overset{*}{X}$$

where

$$\begin{aligned} &\breve{G}_{k+1|k}[\breve{h}|X, \overset{*}{X}, U] \\ &= \int\int_{|\overset{*}{X}|=\overset{*}{n}} \breve{h}^Y \cdot \breve{h}^{\overset{*}{Y}} \cdot f_{k+1|k}(Y, \overset{*}{Y}|X, \overset{*}{X}, U)\delta Y \delta \overset{*}{Y} \\ &= \int\int_{|\overset{*}{X}|=\overset{*}{n}} \breve{h}^Y \cdot \breve{h}^{\overset{*}{Y}} \cdot f_{k+1|k}(Y|X) \cdot f_{k+1|k}(\overset{*}{Y}|\overset{*}{X}, U)\delta Y \delta \overset{*}{Y} \\ &= \breve{G}_{k+1|k}[\breve{h}|X] \cdot \breve{G}_{k+1|k}[\breve{h}|\overset{*}{X}, U] \end{aligned} \tag{203}$$

and where

$$\breve{G}_{k+1|k}[\breve{h}|X] = \int \breve{h}^Y \cdot f_{k+1|k}(Y|X)\delta Y$$

$$\breve{G}_{k+1|k}[\breve{h}|\overset{*}{X}, U] = \int_{|\overset{*}{X}|=\overset{*}{n}} \breve{h}^{\overset{*}{Y}} \cdot f_{k+1|k}(\overset{*}{Y}|\overset{*}{X}, U)\delta \overset{*}{Y}$$

In our particular case, $f_{k+1|k}(\overset{*}{Y}|\overset{*}{X}, U) = 0$ unless

$$\begin{aligned} &\overset{*}{f}_{k+1|k}(\{\overset{*1}{\mathbf{y}}, ..., \overset{*e}{\mathbf{y}}\}|\{(\overset{*1}{\mathbf{x}}, \overset{1}{\mathbf{u}}_1), ..., (\overset{*e}{\mathbf{x}}, \overset{e}{\mathbf{u}}_e)\}) \\ &= \sum_\sigma \overset{*1}{f}_{k+1|k}(\overset{*1}{\mathbf{y}}|\overset{*\sigma 1}{\mathbf{x}}, \overset{\sigma 1}{\mathbf{u}}_{\sigma 1}) \cdots \overset{*e}{f}_{k+1|k}(\overset{*e}{\mathbf{y}}|\overset{*\sigma e}{\mathbf{x}}, \overset{\sigma e}{\mathbf{u}}_{\sigma e}) \end{aligned}$$

Consequently,

$$
\begin{aligned}
&\breve{G}_{k+1|k}[\breve{h}|\mathring{X},U] \\
&= \int_{|\mathring{X}|=\mathring{n}} \breve{h}^{\mathring{Y}} \cdot f_{k+1|k}(\mathring{Y}|\mathring{X},U)\delta\mathring{Y} \\
&= \frac{1}{e!}\int \breve{h}(\mathring{\mathbf{y}}^{1})\cdots\breve{h}(\mathring{\mathbf{y}}^{e}) \\
&\cdot\left(\sum_{\sigma} \mathring{f}^{1}_{k+1|k}(\mathring{\mathbf{y}}^{1}|\mathring{\mathbf{x}}^{\sigma 1},\mathbf{u}_{\sigma 1})\cdots \mathring{f}^{e}_{k+1|k}(\mathring{\mathbf{y}}^{e}|\mathring{\mathbf{x}}^{\sigma e},\mathbf{u}_{\sigma e})\right) d\mathring{\mathbf{y}}^{1}\cdots d\mathring{\mathbf{y}}^{e} \\
&= \frac{1}{e!}\left(\begin{array}{c} \sum_{\sigma}\left(\int \breve{h}(\mathring{\mathbf{y}}^{1})\cdot \mathring{f}^{1}_{k+1|k}(\mathring{\mathbf{y}}^{1}|\mathring{\mathbf{x}}^{\sigma 1},\mathbf{u}_{\sigma 1})d\mathring{\mathbf{y}}^{1}\right) \\ \cdots\left(\int \breve{h}(\mathring{\mathbf{y}}^{e})\cdot \mathring{f}^{e}_{k+1|k}(\mathring{\mathbf{y}}^{e}|\mathring{\mathbf{x}}^{\sigma e},\mathbf{u}_{\sigma e})d\mathring{\mathbf{y}}^{e}\right)\end{array}\right) \\
&= \left(\int \breve{h}(\mathring{\mathbf{y}}^{1})\cdot \mathring{f}^{1}_{k+1|k}(\mathring{\mathbf{y}}^{1}|\mathring{\mathbf{x}}^{1},\mathbf{u}_{1})d\mathring{\mathbf{y}}^{1}\right) \\
&\cdots\left(\int \breve{h}(\mathring{\mathbf{y}}^{e})\cdot \mathring{f}^{e}_{k+1|k}(\mathring{\mathbf{y}}^{e}|\mathring{\mathbf{x}}^{e},\mathbf{u}_{e})d\mathring{\mathbf{y}}^{e}\right) \\
&= (\mathring{p}_{\breve{h}})^{(\mathring{X},U)}
\end{aligned}
$$

where

$$
\begin{aligned}
\mathring{p}_{\breve{h}}(\mathring{\mathbf{x}},\mathbf{u}) &= \int \breve{h}(\mathring{\mathbf{y}}^{j})\cdot \mathring{f}^{j}_{k+1|k}(\mathring{\mathbf{y}}^{j}|\mathring{\mathbf{x}}^{j},\mathbf{u}_{j})d\mathring{\mathbf{y}}^{j} \\
\sigma^{(\mathring{X},U)} &= \prod_{(\mathring{\mathbf{x}},\mathbf{u})\in(\mathring{X},U)} \sigma(\mathring{\mathbf{x}},\mathbf{u})
\end{aligned}
$$

So

$$
\mathring{G}_{k+1|k}[\breve{h}|\mathring{X},U] = (\mathring{p}_{\breve{h}})^{(\mathring{X},U)}
$$

On the other hand, from section VI-J of [12] we know that

$$
G_{k+1|k}[\breve{h}|X] = e_{\breve{h}}\cdot\left(1-p_S+p_S p_{\breve{h}}\right)^{X}\cdot b_{\breve{h}}^{X} \tag{204}
$$

We now turn to the derivation of formulas for the joint PHD. First note that the first functional derivative of the predicted p.g.fl. $\breve{G}_{k+1|k}[\breve{h}]$ is

$$
\begin{aligned}
&\frac{\delta}{\delta\breve{\mathbf{x}}}\breve{G}_{k+1|k}[\breve{h}] \\
&\quad= \int\int_{|\mathring{X}|=\mathring{n}} \frac{\delta}{\delta\breve{\mathbf{x}}}\left(G_{k+1|k}[\breve{h}|X]\cdot \mathring{G}_{k+1|k}[\breve{h}|\mathring{X},U]\right)\cdot f_{k|k}(X,\mathring{X})\delta X\delta\mathring{X} \\
&\quad= \int\int_{|\mathring{X}|=\mathring{n}} \left(\frac{\delta G_{k+1|k}}{\delta\breve{\mathbf{x}}}[\breve{h}|X]\cdot \mathring{G}_{k+1|k}[\breve{h}|\mathring{X},U]\right)\cdot f_{k|k}(X,\mathring{X})\delta X\delta\mathring{X} \\
&\quad+ \int\int_{|\mathring{X}|=\mathring{n}} \left(G_{k+1|k}[\breve{h}|X]\cdot \frac{\delta \mathring{G}_{k+1|k}}{\delta\breve{\mathbf{x}}}[\breve{h}|\mathring{X},U]\right)\cdot f_{k|k}(X,\mathring{X})\delta X\delta\mathring{X}
\end{aligned}
$$

and so the joint PHD is

$$
\begin{aligned}
\breve{D}_{k+1|k}(\breve{\mathbf{x}}) &= \frac{\delta}{\delta\mathbf{x}}\breve{G}_{k+1|k}[1] \qquad (205)\\
&= \int\int_{|\mathring{X}|=\mathring{n}} \left[\frac{\delta G_{k+1|k}}{\delta\breve{\mathbf{x}}}[\breve{h}|X]\right]_{\breve{h}=1}\cdot f_{k|k}(X,\mathring{X})\delta X\delta\mathring{X} \\
&+ \int\int_{|\mathring{X}|=\mathring{n}} \left[\frac{\delta \mathring{G}_{k+1|k}}{\delta\breve{\mathbf{x}}}[\breve{h}|\mathring{X},U]\right]_{\breve{h}=1}\cdot f_{k|k}(X,\mathring{X})\delta X\delta\mathring{X}
\end{aligned}
$$

Given our assumptions, for target states $\breve{\mathbf{x}}=\mathbf{x}$ this becomes

$$
\breve{D}_{k+1|k}(\mathbf{x}) = \int\int_{|\mathring{X}|=\mathring{n}} \left[\frac{\delta G_{k+1|k}}{\delta\mathbf{x}}[\breve{h}|X]\right]_{\breve{h}=1}\cdot f_{k|k}(X,\mathring{X})\delta X\delta\mathring{X} \qquad (206)
$$

whereas for sensor states $\breve{\mathbf{x}}=\mathring{\mathbf{x}}$ it becomes

$$
\breve{D}_{k+1|k}(\mathring{\mathbf{x}}) = \int\int_{|\mathring{X}|=\mathring{n}} \left[\frac{\delta \mathring{G}_{k+1|k}}{\delta\mathring{\mathbf{x}}}[\breve{h}|\mathring{X},U]\right]_{\breve{h}=1}\cdot f_{k|k}(X,\mathring{X})\delta X\delta\mathring{X} \qquad (207)
$$

PHD for Target States: If $\breve{\mathbf{x}}=\mathbf{x}$ (target states) then Eq. (206) is determined entirely by Eq. (204) and we can just apply the PHD predictor derivation presented in section VI-J of [12]. This gives us the claimed formula.

PHD for Sensor States: If $\breve{\mathbf{x}}=\mathring{\mathbf{y}}^{i}$ (state of ith sensor) then Eq. (207)

becomes

$$\begin{aligned}
&\left[\frac{\delta \overset{*}{G}_{k+1|k}}{\delta \overset{*i}{\mathbf{y}}}[\breve{h}|\overset{*}{X},U]\right]_{\breve{h}=1} \\
&= \left[\frac{\delta}{\delta \overset{*i}{\mathbf{y}}}(\overset{*}{p}_{\breve{h}})^{(\overset{*}{X},U)}\right]_{\breve{h}=1} \\
&= \left[\frac{\delta}{\delta \overset{*i}{\mathbf{y}}}\left(\begin{array}{c} \left(\int \breve{h}(\overset{*1}{\mathbf{y}}) \cdot \overset{*1}{f}_{k+1|k}(\overset{*1}{\mathbf{y}}|\overset{*1}{\mathbf{x}},\mathbf{u}_1)d\overset{*1}{\mathbf{y}}\right) \\ \cdots \left(\int \breve{h}(\overset{*e}{\mathbf{y}}) \cdot \overset{*e}{f}_{k+1|k}(\overset{*e}{\mathbf{y}}|\overset{*e}{\mathbf{x}},\mathbf{u}_e)d\overset{*e}{\mathbf{y}}\right)\end{array}\right)\right]_{\breve{h}=1}
\end{aligned}$$

and so

$$\begin{aligned}
&\left[\frac{\delta \overset{*}{G}_{k+1|k}}{\delta \overset{*i}{\mathbf{y}}}[\breve{h}|\overset{*}{X},U]\right]_{\breve{h}=1} \\
&= \left[\begin{array}{c} \sum_{j=1}^{e} \left(\int \breve{h}(\overset{*1}{\mathbf{y}}) \cdot \overset{*1}{f}_{k+1|k}(\overset{*1}{\mathbf{y}}|\overset{*1}{\mathbf{x}},\mathbf{u}_1)d\overset{*1}{\mathbf{y}}\right) \cdots \frac{\delta}{\delta \overset{*i}{\mathbf{y}}} \int \breve{h}(\overset{*j}{\mathbf{y}}) \cdot \overset{*j}{f}_{k+1|k}(\overset{*j}{\mathbf{y}}|\overset{*j}{\mathbf{x}},\mathbf{u}_j)d\overset{*j}{\mathbf{y}} \\ \cdots \left(\int \breve{h}(\overset{*e}{\mathbf{y}}) \cdot \overset{*e}{f}_{k+1|k}(\overset{*e}{\mathbf{y}}|\overset{*e}{\mathbf{x}},\mathbf{u}_e)d\overset{*e}{\mathbf{y}}\right)\end{array}\right]_{\breve{h}=1} \\
&= \left[\begin{array}{c} \left(\int \breve{h}(\overset{*1}{\mathbf{y}}) \cdot \overset{*1}{f}_{k+1|k}(\overset{*1}{\mathbf{y}}|\overset{*1}{\mathbf{x}},\mathbf{u}_1)d\overset{*1}{\mathbf{y}}\right) \cdots \overset{*i}{f}_{k+1|k}(\overset{*i}{\mathbf{y}}|\overset{*i}{\mathbf{x}},\mathbf{u}_i) \\ \cdots \left(\int \breve{h}(\overset{*e}{\mathbf{y}}) \cdot \overset{*e}{f}_{k+1|k}(\overset{*e}{\mathbf{y}}|\overset{*e}{\mathbf{x}},\mathbf{u}_e)d\overset{*e}{\mathbf{y}}\right)\end{array}\right]_{\breve{h}=1} \\
&= \overset{*i}{f}_{k+1|k}(\overset{*i}{\mathbf{y}}|\overset{*i}{\mathbf{x}},\mathbf{u}_i)
\end{aligned}$$

So

$$\begin{aligned}
&\breve{D}_{k+1|k}(\overset{*i}{\mathbf{y}}) \\
&= \int \int_{|\overset{*}{X}|=e} \left[\frac{\delta \overset{*}{G}_{k+1|k}}{\delta \overset{*i}{\mathbf{y}}}[\breve{h}|\overset{*}{X},U]\right]_{\breve{h}=1} \cdot f_{k|k}(X,\overset{*}{X})\delta X \delta \overset{*}{X} \\
&= \frac{1}{e!}\int \int \left[\frac{\delta \overset{*}{G}_{k+1|k}}{\delta \overset{*i}{\mathbf{y}}}[\breve{h}|\overset{*}{X},U]\right]_{\breve{h}=1} \cdot f_{k|k}(X,\{\overset{*1}{\mathbf{x}},...,\overset{*i}{\mathbf{x}},...,\overset{*e}{\mathbf{x}}\})\delta X d\overset{*1}{\mathbf{x}} \cdots d\overset{*i}{\mathbf{x}} \cdots d\overset{*e}{\mathbf{x}}
\end{aligned}$$

and so, as claimed,

$$\begin{aligned}
&\breve{D}_{k+1|k}(\overset{*i}{\mathbf{y}}) \\
&= \frac{1}{(e-1)!}\int\int \overset{*i}{f}_{k+1|k}(\overset{*i}{\mathbf{y}}|\overset{*i}{\mathbf{x}},\mathbf{u}_i) \\
&\cdot f_{k|k}(X,\{\overset{*1}{\mathbf{x}},...,\overset{*i-1}{\mathbf{x}},\overset{*i+1}{\mathbf{x}},...,\overset{*e}{\mathbf{x}}\}\cup\{\overset{*i}{\mathbf{x}}\})\delta X d\overset{*1}{\mathbf{x}}\cdots d\overset{*i}{\mathbf{x}}\cdots d\overset{*e}{\mathbf{x}} \\
&= \int \overset{*i}{f}_{k+1|k}(\overset{*i}{\mathbf{y}}|\overset{*i}{\mathbf{x}},\mathbf{u}_i)\cdot\left(\int\int_{|\overset{*}{X}|=\overset{*}{n}-1} f_{k|k}(X,\overset{*}{X}\cup\{\overset{*i}{\mathbf{x}}\})\delta X\delta\overset{*}{X}\right)d\overset{*i}{\mathbf{x}} \\
&= \int \overset{*i}{f}_{k+1|k}(\overset{*i}{\mathbf{y}}|\overset{*i}{\mathbf{x}},\mathbf{u}_i)\cdot\breve{D}_{k|k}(\overset{*i}{\mathbf{x}})d\overset{*i}{\mathbf{x}}
\end{aligned}$$

8.4. *Derivation of Eqs. (170) and (171)*

We are to derive the joint corrector equations

$$\begin{aligned}
&\breve{D}_{k+1|k+1}(\mathbf{x}) \\
&= \overset{*}{D}_{k+1|k}[1-\overset{*}{p}_D p_{D,\mathbf{x}}]\cdot\breve{D}_{k+1|k}(\mathbf{x}) \\
&+\sum_{i=1}^{m}\frac{\overset{*}{f}_{k+1|k}[p_{D,\mathbf{x}}\overset{*}{p}_D L_{\mathbf{z}_i,\mathbf{x}}]\cdot\breve{D}_{k+1|k}(\mathbf{x})}{\overset{*}{D}_{k|k}[\overset{*}{p}_D]\cdot\lambda_{k+1}c_{k+1}(\mathbf{z}_i)+(D_{k+1|k}\times\overset{*}{D}_{k+1|k})[p_D\overset{*}{p}_D L_{\mathbf{z}_i}]} \\
&\breve{D}_{k+1|k+1}(\overset{*}{\mathbf{x}}) \\
&= (1-\overset{*}{p}_D(\overset{*}{\mathbf{x}}))\cdot\breve{D}_{k+1|k}(\overset{*}{\mathbf{x}})+\frac{\breve{D}_{k+1|k}(\overset{*}{\mathbf{x}})\cdot\overset{*}{p}_D(\overset{*}{\mathbf{x}})\overset{*}{L}_{\overset{*}{\mathbf{z}}}(\overset{*}{\mathbf{x}})]}{\breve{D}_{k+1|k}[\overset{*}{p}_D\overset{*}{L}_{\overset{*}{\mathbf{z}}}]}
\end{aligned}$$

We use the notation of Eqs. (185) and (186). First take the functional derivatives of Eq. (191) with respect to the observation variable $\breve{g}$:

$$\begin{aligned}
\frac{\delta\breve{F}_{k+1}}{\delta Z_{k+1}\delta\overset{*}{\mathbf{z}}_{k+1}}[\breve{g},\breve{h}] &\cong \exp\left(\begin{array}{c}-\overset{*}{\mu}-\lambda-\mu+\overset{*}{\mu}\overset{*}{s}[\breve{h}(1-\overset{*}{p}_D+\overset{*}{p}_D\overset{*}{p}_{\breve{g}})] \\ +\lambda\overset{*}{s}[1-\overset{*}{p}_D+\overset{*}{p}_D c_{\breve{g}}] \\ +\mu(s\times\overset{*}{s})[(\breve{h}\times 1)(1-\overset{*}{p}_D p_D+p_D\overset{*}{p}_D p_{\breve{g}})]\end{array}\right) \\
&\cdot\prod_{i=1}^{m}\left(\overset{*}{s}[\overset{*}{p}_D]\cdot\lambda c(\mathbf{z}_i)+\mu(s\times\overset{*}{s})[(\breve{h}\times 1)p_D\overset{*}{p}_D L_{\mathbf{z}_i}]\right) \\
&\cdot\overset{*}{\mu}\overset{*}{s}[\breve{h}\overset{*}{p}_D\overset{*}{L}_{\overset{*}{\mathbf{z}}_{k+1}}] \qquad (208)
\end{aligned}$$

By Eq. (89) the posterior p.g.fl. corresponding to the collection of Z_{k+1}

and $\overset{*}{\mathbf{z}}_{k+1}$ is:

$$\begin{aligned}
&\breve{G}_{k+1|k+1}[\breve{h}] \\
&\quad= \frac{\frac{\delta \breve{F}_{k+1}}{\delta Z_{k+1}\delta \overset{*}{\mathbf{z}}_{k+1}}[0,\breve{h}]}{\frac{\delta \breve{F}_{k+1}}{\delta Z_{k+1}\delta \overset{*}{\mathbf{z}}_{k+1}}[0,1]} \\
&\quad= \exp\left(\overset{*}{\mu}\overset{*}{s}[(\breve{h}-1)(1-\overset{*}{p}_D)] + \mu(s\times\overset{*}{s})[(\breve{h}\times 1-1)(1-\overset{*}{p}_D p_D)]\right) \\
&\quad\cdot \prod_{i=1}^{m} \left(\frac{\overset{*}{s}[\overset{*}{p}_D]\cdot\lambda c(\mathbf{z}_i) + \mu(s\times\overset{*}{s})[(\breve{h}\times 1)p_D\overset{*}{p}_D L_{\mathbf{z}_i}]}{\overset{*}{s}[\overset{*}{p}_D]\cdot \lambda c(\mathbf{z}_i)+\mu(s\times\overset{*}{s})[p_D\overset{*}{p}_D L_{\mathbf{z}_i}]}\right) \\
&\quad\cdot \left(\frac{\overset{*}{s}[\breve{h}\overset{*}{p}_D \overset{*}{L}_{\overset{*}{\mathbf{z}}_{k+1}}]}{\overset{*}{s}[\overset{*}{p}_D\overset{*}{L}_{\overset{*}{\mathbf{z}}_{k+1}}]}\right)
\end{aligned} \tag{209}$$

Then by Eq. (48) or Eq. (104), the joint PHD is given by the formula

$$\breve{D}_{k+1|k+1}(\breve{\mathbf{y}}) = \frac{\delta \log \breve{G}_{k+1|k+1}}{\delta \breve{\mathbf{y}}}[1] \tag{210}$$

where in our present case

$$\begin{aligned}
&\log \breve{G}_{k+1|k+1}[h] \\
&\quad= \overset{*}{\mu}\overset{*}{s}[(\breve{h}-1)(1-\overset{*}{p}_D)] + \mu(s\times\overset{*}{s})[(\breve{h}\times 1-1)(1-\overset{*}{p}_D p_D)] \\
&\quad+ \sum_{i=1}^{m} \log\left(\frac{\overset{*}{s}[\overset{*}{p}_D]\cdot\lambda c(\mathbf{z}_i) + \mu(s\times\overset{*}{s})[(\breve{h}\times 1)p_D\overset{*}{p}_D L_{\mathbf{z}_i}]}{\overset{*}{s}[\overset{*}{p}_D]\cdot \lambda c(\mathbf{z}_i)+\mu(s\times\overset{*}{s})[p_D\overset{*}{p}_D L_{\mathbf{z}_i}]}\right) \\
&\quad+ \log\left(\frac{\overset{*}{s}[\breve{h}\cdot\overset{*}{p}_D \overset{*}{L}_{\overset{*}{\mathbf{z}}_{k+1}}]}{\overset{*}{s}[\overset{*}{p}_D\overset{*}{L}_{\overset{*}{\mathbf{z}}_{k+1}}]}\right)
\end{aligned} \tag{211}$$

To compute the functional derivative in Eq. (210), first note that if $\breve{F}_{\mathbf{x},\overset{*}{\mathbf{x}}}$ is the functional defined by

$$\breve{F}_{\mathbf{x},\overset{*}{\mathbf{x}}}[\breve{h}] = (\breve{h}\times 1)(\mathbf{x},\overset{*}{\mathbf{x}}) = \breve{h}(\mathbf{x})$$

for all $\breve{h}$ and fixed $\mathbf{x},\overset{*}{\mathbf{x}}$, then for a target state $\breve{\mathbf{y}} = \mathbf{y}$,

$$\begin{aligned}
\frac{\delta \breve{F}_{\mathbf{x},\overset{*}{\mathbf{x}}}}{\delta \mathbf{y}}[\breve{h}] &= \lim_{\varepsilon\to 0}\frac{\breve{F}_{\mathbf{x},\overset{*}{\mathbf{x}}}[\breve{h}+\varepsilon\breve{\delta}_{\mathbf{y}}] - \breve{F}_{\mathbf{x},\overset{*}{\mathbf{x}}}[\breve{h}]}{\varepsilon} = \lim_{\varepsilon\to 0}\frac{\breve{h}(\mathbf{x})+\varepsilon\breve{\delta}_{\mathbf{y}}(\mathbf{x}) - \breve{h}(\mathbf{x})}{\varepsilon} \\
&= \delta_{\mathbf{y}}(\mathbf{x})
\end{aligned}$$

whereas for a sensor state $\breve{\mathbf{y}} = \mathring{\mathbf{y}}$,

$$\frac{\delta \breve{F}_{\mathbf{x},\mathring{\mathbf{x}}}}{\delta \mathring{\mathbf{y}}}[\breve{h}] = \lim_{\varepsilon \to 0} \frac{\breve{h}(\mathbf{x}) + \varepsilon \breve{\delta}_{\mathring{\mathbf{y}}}(\mathbf{x}) - \breve{h}(\mathbf{x})}{\varepsilon} = \breve{\delta}_{\mathring{\mathbf{y}}}(\mathbf{x}) = 0$$

(Here, $\breve{\delta}_{\breve{\mathbf{y}}}(\breve{\mathbf{x}})$ is the joint Dirac delta defined in Eq. (22).) So for a target state $\breve{\mathbf{y}} = \mathbf{y}$

$$\frac{\delta \log \breve{G}_{k+1|k+1}}{\delta \mathbf{y}}[h] = \mathring{s}[1 - \mathring{p}_D p_{D,\mathbf{y}}] \cdot \mu s(\mathbf{y}) + \sum_{i=1}^{m} \frac{\mathring{s}[p_{D,\mathbf{y}} \mathring{p}_D L_{\mathbf{z}_i,\mathbf{y}}] \cdot \mu s(\mathbf{y})}{\mathring{s}[\mathring{p}_D] \cdot \lambda c(\mathbf{z}_i) + \mu(s \times \mathring{s})[(\breve{h} \times 1) p_D \mathring{p}_D L_{\mathbf{z}_i}]} \tag{212}$$

where $p_{D,\mathbf{y}}(\mathring{\mathbf{x}}) \overset{\text{abbr.}}{=} p_D(\mathbf{y}, \mathring{\mathbf{x}})$ and $L_{\mathbf{z},\mathbf{y}}(\mathbf{x}) \overset{\text{abbr.}}{=} L_{\mathbf{z}}(\mathbf{z}|\mathbf{y}, \mathring{\mathbf{x}})$ and $\mathring{s}[\breve{h}] = \int \breve{h}(\mathring{\mathbf{x}}) \cdot s(\mathring{\mathbf{x}}) d\mathring{\mathbf{x}}$.

For a sensor state $\breve{\mathbf{y}} = \mathring{\mathbf{y}}$, on the other hand,

$$\frac{\delta \log \mathring{G}_{k+1|k+1}}{\delta \mathring{\mathbf{y}}}[h] = (1 - \mathring{p}_D(\mathring{\mathbf{y}})) \cdot \mathring{\mu}\mathring{s}(\mathring{\mathbf{y}}) + \frac{\mathring{s}(\mathring{\mathbf{y}}) \cdot \mathring{p}_D(\mathring{\mathbf{y}}) \mathring{L}_{\mathring{\mathbf{z}}}(\mathring{\mathbf{y}})]}{\mathring{s}[\breve{h} \mathring{p}_D \mathring{L}_{\mathring{\mathbf{z}})}]} \tag{213}$$

Consequently, for a target state $\breve{\mathbf{y}} = \mathbf{y}$, by Eq. (210) the PHD is, as claimed,

$$\breve{D}_{k+1|k+1}(\mathbf{y}) = \mathring{s}[1 - \mathring{p}_D p_{D,\mathbf{y}}] \cdot \mu s(\mathbf{y}) + \sum_{i=1}^{m} \frac{\mathring{s}[p_{D,\mathbf{y}} \mathring{p}_D L_{\mathbf{z}_i,\mathbf{y}}] \cdot \mu s(\mathbf{y})}{\mathring{s}[\mathring{p}_D] \cdot \lambda c(\mathbf{z}_i) + \mu(s \times \mathring{s})[p_D \mathring{p}_D L_{\mathbf{z}_i}]} \tag{214}$$

For a sensor state $\breve{\mathbf{y}} = \mathring{\mathbf{y}}$ the PHD is, as claimed,

$$\breve{D}_{k+1|k+1}(\mathring{\mathbf{y}}) = (1 - \mathring{p}_D(\mathring{\mathbf{y}})) \cdot \mathring{\mu}\mathring{s}(\mathring{\mathbf{y}}) + \frac{\mathring{s}(\mathring{\mathbf{y}}) \cdot \mathring{p}_D(\mathring{\mathbf{y}}) \mathring{L}_{\mathring{\mathbf{x}}}(\mathring{\mathbf{y}})]}{\mathring{s}[\mathring{p}_D \mathring{L}_{\mathring{\mathbf{x}}}]} \tag{215}$$

9. Conclusions

In this chapter we have developed a general approach for approximate control-theoretic multisensor-multitarget sensor management. Our refined approach now encompasses: (1) targets of current or potential tactical interest; (2) multistep look-ahead (control of sensor resources throughout a future time-window); (3) sensors with non-ideal dynamics, including sensors residing on moving platforms such as UAVs; (4) sensors whose states are observed indirectly by internal actuator sensors; and (5) possible communication drop-outs. Our approach also addresses the impossibility of deciding between an infinitude of plausible objective functions by concentrating on "probabilistically natural" core goals of sensor management, such as maximizing $\mathring{N}_{k+1|k+1}$.

Despite this progress, our work still has significant limitations. We must assume that the sensors are fixed in number. This precludes the possibility of sensors entering or leaving a scenario. We must assume that each platform carries exactly one sensor. Our basic scheme is still centralized: observations collected by all sensors must be transmitted to a single data fusion engine for processing; and this same site constructs a centralized control decision. Future work must address all of these issues.

Acknowledgments

The work reported in this chapter was supported by the U.S. Air Force Office of Scientific Research under contract F49620-01-C-0031. The content does not necessarily reflect the position or the policy of the Government. No official endorsement should be inferred.

References

[1] Y. Bar-Shalom and X.-R. Li, *Estimation and Tracking: Principles, Techniques, and Software*, Ann Arbor: Artech House, 1993.

[2] D.J. Daley and D. Vere-Jones, *An Introduction to the Theory of Point Processes*, Springer-Verlag, 1988.

[3] I.R. Goodman, R.P.S. Mahler, and H.T. Nguyen, *Mathematics of Data Fusion*, New York: Kluwer Academic Publishers, 1997.

[4] R. Mahler (2003) "Comments on Generalized Probability Generating Functionals," personal communication to Prof. B.-N. Vo, dated Sept. 25, 2003.

[5] R. Mahler, "Engineering Statistics for Multi-Object Tracking," Proc. 2001 IEEE Workshop on Multi-Object Tracking, July 8, 2001, Vancouver, pages 53-60, 2001.

[6] R. Mahler, "An extended first-order Bayes filter for force aggregation," in O. Drummond (ed.), *Signal and Data Processing of Small Targets 2002*, SPIE Vol. 4728, pages 196-207, 2002.

[7] R. Mahler, "Global posterior densities for sensor management," in M.K. Kasten and L.A. Stockum (eds.), *Acquisition, Tracking, and Pointing XII*, SPIE Vol. 3365, pages 252-263, 1998.

[8] R. Mahler, "Global Optimal Sensor Allocation," Proc. Ninth Nat'l Symp. on Sensor Fusion, Vol. I (Unclassified), Mar. 12-14 1996, Naval Postgraduate School, Monterey CA, 347-366.

[9] R. Mahler, *An Introduction to Multisource-Multitarget Statistics and Its Applications*, Lockheed Martin Technical Monograph, Mar. 15, 2000, 114 pages.
[10] R. Mahler, "Multisensor-Multitarget Sensor Management: A Unified Bayesian Approach," in I. Kadar (ed.), *Signal Processing, Sensor Fusion, and Target Recognition XII*, SPIE Proc. vol. 5096, pages 222-233, 2003.
[11] R. Mahler, "Multitarget moments and their application to multitarget tracking," *Proc. Workshop on Estimation, Tracking, and Fusion: A Tribute to Yaakov Bar-Shalom*, May 17, 2001, Naval Postgraduate School, Monterey, CA, pages 134-166, 2001.
[12] R. Mahler, "Multitarget Filtering via First-Order Multitarget Moments," *IEEE Trans. Aerospace and Electronic Systems*, Vol. 39 No. 4, pages 1152-1178, 2003.
[13] R. Mahler, "Objective Functions for Bayesian Control-Theoretic Sensor Management, I: Multitarget First-Moment Approximation," *Proc. 2003 IEEE Aerospace Conference*, Big Sky MT, March 8-15 2003.
[14] R. Mahler, "Objective Functions for Bayesian Control-Theoretic Sensor Management, II: MHC-Type Approximation," in S. Butenko, R. Murphey, and P. Paralos (eds.), *New Developments in Cooperative Control and Optimization*, Kluwer Academic Publishers, to appear.
[15] R. Mahler, "Random Set Theory for Target Tracking and Identification," in D.L. Hall and J. Llinas (eds.), *Handbook of Multisensor Data Fusion*, Boca Raton FL: CRC Press, Chapter 14, 2002.
[16] R. Mahler, "'Statistics 101' for Multisensor, Multitarget Data Fusion," *IEEE Aerospace and Electronic Systems Magazine, Part 2: Tutorials*, Vol. 19 No. 1, January 2004, pages 53-64.
[17] R. Mahler, "Tractable Multistep Sensor Management via MHT," *Proceedings of the Workshop on Multi-Hypothesis Tracking: A Tribute to Samuel Blackman*, San Diego CA, May 30 2003, to appear.
[18] R. Mahler and R. Prasanth, "Technologies Leading to Unified Multi-Agent Collection and Coordination," in S. Butenko, R. Murphey, and P.M. Pardalos (eds.), *Cooperative Control: Models, Applications, and Algorithms*, Kluwer Academic Publisheres, pages 215-251, 2003.
[19] G. Mathéron, *Random Sets and Integral Geometry*, J. Wiley, 1975.
[20] L.H. Ryder, *Quantum Field Theory,* 2nd Edition, Cambridge U. Press, 1996.
[21] D. Stoyan, W.S. Kendall, and J. Meche, *Stochastic Geometry and Its Applications,* Second Edition, John Wiley & Sons, 1995.
[22] B.-N. Vo (2003) Personal communication to R. Mahler, Sept. 22, 2003.

CHAPTER 13

COMMUNICATION REQUIREMENTS IN THE COOPERATIVE CONTROL OF WIDE AREA SEARCH MUNITIONS USING ITERATIVE NETWORK FLOW

Jason W. Mitchell,[a] Steven J. Rasmussen[b]
General Dynamics Advanced Information Systems
Wright-Patterson AFB, OH 45433-7531
{Jason.Mitchell,Steve.Rasmussen}@wpafb.af.mil

Andrew G. Sparks[c]
Air Force Research Laboratory
Wright-Patterson AFB, OH 45433-7531
Andrew.Sparks@wpafb.af.mil

Communication requirements are considered for the cooperative control of wide area search munitions where resource allocation is performed by an iterative network flow. We briefly outline both the single and iterative network flow assignment algorithms and their communication requirements. Then, using the abstracted communication framework recently incorporated into AFRL's MultiUAV simulation package, a model is constructed to investigate the peak and average data rates occurring in a sequence of vehicle-target scenarios using an iterative network flow for task allocation, implemented as a redundant, centralized optimization, that assumes perfect communication.[d]

Keywords:
Cooperative control, uninhabited aerial vehicles, information flow, communication requirements, data rates, MultiUAV

---

[a]Aerospace Scientist
[b]Senior Aerospace Engineer
[c]Aerospace Scientist
[d]This material is declared a work of the U.S. Government and is not subject to copyright protection in the United States.

1. Introduction

Coordination and cooperation between uninhabited aerial vehicles (UAV) has the potential to significantly improve their effectiveness in many situations. For the typical tasks that these vehicles must perform, i.e. detection, classification, attack, and verification, explicit vehicle cooperation may be required to meet specific objectives. Thus, the ability to communicate information between vehicles becomes mission essential and provides an opportunity to enhance overall capability.

While vehicle communications may provide the opportunity to enhance performance, it is likely not without cost. Frequently, control algorithms are designed without regard to their associated communication needs or effects. For the control system designer, such treatment is undertaken to reduce algorithmic complexity and obtain a manageable result. Consequently, it becomes necessary to quantify the communicated data driving the control algorithms ex post facto. As an example of this design strategy, consider several methods that have been previously studied to produce near-optimal single task assignments [10, 6], and more recently, the near-optimal assignment of a sequence of tasks using an iterative network flow model [11]. In these cases, the amount of information necessary to drive these cooperative control algorithms was not considered.

In this work, communication requirements are considered for the cooperative control of uninhabited aerial vehicles with resource allocation performed by an iterative network flow. In the following, we briefly outline the single and iterative network flow assignment algorithms and their communication requirements. Then, we briefly describe the MultiUAV simulation package [7, 9], and the framework recently incorporated to model vehicle-to-vehicle communication. Using this framework, a model is constructed to investigate the peak and average data rates occurring in a sequence of vehicle-target scenarios using an iterative network flow for task allocation, implemented as a redundant, centralized optimization, that assumes perfect communication.

2. Background

We begin with a short description of a typical MultiUAV simulation scenario and a brief outline of the network flow task allocation models.

The current configuration of MultiUAV simulates, but is not limited to, autonomous wide area search munitions (WASM), which are small UAVs powered by a turbojet engine with sufficient fuel to fly for a short period of time. They are deployed in groups from larger aircraft flying at higher altitudes. Individually, they are capable of searching for, recognizing, attacking, and verifying targets.

2.1. *Scenario*

We begin with a set of N vehicles, deployed simultaneously, each with a life span of approximately thirty (30) minutes, that are indexed by $i \in \mathbb{Z}[1, N]$. Targets that may be found by searching fall into known classes according to the value or score associated with their destruction. These targets are indexed by j as they are found, thus we find $j \in \mathbb{Z}[1, M]$ with V_j as the value of target j. The individual vehicles assume no precise a priori information about the total number of targets or their initial locations. This information can only be obtained by the vehicles searching for and finding potential targets via Automatic Target Recognition (ATR) methodologies. The ATR process is modeled using a system that provides a probability that the target has been correctly classified. The probability of a successful classification is based on the viewing angle of the vehicle relative to the target, Rasmussen et al. [9]. For this exercise, the possibility of incorrect identification is not modeled, however targets are not attacked unless a 90% probability of correct identification is achieved. Further details of the ATR methodology can be found in Chandler and Pachter [2], with a detailed discussion available in Chandler and Pachter [1]. Once successfully classified as a target, the attack vehicle is selected. Upon reaching the selected target, the vehicle releases its munition and is subsequently declared an unavailable asset, i.e. attack is a terminal task for WASM. Finally, the selected target must be verified as destroyed to complete the target specific task chain.

Throughout the simulation, at each target state change or task failure, a resource allocation algorithm is executed to compute task assignments. The resulting assignment is sub-optimal. Fortunately, Rasmussen et al. [8] has shown that these assignments are frequently near-optimal in an average sense.

2.2. *Task Allocation: Network Optimization Model*

The weapon system allocation is treated as follows: individual vehicles are discrete supplies of single units, executing tasks corresponding to flows on arcs through the network, with the ultimate disposition of the vehicles representing the demand. Thus, the flows are zero (0) or one (1). We assume that each vehicle operates independently, and makes decisions when new

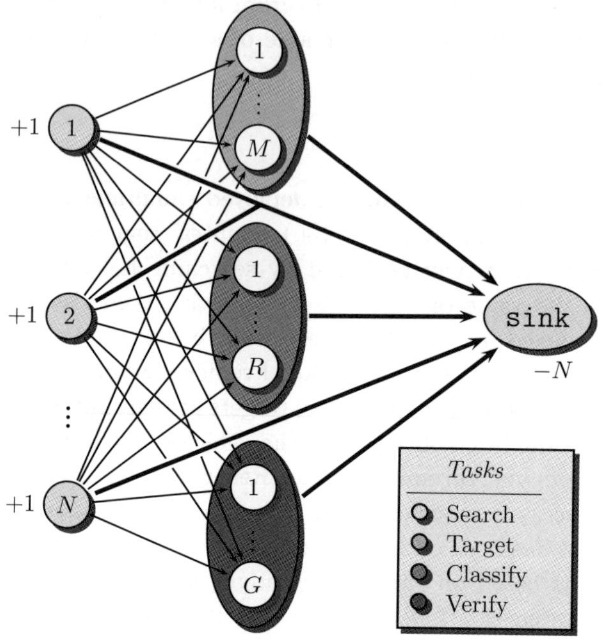

Fig. 1. Network flow diagram.

information is received. These decisions are determined by the solution of the network optimization model. The receipt of new target information triggers the formulation and solving of a fresh optimization problem that reflects current conditions, thus achieving feedback action. At any point in time, the database on-board each vehicle contains a *target* set, consisting of indices, types and locations for targets that have been classified above the probability threshold. There is also a *speculative* set, consisting of indices, types and locations for potential targets that have been detected, but are classified below the probability threshold and thus require further inspection.

The network flow model, seen in Figure 1, is demand driven. The `sink` node at the right exerts a demand-pull of N units, causing the nodes on the left to flow through the network. In the middle layer, the top M nodes represent all of the successfully classified targets, and thus are ready to be attacked. An arc exists from a specific vehicle node to a target node if and only if it is a feasible vehicle/target pair. At a minimum, the feasibility requirement would mean that there is sufficient fuel remaining to strike the target if so tasked. Other feasibility conditions could also be considered, e.g. heterogeneous weapons or sensing platforms, poor look-angles. The center R nodes of the middle layer represent potential targets that have been detected, but do not meet the minimum classification probability. We call them *speculatives*. The minimum feasibility requirement to connect a vehicle/speculative pair is sufficient fuel for the vehicle to deploy its sensor to elevate the classification probability. The lower-tier G nodes model alternatives for verification of targets that have been struck. Finally, each node in the vehicle set on the left has a direct arc to the far right node labelled `sink`, modeling the option of continuing to search. The capacities on the arcs from the target and speculative sets are fixed at one (1). From the integrality property, flow values are constrained to be either zero (0) or one (1). Each unit of flow along an arc has a benefit which is an expected future value. The optimal solution maximizes total value. For a more detailed discussion, including the issue of the *benefit calculation*, see Schumacher et al. [11].

2.2.1. *Single Pass Network Flow*

Single task assignment in MultiUAV is formulated as the capacitated transshipment problem (CTP) [10]. Due to the special structure of the problem, there will always be an optimal solution that is all integer [6]. Thus, solutions to this problem pose a small computational burden, making it feasible for implementation on the processors likely to be available on inexpensive wide area search munitions.

2.2.2. *Iterative Network Flow*

Due to the integrality property, it is not normally possible to simultaneously assign multiple vehicles to a single target, or multiple targets to a single vehicle. However, using the network assignment iteratively, *tours* of multiple assignments can be determined [11]. This is done by solving the initial assignment problem once, and only finalizing the assignment with the shortest estimated arrival time. The assignment problem can then be updated assuming that assignment is performed, updating target and vehicle states, and running the assignment again. This iteration can be repeated until all of the vehicles have been assigned terminal tasks, or until all of the target assignments have been fully distributed. The target assignments are complete when classification, attack, and verification tasks have been assigned for all known targets. Assignments must be recomputed if a new target is found or a munition fails to complete an assigned task.

2.3. *Information Requirements*

The implementation of the task allocation algorithms outlined above requires communication of information between vehicles. As with several previous studies where MultiUAV was used to investigate optimal task allocation, we assume perfect and error-free access to information about vehicle and target states. From many perspectives, these assumptions are clearly unrealistic, particularly when considering physical communication and processing constraints. However, to determine the requirements of a physically realizable system, we must also understand what information is necessary and the quantity needed to drive the algorithms under ideal conditions.

Since both algorithms discussed here make use of network flow, the necessary information is common between them. The overarching optimization problem can be characterized as both centralized and redundant, i.e. each vehicle computes its own network flow. Momentarily disregarding communication issues, the problem, in general, requires a synchronized database of target and vehicle state information. With this, each vehicle computes the benefits for the arcs in the network, and solves the optimization problem to maximize the total benefit. From Mitchell et al. [5], the MultiUAV network flow implementation requires the following communicated information: ATR data; target and vehicle positions; target, vehicle, and task status; and vehicle trajectory waypoints.

Having identified the information necessary, we can begin to consider the volume of information communicated between vehicles. To do this, we turn to the MultiUAV simulation package.

3. Simulation Framework

The MultiUAV simulation package [9] is capable of simulating multiple uninhabited aerospace vehicles which cooperate to accomplish a predefined mission. The purpose of the package is to provide a simulation environment that researchers can use to implement and analyze cooperative control algorithms. The simulation is built using a hierarchical decomposition where inter-vehicle communication is explicitly modeled. The package includes plotting tools and provides links to external programs for post-processing analysis. Each of the vehicle simulations include six-degree-of-freedom dynamics and embedded flight software (EFS). The EFS consists of a collection of *managers* or *agents* that control situational awareness and responses of the vehicles. In addition, the vehicle model includes an autopilot that provides waypoint navigation capability. In its original form, MultiUAV [7] could simulate a maximum of eight (8) vehicles and ten (10) targets, however recent work eases the previous burden of extending these limits. The EFS managers implement the cooperative control algorithms, including the iteratively applied CTP algorithm previously discussed. The individual managers contained within the vehicles include: Tactical Maneuvering, Sensor, Target, Cooperation, Route, and Weapons. At the top level, these managers are coded as SIMULINK models, with supporting code written in both MATLAB script and C++.

3.1. *Communication Model*

The communication simulation used in this work is very similar to that used in Mitchell et al. [5]. However, in this instance, communication is not delayed, so that the messages,[e] generated by the simulated vehicle communication at each major model update, arrive in the *in-box* of a given vehicle at the completion of the current update, and are available for use at the next major update. At the present time, the major model occurs at 10 Hz. This fairly course grained update is necessary to maintain a reasonable runtime for individual scenarios to complete, in a larger Monte-Carlo sense, on a desktop/personal computer. The minor model update, which controls the vehicle dynamics and other underlying subsystems, is scheduled at 100 Hz.

As a consequence of the model update rates, we define the *data rate* necessary at a given major model step as the total size of the messages collected, in bits, divided by the duration of the model update, yielding a rate in bits/s. This simplistic definition is a result of the elementary requirement that each vehicle must have access to all the currently generated messages by the next major update in order to function. Currently, all message data is represented in MATLAB using double-precision floating-point numbers, and in the computation of data rate, the message overhead is not considered, only the message payload. In a physical communication implementation there would be considerably more overhead, including redundancy, error correction, encryption, etc. Thus, retaining double-precision in the ideal communication model remains a reasonable indicator of real-world data rates, particularly since we are interested only in an initial estimate and

[e]The use of *message* here refers to the information format dictated by the MultiUAV package, rather than to messages related to a specific communication system model or protocol.

perhaps a relative comparison of communication necessary in executing various scenarios. Furthermore, a *broadcast* communication model is implicitly assumed, so that generated messages are counted only once. While not specifically targeted to address a particular physical implementation, such a model encompasses the typical view that the communications are time-division multiplexed.

4. Simulation

In this work, we investigate the communication data rate requirements for the cooperative control of wide area search munitions using a iterative network flow of depth three (3). To study this, a Monte-Carlo approach is taken, consisting of one hundred (100) individual simulations, each with a maximum mission time of $t_f = 200\,\text{s}$.

Individual scenarios are composed of eight (8) vehicles with four (4) targets distributed over an area of approximately $16\,\text{mi}^2$. The vehicle properties are: constant velocity of 370 ft/s or approximately mach 0.33, constant altitude of 675 ft, minimum turn radius of 2000 ft, and fuel for a maximum of 30 min of search operation. Since search is not the focus of this study, vehicles begin in a line formation, and initially follow a preprogrammed *zamboni race* search pattern. The targets are uniformly distributed throughout the domain and oriented with uniformly random pose-angles.

5. Results

As a simple measure to convince ourselves that the Monte-Carlo data collected has sufficient statistical weight, we plot the maximum data rate for all 100 simulations, and compute the cumulative average, seen in Figure 2. Surprisingly, we see that there is considerable variation in the maximum data rate. Fortunately, in terms of statistical weight, we see that the average maximum data rate is within 0.2 % of the final cumulative average after just 50 simulations. This is not surprising based on previous work in performing Monte-Carlo simulation with MultiUAV [8, 4, 5]. From the distribution of maximum data rates seen in Figure 3, it appears that the largest number fall between 120 –150 kbit/s. Most of the remaining data is distributed at a lower maximum data rate centered around 105 kbit/s. The single remaining maximum rate is centered at 170 kbit/s.

From this information, we see that, for the given model update resolution and iterative network flow cooperative control algorithm, a significant data rate is required for operation. This obviously ignores consideration of

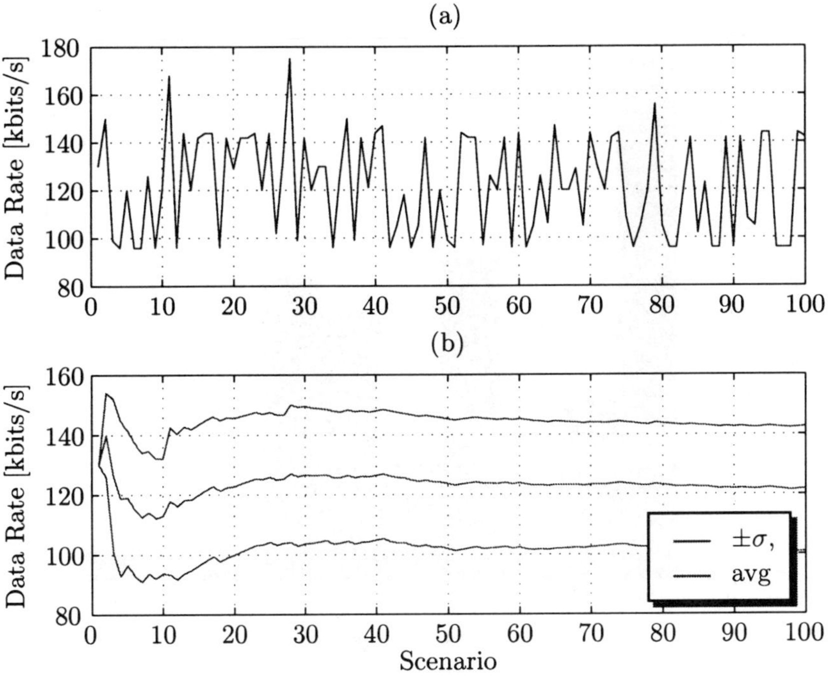

Fig. 2. Maximum data rate (a) and cumulative average (b) over 100 simulations.

any hardware or software to mitigate communication delay effects or insure information integrity that are likely to be included in a physical implementation. Nevertheless, by disregarding the actual magnitude of the maximum data rate, and considering only a relative measure between scenarios, we find that the largest data rate necessary is nearly twice the smallest maximum data rate.

Rather than attempt to analyze each individual simulation run, it is more interesting to compare the scenarios representing the smallest, average, and largest maximum data rates: 96 kbit/s, 120 kbit/s, and 175 kbit/s, respectively. The corresponding communication data rate histories can be seen in Figures 4–6, respectively.

For the smallest maximum data rate, seen in Figure 4, the peaks are well spaced, and decrease as targets are destroyed. Based on the distribution of data rates, Figure 3, this appears to be the less frequent of two typical operational modes. For the average maximum data rate, given by Figure 5, we find the more typical communication situation. For this scenario, the

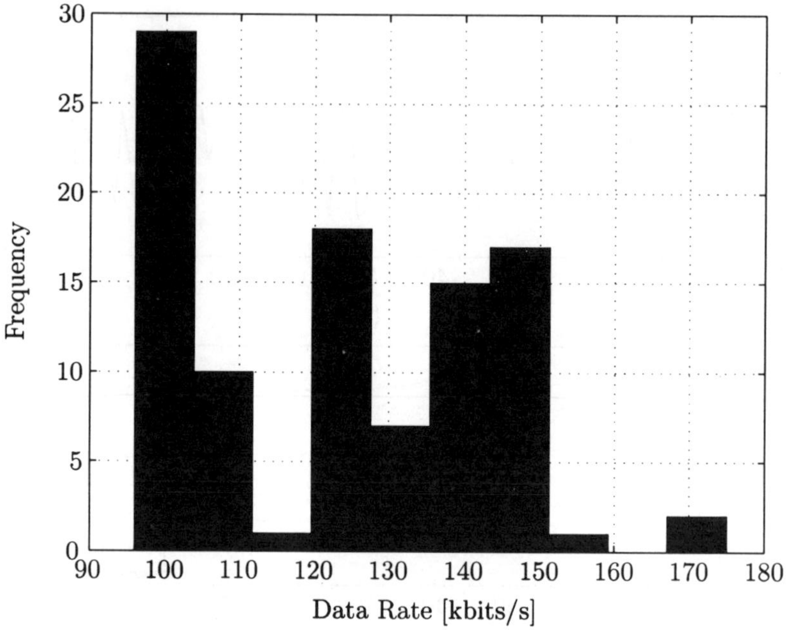

Fig. 3. Maximum data rate frequency distribution of 100 simulations.

rate peaks are much more closely spaced, and do not always decrease as targets are destroyed due to the spike at $t \approx 35$ s. There is also considerable communication activity for $t \in [80, 100]$ s. Lastly, for the largest data rate, found in Figure 6, the magnitude of the largest peak is nearly twice that of the other rate peaks occurring. Given this information, it is instructive to study the vehicle trajectories for the corresponding scenarios. These trajectories appear in Figures 7–9, where vehicles are identified by a `typewriter` style, e.g. 2, and targets are identified by an *italics* style, e.g. *2*, so that they may be more easily distinguished.

The vehicle trajectories for the smallest maximum data rate are found in Figure 7. The trajectory traces are relatively simple, particularly since targets appear in two clusters: *1,3* and *2,4*. For the communication burst around $t \approx 40$ s, we find that a target classification has failed, requiring further classification. For the average data rate seen in Figure 8, the vehicle trajectories are much more complex, with considerable looping and backtracking. Again, we notice that targets appear in two clusters: *1* and *2,3,4*. However, the second cluster contains three targets. The spike at $t \approx 35$ s

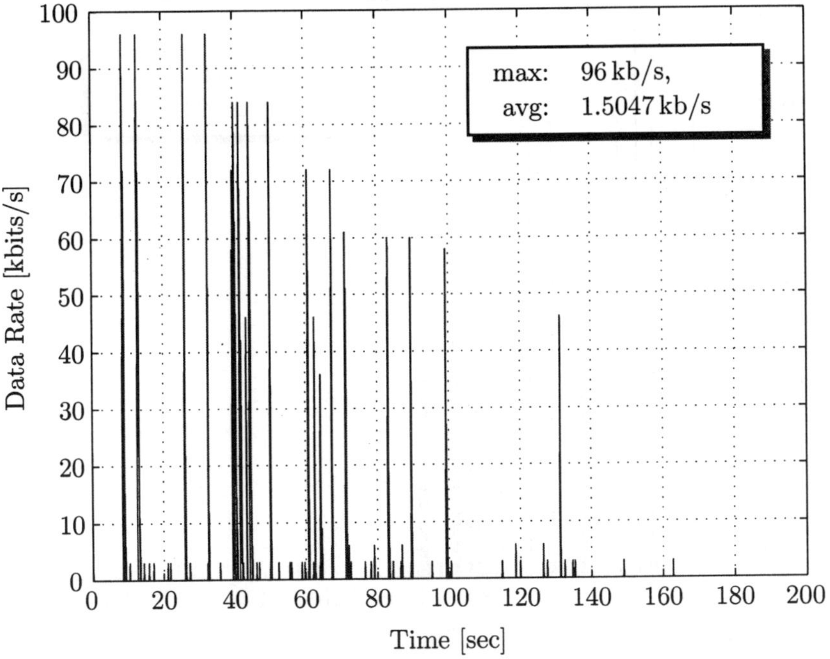

Fig. 4. Communication history: smallest maximum data rate.

results from a failed classification attempt, while the end communication bursts are a result of the three-target cluster. Lastly, for the largest data rate, given by Figure 9, the vehicle trajectories are extremely convoluted. The target clustering is similar to the smallest maximum data rate case, but with the target clusters placed closer together. This explains the communication burst at $t \approx 75\,\mathrm{s}$. In addition, at the time of the largest spike, a number of failures occur. At $t = 74.3\,\mathrm{s}$, classification of target *2* fails. Then, at $t = 74.9\,\mathrm{s}$, two classifications, viz. targets *3* and *4*, fail simultaneously. At $t = 75.3\,\mathrm{s}$, target *4* is successfully classified. Following this, at $t = 76.3\,\mathrm{s}$, target *2* is discovered, then viewed a second time and successfully classified, at $t = 76.6\,\mathrm{s}$. The second classification resulted in a task being completed by a vehicle not assigned to that task, producing an additional task failure. Overall, this particular scenario appears to be a quite pathological case of task sequencing.

In summary, the Monte-Carlo data indicates that there were two primary operational communication modes. In a relative comparison sense,

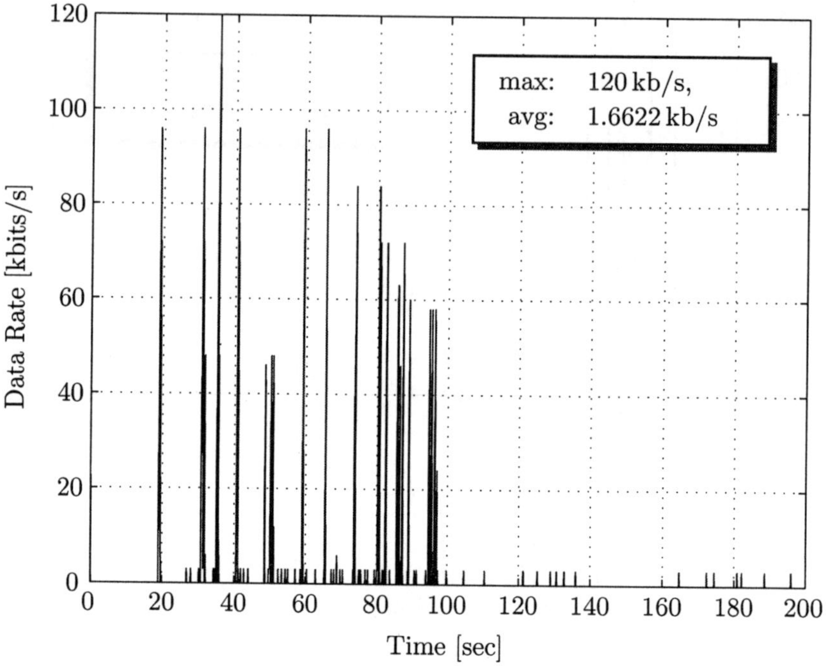

Fig. 5. Communication history: average maximum data rate.

the mode corresponding to the smaller maximum data rate, centered at 105 kbit/s, represents scenarios with a lower incidence of task failure, or lower target density. As the number of task failures or target density increases, the maximum data rate increases to accommodate the additional information necessary to make more frequent decisions; ranging between 120 kbit/s and 150 kbit/s. For the remaining case, we see that pathological task sequencing composed of both simultaneous task failures and simultaneous events generates the largest maximum data rate of 175 kbit/s.

6. Conclusions

In this work, the communication requirements were considered for the cooperative control of wide area search munitions. Using the MultiUAV simulation package, a model was constructed to investigate the peak and average data rates occurring in a sequence of vehicle-target scenarios using an iterative network flow for task allocation, which was implemented as a redundant, centralized optimization. This model assumed perfect vehicle-

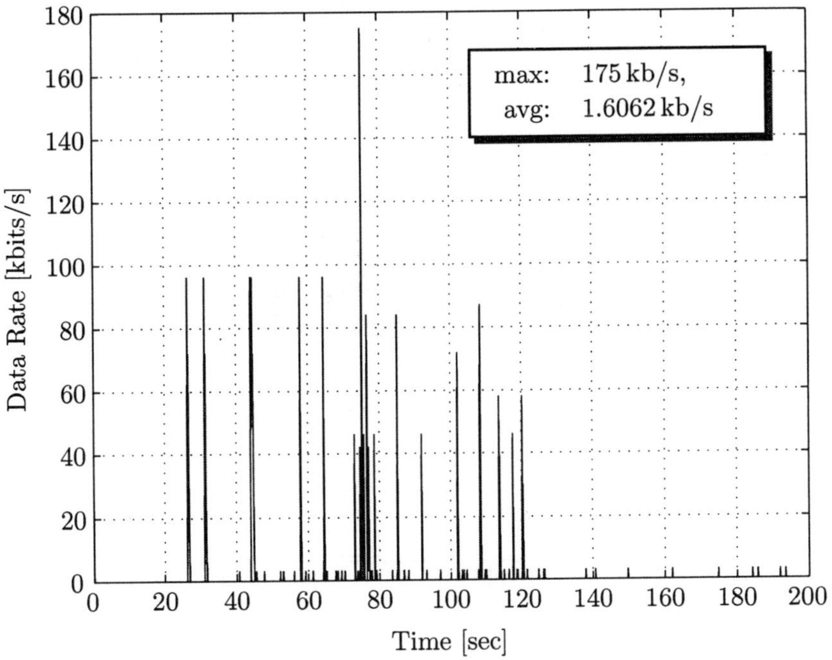

Fig. 6. Communication history: largest maximum data rate.

to-vehicle communication. The data rate was defined to be the amount of data communicated during a major model update divided by the major update duration, were each element was represented by a double-precision floating-point value.

The communication data rate indicated that when a mission scenario suffered setbacks, such as failed tasks, event accumulation bursts, or other difficulty, mission performance suffered, even with perfect communication. This information clearly represented a relative measure of mission health, even during execution.

Having observed that the structure of the communication data rate history correlated well with the likelihood that a particular scenario suffered some difficulty, we hope this quantification of cooperation may be used as a measure to help maintain a desired level of coordination. Such a measure could be used, for example, to ensure the graceful degradation of mission performance in the presence of constrained information flow between vehicles.

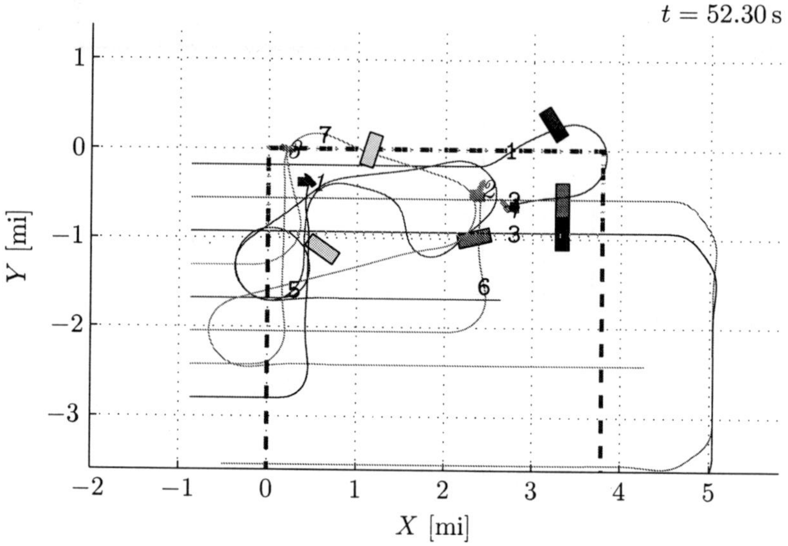

Fig. 7. Vehicle Trajectories: smallest maximum data rate.

Regarding the actual magnitudes of the maximum data rates, these should not be taken as exact requirements or measures, particularly because no specific communication protocol or hardware implementation has been defined. Rather, the magnitudes should be seen to represent traditional engineering estimates that say more in their relative significance than individual significance. With that said, these values do indicate the amount of raw data necessary to drive the cooperative control algorithms, allowing for comparisons between individual implementations of an algorithm.

Acknowledgments

A portion of this work was performed while the first named author held a National Research Council Research Associateship Award at the Air Force Research Laboratory (AFRL) in the Air Vehicles Directorate Control Theory Optimization Branch (VACA) located at the Wright-Patterson Air Force Base.

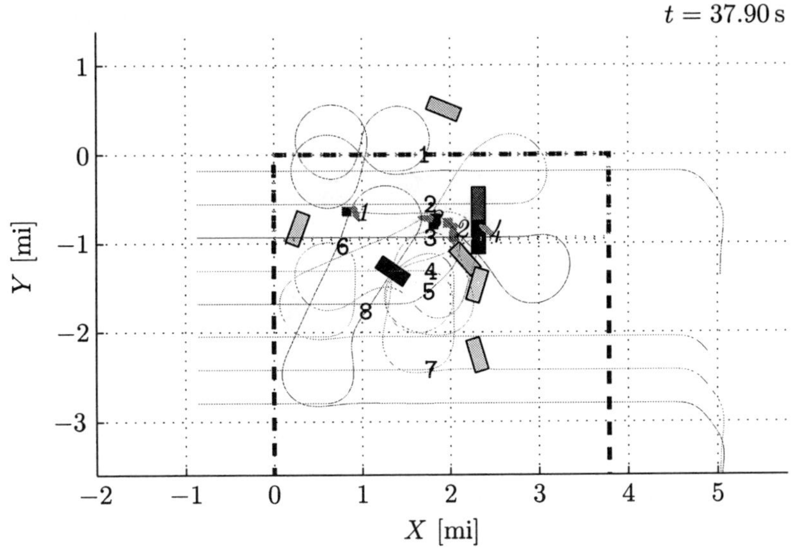

Fig. 8. Vehicle Trajectories: average maximum data rate.

References

[1] Phillip R. Chandler and Meir N. Pachter. Hierarchical control of autonomous teams. In *Proceedings of the AIAA Guidance, Navigation, and Control Conference*, 2001.

[2] Phillip R. Chandler and Meir N. Pachter. UAV cooperative classification. In *Workshop on Cooperative Control and Optimization*. Kluwer Academic Publishers, 2001.

[3] L.R. Ford, Jr. and D.R. Fulkerson. *Flows in Networks*. Princeton University Press, Princeton, NJ, 1962.

[4] Jason W. Mitchell, C. Schumacher, Phillip R. Chandler, and Steven J. Rasmussen. Communication delays in the cooperative control of wide area search munitions via iterative network flow. In *Proceedings of the AIAA Guidance, Navigation, and Control Conference*, 2003.

[5] Jason W. Mitchell and Andrew G. Sparks. Communication issues in the cooperative control of unmanned aerial vehicles. In *Proceedings of the Forty-First Annual Allerton Conference on Communication, Control, & Computing*, 2003.

[6] Kendall E. Nygard, Philip R. Chandler, and M. Pachter. Dynamic network flow optimization models for air vehicle resource allocation. In *Proceedings*

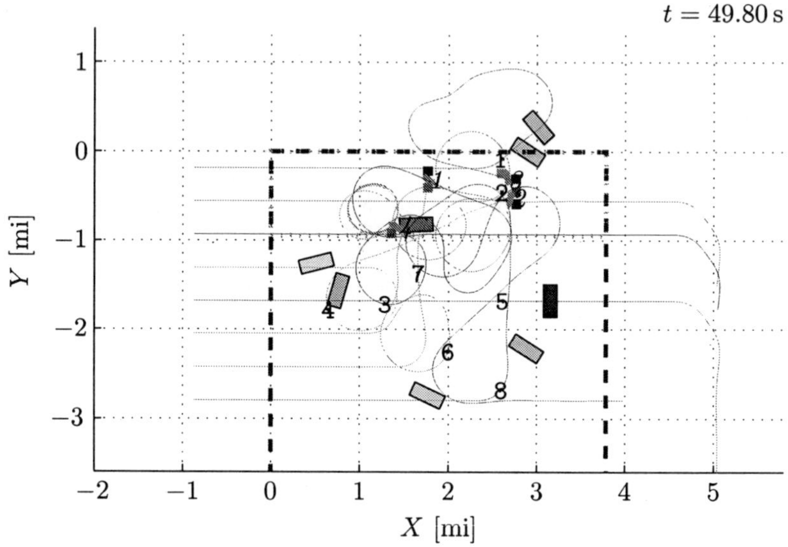

Fig. 9. Vehicle Trajectories: largest maximum data rate.

of the American Control Conference, 2001.

[7] Steven J. Rasmussen and Philip R. Chandler. MultiUAV: A multiple UAV simulation for investigation of cooperative control. In *Proceedings of the Winter Simulation Conference*, 2002.

[8] Steven J. Rasmussen, Phillip R. Chandler, Jason W. Mitchell, C. Schumacher, and Andrew G. Sparks. Optimal vs. heuristic assignment of cooperative autonomous unmanned air vehicles. In *Proceedings of the AIAA Guidance, Navigation, and Control Conference*, 2003.

[9] Steven J. Rasmussen, Jason W. Mitchell, Chris Schulz, C. Schumacher, and Phillip R. Chandler. A multiple UAV simulation for researchers. In *Proceedings of the AIAA Modeling and Simulation Technologies Conference*, 2003.

[10] C. Schumacher, Philip R. Chandler, and Steven J. Rasmussen. Task allocation for wide area search munitions via network flow optimization. In *Proceedings of the AIAA Guidance, Navigation, and Control Conference*, 2001.

[11] C. Schumacher, Philip R. Chandler, and Steven J. Rasmussen. Task allocation for wide area search munitions via iterative network flow optimization. In *Proceedings of the AIAA Guidance, Navigation, and Control Conference*, 2002.

CHAPTER 14

A DECENTRALIZED SWARM APPROACH TO ASSET PATROLLING WITH UNMANNED AIR VEHICLES

Kendall E. Nygard
Department of Computer Science and Operations Research
North Dakota State University
Fargo, ND 58105-5164
Kendall.Nygard@ndsu.nodak.edu

Karl Altenburg
Department of Accounting and Information Systems
North Dakota State University
Fargo, ND 58105-5164
Karl.Altenburg@ndsu.nodak.edu

Jingpeng Tang
Department of Computer Science and Operations Research
North Dakota State University
Fargo, ND 58105-5164
Jingpeng.Tang@ndsu.nodak.edu

Doug Schesvold
Department of Computer Science and Operations Research
North Dakota State University
Fargo, ND 58105-5164
schesvo@web.cs.ndsu.nodak.edu

We present a procedure for controlling a team of Unmanned Air Vehicles (UAVs) for establishing patrol patterns to protect an asset on the ground. The control is decentralized and follows a reactive, behavior-based, emergent intelligent swarm design. The patrol patterns consist of flight tracks with different radii and altitudes around the asset. The multiple tracks help maintain a persistent presence around the asset for the purposes of surveillance and the destruction of hostile intruders. Populating inner tracks is favored over outer tracks, and is accomplished through behaviors that comprise a track switching protocol. Collision avoidance is maintained. Global communication is assumed to be unavailable, and control is established only through passive sensors and minimal short-range radio communication. The model is implemented and successfully demonstrated in an agent-based, simulated urban environment. The simulation establishes that the emergent, behavior-based patrol procedure for UAVs is effective, robust, and scalable. The approach is especially well suited for numerous, small, inexpensive, and expendable UAVs.

Keywords:
swarm, emergent intelligence, decentralized control, patrol

1. Introduction

A bottom-up approach to decentralized control of Unmanned Air Vehicles (UAVs) is investigated in this research. The purpose of the research is to develop a model for emergent formation of UAVs into functional teams that cooperatively complete a mission, such as cooperatively patrolling an asset of high interest and striking any moving or static hostile intruders. Emergent team formation involves the creation of teams without centralized control and based on individual decisions and local information. Most previous UAV mission planning and cooperative control employ global optimization techniques which assume perfect (or near perfect) global communication and complete knowledge sharing. Since reliable global communication in threatening situations is not realistic, systems that rely on it are prone to failure. Other failures that adversely affect global optimization techniques include: loss of global positioning, communication network saturation, lack of battlefield intelligence, highly dynamic battlefield conditions, and the presence of many UAVs within the operational environment [1]. Our simulation shows that mission objectives can be accomplished if all agents follow the same protocols even in the absence of inter-agent communication. The philosophy guiding this research is that of emergent intelligence and its emphasis on bottom-up, decentralized, and behavior-based control. We believe bottom-up approaches are more robust than globally optimized approaches with respect to individual tasks in uncertain and dynamic environments. The emergent intelligence approach relies little on a priori, situational knowledge or high-bandwidth, inter-agent communication. Solutions derived from emergent intelligence are highly adaptive in complex, dynamic, and uncertain environments and they offer a flexibility not easily attained by rigid, globally optimized solutions that assume perfect communication.

2. Simulation Framework

The research builds upon a previously developed agent-based framework to simulate UAVs as virtual agents [2]. This framework is known as ASAS - Autonomous Search and Attack System. By extending the generic, object-oriented agent structure in the framework, we created UAV agents with the intent of simulating the characteristics of small, low cost, expendable UAVs. In an effort to obtain a reasonably high-fidelity model, we assume that agents have limited capabilities. These limitations extend to the UAV's computational processing power, memory, and communication capability. It is also assumed that a UAV's sensors, actuators, and control systems are subject to noise and failures.

Individual UAV agents rely on local information obtained from its sensors and process that information locally. An agent has little or no dependence on another agent's state or presence. However, the agents are opportunistic: if information about another agent is available, that information may be used. Simple signal transmitters and sensors are attached to the agents to allow for limited range, broadcast or directional communication for opportunistic cooperation between agents. Signal reception is often limited to an agent's nearest neighbor. Therefore, an agent may be unaware that its cooperation with a close neighbor may propagate and result in team formation; teams are an epiphenomena of individual agent behavior.

Individual UAV agents are physically simulated with simple actuators allowing for turning (a virtual, coordinated roll and bank), and acceleration based on a simple, discrete set of velocities: slow, cruise, and fast. These capabilities allow the agents to model the rudimentary functionality of operational UAVs.

The control philosophy for the agents is based on task achieving modules with tight sensor-actuator coupling. Providing for the persistence of behavior in the absence of a triggering sensation requires some state information. The agents employ discrete states and may act differently to similar sensations in different states.

3. Asset Patrol Mission

Many situations may arise where it is deemed necessary to protect vital assets in high threat environments. An example of this is in the area of homeland security in which intelligence indicates that particular assets could be at risk from terrorist attacks. Mission goals for protecting such assets may include maintenance of a persistent presence around the asset, surveillance, and destruction of hostile intruders. A UAV is an ideal choice for carrying out this type of mission. A persistent presence around the asset could be maintained by establishing flight patrol patterns around the asset. Multiple UAVs in these patrol patterns at any point in time would ensure complete surveillance coverage and provide redundancy that would minimize the impact of individual UAV failures in the overall mission objective.

4. Asset Patrol Algorithm

4.1. *Patrol Structure*

The patrol patterns consist of flight tracks with different radii and altitudes around the asset. The multiple tracks help maintain a persistent presence around the asset for the purposes of surveillance and the destruction of hostile intruders. They also provide multiple viewpoints for surveillance as well as multiple layers of protection. Populating inner tracks is favored over outer tracks and is accomplished through behaviors that comprise a track switching protocol. Collisions in an urban area, especially around the asset being protected, would be extremely hazardous. Therefore, one of the main objectives of the protocols is that of collision avoidance. The altitude of the patrol tracks is proportional to the radius - lower tracks are smaller. Each patrol track consists of a fixed number of waypoints that form a regular polygon with the asset at its center.

4.2. *UAV Sensor/Communication Capability*

Global communication is assumed to be unavailable, and control is established only through passive sensors and minimal, short-range radio communication. One of the main objectives in the design philosophy is to determine what can be accomplished with minimal inter-agent communication. The motivation for this is to build systems that are highly robust. Systems that do not rely on capabilities that are prone to failure, such as global communication, are inherently more robust. Greater communication capabilities may be considered later to increase performance. The advantage of our design philosophy is that if these added communication capabilities fail, the system will still function reliably because it was designed to work without them.

4.3. *UAV Behaviors*

The high level objective of asset patrol is accomplished (emerges) from the more local UAV behaviors of *collision avoidance, patrolling*, and *attacking.* The *collision avoidance* and *attacking* behaviors are similar to those used in the sweep search described in [1]. The focus here is on the *patrolling* behavior.

The high-level control structure is illustrated in the state chart of Figure 1. The control is hierarchical, with the *Choose* module of Figure 1 being a high-level construct charged with identifying which lower-level state chart should appropriately be in control in the current situation. The current situation is assessed at regular time intervals by the *Choose* module. At each cycle, sensory input is processed to determine the best choice of action. Figure 2 illustrates an expansion of the *Patrol Asset* module into its lower-level state chart consisting of the behaviors *enter patrol, patrol, seek gap*, and *exit patrol.*

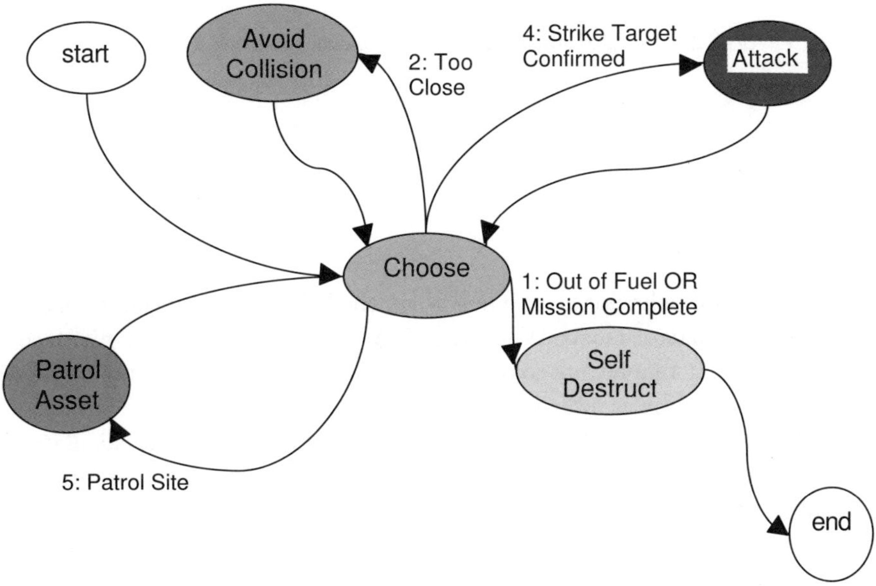

Fig. 1. Hierarchical State charts of UAV behavior.

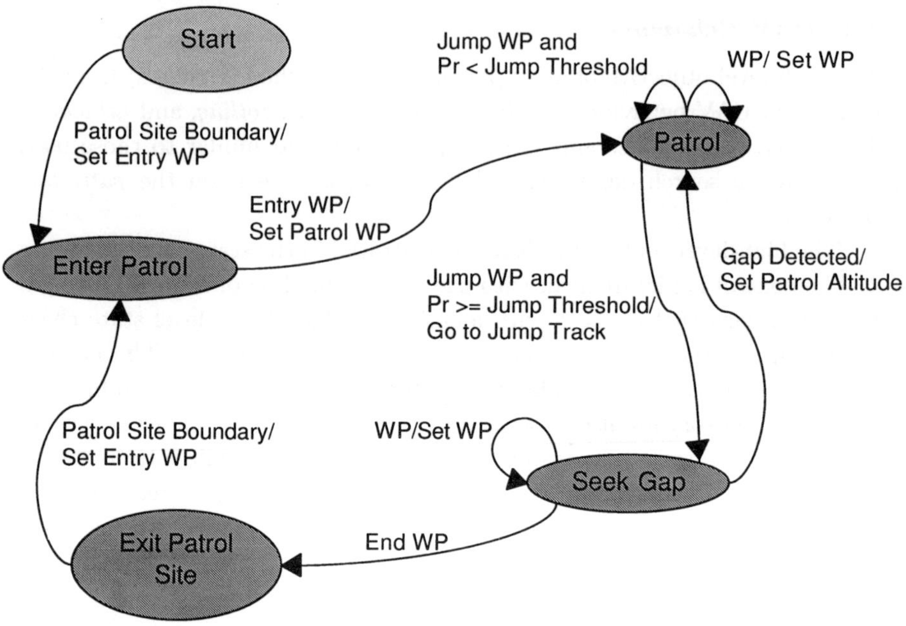

Fig. 2. Detailed state charts of UAV patrol asset behavior.

4.4. *Enter Patrol Behavior*

The enter patrol behavior is executed when the UAV is attempting to enter the outermost patrol track. When the UAV reaches a particular distance from the asset, it maneuvers to orient itself in the direction of patrol flight. This patrol direction is known in advance and is either clockwise or counterclockwise. Once the UAV is oriented in the patrol direction, it calculates which of the pre-specified entry points of the outer track is the closest to its current heading. If the UAV doesn't encounter any obstacles, such as other UAVs, on its way to the entry point, it will enter the outer patrol track. If another UAV is encountered, it will fly away from the asset for some distance before repeating the enter patrol behavior. The behavior of flying away from the asset when encountering other UAVs in close proximity provides congestion control for the outer track.

4.5. *Patrol Behavior*

The patrol behavior consists of orbiting around the asset in the current track by flying from waypoint to waypoint while scanning for possible intruders. UAVs maintain cruise speed while in a patrol track. While patrolling in an outer track, a UAV use a probability calculation to decide wether to attempt to switch to the next inner track. This decision is always made at a pre-specified waypoint. Limiting track switching attempts to a pre-specified point minimizes potential collisions.

4.6. *Track Switching Protocol*

The decision to switch tracks is based on the UAV's perception of congestion of the target track. If a UAV tries unsuccessfully to switch to a particular track, it will remember this and lower its probability of trying the next time. Initially, the UAVs attempt track switches with 100% probability. A track switch attempt consists of three steps, as depicted in Figure 3: 1) Jump from patrol track to jump track at the pre-specified waypoint, 2) Jump from jump track to patrol track if a gap is detected, and 3) Start over if a gap is not detected before the pre-specified exit point is reached.

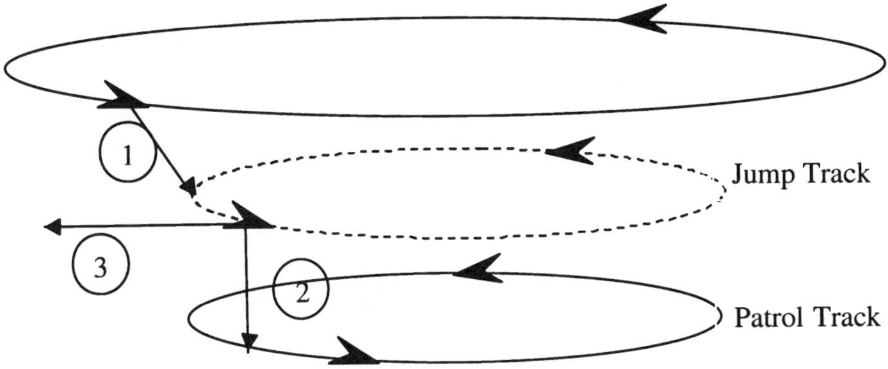

Fig. 3. Track switch protocol.

The jump track for a particular patrol track is the same radius as the desired patrol track but at an altitude that is half way between the two tracks involved in the switch. Once the UAV enters the jump track at the pre-specified point, it executes the *gap seeking* behavior.

4.7. *Gap Seeking Behavior*

After entering the jump track, the UAV accelerates to fast speed and begins looking for a point where it can fit in the desired patrol track . This activity is known as the gap seeking behavior. The point that the UAV seeks is such that a minimum separation distance between UAVs is maintained. It is assumed that the UAV has only forward scanning visual sensors. A UAV in the jump track uses a timer to determine if there is enough room behind it in the target patrol track. This timer is set to zero when the UAV first enters the jump track. Each time the jump UAV observes a patrol UAV directly below it, the timer is reset to zero. Given the difference in speed of UAVs in the jump track and UAVs in the patrol track, the jump UAV determines that there is enough room behind when the timer reaches a certain value. If the timer reaches this value, the jump UAV simply scans to see if there is enough room ahead as well. If so, the UAV has found a gap. The timer reset is illustrated in Figure 4. The gap calculation is illustrated in Figure 5.

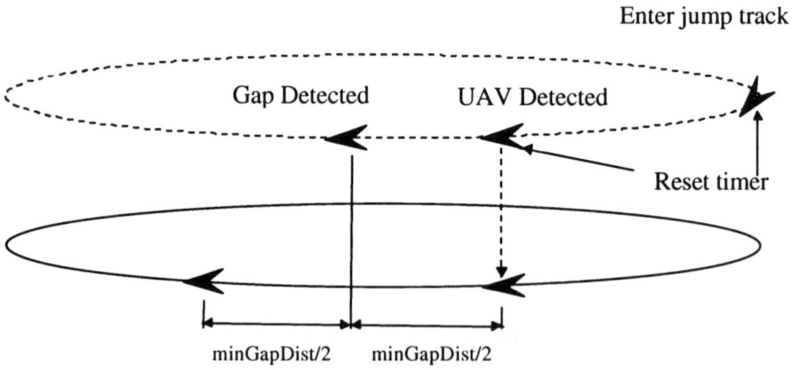

Fig. 4. Gap detection.

$$t(v_{fast} - v_{cruise}) \geq \frac{minGapDist}{2} \quad (1)$$

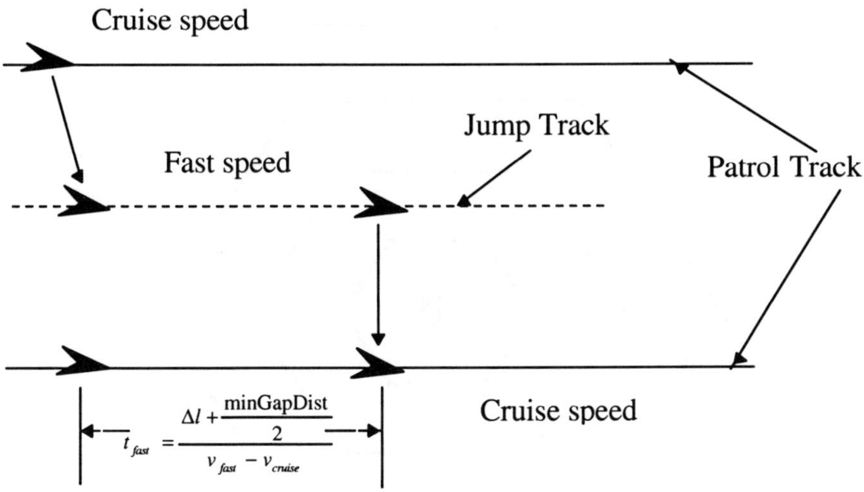

Fig. 5. Gap timer calculation.

$$t_{fast} = \frac{\Delta l + \frac{minGapDist}{2}}{v_{fast} - v_{cruise}} \tag{2}$$

If the UAV doesn't find a gap large enough before it reaches the exit point, it will exit the patrol area by executing the exit patrol behavior. The entry and exit points of the jump track are such that a UAV is in the jump track slightly less than one complete orbit. This restriction eliminates possible collisions in the jump track.

4.8. *Exit Patrol Behavior*

The purpose of the exit patrol behavior is to exit the patrol area after a failed track switch to avoid collisions with other patrolling UAVs. Starting at the exit point of the jump track, the UAV flies away from the asset until it is well beyond the outer most patrol track. Until it reaches this point, it maintains the altitude of the jump track it just exited since no other patrolling UAVs are at this altitude. It then begins climbing to the altitude of the outer most patrol track. Then the enter patrol behavior is invoked. The exit patrol behavior may also be used when UAVs are low on fuel and need to return to base.

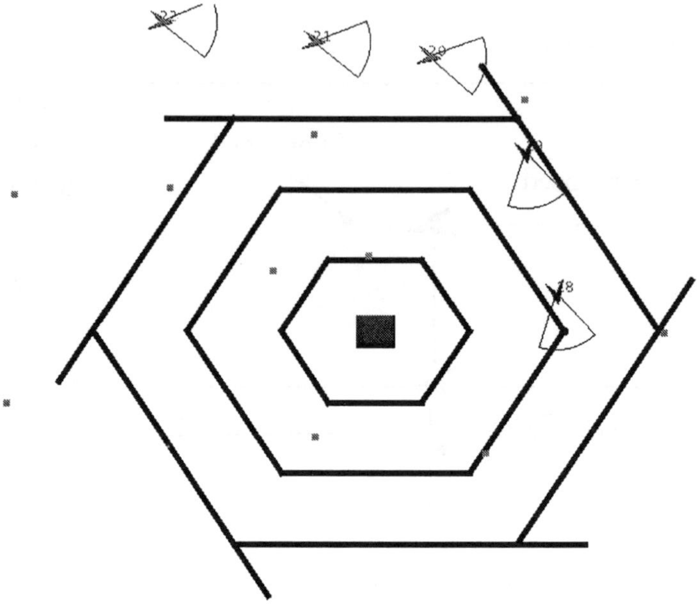

Fig. 6. View of patrolling simulation, entering patrol.

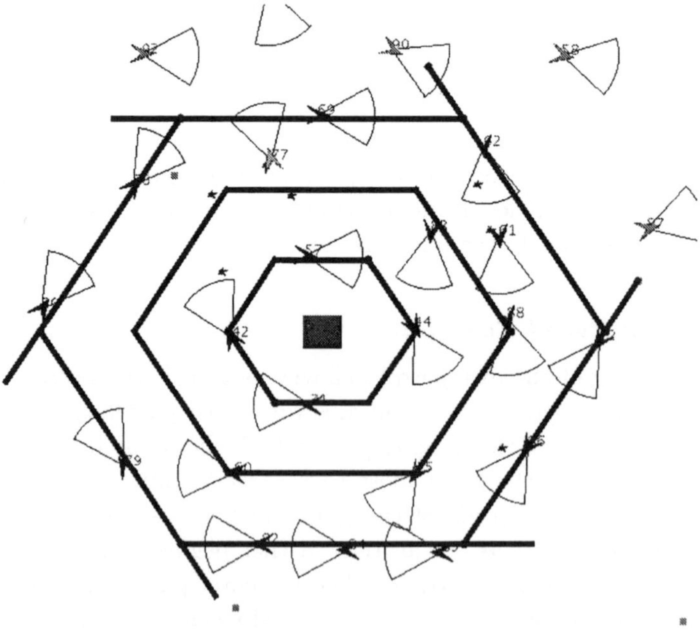

Fig. 7. View of patrolling simulation, patrolling.

Table 1. System performance under varying conditions using hexagonal tracks.

Shape of track	Number of UAVs	Threat density	System performance
Hexagon	1∼5	None	4 to 5 UAVs populate the inner most track
		Low	UAVs strike threats before tracks are populated
		High	(Same as above)
	16	None	UAVs populate inner two tracks, no evasive action required
		Low	UAVs populate inner two tracks, most threats destroyed, no evasive action required
		High	UAVs populate inner most track, most threats destroyed, no evasive action required
	32	None	UAVs populate all the three tracks, evasive action required
		Low	UAVs populate all tracks, most threats destroyed, evasive action required
		High	(Same as above)
	> 32	None	UAVs populate all tracks, evasive action required, collisions occurred
		Low	UAVs populate all tracks, most threats destroyed, evasive action required, collisions occurred
		High	UAVs populate all tracks, most threats destroyed, evasive action required

5. Experimental Results and Observations

The system has been tested under varying experimental conditions which include: varying number of UAVs, varying number of threats, and differing track shapes. Threat density was varied from none, low, and high with 0, 5 to 10, and 10 to 15 threats respectively. The experimental results with varying numbers of UAVs and threats using hexagonal tracks are shown in Table 1, and the view of the patrolling system is shown in Figures 6 and 7.

As the shapes of the tracks increase from hexagon to octagon and 16-gon, there were fewer evasive action maneuvers required when attempting to enter the outer track. This is due to the increased number of entry points making it less likely that two UAVs would seek the same entry point simultaneously. With 32-gon tracks, there were more evasive action maneuvers required when attempting to enter the outer track. This is due to the entry points being too close together. The system becomes unstable due to cascading evasive action maneuvers when more than 32 UAVs are in the patrol area.

6. Conclusions and Future Works

The asset patrol and protection model is implemented and successfully demonstrated in an agent-based, simulated urban environment. The simulation establishes that the emergent, behavior-based patrol procedure for UAVs is effective, robust, and scalable. The approach is especially well suited for numerous, small, inexpensive, and expendable UAVs. The use of virtual beacons (waypoints), signal-based communication, and simple rules provide a robust and effective method for cooperative control among n UAVs to patrol an asset. The model presented demonstrates that neither high-level control nor high-bandwidth communication is necessary for this complex cooperative control task. The simulation shows that communication is not necessary if all the agents follow the prescribed protocols.

There are several areas that are being explored to expand and extend our current multi-agent model. High level decision layers, based on a Partially Observable Markov Decision Process (POMDP) and a Bayesian Network, are under development to function on the top of the reactive behavior-based agent control. This would allow agents to function more intelligently if more global information is available. In the absence of this global information, agents can fall back on the reactive behavior-based control. The agents may be augmented with a greater behavioral repertoire allowing them to perform a variety of tactics as well as other coordinated movements.

References

[1] Schlecht J, Altenburg K, Ahmed BM, Nygard KE., "Decentralized Search by Unmanned Air Vehicles using Local Communication", Las Vegas, NV: *Proceedings of the International Conference on Artificial Intelligence*, Volume II, pages 757-762, 2003.

[2] Altenburg K, Schlecht J, Nygard KE., "An Agent-based Simulation for Modeling Intelligent Munitions", Athens, Greece: *Advances in Communications and Software Technologies*, pages 60-65, 2002.

CHAPTER 15

K-MEANS CLUSTERING USING ENTROPY MINIMIZATION

Anthony Okafor and Panos M. Pardalos
Department of Industrial and Systems Engineering
University of Florida, Gainesville, FL

Associated with use of the K-means Algorithm for data partitioning is the problem of initializing the number of clusters and their centers. In this paper, we propose to integrate as a variable the number of clusters in the optimization problem. By using entropy minimization via Bayesian inference, the number of optimum clusters can easily be found. Depending on the clustering requirements of the data, the entropy constant in our algorithm can be varied in order to obtain different number of clusters.

Keywords:
K-means clustering, entropy, Bayesian inference

1. Introduction

Data clustering and classification arise in many different applications such as pattern recognition and pattern classification, data mining and knowledge discovery, data compression and vector quantization [9]. The quality of a good cluster is application dependent since there are many methods for finding clusters subject to various criteria which are both ad hoc and systematic [9]. The different clustering methods are usually referred to as unsupervised. Unsupervised methods are also referred to as automatic data partition methods. For these methods, the user intervention is simply reduced to initializing the process (for instance in the K-means algorithm, defining the number of clusters and their centers) and interpreting the results. The results obtained are user independent as opposed to supervised methods. Unsupervised methods include the K-means [18], isodata [5], fuzzy c-means [12], and the maximum likelihood with expectation maximization (EM) [10] or sometimes called maximum likelihood estimation.

One of the difficulties in using unsupervised methods is the need for input parameters. Many algorithms, especially the K-means and other hierarchical methods [7] require that the initial number of clusters be specified. Several authors have proposed methods that automatically determine the number of clusters in the data [5, 10, 6]. These methods use some form of cluster validity measures like variance, *a priori* probabilities and the difference of cluster centers. The obtained results are not always as expected and are data dependent [19]. Some criteria from information theory have also been proposed. The Minimum Descriptive Length (MDL) criteria evaluates the compromise between the likelihood of the classification and the complexity of the model [17].

In this chapter, we propose to incorporate the clustering problem into a Bayesian inference to automatically detect the number of clusters. Entropy is used to derive prior probability in the proposed model. Some automatic thresholding methods have been proposed using entropy either by maximizing the information between the two clusters derived from the Renyi's entropy [11, 12] or by minimizing the cross entropy [6]. In this chapter, we consider a problem of partitioning a data. To accomplish this, we therefore minimize the entropy associated with the data clustering histogram.

The chapter is organized as follows. In the next section, we provide some background on the K-means algorithm. A brief introduction of entropy is presented in section 3. The proposed model is derived in section 4. With the addition of Gaussian likelihood, the proposed model extends to the K-means algorithm. The results of our algorithm are discussed in section 5. We conclude briefly in section 6.

2. K-Means Clustering

The K-means clustering [13] is a method commonly used to partition a data set into k groups. In the K-means clustering, we are given a set of n data points in d dimensional space R^d and an integer k and the problem is to determine a set of points (centers) in R^d so as to minimize the distance from each data point to its nearest center. The K-means consists of primarily two steps:

1) The assignment step where based on initial k cluster centers (classes), instances are assigned to the closest class.

2) The re-estimation step where the class centers are recalculated from the instances assigned to the class.

These steps are repeated until convergence occurs; that is when the re-estimation step leads to minimal change in the class center. The algorithm is outlined in Figure 1.

The K-means Algorithm
Input
$P = \{p_1, \ldots, p_n\}$ (points to be clustered)
k (number of clusters)
Output
$C = \{c_1, \ldots c_k\}$ (cluster centers)
$m : P \rightarrow \{1, \ldots k\}$ (cluster membership)
Procedure K-means
1. Initialize C (random selection of P).
2. For each $p_i \in P$, $m(p_i) = \arg\min_{j \in k} \text{distance}(p_i, c_j)$.
3. If m has not changed, stop, else proceed.
4. For each $i \in \{1, \ldots k\}$, recompute c_i as a center of $\{p \mid m(p) = i\}$.
5. Go to step 2.

Fig. 1. The K-Means Algorithm

Several distance metrics like the Manhattan or the Euclidean are commonly used. In this chapter, we consider the Euclidean distance metric. Issues that arise in using the K-means include: choosing the number of clusters and degeneracy. Degeneracy arises when the algorithm is trapped in a local minimum thereby resulting in some empty clusters. These two problems are addressed in our approach by using entropy minimization.

3. A Brief Overview of Entropy Optimization

The concept of entropy was originally developed by the physicist Rudolf Clausius around 1865 as a measure of the amount of energy in a thermodynamic system [2]. This concept was later extended through the development of statistical mechanics. It was first introduced into information theory in 1948 by Claude Shannon [16]. Entropy can be understood as the degree of disorder of information contents. It is also a measure of uncertainty about a partition [16, 10].

The philosophy of entropy minimization in the pattern recognition field can be applied to classification, data analysis, and data mining where one of the tasks is to discover patterns or regularities in a large data set. The regularities of the data structure are characterized by small entropy values, where randomness is characterized by large entropy values [10]. In the data mining field, the most well known application of entropy is information gain of decision trees. Entropy based discretization recursively partitions the values of a numeric attribute to a hierarchy discretization using entropy as information measure evaluates attribute importance by examining information theoretic measures [10].

In this chapter entropy minimization is used to determine the number of clusters and to overcome degeneracy. Entropy is used as an information measure of cluster distribution of data in that cluster. We can represent data belonging to a cluster as one bin. Thus a histogram represents cluster distribution of data. From the entropy theory, a histogram of cluster label with low entropy shows a classification with high confidence, while a histogram with high entropy shows a classification with low confidence.

3.1. *Minimum Entropy and Its Properties*

Shannon Entropy is defined as:

$$H(X) = -\sum_{i=1}^{n}(p_i \ln p_i) \tag{1}$$

where X is a random variable with outcomes $1, 2, ..., n$ and associated probabilities $p_1, p_2, ..., p_n$.

Since $-p_i \ln p_i \geq 0$ for $0 \leq p_i \leq 1$ it follows from (1) that $H(X) \geq 0$, where $H(X) = 0$ iff one of the p_i equals 1; all others are then equal to zero. Hence the notation $0 \ln 0 = 0$. For continuous random variable with probability density function $p(x)$, entropy is defined as

$$H(X) = -\int p(x) \ln p(x) dx \tag{2}$$

This entropy measure tells us whether one probability distribution is more informative than the other. The minimum entropy provides us with minimum uncertainty, which the limit of the knowledge we have about a system and its structure [16]. In pattern recognition for example the quest is finding minimum entropy [16]. The problem of evaluating minimal entropy probability distribution is the global minimization of the Shannon entropy measure subject to the given constraints. This problem is known to be NP-hard [16].

Two properties of minimal entropy which will be fundamental in the development of our model are *concentration* and *grouping* [16].

Grouping implies moving all the probability mass from one state to another, that is reduce the number of states. This reduction can decrease entropy.

Proposition 1: *Given a partition* $ß = [B_a, B_b, A_2, A_3, ...A_N]$, *we form the partition* $\AA = [A_1, A_2, A_3, ...A_N]$ *obtained by merging* B_a *and* B_b *into* A_1, *where* $p_a = P(B_a)$, $p_b = P(B_b)$ *and* $p_i = P(A_i)$, *we maintain that:*

$$H(\AA) \leq H(ß) \tag{3}$$

Proof: The function $\varphi(p) = -p \ln p$ is convex. Therefore for $\lambda > 0$ and $p_1 - \lambda < p_1 < p_2 < p_2 + \lambda$
we have:

$$\varphi(p_1 + p_2) < \varphi(p_1 - \lambda) + \varphi(p_2 + \lambda) < \varphi(p_1) + \varphi(p_2) \tag{4}$$

Clearly,

$$H(ß) - \varphi(p_a) - \varphi(p_b) = H(\AA) - \varphi(p_a + p_b)$$

because each side equals the contribution to $H(ß)$ and $H(\AA)$ respectively due the to common elements of ß and $\AA$
Hence, (3) follows from (4). □

Concentration implies moving probability mass from a state with low probability to a state with high probability. Whenever this move occurs, the system becomes less uniform and thus entropy decreases.

Proposition 2: *Given two partitions* $ß = [b_1, b_2, A_3, A_4, ...A_N]$ *and* $\AA = [A_1, A_2, A_3, ...A_N]$ *that have the same elements except the first two. We maintain that if*
$p_1 = P(A_1)$, $p_2 = P(A_2)$ *with* $p_1 < p_2$ *and* $(p_1 - \lambda) = P(b_1) \leq (p_2 + \lambda) = P(b_2)$,
then

$$H(ß) \leq H(\AA) \tag{5}$$

Proof:
Clearly,

$$H(\mathring{A}) - \varphi(p_1) - \varphi(p_2) = H(\text{ß}) - \varphi(p_1 - \lambda) - \varphi(p_2 + \lambda)$$

because each side equals the contribution to $H(\text{ß})$ and $H(\mathring{A})$ respectively due to the common elements of $\mathring{A}$ and ß
Hence, (5) follows from (4). □

3.2. *The Entropy Decomposition Theorem*

Another attractive property of entropy is the way in which aggregation and disaggregation are handled [4]. This is because of the property of additivity of entropy. Suppose we have n outcomes denoted by $X = \{x_1,...,x_n\}$, with probability $p_1, ..., p_n$. Assume that these outcomes can be aggregated into a smaller number of sets $C_1, ..., C_K$ in such a way that each outcome is in only one set C_k where $k = 1, ...K$. The probability that outcomes are in set C_k is

$$p_k = \sum_{i \in C_k} p_i \tag{6}$$

The entropy decomposition theorem gives the relationship between the entropy $H(X)$ at level of the outcomes as given in (1) and the entropy $H_0(X)$ at the level of sets. $H_0(X)$ is the between group entropy and is given by:

$$H_0(X) = -\sum_{k=1}^{K} (p_k \ln p_k) \tag{7}$$

Shannon entropy (1) can then be written as:

$$
\begin{aligned}
H &= -\sum_{i=1}^{n} p_i \ln p_i \\
&= -\sum_{k=1}^{K}\sum_{i\in C_k} p_i \ln p_i \\
&= -\sum_{k=1}^{K} p_k \sum_{i\in C_k} \frac{p_i}{p_k}\left(\ln p_i + \ln \frac{p_k}{p_i}\right) \\
&= -\sum_{k=1}^{K}(p_k \ln p_k) - \sum_{k=1}^{K} p_k \sum_{i\in C_k} \frac{p_i}{p_k} \ln \frac{p_i}{p_k} \\
&= H_0(X) + \sum_{k=1}^{K} p_k H_k(X) \qquad (8)
\end{aligned}
$$

where

$$H_k(X) = -\sum_{i\in C_k} \frac{p_i}{p_k} \ln \frac{p_i}{p_k} \qquad (9)$$

A property of this relationship is that $H(X) \geq H_0(X)$ because p_k and $H_k(X)$ are nonnegative. This means that after data grouping, there cannot be more uncertainty (entropy) than there was before grouping.

4. The Proposed Model

In this section we outline our proposed approach and show how it extends to the K-means algorithm.

4.1. *Entropy as a Prior Via Bayesian Inference*

Given a data set represented as $X = \{x_1,...,x_j,...,x_n\}$, a clustering is the partitioning of the data set to get the clusters $\{C_i,\ i = 1,...K\}$, where K is usually less than n. Thus a partition of the data corresponds to a cluster defined by $C_i = \{j \in l | x_j = i\}$ for $l \in (1, 2, ...n)$.

The K-means clustering algorithms aims to find the partition that minimizes the square distance between the data and the classification (cluster center). Similarly the Bayesian approach takes into account the distance between the data and the classification. In addition, it also considers the prior on the classification. Thus by the Bayes approach, a classified data is obtained by maximizing the posterior probability. Since the number of clusters is unknown and must be specified to use the K-means, we propose to find it by using the Bayesian inference. This can be done by using an entropy prior in the Bayes rule and incorporating the number of clusters into this prior and estimating it. Suppose that resulting data from the clustering is $\Theta = \{\theta_1, \theta_2, ..., \theta_N\}$, by Bayes rule, the posterior probability $P(\Theta|X)$ is given as;

$$P(\Theta|X) = \frac{P(X|\Theta)P(\Theta)}{P(X)} \propto P(X|\Theta)P(\Theta) \tag{10}$$

where $P(X|\Theta)$ is the likelihood and measures the accuracy in clustering the data and the prior $P(\Theta)$ measures consistency with our background knowledge. The likelihood has the following form:

$$P(X|\Theta) = \prod_{j\in l} P(x_j|\theta_j) = e^{\sum_{j\in l} \ln p(x_j|\theta_j)} \tag{11}$$

To find the number of clusters, we proceed as follows: we initially select an arbitrary large number of clusters K. A rule of thumb value $K = \sqrt{n}$ may be used [3]. To reduce this number, we have to sharpen the histogram associated with the clustering. We propose to minimize the entropy of the classified data histogram. The entropy decreases as the number of bins with probability zero increases.

From equation (1), we write Shannon Entropy as:

$$H(X) = -\sum_{i=1}^{K} p_i \ln p_i \tag{12}$$

If we consider a clustering that has k clusters, we have:

$$H(X) = -\sum_{i=1}^{K} p_i \ln p_i = -\sum_{i=1}^{k} p_i \ln p_i \tag{13}$$

Defining the prior as an exponential distribution, we have;

$$P(\Theta) \propto e^{\beta \sum_{i=1}^{k} p_i \ln p_i} \tag{14}$$

where $p_i = |C_i|/n$ is the prior probability of cluster i, and β (entropy constant) refers to a weighting of the a prior knowledge. The posterior probability now becomes:

$$P(\Theta|X) \propto \exp \sum_{j\in l} (p_i \ln p_i) \left(\exp \left[\beta \sum_{i=1}^{k} p_i \ln p_i \right] \right) \propto \exp(-E) \tag{15}$$

where E is written as follows:

$$E = -\sum_{j \in L} \ln p(x_j|\theta_j) - \beta \sum_{i=1}^{k} p_i \ln p_i \tag{16}$$

If we assume that the x_j have Gaussian distribution with mean values $\theta_i, i = (1, ..., k)$ and constant cluster variance. Then,

$$P(x_j|\theta_i) = \frac{1}{\sqrt{2\pi\sigma^2}} e^{\left(-\frac{(x_j-\theta_i)^2}{2\sigma^2}\right)} \tag{17}$$

Taking natural log and omitting constants, we have

$$\ln P(x_j|\theta_i) = -\frac{(x_j-\theta_i)^2}{2\sigma^2} \tag{18}$$

Equation (16) becomes

$$E = \sum_{i=1}^{k} \sum_{j \in C_i} \frac{(x_j-\theta_i)^2}{2\sigma^2} - \beta \sum_{i=1}^{k} p_i \ln p_i \tag{19}$$

or

$$E = \sum_{i=1}^{k} \sum_{j \in C_i} \left[\frac{(x_j-\theta_i)^2}{2\sigma^2} - \frac{\beta}{N} \ln p_i) \right] \tag{20}$$

We note that when $\beta = 0$, E is the cost of the K-means clustering algorithm.

The Entropy K-means algorithm is given in Figure 2.

This algorithm iteratively reduces the numbers of clusters because some of the bins (clusters) will vanish where $p_i = 0$.

5. Results

The entropy K-means algorithm was tested on some synthetic images and on the Iris data set. The results are given in 5.1 and 5.2 respectively.

5.1. *Image Clustering*

For the synthetic images, the objective is to reduce the complexity of the grey levels. Our algorithm was implemented with synthetic images for which the ideal clustering is known. A total of three test images were used with varying numbers of clusters. The first two images, test1 and test2, have four clusters. Three of the clusters had uniformly distributed values with a range of 255, and the other had a constant value. Test1 had clusters of varying size while test2 had equal sized clusters. The third synthetic image, test3, has nine clusters each of the same size and each having values uniformly distributed with a range of 255. We initialized the algorithm with the number of clusters equal to the number of grey levels and the value of cluster centers equal to the grey values. The initial probabilities (p_i) were computed from the image histogram. The algorithm was able to correctly detect the number of clusters. Different clustering results were obtained as the value of the entropy constant was changed as is shown in Table 1. For the image test3, the correct number of clusters was obtained using a β of 1.5. For the images test1 and test2, a β value of 5.5 yielded the correct number of clusters. In Table 1, the optimum number of clusters for each synthetic image are bolded.

Table 1. The number of clusters for different values of β

β	Images		
	test1	test2	test3
1.0	10	10	13
1.5	6	8	**9**
3.5	5	5	6
5.5	**4**	**4**	5

5.2. *Iris Data*

Next we tested the algorithm on the Iris Data. The Iris data is well known [1, 8] and serves as a benchmark for supervised learning techniques. It consists of three types of Iris plants: *Iris Versicolor, Iris Virginica, and Iris Setosa* with 50 instances per class. Each datum is four dimensional and consists of plants' morphology namely *sepal width, sepal length, petal width, and petal length.* One class *Iris Setosa* is well separated from the other two. Our algorithm was able to obtain the three-cluster solution when using entropy constant β of 4.0 and 4.5. Two cluster solutions were also obtained using entropy constants of 5.0, 5.5, 6.0, and 6.5. Table 2 shows the results of the clustering.

To evaluate the performance of our algorithm, we determined the percentage of data that were correctly classified for three cluster solution. We compared it to the results of direct K-means. Our algorithm had a 91%

correct classification while the direct K-means achieved only 68% percent correct classification, see Table 3.

Table 2. The number of clusters as a function of β for the Iris Data

β	4.0	4.5	5.0	5.5	6.0	6.5
k	3	3	2	2	2	2

Table 3. Percentage of correct classification of Iris Data

k	3.0	3.0	2.0	2.0	2.0	2.0
%	90	91	69	68	68	68

6. Conclusion

By incorporating entropy as a prior in the Bayesian inference, the number of clusters present in a data set can be determined automatically. Varying the entropy constant (β) allows us to vary the final number of clusters. The approach worked well with test images and Iris data producing the expected number of clusters. Further work will the extension of this method to multidimensional data and large data sets.

References

[1] R.O. Duba, and P.E. Hart, *Pattern Classification and Scene Analysis*, Wiley-Interscience, New York, NY, 1974.

[2] S. Fang, J.R. Rajasekera., and H.-S. J. Tsao, *Entropy Optimization and Mathematical Programming*, Kluwer's Academic Publishers, 1997.

[3] M. Figueiredo and A.K. Jain. *Unsupervised learning of finite mixture models,* IEEE Trans. Pattern Analysis and Machine Intelligence, 24(3): 381-396, 2002.

[4] K. Frenken, *Entropy Statistics and Information Theory* ,The Elgar Companion to Neo-Schumpeterian Economics, Cheltenham, UK and Northampton MA: Edward Elgar Publishing, (Submitted for Publication.), 2003.

[5] D. Hall and G Ball, ISODATA *A Novel Method of Data Analysis and Pattern classification, Tech Report.,* Stanford Research institute, Menlo Park, CA, 1965.

[6] G. Iyengar, and A. Lippman, *Clustering Images using Relative Entropy for Efficient retrieval*, IEEE Computer Magazine, 28(9): 23-32, Sept. 1995.

[7] A. Jain and M. Kamber, *Algorithm for Clustering*, Prentice Hall, 1998

[8] M. James, *Classification Algorithm*, Wiley-Interscience, New York, NY, 1985.

[9] T. Kanungo, D.M. Mount, N.S. Netayahu, C.D. Piako, R. Silverman, and A.Y. Wu. *An Efficient K-Means Clustering Algorithm: Analysis and Implementation,* IEEE Trans. Pattern Analysis and Machine Intelligence, 24(7): 881-892, 2002.

[10] J.N. Kapur and H.K. Kesaven, *Entropy Optimization Principle with Applications*, London Academic, 1997, Ch.1.

[11] Nailong Wu, *The Maximum Entropy Method*, Springer, 1997, Ch.5.

[12] Y.W. Lim and S.U. Lee, *On the Color Image Segmentation Algorithm based on Thresholding and Fuzzy C-means Techniques,* Pattern Recognition, vol. 23, pp. 935-952, 1990.

[13] J.B. McQueen (1967), *Some Methods for Classification and Analysis of Multivariate Observations,* Proceedings of the Fitfth Symposium on Math, Statistics, and Probability, pp 281-297. Berkely, CA: University of California Press.

[14] D. Miller, A. Rao, K. Rose and A. Gersho, *An Information Theoretic Framework for Optimization with Application to Supervised Learning*, IEEE International Symposium on Information Theory, Whistler B.C., Canada, September 1995.

[15] B. Mirkin, *Mathematical Classification and Clustering – Nonconvex Optimization and its Applications,* v11, Kluwer, 1996.

[16] D. Ren, *An An Adaptive Nearest Neighbor Classification Algorithm,* Available at `www.cs.ndsu.nodak.edu/~dren/papers/CS785finalPaper.doc`

[17] J. Rissanen, *A Universal prior for integers and Estimation by Minimum Description Length*, Annals of Statistics, 1983.

[18] J.T. Tou and R.C. Gonzalez, *Pattern Recognition Principles* , Massachusetts: Addison-Wesley, 1994.

[19] M.M. Trivedi and J.C. Bezdeck, *Low-level segmentation of aerial with fuzzy clustering,* IEEE Trans. Syst. Man, Cybern., vol. SMC-16, pp 589-598, 1986.

Entropy K-means Algorithm

1. Select the initial number of clusters k and a value for the stopping criteria ε.

2. Randomly initialize the cluster centers $\theta_i(t)$, and the *a priori* probabilities p_i, $i = 1, 2, ..., k$, β, and the counter $t = 0$.

3. Classify each input vector x_j, $j = 1, 2, ..., n$ to get the partition C_i such that for each $x_j \in C_r$, $r = 1, 2, ..., k$
$$[x_j - \theta_r(t)]^2 - \frac{\beta}{n}\ln(p_r) \le [x_j - \theta_i(t)]^2 - \frac{\beta}{n}\ln(p_i)$$

4. Update the cluster centers
$$\theta_i(t+1) = \frac{1}{|C_i|}\sum_{j \in C_i} x_j$$
and the *a priori* probabilities of clusters
$$p_i(t+1) = \frac{|C_i|}{n}$$

5. Check for convergence; that is see if
$$\max_i |\theta_i(t+1) - \theta_i(t)| < \varepsilon$$
if it is not, update $t = t + 1$ and go to step 3.

Fig. 2. The Entropy K-means Algorithm

CHAPTER 16

INTEGER FORMULATIONS FOR THE MESSAGE SCHEDULING PROBLEM ON CONTROLLER AREA NETWORKS

Carlos A.S. Oliveira
Department of Industrial and Systems Engineering
University of Florida, Gainesville, FL
oliveira@ufl.edu

Panos M. Pardalos
Center for Applied Optimization
Department of Industrial and Systems Engineering
University of Florida, Gainesville, FL
pardalos@ufl.edu

Tania M. Querido
CEFET-RJ, Av. Maracanã
229, Rio de Janeiro, RJ, Brazil
Supported by the Brazilian National Research Council (CNPQ)
tania.querido@att.net

In this work, the problem of scheduling messages in a controller area network (CAN) is presented. CAN is an important type of hard real-time distributed system, which is used to control embedded devices, connected to a main processor through a serial communication infrastructure. The main problem in CAN concerns the optimal allocation of messages in the bus field connecting processor nodes. We propose linear integer programming formulations for this problem. Our objective is to find a message schedule minimizing the time for dispatching of messages with high priority. We show that the problem is NP-hard, and present results of the mathematical programming models for a set of instances defined over subsets of the SAE Benchmark for Automotive Systems.

Keywords:
Controller area network, integer programming, computational complexity

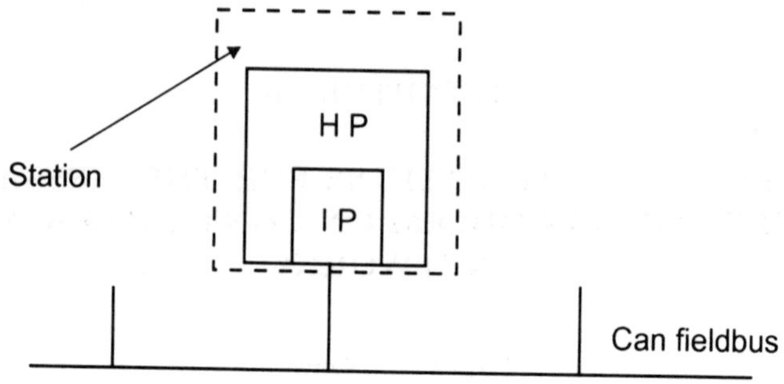

Fig. 1. CAN architecture

1. Introduction

Applications of real-time distributed systems appear in many industrial areas. For such applications the use of computer networks has increased in the last years, due to the trend of component automation through the use of embedded processors. In automobile industries, the controller area network (CAN) is a type of hard real-time distributed system designed to coordinate the demand of messages among in-vehicle electrical resources integrated with computational devices. CAN has been adapted by automotive manufacturers to operate the ever growing number of vehicle's electrical accessories and, essentially, to deal with security components. Examples of automobile subsystems connected through CAN are brakes, engine, lubrication, etc.

In a CAN, processor nodes are connected by a serial communication bus, also known as *fieldbus*. A scheme of CAN architecture is shown in Figure 1. Each processor uses preemptive scheduling to select running tasks in the form of short messages (the maximum data length in the network is restricted to 8 bytes). Stations on a CAN fieldbus receive messages based on the message identifier, which is used to filter messages and assign priorities. The component responsible for identifying messages arriving to a device is called *interface processor* (IP). The IP presents to the main *host processor* (HP) only messages with desired identifiers.

Messages in a CAN system are typically periodic. The different periods are designed according to control specifications of the distributed system. However, due to interference of different messages transmission periods, the

Table 1. Sample of SAE requirements for CAN messages.

Signal Num.	Signal Description	Size (bits)	Period (ms)	Deadline (ms)	Priority
1	Traction Battery Voltage	8	100	100	2
2	Traction Battery Current	8	100	100	2
3	Traction Battery Temp, Average	8	1000	1000	1
4	Auxiliary Battery Voltage	8	100	100	2
5	Traction Battery Temp, Max.	8	1000	1000	1
6	Auxiliary Battery Current	8	100	100	2
7	Accelerator Position	8	5	5	5
8	Brake Pressure, Master Cylinder	8	5	5	5
9	Brake Pressure, Line	8	5	5	5
10	Transaxle Lubrication Pressure	8	100	100	2
11	Transaction Clutch Line Pressure	8	5	5	5
12	Vehicle Speed	8	100	100	2
13	Traction Battery Ground Fault	1	1000	1000	1
14	Hi&Lo Contactor Open/Close	4	50	5	5
15	Key Switch Run	1	50	20	3

time intervals between successive instances of the same periodic message may suffer some fluctuations. The resulting time interval between instances of messages is commonly called *jitter*. To reduce message transmission delay and thus obtain higher communication channel utilization, the controller network employs a message priority scheme, i.e., highest priorities are assigned to messages with shortest deadlines. As messages compete for the exclusive use of a transmission, a policy is needed for determining which message should be sent next when the network is available.

Most companies working on CAN use the standard recommendations developed by the Society of Automotive Engineering (SAE). In particular, the benchmark for class C automotive systems concerning safety critical control applications [5] will be considered. Table 1 reports some of the SAE specifications, where the number of processing nodes, sizes, periods and deadlines of messages are the given parameters. Classes of messages are derived from their priorities. According to the specification, the SAE protocol identifies 6 different classes among the total of 53 messages. In this work, the SAE recommendations are used for modeling purposes, and also to help in creating realistic instances for the CAN system.

1.1. *Previous Work*

In CAN systems, the main optimization problem consists in studying the optimal ordering of messages in the network. This is motivated by the possibly serious consequences that a bad message ordering can have in the real-time system. The correct understanding of best and worst scenario have been the main reasons for studying CAN from the simulation as well as the optimization point of view.

Due to its combinatorial nature, message scheduling in CAN is a challenge for those who deal with the analysis of optimal message ordering. Jitter minimization has been studied as a combinatorial problem leading to an enormous number of different combinations. Several works have partially included some effects of precedence constraints in a limited heuristic approach [1, 7, 8]. For example, in [1] a modification of the genetic algorithm was used to support a simulation of the process. The performance of control loops, considering jitter interference, was studied by [7] and [1]. An interesting worst-case analysis of the problem was made by Tindell and Burns [8].

Although simulation has been successfully used to conduct studies on CAN [2, 6, 7, 9], a mathematical formulation is fundamental to provide an intelligent message strategy and to increase reliability of the overall system. Such a mathematical formulation would allow, on the other hand, a reduction on the reliance on costly prototypes for the product development process.

In this work, a mathematical model for scheduling messages on a CAN network is presented, based on integer linear programming. Our objective is to allow the delivery of the maximum number of messages, giving precedence to messages with higher priority.

The chapter is organized as follows. Section 2 introduces the message scheduling problem on CAN and gives a sequence of linear IP formulations. These are used to define clearly the problem, and provide some of its formal properties. Experimental studies were conducted to determine the quality and computational performance of the formulations proposed. These results are reported in Section 3. Finally, in Section 4 we give some concluding remarks and future directions for this work.

2. Problem Definition

The scheduling of messages in a CAN network is a hard real time process. This means that the delivery of messages cannot be postponed by more than a very small fraction of time. The well functioning of diverse sensitive components, such as the brakes in an automobile, depend on the correct delivery of messages, and is the main motivation for optimizing the scheduling of messages in CAN.

To define optimal message scheduling in a CAN network, we introduce a mathematical notation for the message delivery system. In a CAN, a set of messages is defined that can be used to control the different embedded devices. We assume that there are m different types of messages. Messages are periodic, and each message has period T_i, for $i \in \{1, \ldots, m\}$. This means that the j-th occurrence of message i happens anywhere between times $(j-1)T_i + 1$ and jT_i, inclusive, due to uncertainties in the system. Messages have attached attributes, such as priority p_i, and time duration σ_i, for $i \in \{1, \ldots, m\}$.

Assume that the system is simulated in the time interval $[0, t]$. The total time is divided into slots, and the size T_s of each slot is given by the maximum common divisor of the periods of all messages, i.e., $T_s = \gcd\{T_1, \ldots, T_m\}$. Thus, there are $n = t/T_s$ slots in the simulated system. Similarly, the number of occurrences of a message of type i is given by $n_i = t/T_i$. To simplify calculations, we assume that $T_s = 1$, which can always be enforced by changing the unit of time to be equal to T_s. Another assumption used in the following models is that each slot of time can have only an integer number of messages. This means that a message cannot be assigned at the same time to two different slots, i.e. it cannot be "split" between the boundaries of the slots. This assumption is not completely true in practice, but is a good approximation for most systems, since messages are scheduled to be dispatched at the beginning of a particular time slot.

With these definitions we can formulate a problem that represents the optimal scheduling of messages. This will be done in the next subsection using three models that represent, with increasing complexity, different aspects of the problem.

2.1. *Integer Programming Formulations*

We propose a mathematical programming model for the scheduling of messages in a controller area network. Let $x_{i,j,k}$ be defined as

$$x_{i,j,k} = \begin{cases} 1 \text{ if the } j\text{-th occurrence of message } i \text{ appears in slot } k \\ 0 \text{ otherwise.} \end{cases}$$

It is assumed initially that all messages appear at the beginning of its period. Then, we can formulate the CAN message scheduling problem in the following way:

$$\text{CANMS1}: \qquad \min \sum_{k=1}^{n} \sum_{i=1}^{m} \sum_{j=1}^{n_i} (k - T_i(j-1))(P - p_i) x_{ijk} \tag{1}$$

subject to

$$x_{ijk} = 0 \qquad 1 \le i \le m,\ 1 \le j \le n_i,\ 1 \le k \le (j-1)T_i \tag{2}$$

$$x_{ijk} = 0 \qquad 1 \le i \le m,\ 1 \le j \le n_i,\ jT_i + 1 \le k \le n \tag{3}$$

$$\sum_{k=(j-1)T_i+1}^{jT_i} x_{ijk} = 1 \qquad 1 \le i \le m,\ 1 \le j \le n_i \tag{4}$$

$$\sum_{i=1}^{m} \sum_{j=1}^{n_i} \sigma_i x_{ijk} \le T_s = 1 \qquad 1 \le k \le n \tag{5}$$

$$x_{ijk} \in \{0, 1\} \qquad 1 \le i \le m,\ 1 \le j \le n_i,\ 1 \le k \le n, \tag{6}$$

where $P = \max_i\{p_i\}$.

In this formulation, the objective function (1) focus on minimizing the total displacement of messages from the beginning of its cycle. The aim is to make messages with higher priority (smaller value of p_i) to appear before messages with lower priority. Constraints (2) and (3) state that the j-th occurrence of a message cannot appear before or after its period. The correct appearance of a message is then established by Constraint (4). The set of inequalities (5) is used to constrain the total time of messages assigned to one slot to be at most T_s. Finally, Constraints 6 are used to define the correct domain for the variables used.

The formulation above can be used to determine the best ordering of messages in an interval $[0, t]$ of time. However, if $t \ge T_{\max} = \text{lcm}_i(T_i)$ (where lcm is the least common multiplier), then it is easy to check that the solutions will repeat the pattern of the initial $T_{\max}$ period. This happens because we assume that all messages are sent exactly at the beginning of their respective periods.

Proposition 1: *If $t > T_{\max}$, where $T_{\max} = \text{lcm}_i(T_i)$, given an optimal solution from time 0 to $T_{\max}$, then there is an optimum solution from 0 to t that is a repetition of the pattern from 0 to $T_{\max}$, i.e., $x_{i,j,k} = x_{i,j,k+lT_{max}}$, for $l \in \mathrm{N}$, $1 \le l \le t/T_{\max}$.*

Proof: An optimal solution for CANMS1 can be constructed that has the stated property. Suppose that a solution with optimal cost z^* does not have this property, i.e. there is a $k < T_{max}$ such that $x_{i,j,k} \ne x_{i,j,k+lT_{max}}$, for

$l > 1$. Let k' be the smallest such k. Assuming that all messages arrive at the beginning of the period, there is a symmetry in the problem, and the set of messages waiting to be sent at time slot k' are the same as the messages waiting at time $lT_{max} + k'$, for $l > 1$. Thus, the solution given by $x_{i,j,lT_{max}+k} = x_{i,j,k}$ for $k' \leq k \leq T_{max}$ must have objective cost $z' \leq z^*$. However, z^* is the optimum and therefore $z' = z^*$. Thus, there is an optimal solution with the stated property. □

2.1.1. *Modeling Message Delays*

To make the model more realistic, we define a generalization of the given formulation that includes variations in the arrival time of messages. In this case we assume that, with each message i in its j-th appearance, there is an associated delay $d_{ij} \in \mathrm{R}$. The delay d_{ij} represents the actual instant, within the current period, when the message was sent. The modified formulation then becomes

$$\text{CANMS2}: \qquad \min \sum_{k=1}^{n} \sum_{i=1}^{m} \sum_{j=1}^{n_i} (k - T_i(j-1) - d_{ij})(P - p_i) x_{ijk} \qquad (7)$$

subject to

$$x_{ijk} = 0 \qquad 1 \leq i \leq m,\ 1 \leq j \leq n_i,\ 1 \leq k \leq (j-1)T_i \qquad (8)$$

$$x_{ijk} = 0 \qquad 1 \leq i \leq m,\ 1 \leq j \leq n_i,\ jT_i + 1 \leq k \leq n \qquad (9)$$

$$\sum_{k=(j-1)T_i+1}^{jT_i} x_{ijk} = 1 \qquad 1 \leq i \leq m,\ 1 \leq j \leq n_i \qquad (10)$$

$$\sum_{i=1}^{m} \sum_{j=1}^{n_i} \sigma_i x_{ijk} \leq T_s = 1 \qquad 1 \leq k \leq n \qquad (11)$$

$$x_{ijk} \in \{0, 1\} \qquad 1 \leq i \leq m,\ 1 \leq j \leq n_i,\ 1 \leq k \leq n \qquad (12)$$

Another problem that arises in the model CANMS1 (and CANMS2 as well) is that it may be impossible to find feasible solutions, due to Constraint (5) (the same as (11)). A third integer linear model, must be proposed to account for situations where the bandwidth is insufficient to satisfy all message requests. To see this, remember that IP formulations CANMS1 and CANMS2 require that all messages be allocated to some time slot (according to Constraint (4)). However, as the size of the slots are fixed, it is possible to have more messages than bandwidth to satisfy requests.

This may cause some instances of the problem to become infeasible, even when just a small number of messages with low priority cannot be sent. A more general goal is to allow sub-optimal schedules, with a minimum set of messages $\mathcal{L}$ that cannot be dispatched due to time constraints.

A way to solve this problem consists of using a Lagrangian relaxation of the given formulation. By relaxing Constraints (5), (11) and adding them, instead, as a penalty in the objective function, we can remove the infeasibility associated to this constraint. The objective of the formulation becomes now to schedule all messages, minimizing at the same time the delay incurred by high priority messages and the penalty incurred by using more time than what is available in the time slots. The updated ILP, which will be called CANMS3, is presented bellow:

$$\min \sum_{k=1}^{n}\sum_{i=1}^{m}\sum_{j=1}^{n_i}(k - T_i(j-1) - d_{ij})(P - p_i)x_{ijk} + \rho\sum_{k=1}^{n}(\sum_{i=1}^{m}\sum_{j=1}^{n_i}\sigma_i\, x_{ijk} - T_s) \tag{13}$$

subject to

$$x_{ijk} = 0 \qquad 1 \le i \le m,\ 1 \le j \le n_i,\ 1 \le k \le (j-1)T_i \tag{14}$$

$$x_{ijk} = 0 \qquad 1 \le i \le m,\ 1 \le j \le n_i,\ jT_i + 1 \le k \le n \tag{15}$$

$$\sum_{k=(j-1)T_i+1}^{jT_i} x_{ijk} = 1 \qquad 1 \le i \le m,\ 1 \le j \le n_i \tag{16}$$

$$x_{ijk} \in \{0,1\} \qquad 1 \le i \le m,\ 1 \le j \le n_i,\ 1 \le k \le n, \tag{17}$$

where $\rho \in R_+$ is a positive value expressing the relative importance of the feasibility constraints. Note that in this formulation, for instances with feasible optimal solutions where Constraint (5) is not tight, it is possible to have negative optimal values.

The models introduced above represent different ways of looking at the scheduling problem on CAN systems. The first model (CANMS1) is useful when we are interested in solving the ordering problem for a generic system when fluctuations in the arrival time of messages are not being considered. We call the problem defined by the first model the *cycle scheduling problem for CAN*. The second formulation gives rise to more specific instances of the scheduling problem. In this case, one must consider the exact period at which messages are delivered, which can have different values across the modeled periods. The resulting problem will be refereed to as the *message scheduling problem on CAN*. Lastly, the third model proposed is able to cope with infeasibility situations, thus it can be a useful tool for determining the amount of infeasibility of a proposed system given some practical instances. This formulation will be called the *infeasibility minimization version* of the message scheduling problem on CAN systems.

2.2. *Computational Complexity*

The CAN message scheduling problem represents the set of combinatorial decisions that must be made in real time by a CAN system. The difficulty consists of finding the exact ordering of messages with the goal of minimizing the total delay incurred. It is not surprising that the resulting problem is in the NP-hard class, as shown bellow.

Proposition 2: *The CAN message scheduling problem is NP-hard.*

Proof: The proof is by reduction from the bin packing problem, which is well known to be NP-hard [4]. In the bin packing problem, we are given a set U of objects and a set $\mathcal{B}$ of bins, each with size B, where objects can be stored. Each object $u \in U$ has a size $s(u)$. The objective of the problem is to minimize the number of bins needed to store all objects in U, i.e.,

$$\min x$$

subject to

$$i y_{ui} \leq x \qquad \text{for all } u \in U,\ b_i \in \mathcal{B} \tag{18}$$

$$\sum_{b_i \in \mathcal{B}} y_{ui} = 1 \qquad \text{for all } u \in U \tag{19}$$

$$\sum_{u \in U} s(u) y_{ui} \leq B \qquad \text{for all } b_i \in \mathcal{B} \tag{20}$$

$$y_{ui} \in \{0, 1\} \qquad \text{for all } u \in U,\ b_i \in \mathcal{B} \tag{21}$$

$$x \in \mathrm{Z}, \tag{22}$$

where y_{ui} is a decision variable such that $y_{ui} = 1$ if and only if object $u \in U$ is assigned to bin $b_i \in \mathcal{B}$.

A general instance of this problem can be readily translated into an equivalent instance of the CAN message scheduling problem. First, each object $u \in U$ corresponds to a message. Messages do not need to be periodic in this case, which is also equivalent to give a very large periodicity T_i for each message $i \in \{1, \ldots, m\}$. Second, each new bin used in the solution corresponds to a slot in the CAN system.

We claim that if the priority p_i of every message m_i is equal to 1, then minimizing the objective function (1) is equivalent to minimizing the number of bins used in the solution of the bin packing problem. This is true because, as there is no difference between priorities, messages are interchangeable in terms of their contributions to the objective function, and the goal turns into put the maximum number of messages close to the beginning of the cycle. Moreover, to minimize (1) no unused spaces capable of receiving another message can be left between allocated time slots. Thus, the slots used in the optimal solution to the CANMS problem, when taken in any order, correspond to an optimal solution to the bin packing problem. Conversely, an optimal solution to the bin packing problem can be easily translated into a solution to the CANMS problem by listing the contents of the used bins in increasing order of the number of objects contained in the bin. Thus, making $m = |U|$, $\min_i T_i = \infty$ and $p_i = 1$ for all messages $i \in \{1, \ldots, m\}$, we can reduce any instance of the bin packing problem to an equivalent instance of the CANMS problem. We conclude that the CANMS problem is NP-hard. □

3. Experimental Results

To determine the efficiency of the integer programming model proposed on the previous sections, a set of computational experiments were proposed. In the first step, we designed a set of instances of the CANMS problem that could be representative of problems encountered in real life situations. With this purpose, we generated a set of random instances, based on specifications for CAN proposed on the SAE benchmark (Table 1). In all instances the number of messages is the same, and their parameters are defined according to the SAE specifications.

3.1. *Experimental Settings*

The integer programming models were implemented using the XPress Mosel^{TM} modeling system [3], which uses its own simplex implementation. It uses a branch and bound code, with the addition of common families of cuts, such as Gomory cuts. The computer used was a PC with 512MB of main memory and a 2.7GHz processor. The instances were created with random values of the displacement parameter d_{ij}, for each message $i \in \{1, \ldots, m\}$ and occurrence $j \in \{1, \ldots, n_i\}$. For each instance, a different value for the time period was defined, ranging from 0.1 to 100 seconds.

Table 2. Result from the IP models on randomly generated instances of the CANMS.

instance	simulated time (ms)	num. of variables	execution time (s)	cost
1	100	940	0.031	-474212
2	500	4700	0.141	-2.37e+6
3	1000	10600	0.484	-5.35e+6
4	2000	21200	1.64	-1.07e+7
5	3000	31800	3.609	-1.60e+7
6	4000	42400	8.141	-2.14e+7
7	5000	53000	14.281	-2.67e+7
8	6000	63600	21.875	-3.21e+7
9	7000	74200	30.813	-3.75e+7
10	8000	84800	40.922	-4.28e+7
11	9000	95400	52.422	-4.82e+7
12	10000	106000	65.172	-5.35e+7

3.2. *Results*

Table 2 gives a summary of results found using the proposed model. The first column gives the number of the instance, while the second column shows the time of the total simulation represented by the instance. In the third column the number of variables in the formulation is given. This number shows that the size of the formulation is approximately a linear function of the total time. The fourth column gives the execution time for the linear relaxation of the integer programming model. Finally, the last column shows the optimal objective value for the linear relaxation.

The first set of instances were created using the standard definitions. However, the resulting slot size has a comparatively large number (5 ms). This resulted in instances where the solutions had enough space to place all occurrences of messages. This is reflected in the large small numbers in the solutions reported in Table 2.

A second set of instances were generated, this time with an extended collection of messages. Some of the newly added messages had very small deadlines, which induced a small slot size (0.1 ms). Running the optimization code a second time, the results showed that it was more difficult to find optimal solutions, as displayed in Table 3.

4. Concluding Remarks

A new mathematical model for hard real-time distributed systems in the form of CAN network is presented in this work. The properties of the model

Table 3. Result from the IP models on randomly generated instances of the CANMS, with slot size equal to 0.1 seconds.

instance	simulated time (ms)	num. of variables	execution time (s)	cost
1	100	940	0.038	972.4
2	500	4700	0.239	4212.9
3	1000	10600	0.532	7528.3
4	2000	21200	1.81	12551.4
5	3000	31800	3.973	22152.7
6	4000	42400	9.426	29753.0
7	5000	53000	18.962	36233.5
8	6000	63600	28.821	45231.7
9	7000	74200	37.612	52513.8
10	8000	84800	49.269	65231.7
11	9000	95400	61.823	76975.4
12	10000	106000	72.814	83016.9

are proposed, as well as an efficient method for computing near-optimal solutions. The problem of CAN message scheduling is defined as the minimization of message delays, weighted by priorities of the classes of messages in the system. The resulting problem is shown to be NP-hard, by reduction from the bin packing problem. This fact has prompted the proposal of an efficient heuristic algorithm, which can be used to give high quality solutions in practical instances.

Given the importance of CAN in many industrial areas for controlling embedded devices, it is highly desirable to have good methods giving analytical solutions for the scheduling problem. Such methods can be used to guide the development of new systems, as well as to improve the understanding of current CAN systems. The use of the model proposed here can potentially bring large improvements to the processes of designing and managing communication systems based on controller area networks.

References

[1] J. Barreiros, E. Costa, J. Fonseca, and F. Coutinho. Jitter reduction in a real-time message transmission system using genetic algorithms. In *Proceedings of the CEC 2000 - Conference of Evolutionary Computation*, 2000.

[2] A. Burns, K. Tindell, and A. Wellings. Fixed priority scheduling with deadlines prior to completion. In *Proceedings Sixth Euromicro Workshop on Real-time Systems*. IEEE Computer Society Press, 1994.

[3] Dash Optimization Inc. *Xpress-Optimizer Reference Manual*, 2003.

[4] M. R. Garey and D. S. Johnson. *Computers and Intractability: A Guide to the Theory of NP-completeness*. W.H. Freeman and Company, 1979.

[5] SAE. Class C application requirement considerations. Technical report, SAE Technical Report J2056/1, June 1993.

[6] L. Sha, J.P. Lehoczky, and R. Rajkumar. Solutions for some practical problems in prioritized pre-emptive scheduling. In *IEEE Real-time Systems Symposium*, pages 181–191, 1986.

[7] A. Stothert and I. MacLeod. Effect of timing jitter on distributed computer control system performance. In *Proceedings of DCCS'98 - IFAC Workshop on Distributed Computer Control Systems*, Sep 1998.

[8] K. Tindell and A Burns. Guaranteed message latencies for distribute safety-critical hard real-time networks. Technical report, YCS 229, Department of Computer Science, Univ. of York, 1994.

[9] K.M. Zuberi and K.G. Shin. Design and implementation of efficient message scheduling for controller area network. *IEEE Transactions on Computers*, 8(2):182–188, 2000.

CHAPTER 17

MULTIPLE RADAR PHANTOM TRACKS FROM COOPERATING VEHICLES USING RANGE-DELAY DECEPTION

Meir Pachter
Dept. of Electrical Engineering
AF Institute of Technology
Wright-Patterson AFB, OH
meir.pachter@afit.edu

Phillip R. Chandler
Air Force Research Laboratory, Air Vehicles Directorate
Wright-Patterson AFB, OH
phillip.chandler@wpafb.af.mil

Keith B. Purvis
Dept. of Mechanical Engineering
UC-Santa Barbara
Santa Barbara, CA

Scott D. Waun
Dept. of Electrical Engineering
Ohio State University
Columbus, OH

Reid A. Larson, 2d Lt, USAF
Air Force Research Laboratory, Air Vehicles Directorate
Wright-Patterson AFB, OH
reid.larson@wpafb.af.mil

Multiple cooperating Electronic Combat Air Vehicles (ECAV) are used to generate phantom radar tracks in a multiple radar air defense network. The vehicles use a range delay deception transponder, which delays the radar pulses received by the ECAV and sends them back to the radar. This results in the radar calculating an erroneous target range. A radar network will correlate tracks to identify phantom targets. The ECAV team, however, precisely positions and dynamically coordinates the motion of the vehicles so that all radars see the same phantom track. This chapter presents the two-dimensional mathematical relationships between the motion of the vehicle and the motion of the phantom track. Phantom tracks investigated include: a) constant heading and constant velocity; b) circular trajectory with constant velocity; and c) arbitrary trajectory and arbitrary velocity. Closed form solutions are obtained for the ECAV trajectory given a specified phantom track. Parametric analyses are performed with constraints on the vehicle dynamics, constraints on the transponder, vehicle initial position and state, and phantom track state. Results are presented for a single vehicle and a single radar, and for up to four vehicles generating a single coherent phantom track for up to four radars correlating returns. The vehicles must tightly coordinate their highly coupled actions to generate an effective phantom track using range delay deception. Also addressed is the generation of multiple phantom tracks through exploitation of sidelobes in a radar's antenna pattern.

1. Introduction

This chapter addresses the control of Electronic Combat Air Vehicles (ECAVs) generating phantom tracks using range delay deception. In range delay deception, the received pulse from a radar is delayed by the ECAV and retransmitted back to the targeted radar. This causes the victim radar to calculate an erroneous target range. Multiple tracking radars cooperate in a defense network by correlating the targets' tracks, as a counter-measure to range delay deception. As a counter-counter measure, multiple ECAVs cooperate in deceiving the networked radars by generating a coherent phantom track. This chapter addresses the cooperative control of ECAVs using range delay deception against a track-correlating radar network.

Figure 1 shows the phantom track scenario. In this example, there are four radars that share track files around the network. There are four ECAVs, one for each radar. At time t_1, the ECAVs are in the radars' line of sight to the phantom target's position T_1. The radar pulses are delayed by the ECAVs so that the perceived range vectors all intersect at T_1. The ECAVs are repositioned to continuously stay in the radars' line of sight at t_2. This generates the phantom track of the desired speed and heading. Since each of the radars confirms the other's target track, the track is considered valid.

A formal description of range-delay based deception is given. Using this

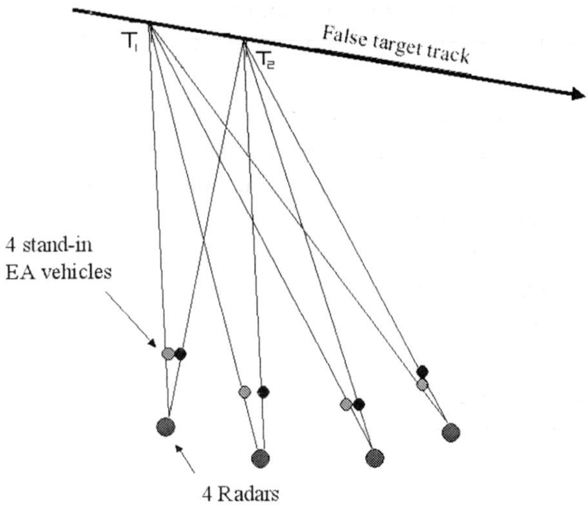

Fig. 1. Representation of multiple ECAVs deceiving an integrated radar network by generating a single, coherent phantom track using range delay deception.

description of the problem, generalized mathematical formulations are made based on the kinematics of a single vehicle generating a phantom track against a single radar. The formulas are extended to two specific cases of interest: a) an ECAV generating a phantom track of constant heading and velocity; and b) an ECAV generating a phantom track of circular trajectory. Additionally, numerical equations are developed to conveniently address an ECAV creating a phantom track of arbitrary heading and velocity, which leads to the initial development of multiple ECAV's cooperatively forming a single, coherent phantom target and track. Finally, this investigation ends with a discussion of a single vehicle generating multiple phantom targets by exploitation of sidelobes in a radar's antenna pattern among a network of integrated radars.

2. Range Delay-Based Deception

Here, we describe in more detail how an ECAV would deceive a radar, or network of radars, using range-delay deception techniques. The ECAV employs a transponder or a repeater for false target generation. The former consists of a receiver, a variable delay circuit, a signal generator, a power amplifier, and an antenna. Upon receiving a pulse from a threat radar, the transponder waits for a period corresponding to the desired additional range of the false target; then, transmits back to the radar an internally generated signal simulating a target echo. A repeater for generating false targets generally includes a memory, enabling it to produce much more realistic "targets". The digital RF memory stores the actual pulse received from the radar. After the desired delay, the pulse is read out, amplified, and transmitted back to the radar.

Using either a repeater or a transponder, and by making the time delay longer than the radar's interpulse period, false targets may be made to appear on the victim's radar at greater ranges than that of the ECAV.

Range delay deception causes the tracking radar to calculate an erroneous target range R. Hence, a phantom target is presented to the victim radar. Moreover, the tracking radar invariably assumes that the phantom target is in its main lobe. Hence, if the ECAV is in the radar's main lobe, the phantom target will be positioned on the Line Of Sight (LOS) from the ECAV to the radar. Thus, the phantom target is placed on a radial connecting the radar and the ECAV and both the ECAV and the phantom target share the same bearing θ.

Delaying the returned pulse causes the phantom target to be on the LOS to the radar, further away from the ECAV: $R > r$, where r is the ECAV's distance from the radar. This places a constraint on the ECAV's range r. If however the radar is not pulse-to-pulse agile, the ECAV also has the option of advancing the returned pulse, thus placing the phantom target between the ECAV and the radar such that $R < r$. In this case r is not constrained and moreover, in view of the "radar equation", the ECAV could then operate outside the operational range of the radar, such that $r > R_{max}$. The amount of "delay" generated by the ECAV electronics is bounded by the known Pulse Repetition Frequency (PRF) of the victim radar; the PRF, in turn, is commensurate with the radar's power-constrained maximum range R_{max}. The phantom target is invariably placed in the operational envelope of the radar, i.e., $0 < R < R_{max}$. Over time, an ECAV engaged in range delay deception will generate a phantom target track. The phantom target's track is determined by both the range "delay" action and the position of the ECAV relative to the targeted radar because both the ECAV and the phantom target have the same bearing. The ECAV's motion and the range "delay" determine the phantom target's track.

3. Single ECAV, Single Radar Engagement

As an initial step toward the goal of cooperatively deceiving a network of radars, we address the one-on-one engagement of a single ECAV spoofing a single radar. The engagement is initially treated analytically by resolving the vehicle and phantom kinematics using two methods: (1) solution of the "direct problem", i.e., calculate the phantom target trajectory based on a pre-determined ECAV trajectory; and (2) solution of the "inverse problem", i.e., calculate the ECAV trajectory based on a pre-determined phantom target trajectory.

3.1. *The Direct Problem*

Here, the ECAV is tasked with creating a phantom target track within the operational range of a single radar. Refer to Figure 2. The ECAV (E) has "simple motion", viz., its speed v_E is constant and its control, the course $\phi_E(t)$, is piecewise continuously differentiable. This is consistent with controlling the ECAV's turning rate. More elaborate kinematic models, e.g., the imposition of an upper bound on the ECAV's turning rate and/or the control of its speed $v_E(t)$, are considered later. Using polar coordinates for convenience, we locate the victim radar at the origin. The range of the ECAV from the radar is r and its bearing is θ; without loss of generality, we set $\theta(0) = 0$ and we also assume $\frac{d\theta}{dt}(0) > 0$. The range of the phantom target T is R ($\leq R_{max}$). Moreover, the ECAV also controls the phantom target's range rate $\dot{R}$. Thus, the state variables are r, θ, and R, and the (ECAV's) control variables are $\phi_E(t) \in [0, 2\pi)$ and $v(t) \in R$.

The relationship between the ECAV's trajectory and the ECAV-generated phantom target's trajectory is now elucidated. In polar coordinates, the ECAV's trajectory is given by the functions $r(t)$ and $\theta(t)$. The phantom target's track is given by the functions $R(t)$ and $\theta(t)$. Alternatively, one might be interested in the phantom target's speed v_T and its course ϕ_T - see, e.g., Figure 2. The phantom target's track is then determined by $v_T(t)$, $\phi_T(t)$, $0 \leq t$, and the initial conditions $\theta(0) = 0$ and $R(0) = R_0$ - the phantom target's initial range, R_0, being set by the ECAV. Thus, the output variables are $R(t)$ and $\theta(t)$, or, $v_T(t)$ and $\phi_T(t)$.

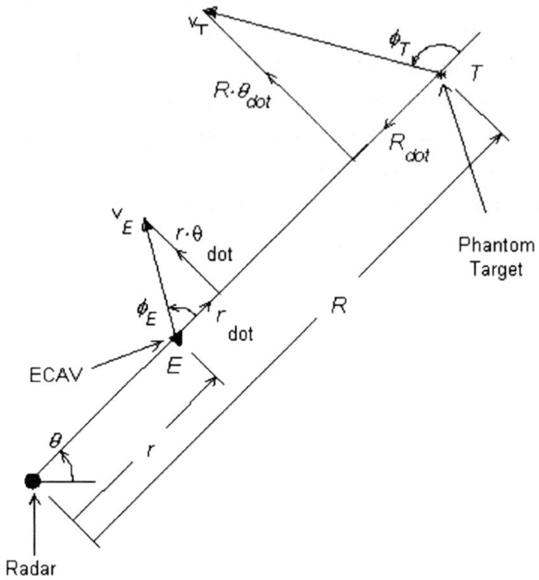

Fig. 2. Kinematic representation of the single ECAV, single radar engagement.

It is convenient to use non-dimensional variables:

$$
\begin{aligned}
r &\rightarrow \frac{r}{R_0} \\
R &\rightarrow \frac{R}{R_0} \\
R_{max} &\rightarrow \frac{R_{max}}{R_0} \\
v &\rightarrow \frac{v}{v_E} \\
t &\rightarrow \frac{v_E}{R_0} t \\
\alpha &\rightarrow \frac{v_T}{v_E}
\end{aligned}
$$

The non-dimensional variables t, r, R and α are non-negative and $R_{max} > 1$. The phantom target's speed is now quantified by the speed ratio α, which is time - dependent.

The equations of motion of the ECAV are

$$\dot{r} = \cos\phi_E \;, \qquad r(0) = r_0 \tag{1}$$

$$\dot{\theta} = \frac{1}{r}\sin\phi_E \;, \quad \theta(0) = 0 \tag{2}$$

and the phantom target's range is governed by

$$\dot{R} = v \;, \quad R(0) = 1 \tag{3}$$

The ECAV's controls v and ϕ_E determine the phantom target's track. Indeed, $r(t)$ and $\theta(t)$ are governed by the control $\phi_E(t)$, and are given by the solutions of eqs. (1) and (2), respectively. $R(t)$ is governed by the control $v(t)$, and is given by the solution of eq. (3). Hence, the phantom target's track $(R(t),\ \theta(t))$ is completely characterized in terms of the ECAV's controls $\phi_E(t)$ and $v(t)$. Strictly speaking, the kinematics of range delay-based deception are modeled as a nonlinear control system in terms of a three states, two inputs, and two outputs. The system's states are r, θ and R; the controls are ϕ_E and v; and the outputs are R and θ. Thus, the equations of motion of the phantom target are (1) - (3) and the ECAV's controls $v(t)$ and $\phi_E(t)$ determine the phantom target's trajectory $R(t)$ and $\theta(t)$.

3.2. *The Inverse Problem*

While the direct problem can be useful, in general we are more interested in the inverse problem, where it is required to synthesize an ECAV trajectory on the time interval $0 \leq t \leq t_f$, $\forall t$, for a specified phantom trajectory.

Suppose that the phantom target's path is given in parametric form: $R(t)$ and $\theta(t)$ are specified $\forall\ t$, $0 \leq t \leq t_f$. It is required to determine the ECAV controls $v(t)$ and $\phi_E(t)$, and the ECAV's trajectory $r(t)$ and/or $r(\theta)$.

From eq. (3) we directly calculate

$$v(t) = \frac{dR}{dt} \tag{4}$$

From eq. (2) we obtain

$$\sin\phi_E = r\frac{d\theta}{dt} \tag{5}$$

so that

$$\cos\phi_E = \sqrt{1 - (\frac{d\theta}{dt})^2 r^2} \tag{6}$$

or

$$\cos\phi_E = -\sqrt{1 - (\frac{d\theta}{dt})^2 r^2} \tag{7}$$

Inserting eqs. (6) or (7) into eq. (1) then yields the differential equations

$$\dot{r} = \sqrt{1 - (\frac{d\theta}{dt})^2 r^2}\ , \quad r(0) = r_0\ , \quad 0 \le t \tag{8}$$

or

$$\dot{r} = -\sqrt{1 - (\frac{d\theta}{dt})^2 r^2}\ , \quad r(0) = r_0\ , \quad 0 \le t\ , \tag{9}$$

respectively. Evidently, the existence of an ECAV trajectory on a time interval $0 \le t \le t_s$, where $t_s \le t_f$, requires that the ECAV's initial range satisfies

$$r_0 < 1/\frac{d\theta}{dt}(0) \tag{10}$$

The construction of an ECAV trajectory entails the solution of the differential equations (8) and/or (9). One may begin solving either the differential equation (8), in which case r is monotonically increasing and

$$0 \le \phi_E \le \frac{\pi}{2} \ \ if \ \ \frac{d\theta}{dt} \ge 0 \qquad and \qquad \frac{3\pi}{2} \le \phi_E \le 2\pi \ \ if \ \ \frac{d\theta}{dt} < 0\ ;$$

or, one might solve the differential equation (9), in which case r is monotonically decreasing and

$$\pi \ge \phi_E \ge \frac{\pi}{2} \ \ if \ \ \frac{d\theta}{dt} \ge 0 \qquad and \qquad \frac{3\pi}{2} \ge \phi_E > \pi \ \ if \ \ \frac{d\theta}{dt} < 0$$

Note that since $\frac{d\theta}{dt}(0) > 0$, if one initially embarks on solving eq. (8) then $0 \le \phi_E(0) \le \frac{\pi}{2}$ and if one initially solves eq. (9) then $\pi \ge \phi_E(0) \ge \frac{\pi}{2}$.

Since $\phi_E(t)$ is continuous, as long as the solution $r(t)$ satisfies the condition

$$\mid \frac{d\theta}{dt}(t) \mid \cdot r(t) \le 1\ , \tag{11}$$

one stays with the differential equation under consideration. Condition (10) and the hypothesis $\frac{d\theta}{dt}(0) > 0$ guarantee that initially, condition (11) holds. In other words, there exists a time interval $0 \le t \le t_s$, $0 < t_s$, for which an initial ECAV trajectory segment can be constructed. Obviously, we are interested in extending t_s s.t. $t_s = t_f$.

We have no control over the function $\mid \frac{d\theta}{dt}(t) \mid$, for the latter is specified. At the same time, the positive function $r(t)$ given by the solution of the differential equations (8) or (9) is, in part, determined by our course of action: choosing to integrate the differential equation (9) and thus reducing $r(t)$, helps us meet the requirement (11) - provided, of course, that $r(t)$ is not reduced to zero. Hence, if during the integration of eq. (8), at some time $t = t_s < t_f$

$$\mid \frac{d\theta}{dt}(t_s) \mid \cdot r(t_s) = 1 \ ,$$

switching to the solution of the differential eq. (9) may be warranted.

The switching action is now analyzed. Specifically, the switching function

$$s(t) \ \equiv \ 1 - (\frac{d\theta}{dt})^2 r^2 \quad (\leq 1)$$

is defined. Using the $s(t)$ definition, we see that

$$\dot{r} = \pm\sqrt{s(t)}$$

and we therefore require

$$s(t) \geq 0 \quad \forall\, t \ , \quad 0 \leq t \leq t_f$$

Now, condition (10) renders the switching function positive at $t = 0$:

$$s(0) > 0$$

and therefore there exists a time interval $0 \leq t \leq t_s$ s.t. $s(t) > 0 \ \forall\ t$, $0 \leq t < t_s$.

Attention is now given to the case where, upon integrating the differential eqs. (8) or (9), the switching function vanishes at time $t = t_s$ and $t_s < t_f$. Let t_s be the first time instant where the switching function vanishes. Thus

$$s(t_s) = 0 \ ,$$

i.e.,

$$\frac{d\theta}{dt}(t_s) \neq 0$$

and

$$r(t_s) = 1/ \mid \frac{d\theta}{dt}(t_s) \mid \tag{12}$$

Now, at $t = t_s$, $\frac{d\theta}{dt}(t_s) \neq 0$. Assume $\frac{d\theta}{dt}(t_s) > 0$. The continuity of $\phi_E(t)$ renders that $\frac{d\theta}{dt}(t)$ is continuous. Thus, also $\frac{d\theta}{dt}(t_s^-) > 0$. Hence, if initially the differential equation being integrated is eq. (8), then $\frac{d\theta}{dt}(t_s^-) > 0$ and eqs. (5) and (6) yield the control $\phi_E(t_s^-) = \frac{\pi}{2} - \epsilon$, where ϵ is a small positive number. Assume that at $t = t_s$ one switches to the integration of the differential eq. (9). The continuity of $\frac{d\theta}{dt}(t)$ implies that also $\frac{d\theta}{dt}(t_s^+) > 0$.

Hence, if the differential eq. (9) is integrated then $\frac{d\theta}{dt}(t_s^+) > 0$ and eqs. (5) and (7) yield the control $\phi_E(t_s^+) = \frac{\pi}{2} + \epsilon$, where ϵ is a small positive number. Hence, if $\frac{d\theta}{dt}(t_s) > 0$ and at $t = t_s$ one switches from the integration of eq. (8) to the integration of eq. (9), the control $\phi_E(t)$ is continuous, monotonically increasing and $\phi_E(t_s) = \frac{\pi}{2}$. Similarly, if $\frac{d\theta}{dt}(t_s) > 0$ and at $t = t_s$ one switches from eq. (9) to eq. (8), the control $\phi_E(t)$ is continuous, monotonically decreasing and $\phi_E(t_s) = \frac{\pi}{2}$. If $\frac{d\theta}{dt}(t_s) < 0$ and at $t = t_s$ one switches from eq. (8) to eq. (9), the control $\phi_E(t)$ is continuous, is monotonically decreasing, and $\phi_E(t_s) = \frac{3\pi}{2}$. If $\frac{d\theta}{dt}(t_s) < 0$ and at $t = t_s$ one switches from eq. (9) to eq. (8), the control $\phi_E(t)$ is continuous, is monotonically increasing, and $\phi_E(t_s) = \frac{3\pi}{2}$. Thus, we have shown that during a switch, at $t = t_s$, the continuity of the control $\phi_E(t)$ is preserved.

4. Phantom Target Scenarios

Having formulated the general equations of motion for the one-on-one ECAV and radar engagement, two specific cases are addressed: a) a phantom target with constant velocity and constant heading; and b) a phantom target with constant velocity and circular trajectory.

4.1. *Constant Heading, Constant Velocity Phantom Target*

We now investigate the ECAV maneuver required for generating a phantom target which appears to hold a constant course and a constant speed. Thus,

$$\alpha(t) = \alpha$$

and, without loss of generality, the phantom target's course is $0 < \psi < \pi$. A kinematic diagram is illustrated in Figure 3. We confine our attention to the time interval $0 \leq t \leq t_f$, where

$$t_f < \frac{1}{\alpha}(\cos\psi + \sqrt{R_{max}^2 - \sin^2\psi})$$

Given the phantom target's track, we need to calculate $\dot{\theta}(t)$. To this end we solve the $\triangle\, OT_0T$ (ref. Figure 3) and we calculate

$$R(t) = \sqrt{1 + \alpha^2 t^2 - 2\alpha t \cos\psi} \tag{13}$$

$$\cos\theta = \frac{1 - \alpha t \cos\psi}{R} \tag{14}$$

$$\sin\theta = \frac{\alpha t \sin\psi}{R} \tag{15}$$

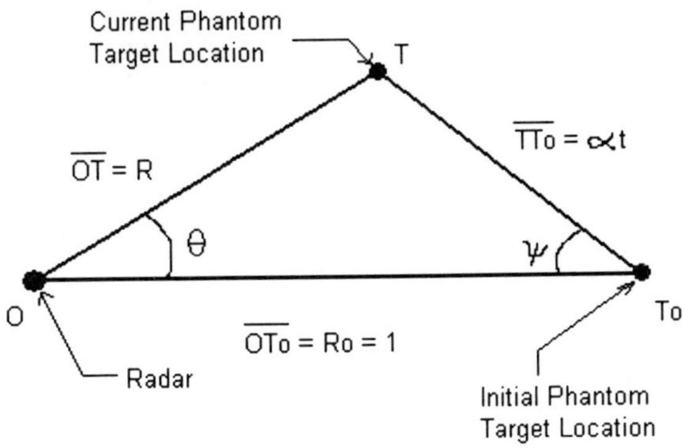

Fig. 3. $\triangle OT_0T$ represents the geometric relation between the radar and the locations of the phantom target at time $t_0 = 0$ (Point T_0) and time t (Point T). The segment $\overline{TT_0}$ represents the constant course phantom track.

Differentiating eq. (15) we obtain

$$\cos\theta\ \dot{\theta} = \alpha \sin\psi\left(\frac{1}{R} - t\frac{\dot{R}}{R^2}\right) \tag{16}$$

Differentiating eq. (13) yields

$$\dot{R} = \frac{\alpha}{R}(\alpha t - \cos\psi) \tag{17}$$

and inserting eqs. (13), (14) and (17) into eq. (16) gives

$$\dot{\theta}(t) = \frac{\alpha \sin\psi}{1 + \alpha^2 t^2 - 2\alpha t \cos\psi} \tag{18}$$

The expression (18) is inserted into eqs. (8) and (9). We obtain the differential equations for the ECAV's path

$$\dot{r} = \sqrt{1 - \left(\frac{\alpha \sin\psi}{1 + \alpha^2 t^2 - 2\alpha t \cos\psi}\right)^2 r^2} \tag{19}$$

and

$$\dot{r} = -\sqrt{1 - \left(\frac{\alpha \sin\psi}{1 + \alpha^2 t^2 - 2\alpha t \cos\psi}\right)^2 r^2} \tag{20}$$

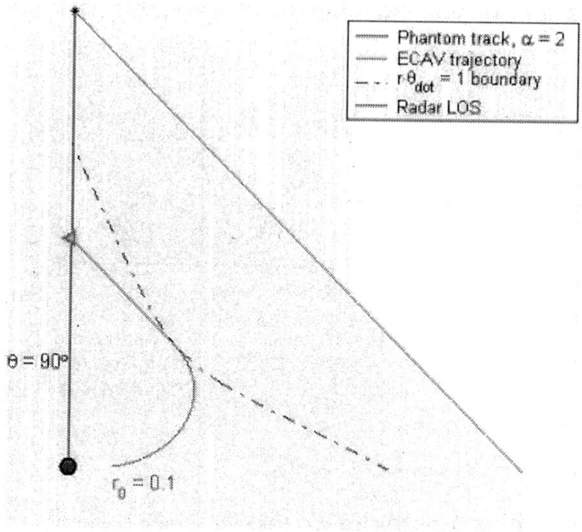

Fig. 4. ECAV trajectory for generating a constant velocity, constant course phantom track. In this example, the victim radar follows the ECAV for a 90-degree sector of its tracking area. The ECAV begins at position $r_0 = 0.1(\alpha = 2)$.

and we proceed to establish the solution $r(t)$.

Two solutions of the series of differential equations (18)-(20) are graphically represented in Figures 4 and 5. The ECAV is given a different starting point in each of the two examples; $r_0 = 0.1$ and $r_0 = 0.5$ respectively. In each case, the ECAV approaches the $r\dot{\theta} = 1$ boundary. We recall from equations (8) and (9) that when $r\dot{\theta}$ is greater than one, the differential equation(s) transition into the imaginary realm and thus cannot be used to determine the ECAV's trajectory. In each figure, the $r\dot{\theta}$ boundary is represented by a dotted line. If the ECAV's trajectory intersects with this boundary, either the switching condition (switching between solving eqn. (8) to (9) or vice versa) must be employed or the phantom track will be compromised. In both Figures 4 and 5, it can be observed that the ECAV touched the boundary. The switching condition was immediately applied in each simulation, allowing the phantom target tracks to remain intact.

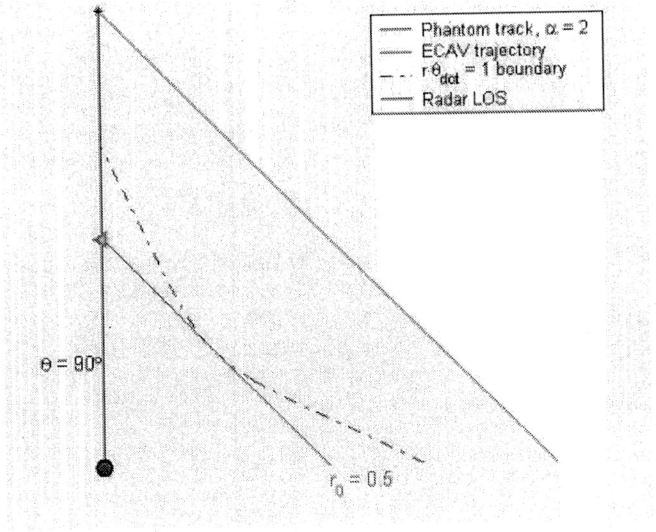

Fig. 5. ECAV trajectory for generating a constant velocity, constant course phantom track. In this example, the victim radar tracks the ECAV for a 90-degree sector of its tracking area. The ECAV begins at position $r_0 = 0.5(\alpha = 2)$. In general, an ECAV will have a trajectory parallel to the constant heading phantom track if $r_0 = 1/\alpha$.

4.2. *Circular Trajectory, Constant Velocity Phantom Target*

We now consider the special case of the ECAV maneuver required for generating a phantom target which holds a constant speed v_T and which follows a circular path of radius R_0, centered at the radar's position.

$$R(t) \equiv 1 \tag{21}$$

and

$$\alpha \equiv \frac{v_T}{v_E} \ (= const.) \ \ \forall\, 0 \leq t \tag{22}$$

Obviously,

$$v \equiv 0 \tag{23}$$

and

$$\phi_T = \frac{\pi}{2}$$

Also,

$$\frac{d\theta}{dt} \equiv \alpha \ (= const.) \ \ \forall\, 0 \leq t \tag{24}$$

Hence, the differential equations (8) and (9) are autonomous:

$$\dot{r} = \sqrt{1-\alpha^2 r^2}\ , \quad r(0) = r_0\ , \quad 0 \leq t \tag{25}$$

or

$$\dot{r} = -\sqrt{1-\alpha^2 r^2}\ , \quad r(0) = r_0\ , \quad 0 \leq t \tag{26}$$

Evidently, the following must hold:

$$\alpha < \frac{1}{r_0}\ , \tag{27}$$

whereupon the solution of the differential equation (24) is

$$r(t) = \frac{1}{\alpha}\sin(\alpha t + \arcsin(\alpha r_0)) \tag{28}$$

Hence, we calculate

$$\phi_E(t) = \alpha t + \arcsin(\alpha r_0) \tag{29}$$

Finally, the ECAV's trajectory in polar coordinates is

$$r(\theta) = \alpha \sin(\theta + \arcsin(\frac{1}{\alpha} r_0)) \tag{30}$$

and, in dimensional variables,

$$\phi_E = \theta + \arcsin(\frac{v_T}{v_E}\frac{r_0}{R_0})) \tag{31}$$

and

$$r(\theta) = \frac{v_E}{v_T} R_0 \sin(\theta + \arcsin(\frac{v_T}{v_E}\frac{r_0}{R_0})) \ , \tag{32}$$

provided that

$$\frac{r_0}{v_E} < \frac{R_0}{v_T} \ ; \tag{33}$$

i.e., the ECAV can "reach" the radar before the phantom target can reach the radar.

4.3. *Flyable Regions*

In practice, it is desirable to give an ECAV flexibility in making adjustments to its flight path, especially in heading and velocity. These adjustments would need to be made without compromising the integrity of the phantom track. Therefore, there is incentive to compute ECAV "flyable ranges" that allow for minor variations to the velocity, heading, and range of an ECAV with respect to a radar.

An added assumption is made to the ECAV: operational range of the range delay deception antenna. For purposes of this discussion, it is assumed that the deception antennas are positioned and operate on the sides of the ECAV. Thus, imagine an axis perpendicular to $\vec{v}_E(t)$, the vector defining the velocity of the ECAV, that connects the antennas on the sides of the vehicle. We now say that the antennas are effective at an angle $\pm\phi$ to the axis (on both sides of the vehicle). For example, if we set $\phi = 60°$, then each antenna covers 120° on each side of the ECAV ($2 * 60° = 120°$) while 60° to the front of the ECAV and 60° to the rear cannot project phantom pulses ($60° + 120° + 60° + 120° = 360°$). The same analogy would apply to any arbitrary value of ϕ. Of course, for the range delay deception to be effective, the targeted radar must be positioned within one of the antennas' ranges of operation. If the radar was not positioned within the antenna's range, the phantom target track would be compromised.

The equation that defines the magnitude of $\vec{v}_E$ is as follows:

$$v_E = \sqrt{\dot{r}^2 + (r\dot{\theta})^2} \tag{34}$$

Recall that in our previous analyses, v_E was originally set equal to one. Without varying the ratio of the velocity of the phantom target to that of the ECAV (α), we now allow v_E to vary in magnitude. The choice of values is rather arbitrary, however, certain values of v_E can become incompatible with the overall deception technique. Thus, the ECAV now has increased flexibility in its velocity (it is not constrained to a single value of v_E). Additionally, changing the value of v_E will affect the location of the $r\dot{\theta}$ boundary. Thus, the ECAV can also alter its heading more readily (the $r\dot{\theta}$ boundary is now modular with respect to the ECAV). It is noted that while minor fluctuations in heading can be tolerated, a sharp change in heading may not be possible due to inherent flight dynamics. Also noteworthy is the fact that while the ratio α is unchanged, varying the velocity of the ECAV will also affect the velocity of the phantom target. The phantom target will remain on its constant heading, yet its velocity will fluctuate to meet the α requirement.

Upon incorporation of the assumptions of variable v_E and operational antenna range into the ECAV trajectory algorithms, it is possible to map flyable regions for an ECAV-radar engagement. Refer to Figure 6. This figure displays a typical map of flyable regions for an ECAV under the assumptions made above. In addition to the flyable region (labeled in gray), there are two other notable regions of interest: the "black hole" and "off-limits" regions. The "black hole" represents a region that can be entered by an ECAV, but not exited without compromising the phantom track. To attempt an exit from the black hole region would cause one or more of the following: (1) violation of equation (34); (2) interception of the $r\dot{\theta}$ boundary; or (3) an abrupt change in heading that would position the radar outside the deception antenna's operational range. The "off-limits" region is such that an ECAV can exit the section but not enter. The argument supporting the off-limits region is analogous to that used to describe the characteristics of the black hole. Figure 7 is an example of a map for a circular phantom track. These maps were validated through experimentation after the appropriate adjustments to the simulation were made.

5. The Forward Problem

The previous sections dealt with solving the inverse problem in which phantom trajectories are given and the ECAV trajectory with respect to time is calculated. The focus now shifts to the solution of a trajectory in which an arbitrary trajectory and velocity profile has been pre-determined for either the phantom track or the ECAV. Depending on which is given, the other trajectory profile will be solved with respect to time. This case is solved numerically and is well-suited to transition to the case of multiple ECAVs co-

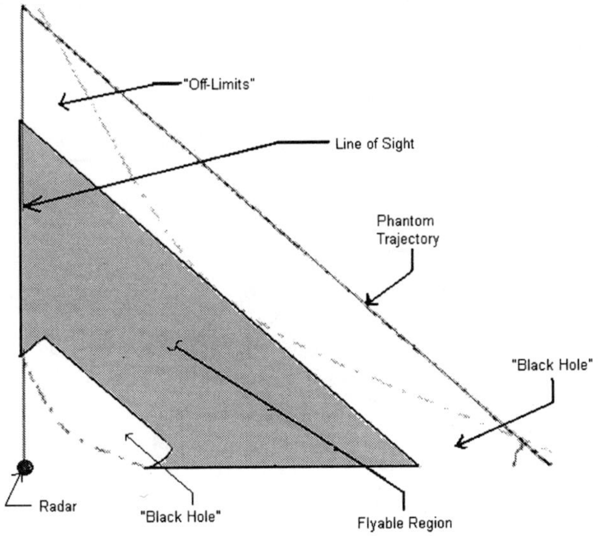

Fig. 6. A visual representation of the flyable regions for a constant velocity, constant heading phantom track. The parameters in this example are as follows: $\alpha = 2; \psi = 45°; v_E = 0.67 \rightarrow 1.5$.

operatively generating a single, coherent phantom track in a radar network. Refer to Figure 8 for the kinematic diagram accompanying this discussion. Note that while most of the variables referred to in this section have the same name and designation as those in previous sections, their respective reference frames are not necessarily identical. Use caution when comparing variables discussed previously with those discussed from this point forward. This section does not account for antenna operating ranges or any possible flight dynamics (minimum velocity, turn radius, etc.).

We begin by allowing for an arbitrary location of a radar with respect to a global origin. $\hat{x}$ and $\hat{y}$ represent the coordinates of an object (ECAV E or phantom T) with respect to a particular radar.

$$\hat{x} = x - x_r$$
$$\hat{y} = y - y_r$$

We now outline a phantom target trajectory with respect to time and solve for the ECAV flight path. We begin by stating velocity and head-

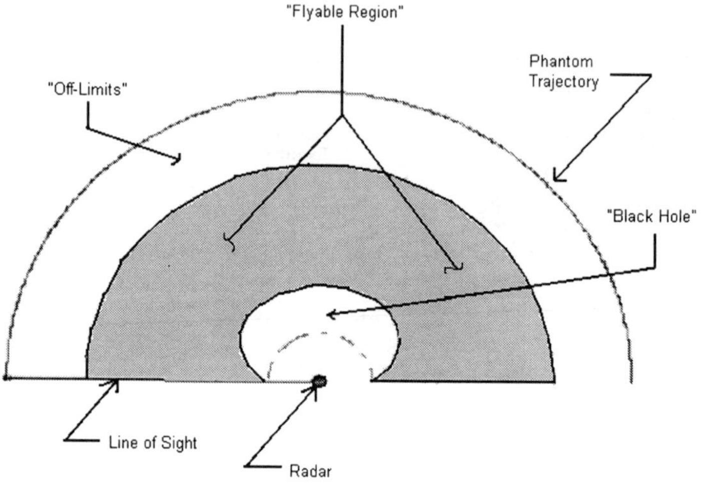

Fig. 7. A visual representation of the flyable regions for a phantom track with a circular trajectory. The parameters in this example are as follows: $\alpha = 2; v_E = 0.67 \rightarrow 1.5$.

ing profiles of the phantom target with respect to time, $V_T(t)$ and $\phi_T(t)$, respectively, and integrate them over time $t = 0$ to $t = t$.

$$\hat{x}_T(t) = \int_0^t V_T(T) \cos \phi_T(T) dT \tag{35}$$

$$\hat{y}_T(t) = \int_0^t V_T(T) \sin \phi_T(T) dT \tag{36}$$

We now relate the ECAV's position with respect to time through the geometry of the system. Recall that the phantom vehicle and the ECAV remain in the radar's line of sight.

$$R(t) = \sqrt{\hat{x}_T^2 + \hat{y}_T^2}$$

$$\beta(t) = \frac{r(t)}{R(t)}$$

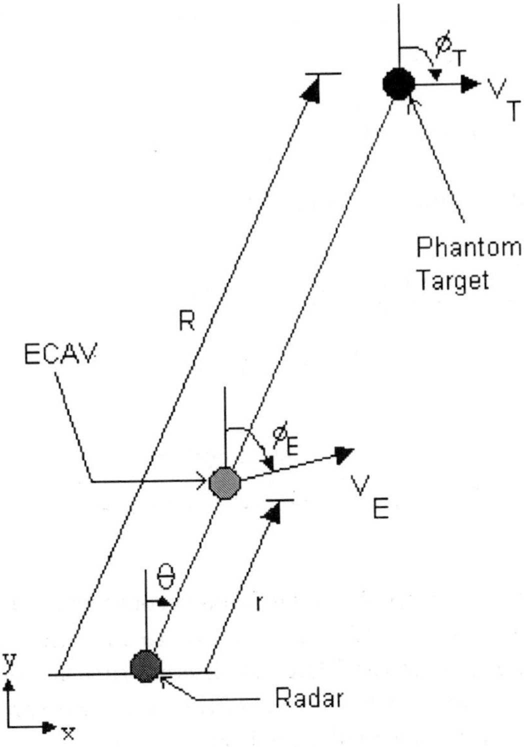

Fig. 8. Re-define the kinematics from Figure 2 to accommodate a numerical formulation for arbitrary velocity and heading of the ECAV and phantom track.

$$\theta(t) = \arctan(\hat{x}_T(t)/\hat{y}_T(t))$$

$$\hat{x}_E(t) = r(t) \sin \theta(t) = \beta(t)\hat{x}_T(t) \tag{37}$$

$$\hat{y}_E(t) = r(t) \cos \theta(t) = \beta(t)\hat{y}_T(t) \tag{38}$$

We rewrite $\hat{x}_E$ and $\hat{y}_E$ in terms of the ECAV velocity and heading; each equivalent to equations (37) and (38) respectively:

$$\hat{x}_E(t) = \int_0^t V_E(t) \sin \phi_E(T) dT \tag{39}$$

$$\hat{y}_E(t) = \int_0^t V_E(t) \cos \phi_E(T) dT \tag{40}$$

Through algebraic/numeric manipulation, we write:

$$\beta(t) = \frac{\int_0^t V_E(T) \sin \phi_E(T) dT}{\hat{x}_T(t)} = \frac{\int_0^t V_E(T) \cos \phi_E(T) dT}{\hat{y}_T(t)}$$

$$\hat{y}_T^{i+1}(\hat{x}_E^i + V_E^i \sin \phi_E^i \Delta t) = \hat{x}_T^{i+1}(\hat{y}_E^i + V_E^i \cos \phi_E^i \Delta t)$$

$$V_E^i(\phi_E^i) = \frac{\hat{x}_T^{i+1} \hat{y}_E^i - \hat{y}_T^{i+1} \hat{x}_E^i}{(\hat{y}_T^{i+1} \sin \phi_E^i - \hat{x}_T^{i+1} \cos \phi_E^i) \Delta t} \tag{41}$$

Thus, V_E is in terms of ϕ_E. A similar formula could be derived to determine V_T in terms of ϕ_T. In order for equation (41) to be of use, we do require knowledge of the position of the ECAV at discrete time increments. This equation can be modified to solve for any of the numerical variables based on the requirements of the particular application. A more likely scenario is that the ECAV velocity and heading is more concrete, and the position of the ECAV at each time increment needs to be resolved.

Having derived the numerical mathematics behind a single vehicle generating an arbitrary phantom track, we are now prepared to address n ECAV's producing a single, coherent phantom target. Mathematically, we set the phantom target coordinates, trajectory, and heading equal for all n ECAVs. We then determine the location of each of the n victim radars and solve the equations accordingly. The result is shown graphically in Figure 9, as generated using a MATLAB simulation. In this case, there are four radars and four ECAVs, each of which has a straight-line trajectory but varying speed. The false target track in the figure has multiple heading and velocity changes.

6. Multiple Phantom Tracks

Multiple phantom tracks is an area of interest as well as having less ECAVs available than tracking radars. One method of generating multiple tracks

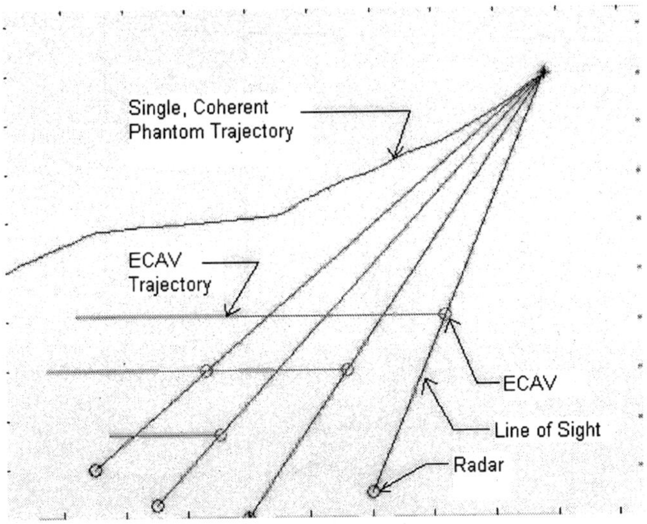

Fig. 9. A single, coherent phantom track is generated through close coordination between four ECAV's spoofing four radars.

is to use a range delay repeater or transponder to generate multiple time delays that could be interpreted by the radar as multiple targets along the same azimuth as the ECAV. Another method would be to have each ECAV generate a single, unique phantom track. These methods are somewhat trivial because they ignore the ability of radar networks to correlate tracks.

An alternative method to producing multiple phantom tracks is to exploit radar sidelobes from a phased array antenna. The antenna produces a radiation pattern consisting of the mainlobe along with residual radiation sidelobes, illustrated in Figure 10. The fact that a sidelobe emits a signal means that a receiver can also register a return from an object in the sidelobe as well, although the return is of much lower intensity than that of a return from the mainlobe.

It is conceivable that an ECAV can send a higher energy pulse in the direction of one or more of the residual sidelobes. The receiver from the radar registers a "return" of energy comparable to the energy emitted from the mainlobe. Thus, the radar interprets the return as being from an object in the mainlobe at the azimuth of the line-of-sight. In fact, the ECAV has created a phantom in the mainlobe line-of-sight while the ECAV is positioned at some angle θ off the main beam. By doing so, an ECAV deceives the radar using angle deception instead of range deception. If range

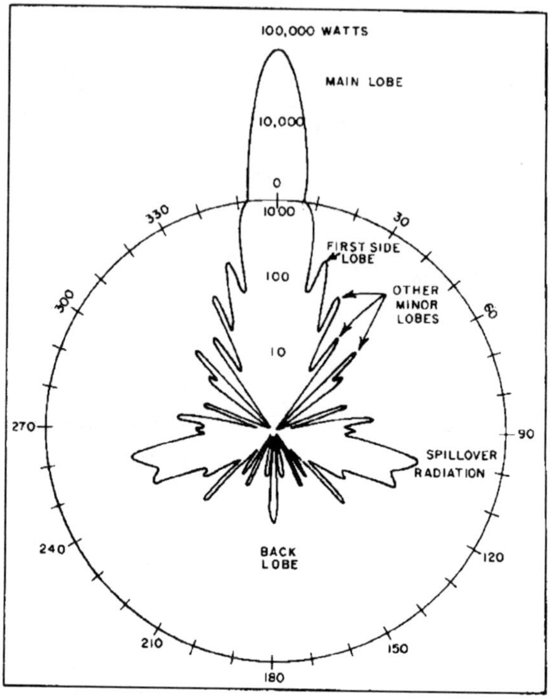

Fig. 10. Typical radar antenna pattern with apparent main lobe and residual side lobes.

delay is incorporated as well, the ECAV deceives a radar with false azimuth and range information simultaneously. We are able to visualize this sidelobe deception technique in the case of one ECAV and two radars in Figure 11.

Sidelobe deception of phased array antennas raises numerous issues. To effectively deceive the radar or radar system, one needs specific knowledge of the victim radar(s). For example, one must have specific knowledge of each individual radar's antenna pattern, operating characteristics, and operational capabilities. This information is needed to maintain the integrity of the phantom tracks while also ensuring that an ECAV can discern when it is operating in sidelobe regions as opposed to the mainlobe beam. Additionally, operating frequency can vary the width and bearing of the mainlobe and sidelobes, making analysis of a radar's output critical (even though it may vary with time). Also, it is vital that one has knowledge of the location of each individual radar in the network with respect to the ECAV's own position. This is important in order to direct energy in the proper direction

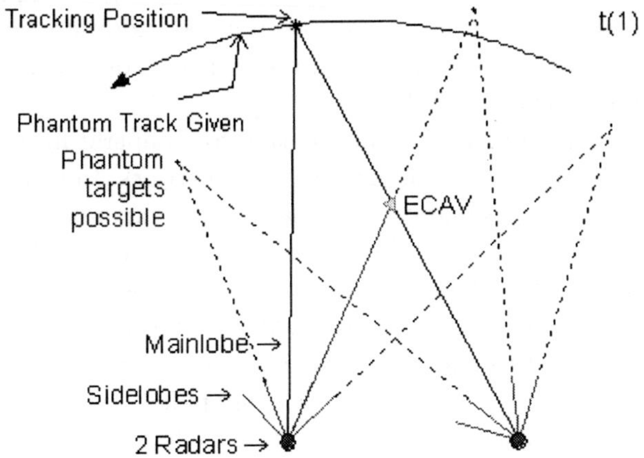

Fig. 11. Representation of side lobe deception used to generate multiple phantom targets simultaneously.

and employ range delay effectively to precisely generate specific phantom trajectories.

7. Conclusions

In this chapter, the task of deceiving an integrated network of radars with multiple ECAVs generating a single, coherent phantom track has been considered. The kinematics and mathematics dictating a single ECAV deceiving a single radar with a phantom target using range-delay deception have been presented. The mathematics are analytically formulated such that a single phantom trajectory is prescribed and an ECAV must calculate its trajectory to produce the prescribed phantom track (the inverse problem). The general case of the one-on-one engagement has been conditioned to two specific cases of interest: a constant velocity, constant course phantom track and a constant velocity, circular trajectory phantom track. These cases represent two likely scenarios of this deception in practice. Next, the mathematics and kinematics of the one-on-one engagement are adjusted to account for calculation of arbitrary trajectories of the phantom target and the ECAV. A discussion regarding the potential of a single ECAV to deceive multiple radars through exploitation of a phased array antenna's sidelobes is presented.

The concepts of range and angle deception investigated appear to be feasible, but many simplifications have been made. Additionally, extensive information is needed about the victim radar or network of radars for the angle deception to be successful.

This chapter has presented a formulation for cooperating ECAVs generating phantom targets in an integrated radar network. In future research, the kinematics and mathematical formulas will be expanded to the three-dimensional case. The 3-D case considers not only range and azimuth of ECAVs and phantom targets, but also altitude of the ECAVs and phantom targets. Additionally, more flight dynamics will be incorporated into the simulations, including restrictions on minimum/maximum velocity, turn radii, and aircraft acceleration. Finally, the simplified radar theory will be expanded to provide insights on the deception's effect against more sophisticated radar technology.

References

[1] Stimson, George W. "Introduction to Airborne Radar" 2d Ed. SciTech Publishing, Raleigh, NC, 1998.

[2] Vakin, S.A., Shustov, L.N., Dunwell, R.H. "Fundamentals of Electronic Warfare." Artech, Norwood, MA, 2001.

CHAPTER 18

POSSIBILITY REASONING AND THE COOPERATIVE PRISONER'S DILEMMA

Henry L. Pfister
Air Force Research Laboratory, Munitions Directorate
Eglin AFB, FL
pfister@eglin.af.mil

Jamie M. Walls
North Carolina A&T State University
Greensboro, NC
walls_jamie@yahoo.com

The problems of reasoning and uncertainty have been studied since the invention of probability three hundred and fifty years ago. This research presents a method for cooperative possibility reasoning with uncertainty developed for logical proposition inferences. The approach explicitly includes "uncertain" as a logic state along with "true" and "false". This leads to a model for possibility variables that can be solved as a linear program. The logical inference from the proposition states can be computed in terms of the possibility variables using the methods from fuzzy set and logic. The Prisoner's Dilemma and Epiminides paradox are used to illustrate the unique features available through the use of possibility reasoning with uncertainty. In addition this research illustrates how variables can be linked with both "AND" and "OR" conjunctions. The Prisoner's Dilemma shows how the two prisoners can cooperate in decision making. This model allows the prisoners to make decisions based on the possible trust for each other. Robert Axelrod made the Prisoner's Dilemma cooperative reasoning problem popular in 1986 and has explored the complexity of cooperation for many years. In contrast the 2000-year-old Epiminides Paradox is solved to illustrate the novel features of possibility reasoning with uncertainty for dealing with contradictory statements.

Keywords:
Boolean Logic Functions, Decision Making, Fuzzy Logic, Linear Programming, Possibility Theory, Logical Paradox, Cooperative Prisoner's Dilemma

1. Introduction

Reasoning with uncertainty has been studied throughout history with solutions ranging from astrology to artificial intelligence. Plato, born in 483 BC, his teacher Socrates, and his student Aristotle laid the foundation for much of western thought and reasoning. Aristotle gave us a formal definition for reasoning when the propositions are certain to be either true or false. He did not believe such logical processes governed all parts of the mind and he also posed a notion of intuition when faced with uncertainty.

The oldest formalism for reasoning under uncertainty is probability theory, which was developed by Pascal and Fermat in an exchange of letters in 1654 about a game of chance and flipping coins. Over the last three and a half centuries the theory has been well defined and its capabilities extensively explored, so that the rules for propagating values are established without question, and may be found in any textbook on probability and statistics. It is less clear what the numbers mean intrinsically. Some maintain that the probability of an event is a measure of its frequency of occurrence in the long term; while others insist that probability is the subjective measure of one's belief in the occurrence of the event. There are convincing arguments for and against both positions.

Rev. Thomas Bayes (1702-1761) wrote:

> "Given the number of times in which an unknown event has happened and failed: Require the chance that the probability of its happening in a single trial lies somewhere between any two degrees or probability and that can be named."

In his famous essay published posthumously, Bayes asked this question, and gave an answer. First, he defined the probability of an event as the ratio of the values at which an expectation depending on the happening of the event ought to be computed, and the value of the thing expected upon its happening. In short, he regarded probability as a rational betting ratio, and he derived the laws of probability from this definition. Then he obtained a solution by means of an ingenious geometrical representation of the problem. He needed such a geometrical representation since he defined probability as a betting ratio, so that a probability be assigned to an event, which either occurs (true) or does not occur (false). Thus, instead of speaking of "the correct value x lies between y and z," he had to talk about a concrete proposition as either coming true or not coming true. His rule is widely used to update probabilities of proposition in light of new evidence.

George Boole introduced a formal language for making logical inference about propositions in 1847 with his three laws of thought:

- Non-contradiction: No proposition is both true and false
- Excluded middle: Every proposition is either true or false
- Identity: No proposition ever changes its truth value

These have been widely accepted in Boolean algebra but failed to address some classic reasoning paradoxes. The first recorded reasoning paradox is that of Epiminides the Cretan, who proclaimed that, "All Cretans are liars." If we believe this proposition, then we should also disbelieve it. But if we disbelieve it, then that is evidence that we should believe it. For centuries such conundrums were considered childish amusements. But around the turn of the century, they presented a formidable problem for philosophers and mathematicians trying to develop formal systems of reasoning. This paradox is analyzed as an illustration of the unique features of possibility reasoning with uncertainty.

Zaheh [2] in 1965 relaxed the requirements for a crisp logic required of probability and formal reasoning by introducing fuzzy sets with membership measures to formulate possibility theory. In essence he stated:

- Every fuzzy proposition is both true and false with membership value
- A fuzzy proposition can change its truth membership value

From this he developed possibility theory and fuzzy logic [1]. Baldwin formulated the use of possibility theory and fuzzy logic with logic programming in 1986 to evaluate uncertain propositions [3].

This research explores the application of possibility theory and fuzzy logic to reasoning with uncertainty [7] along with a computational method to evaluate inferences. The premise is to explicitly include "uncertain" as a logic state along with "true" and "false." This leads to a formulation of possibility variables that can be solved as a linear optimization problem [8]. Inferences on the proposition state can be then computed in terms of the possibility values. Several example cases are presented for different classes of reasoning situations.

The possibility reasoning with uncertainty process is applied to the Prisoner's Dilemma paradox. This cooperation without communication paradox was made popular by Professor Alexrod [5, 6] in 1986, when he conducted a tournament to test different decision-making methods on an iterative computer version of the Prisoner's Dilemma. The Prisoner's Dilemma Tournament became the basis of many of the papers he wrote on analyzing social and political behavior using genetic algorithms (GA).

In addition the 2000-year old Epiminides paradox [4] is analyzed to illustrate the novel features of possibility reasoning to deal with contradictory statements. Similar results are not possible with classical logic. In the paradox, Epiminides the Cretan said "All Cretans are liars." This paradox questions the validity of itself, if Epiminides is truthful then he is a liar, but if he is untruthful, then is he truthful?

2. Reasoning with Uncertainty

An important point about the use of probability theory for reasoning is that it is not truth functional. That is, it is not possible to precisely establish the probability of a combination of two or more logical propositions from the probabilities of the propositions alone. This is in direct contrast to classical logic, which is truth functional. The result of this difference between logic and probability is that attaching probability values to logical propositions is not very fruitful for reasoning. For instance, when two propositions are combined together it is only possible to establish the bounds on the probability of the combination, and these bounds tend to expand to the full range $[0.0, 1.0]$ very quickly rather than reducing the range of uncertainty.

There are two approaches to solving this problem, either to change the representation from logic to something that is more natural from the point of view of probability theory or to use a numerical measure of uncertainty that is truth functional. The first approach leads to Bayes probability networks and decision trees that explicitly record the conditional dependencies between the probabilities of propositions, so that when the probability of a particular proposition is required, it is clear which other probabilities must be taken into account. This provides a means of establishing the precise probabilities of interesting propositions, which is efficient in practice, despite being intractable in the general case. This approach has become very popular, but, despite much recent work, does not have the flexibility of a formal first order logic for reasoning.

The second approach is to use a truth functional value such as possibility as a measure of uncertainty. In this approach, possibilities associated with propositions are combined truth functionally, and may be used to infer the truth of the combined propositions within tighter bounds than with probability. A large number of other methods for reasoning under uncertainty have also been proposed as alternatives to probability theory in numerous published reports in the last three centuries.

Possibility theory is based around a measure, which quantifies the degree to which a proposition might have a particular property. For instance, the possibility that a person is tall is a measure of the degree to which they are 'tall'. In the simplest case the measure of a person being 'tall' comes down to the extent to which they are a member of the set of 'tall' things. From this simple basis a theory has grown. This theory largely parallels probability theory, with different methods of normalizing numerical values distributions and manipulating them to estimate states.

In fact, one of the strengths of possibility theory is that it has a large number of different proposition combination operations, each of which assumes slightly different nuances of meaning for the values. However, from the point of view of providing a means of extending logic to handle uncertain information, possibility theory can go beyond probability theory. Possibility logic quantifies propositions with possibility values, and their dual or necessity values. It provides a means for combining these values truth functionally in the situations that are encountered in logical reasoning. The result is a first order, quantified truth functional logic which is applicable to many instances of possibility reasoning with uncertainty.

2.1. *Possibility and Necessity*

Uncertainty in the truth of propositions leads to the use of possibility reasoning. If information exists then probabilities can be computed but many situations do not lend themselves to probabilistic methods. The possibility of a member of a fuzzy set X in a universe U is defined with membership functions $m(x)$ as:

$$\text{Poss}(x \text{ in } X) = \sup \min(m(x \text{ in } U), m(x \text{ in } X))$$

This defines limiting cases for the possibility of set intersections and unions by:

$$\text{Poss}(X \cup Y) = \max(\text{Poss}(X), \text{Poss}(Y))$$
$$\text{Poss}(X \cap Y) = \min(\text{Poss}(X), \text{Poss}(Y))$$

Another measure is defined that is a dual of the possibility called necessity N defined by:

$$N(X \cap Y) = \min(N(X), N(Y))$$

where, $N(X) = 1$ indicates X is true and the dual relationship is

$$\text{Poss}(X) = 1 - N(\text{NOT } X) \text{ where } \min(N(X), N(\text{NOT } X)) = 0$$

That implies the following relationships:

$$\text{Poss}(X) \geq N(X) \text{ for all } X \text{ in } U$$
$$N(X) > 0 \text{ implies Poss}(X) = 1$$
$$\text{Poss}(X) < \text{ implies } N(X) = 0$$

2.2. *Proposition Uncertainty*

The support for a logical proposition A is given as a value pair [Necessity $= Sn$, Possibility $= Sp$] so that $Sn \leq Sp$ or there exists a necessity support measure less than or equal to a possible support measure and both have negations of $1 - Sn$ and $1 - Sp$. When

$$Sn + (1 - Sp) \leq 1$$

is satisfied and both Sn and Sp are constrained to the numerical range $[0.0, 1.0]$. Then the uncertainty of the support for the proposition can be defined as

$$1 - Sn - Sp$$

The interpretation of these possibility and necessity values for a logic proposition is illustrated by the example:

$$\text{Prop } A : [Sn, Sp] \text{ or example values } [0.2, 0.4]$$

Where the support for the truth of A has necessity $Sn = 0.2$ and possibility $Sp = 0.4$. The necessary support for NOT A is $1 - Sp = 0.6$ and the possible support for NOT A is $1 - Sn = 0.8$. The uncertainty in A is $1 - Sn - Sp = 0.4$ which satisfies $Sn + (1 - Sp) = 0.8 < 1.0$. Given the support pairs for two propositions A and B as A: $[SnA, SpA]$ and B: $[SnB, SpB]$ and considering all logic combinations of the propositions in Table 1 defines the set of variables for the AND conjunction operation of the two propositions necessary support as N_{ij}.

The necessary support variable table entries N_{ij} must satisfy the following seven consistency equations obtained by summing rows, columns, and elements of the table.

Table 1. Necessary support variable table

AND Support Variable N_{ij}	B True is SnB	NOT B is $1 - SpB$	B Uncertain is $SpB - SnB$
A True is SnA	A AND B N_{11}	A AND NOT B N_{12}	A N_{13}
NOT A is $1 - Sp$	NOT A AND B N_{21}	NOT A AND NOT B N_{22}	NOT A N_{23}
A Uncertain is $SpA - SnA$	B N_{31}	NOT B N_{32}	Uncertain N_{33}

$$A \text{ true implies} : SnA = N_{11} + N_{12} + N_{13} \tag{1}$$

$$\text{NOT } A \text{ implies} : 1 - SpA = N_{21} + N_{22} + N_{23} \tag{2}$$

$$\text{Uncertain } A \text{ implies} : SpA - SnA = N_{31} + N_{32} + N_{33} \tag{3}$$

$$B \text{ true implies} : SnB = N_{11} + N_{21} + N_{31} \tag{4}$$

$$\text{NOT } B \text{ implies} : 1 - SpB = N_{12} + N_{22} + N_{32} \tag{5}$$

$$\text{Uncertain } B \text{ implies} : SpB - SnB = N_{13} + N_{23} + N_{33} \tag{6}$$

$$\text{Normalization} : \sum_{\forall i,j} N_{ij} = 1.0 \tag{7}$$

2.3. *AND Boundaries*

These seven equations do not define nine unique values for the N_{ij} variables but they do specify bounds that can be used to derive the following additional constraints. The argument for the lower bound is that the minimum support that can be allocated to any variable is the maximum positive sum of the row and column support less 1.0 or zero. The argument for the upper bound is that the maximum support that can be allocated to any variable is the minimum of the row and column supports:

$$\max(SnA + SnB - 1, 0.0) \le N_{11} \le \min(SnA, SnB) \tag{8}$$

$$\max(1 - SpA - SpB, 0.0) \le N_{22} \le \min(1 - SnA, 1 - SnB) \tag{9}$$

$$\max(SpA - SnA + SpB - SnB - 1, 0.0) \le N_{33} \le \min(SpA - SnA, SpB - SnB) \tag{10}$$

$$\max(1 - SpA + SnB - 1, 0.0) \le N_{21} \le \min(1 - SpA, SnB) \tag{11}$$

$$\max(SpA - SnA + SnB - 1, 0.0) \le N_{31} \le \min(SpA - SnA, SnB) \tag{12}$$

$$\max(SnA + 1 - SpB - 1, 0.0) \le N_{12} \le \min(SnA, 1 - SpB) \tag{13}$$

$$\max(SnA + SpB - SnB - 1, 0.0) \le N_{13} \le \min(SnA, SpB - SnB) \tag{14}$$

$$\max(SpA - SnA + 1 - SpB - 1, 0.0) \le N_{32} \le \min(SpA - SnA, 1 - SpB) \tag{15}$$

$$\max(1 - SpA + SpB - SnB - 1, 0.0) \le N_{23} \le \min(1 - SpA, SpB - SnB) \tag{16}$$

These equations are based on logic of the graph in Figure 1:

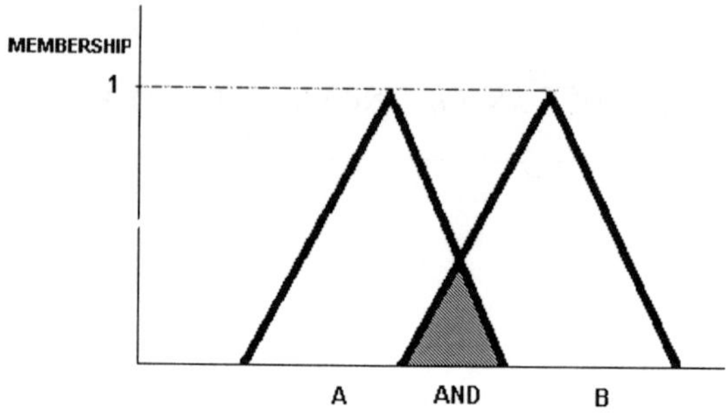

Fig. 1. Membership graph, logic of AND bounds.

2.4. *OR and XOR Boundaries*

Similarly the OR and XOR operations also have set upper and lower bounds that change when the possibility and necessity values change. The AND and OR boundaries are based on Frchet inequalities.
Conjunction

$$\max(0, P(F) + P(G) - 1) \leq P(F \text{ AND } G) \leq \min(P(F), P(G))$$

Disjunction

$$\max(0, P(F) + P(G) - 1) \leq P(F \text{ OR } G) \leq \min(P(F), P(G))$$

The "OR" bounds are shown below

$$\max(SnA, SnB) \leq N_{11} \leq \min(SnA + SnB, 1) \tag{17}$$
$$\max(1 - SpA, 1 - SpB) \leq N_{22} \leq \min(1 - SpA + 1 - SpB, 1) \tag{18}$$
$$\max(SpA - SnA, SpB - SnB) \leq N_{33} \leq \min(SpA - SnA + SpB - SnB, 1) \tag{19}$$
$$\max(1 - SpA, SnB) \leq N_{21} \leq \min(1 - SpA + SnB, 1) \tag{20}$$
$$\max(SpA - SnA, SnB) \leq N_{31} \leq \min(SpA - SnA + SnB, 1) \tag{21}$$
$$\max(SnA, 1 - SpB) \leq N_{12} \leq \min(SnA + 1 - SpB, 1) \tag{22}$$
$$\max(SnA, SpB - SnB) \leq N_{13} \leq \min(SnA + SpB - SnB, 1) \tag{23}$$
$$\max(SpA - SnA, 1 - SpB) \leq N_{32} \leq \min(SpA - SnA + 1 - SpB, 1) \tag{24}$$
$$\max(1 - SpA, SpB - SnB) \leq N_{23} \leq \min(1 - SpA + SpB - SnB, 1) \tag{25}$$

The logic of the "OR" bounds are based on the membership graph (Figure 2) and Table 2:

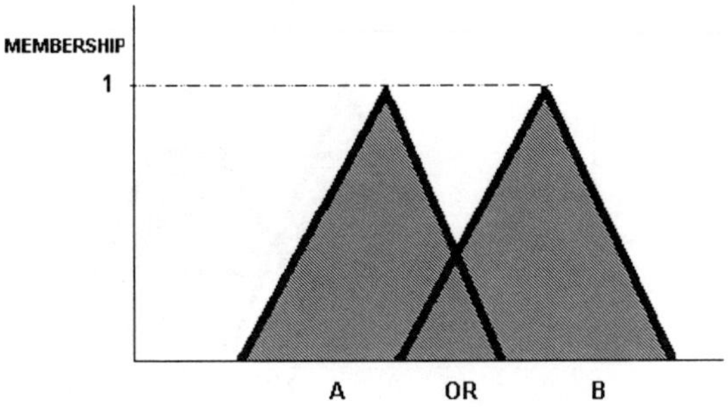

Fig. 2. Membership graph, logic of OR bounds.

Table 2. Necessary support variable table

OR Support Variable N_{ij}	B True is SnB	NOT B is $1 - SpB$	B Uncertain is $SpB - SnB$
A True is SnA	A OR B N_{11}	A OR NOT B N_{12}	A N_{13}
NOT A is $1 - Sp$	NOT A OR B N_{21}	NOT A OR NOT B N_{22}	NOT A N_{23}
A Uncertain is $SpA - SnA$	B N_{31}	NOT B N_{32}	Uncertain N_{33}

The XOR bounds are not as easy to determine as the AND and OR bounds are. Frchet established the equations for bounds on conjunction and disjunction where AND is an example of conjunction and OR is an example of disjunction. The bounds for XOR are not as easily represented. The logic of the "XOR" is based on the graph in Figure 3.
As you can see this is not a direct comparison, the subtraction of the middle of the membership function is not easily represented in terms of simple mathematical equations.

2.5. *Solving for the Support Values*

A fuzzy logic based approach to solving this set of equations and constraints is to use the operations for the union of fuzzy sets (max) and the intersection of fuzzy sets (min) as operations on support functions along with the

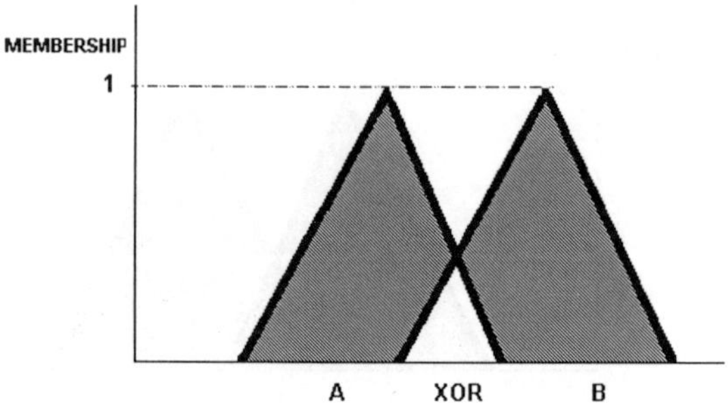

Fig. 3. Membership graph, logic of XOR bounds.

compliment as follows:

$$Sn(A \text{ AND } B) = \min(Sn(A), Sn(B))$$
$$Sn(\text{NOT } A \text{ AND NOT } B) = \min(Sn(\text{NOT } A), Sn(\text{NOT } B))$$
$$Sn(\text{UNCERTAIN } A \text{ AND UNCERTAIN } B)$$
$$= \min(Sn(\text{UNCERTAIN } A), Sn(\text{UNCERTAIN } B))$$
$$Sn(A \text{ OR } B) = \max(Sn(A), Sn(B))$$
$$Sn(\text{NOT } A) = 1 - SpA$$

Other entries in the above tables are computed from the row and column constraints.

2.6. *Support Optimization*

Given the equations and constraints (1) to (16), an initial question is how can this be used to optimize the possibility support values. For the two-proposition comparison table a linear programming optimization can be used to solve for the optimum set of N_{ij} variables as defined by the objectives of the logical propositions. The following numerical example defines two first order propositions for target identification as: target (military): $[Sn = 0.2, Sp = 0.5]$ and target (tank): $[Sn = 0.3, Sp = 0.67]$. These are unit clauses that express a necessary support for a target to be military as 0.2 with possible support as 0.5 and this implies the necessary support for not military is $0.5 = 1.0 - Sp$ and the possible support for not military is $0.8 = 1.0 - Sn$. The values Sn and Sp satisfy the dual relationship $Sn + (1 - Sp) \leq 1$ with a target being military uncertainty of $0.3 = 1 - Sn - Sp$. The target being a tank defines a conjunction predicate logic clause where the target is a tank has necessary support of $0.3 = Sn$ and the necessary support of not being a tank of $0.33 = 1.0 - Sp$. These values can be combined with the resulting support values to form the logical proposition of target (military) AND target (tank). Solving the logical proposition for a target identification instance yields possibility variables and identification (military, tank) defines the following linear equations and constraints.

$$
\begin{aligned}
0.2 &= N_{11} + N_{12} + N_{13} \\
0.5 &= N_{21} + N_{22} + N_{23} \\
0.3 &= N_{31} + N_{32} + N_{33} \\
0.3 &= N_{11} + N_{21} + N_{31} \\
0.33 &= N_{12} + N_{22} + N_{32} \\
0.37 &= N_{13} + N_{23} + N_{33} \\
1.0 &= N_{11} + N_{12} + N_{13} + N_{21} + N_{22} + N_{23} + N_{31} + N_{32} + N_{33}
\end{aligned}
$$

$$
\begin{aligned}
0.0 \le N_{11} &\le 0.3 \\
0.0 \le N_{22} &\le 0.33 \\
0.0 \le N_{33} &\le 0.3 \\
0.0 \le N_{21} &\le 0.3 \\
0.0 \le N_{31} &\le 0.3 \\
0.0 \le N_{12} &\le 0.2 \\
0.0 \le N_{13} &\le 0.2 \\
0.0 \le N_{32} &\le 0.3 \\
0.0 \le N_{23} &\le 0.3
\end{aligned}
$$

Which can be solved by linear programming with an objective of maximizing the logical AND variable N_{11} subject to the constraints and equations. The following data is from a custom computer program implementing these computations where $C(\cdot)$ is the objective variable weight and $X(\cdot)$ is the solution value of the linear optimization program.

```
Input Data
Proposition A: Target is military with Possibility 0.50 and Necessity 0.20
Proposition B: Target is tank with Possibility 0.67 and Necessity 0.30
Solution for 25 equations 43 variables and iterations 21
Maximize → Target is military AND Target is tank
```

$C(1) = 1.00 - X(1) = 0.20$ - Target is military AND Target is tank
$C(2) = 0.00 - X(2) = 0.33$ - NOT (Target is military) AND NOT (Target is tank)
$C(3) = 0.00 - X(3) = 0.30$ - UNCERTAIN (Target is military AND Target is tank)
$C(4) = 0.00 - X(4) = 0.10$ - NOT (Target is military) AND Target is tank
$C(5) = 0.00 - X(5) = 0.00$ - UNCERTAIN (Target is military) AND Target is tank
$C(6) = 0.00 - X(6) = 0.00$ - Target is military AND NOT (Target is tank)
$C(7) = 0.00 - X(7) = 0.00$ - Target is military AND UNCERTAIN (Target is tank)
$C(8) = 0.00 - X(8) = 0.00$ - UNCERTAIN (Target is military) AND NOT (Target is tank)
$C(9) = 0.00 - X(9) = 0.07$ - NOT (Target is military) AND UNCERTAIN (Target is tank)

```
The highest possibility is NOT (Target is military) AND NOT (Target is tank)
with value 0.33

Inferences from the possibilities
Possibility that Target is military is true has support of 0.30
Possibility that Target is tank is true has support of 0.20
Possibility that Target is military is false has support of 0.50
Possibility that Target is tank is false has support of 0.33
Possibility that Target is military is uncertain has support of 0.07
Possibility that Target is tank is uncertain has support of 0.37
Max inference that Target is military is false has support of 0.50
```

The interpretation is that the possibility support is highest at 0.33 that the target is not military and not a tank while the maximum possibility support that the target is military and a tank is only 0.2 which is less than the possibility support of 0.3 for uncertainty that the target is military and is a tank. The inferences about the nine combinations of the two propositions are computed from the solution and yield a maximum possibility support for a false condition that the target is military.

This illustrates a method to do explicit reasoning with uncertainty using fuzzy logic conditions and optimization rather than a probabilistic bounding of classical logic using conditional probabilities. It is clear from this approach that a great number of other objective functions could be maximized and will yield different reasoning results. One example is to choose as an objective to maximize variable N_{22} rather than N_{11}, which is the simultaneous possibility of not military and not a tank, that yields the following solution for the sample problem.

```
Input Data
Proposition A: Target is military with Possibility 0.50 and Necessity 0.20
Proposition B: Target is tank with Possibility 0.67 and Necessity 0.30
Solution for 25 equations 43 variables and iterations 24
Maximize → NOT (Target is military) AND NOT (Target is tank)
C(1) = 0.00 – X(1) = 0.13 - Target is military AND Target is tank
C(2) = 1.00 – X(2) = 0.33 - NOT (Target is military) AND NOT (Target is tank)
C(3) = 0.00 – X(3) = 0.30 - UNCERTAIN (Target is military AND Target is tank)
C(4) = 0.00 – X(4) = 0.17 - NOT (Target is military) AND Target is tank
C(5) = 0.00 – X(5) = 0.00 - UNCERTAIN (Target is military) AND Target is tank
C(6) = 0.00 – X(6) = 0.00 - Target is military AND NOT (Target is tank)
C(7) = 0.00 – X(7) = 0.07 - Target is military AND UNCERTAIN (Target is tank)
C(8) = 0.00 – X(8) = 0.00 - UNCERTAIN (Target is military) AND NOT (Target is
tank)
C(9) = 0.00 – X(9) = 0.00 - NOT (Target is military) AND UNCERTAIN (Target is
tank)

The highest possibility is NOT (Target is military) AND NOT (Target is tank)
with value 0.33

Inferences from the possibilities
Possibility that Target is military is true has support of 0.30
Possibility that Target is tank is true has support of 0.20
Possibility that Target is military is false has support of 0.50
Possibility that Target is tank is false has support of 0.33
Possibility that Target is military is uncertain has support of 0.00
Possibility that Target is tank is uncertain has support of 0.37
Max inference that Target is military is false has support of 0.50
```

The differences in the two solutions is a shift in the values of variable $X(1)$ from 0.2 to 0.13 thus reducing the possibility of being both military and a tank. The confusion variable $X(4)$ goes from 0.1 to 0.17 increasing the impossible condition that it is a nonmilitary tank. The uncertain variable $X(9)$ goes to zero, and variable $X(7)$ goes to 0.07 for a military vehicle and uncertain tank. The change in the inferences is that the possibility of uncertainty that the target is military increases from zero from 0.07. The maximum value inference of the target being military is still false at 0.5 indicating a stable reasoning conclusion under these multiple objectives.

In fact, all of the single variable objectives have the exact same maximum value inference except for the following case.

```
Solution for 25 equations 43 variables and iterations 24
Maximize → NOT (Target is military) AND UNCERTAIN (Target is tank)
```

$C(1) = 0.00 - X(1) = 0.00$ - `Target is military AND Target is tank`

$C(2) = 0.00 - X(2) = 0.13$ - `NOT (Target is military) AND NOT (Target is tank)`

$C(3) = 0.00 - X(3) = 0.00$ - `UNCERTAIN (Target is military AND Target is tank)`

$C(4) = 0.00 - X(4) = 0.00$ - `NOT (Target is military) AND Target is tank`

$C(5) = 0.00 - X(5) = 0.30$ - `UNCERTAIN (Target is military) AND Target is tank`

$C(6) = 0.00 - X(6) = 0.20$ - `Target is military AND NOT (Target is tank)`

$C(7) = 0.00 - X(7) = 0.00$ - `Target is military AND UNCERTAIN (Target is tank)`

$C(8) = 0.00 - X(8) = 0.00$ - `UNCERTAIN (Target is military) AND NOT (Target is tank)`

$C(9) = 1.00 - X(9) = 0.37$ - `NOT (Target is military) AND UNCERTAIN (Target is tank)`

```
The highest possibility is NOT (Target is military) AND UNCERTAIN (Target is
tank) with value 0.37

Inferences from the possibilities
Possibility that Target is military is true has support of 0.30
Possibility that Target is tank is true has support of 0.20
Possibility that Target is military is false has support of 0.50
Possibility that Target is tank is false has support of 0.33
Possibility that Target is military is uncertain has support of 0.67
Possibility that Target is tank is uncertain has support of 0.37
Max inference that Target is military is uncertain has support of 0.67
```

Maximizing the objective of '*not a military target*' and an '*uncertain tank*' increases the inference support from 0.5 to 0.67 and results in a different maximum inference that '*the target is military is uncertain*' with a higher value that the 0.5 for an inference of a '*false military target*'. As is often the case in optimization problems, the objective must be chosen carefully since the algorithm will blindly search for extreme feasible solutions that may not be reasonable. Clearly a number of the potential objective functions are logically confused from a reasoning point of view and not appropriate as objective goals. However, the reasoning inferences from the optimal possibility values are stable and consistent even for the confused objectives. The next section defines some classes of reasoning problems and appropriate objective functions to deal with this problem.

3. Classes of Reasoning Situations

Three major classes of reasoning situations can be stated for the two-proposition model. The first class is conditional propositions so that if proposition B is true then proposition A is always true. The second class is for exclusive propositions where either A is true or B is true but not both and at least one is true. The third class is for independent propositions where A or B can be true or false in any combination.

3.1. *Conditional Propositions*

The tank identification example has a conditional relationship between the propositions. In this example the proposition that the '*target is a tank*' is conditioned on the proposition that the '*target is military*'. No civilian tanks can exist in this logical situation. In this conditional case only the following optimization objectives are logically admissible:

- Target is military AND Target is tank
- NOT (Target is military) AND NOT (Target is tank)
- UNCERTAIN (Target is military AND Target is tank)
- Target is military AND NOT (Target is tank)
- Target is military AND UNCERTAIN (Target is tank)
- UNCERTAIN (Target is military) AND NOT (Target is tank)

This restriction on conditional propositions precludes the erroneous maximum inference result for the proposition '*target is military is uncertain*' as computed in the previous section.

3.2. *Exclusive Propositions*

A detection problem where the propositions are exclusive is presented by example. The propositions are either that a vehicle is military or that it is civilian. In this case only the following optimization objectives are logically admissible:

- NOT (Target is military) AND Target is civilian
- UNCERTAIN (Target is military) AND Target is civilian
- Target is military AND NOT (Target is civilian)

- Target is military AND UNCERTAIN (Target is civilian)

An example case is the following model for the first objective that yields a high possibility that the '*target is civilian is uncertain*' and '*target is military is false*':

Input Data

Proposition A: Target is military with Possibility 0.50 and Necessity 0.30

Proposition B: Target is civilian with Possibility 0.60 and Necessity 0.20

Solution for 25 equations 43 variables and iterations 21

Maximize $\rightarrow$ NOT (Target is military) AND Target is civilian

$C(1) = 0.00 - X(1) = 0.00$ - Target is military AND Target is civilian

$C(2) = 0.00 - X(2) = 0.30$ - NOT (Target is military) AND NOT (Target is civilian)

$C(3) = 0.00 - X(3) = 0.20$ - UNCERTAIN (Target is military AND Target is civilian)

$C(4) = 1.00 - X(4) = 0.20$ - NOT (Target is military) AND Target is civilian

$C(5) = 0.00 - X(5) = 0.00$ - UNCERTAIN (Target is military) AND Target is civilian

$C(6) = 0.00 - X(6) = 0.10$ - Target is military AND NOT (Target is civilian)

$C(7) = 0.00 - X(7) = 0.20$ - Target is military AND UNCERTAIN (Target is civilian)

$C(8) = 0.00 - X(8) = 0.00$ - UNCERTAIN (Target is military) AND NOT (Target is civilian)

$C(9) = 0.00 - X(9) = 0.00$ - NOT (Target is military) AND UNCERTAIN (Target is civilian)

Inferences from the possibilities

Possibility that Target is military is true has support of 0.20

Possibility that Target is civilian is true has support of 0.30

Possibility that Target is military is false has support of 0.50

Possibility that Target is civilian is false has support of 0.40

Possibility that Target is military is uncertain has support of 0.00

Possibility that Target is civilian is uncertain has support of 0.40

Max inference that Target is military is false has support of 0.50

All combinations of these objectives have the same maximum inference proposition that '*target is military is false*' although the optimum decision variables change with the objective function choice.

3.3. *Independent Propositions*

An independent reasoning problem where the objectives are unrestricted is presented by example. The propositions are either that a vehicle is military or that a vehicle is operational. In this case all the optimization variables are logically admissible, all are included in the objective, and are equally weighted:

```
Input Data Proposition
A: Vehicle is military with Possibility 0.60 and Necessity 0.40
Proposition B: Vehicle is operational with Possibility 0.80 and Necessity
0.20
Solution for 25 equations 43 variables and iterations 24
Maximize → EQUAL WEIGHTS
```

$C(1) = 1.00 - X(1) = 0.00$ - Vehicle is military AND Vehicle is operational

$C(2) = 1.00 - X(2) = 0.20$ - NOT (Vehicle is military) AND NOT (Vehicle is operational)

$C(3) = 1.00 - X(3) = 0.20$ - UNCERTAIN (Vehicle is military AND Vehicle is operational)

$C(4) = 1.00 - X(4) = 0.20$ - NOT (Vehicle is military) AND Vehicle is operational

$C(5) = 1.00 - X(5) = 0.00$ - UNCERTAIN (Vehicle is military) AND Vehicle is operational

$C(6) = 1.00 - X(6) = 0.00$ - Vehicle is military AND NOT (Vehicle is operational)

$C(7) = 1.00 - X(7) = 0.40$ - Vehicle is military AND UNCERTAIN (Vehicle is operational)

$C(8) = 1.00 - X(8) = 0.00$ - UNCERTAIN (Vehicle is military) AND NOT (Vehicle is operational)

$C(9) = 1.00 - X(9) = 0.00$ - NOT (Vehicle is military) AND UNCERTAIN (Vehicle is operational)

```
The highest possibility is Vehicle is military AND UNCERTAIN (Vehicle is
operational) with value 0.40

Inferences from the possibilities
Possibility that Vehicle is military is true has support of 0.20
Possibility that Vehicle is operational is true has support of 0.40
Possibility that Vehicle is military is false has support of 0.40
```

```
Possibility that Vehicle is operational is false has support of 0.20
Possibility that Vehicle is military is uncertain has support of 0.00
Possibility that Vehicle is operational is uncertain has support of 0.60
Max inference that Vehicle is operational is uncertain has support of 0.60
```

All combinations of these objectives including the equal weight have the same maximum inference value for the proposition that '*vehicle is operational is uncertain*'.

The Exclusive and Independent Proposition results are shown in Figure 4

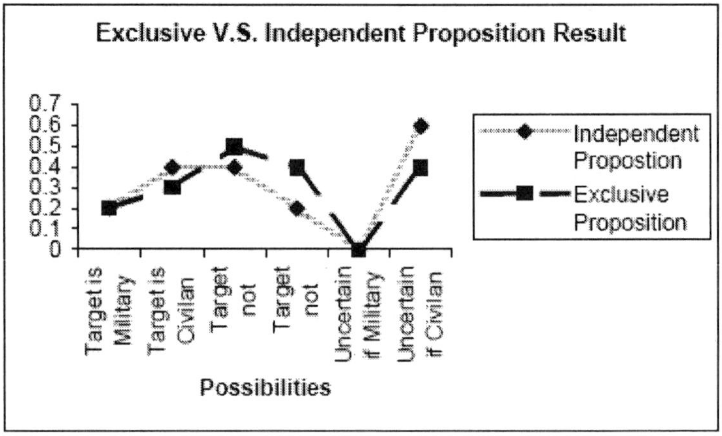

Fig. 4. Exclusive and Independent Proposition results.

When the AND function maximizes the objective of '*not a military target*' and '*an uncertain tank*' increases the inference support from 0.5 to 0.67 the result is a maximum inference that the '*target is military is uncertain*' with a higher value than that when the 0.5 for an inference of '*a false military target.*' This is often the case in optimization problems, the objective must be chosen carefully since the algorithm will blindly search for extreme feasible solutions that may not be reasonable. Clearly, a number of the potential objective functions are logically confused from a reasoning point of view and not appropriate as objective goals. However, the reasoning inferences from the optimal possibility values are stable and consistent even for the confused objectives. The next section defines some classes of reasoning problems and appropriate objective functions to deal with this problem. In order to use the Probability Reasoning for the cooperative control project, the different possible variables need to be optimized with respect to A AND B, A OR B, and A XOR B.

3.4. *AND and OR Comparison*

When the all the coefficients are positive the AND and OR conjunctions give the same or similar resulting possibilities, but not when all the coefficients are negative values. The following two examples demonstrate the difference between using the AND and the OR conjunction with the same set of possibilities and Necessities. In both cases variable A has a Sp of 0.6 and a Sn of 0.3 and variable B has a Sp of 0.7 and a Sn of 0.1.

`Case: AND Example`

`Input Data Proposition A: Variable A with Possibility 0.60 and Necessity 0.30`
`Proposition B: Variable B with Possibility 0.70 and Necessity 0.10`
`Solution for 25 equations 43 variables and iterations 28`
`Maximize` $\longrightarrow$ `Variable A AND Variable B`

$C(1) = 1.00 - X(1) = 0.000$ - `Variable A AND Variable B`
$C(2) = 1.00 - X(2) = 0.300$ - `NOT (Variable A ) AND NOT(Variable B)`
$C(3) = 0.00 - X(3) = 0.000$ - `UNCERTAIN (Variable A AND Variable B)`
$C(4) = 0.00 - X(4) = 0.000$ - `NOT (Variable A ) AND Variable B`
$C(5) = 0.00 - X(5) = 0.100$ - `UNCERTAIN (Variable A ) AND Variable B`
$C(6) = 0.00 - X(6) = 0.000$ - `Variable A AND NOT (Variable B)`
$C(7) = 0.00 - X(7) = 0.300$ - `Variable A AND UNCERTAIN (Variable B)`
$C(8) = 0.00 - X(8) = 0.000$ - `UNCERTAIN (Variable A ) AND NOT (Variable B)`
$C(9) = 0.00 - X(9) = 0.000$ - `NOT (Variable A ) AND UNCERTAIN (Variable B)`

`The highest possibility is NOT (Variable A) AND NOT (Variable B) with value 0.300`

`Inferences from the possibilities`
`Possibility that Variable A is true has support of 0.30`
`Possibility that Variable B is true has support of 0.10`
`Possibility that Variable A is false has support of 0.30`
`Possibility that Variable B is false has support of 0.30`

```
Possibility that Variable A is uncertain has support of 0.10
Possibility that Variable B is uncertain has support of 0.30
Max inference that Variable A is true has support of 0.30
```

When the AND conjunction is used Variable A has true possibility of 0.3, a false possibility of 0.3, and an uncertainty of 0.1.

```
Case: OR Example

Input Data
Proposition A: Variable A with Possibility 0.60 and Necessity 0.30
Proposition B: Variable B with Possibility 0.70 and Necessity 0.10
Solution for 25 equations 43 variables and iterations 17
Maximize → Variable A OR Variable B
```

$C(1) = 1.00 - X(1) = 0.300$ - Variable A OR Variable B
$C(2) = 1.00 - X(2) = 0.400$ - NOT (Variable A) OR NOT(Variable B)
$C(3) = 0.00 - X(3) = 0.000$ - UNCERTAIN (Variable A OR Variable B)
$C(4) = 0.00 - X(4) = 0.000$ - NOT (Variable A) OR Variable B
$C(5) = 0.00 - X(5) = 0.000$ - UNCERTAIN (Variable A) OR Variable B
$C(6) = 0.00 - X(6) = 0.300$ - Variable A OR NOT (Variable B)
$C(7) = 0.00 - X(7) = 0.000$ - Variable A OR UNCERTAIN (Variable B)
$C(8) = 0.00 - X(8) = 0.000$ - UNCERTAIN (Variable A) OR NOT (Variable B)
$C(9) = 0.00 - X(9) = 0.000$ - NOT (Variable A) OR UNCERTAIN (Variable B)

```
The highest possibility is NOT (Variable A) OR NOT (Variable B) with value
0.400
Inferences from the possibilities
Possibility that Variable A is true has support of 0.60
Possibility that Variable B is true has support of 0.30
Possibility that Variable A is false has support of 0.40
Possibility that Variable B is false has support of 0.70
Possibility that Variable A is uncertain has support of 0.00
Possibility that Variable B is uncertain has support of 0.00
Max inference that Variable B is false has support of 0.70
```

The OR conjunction on the other hand, calculates the true possibility for variable A to be 0.6, and the false possibility to be 0.7, but it doesn't have a possibility for uncertainty for either variable A or variable B. All the Optimized Possibilities are shown in Figure 5.

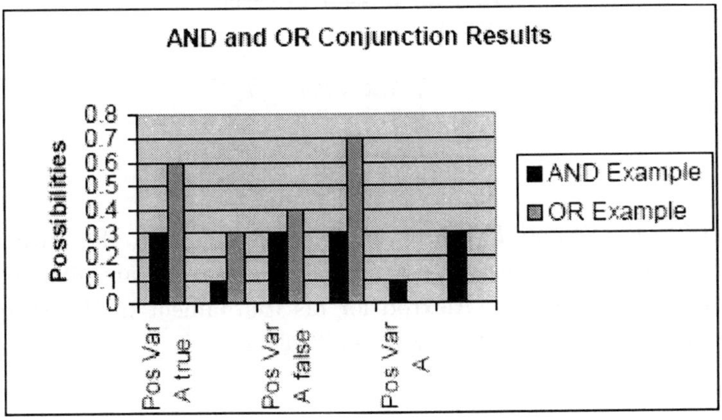

Fig. 5. All the Optimized Possibilities.

4. Prisoner's Dilemma

The Prisoner's Dilemma is the analysis of the decisions that two criminals have to make, it is a common game theory case. The two prisoners are arrested for committing a crime and are placed in two different rooms. Both of the prisoners are then given the same offers. If they confess and their accomplice doesn't they will be released after they testify against their partner. If they both confess they will both get a reduced sentence. Yet if they don't confess and their partner does then they get the book thrown at them, and if they both don't confess, they both get a small sentence for firearm possession.

Robert Axelrod conducted a computer-based tournament in order to analyze different strategies used for the Prisoner's Dilemma. The dilemma was modified so that the prisoners made their decision on whether to co-operate or defect repetitively. That allowed them to use their accomplice's last move in order to make the decision for their current move. In the first round he received fourteen entries, the best of which was the TIT for TAT implementation. In the second round he received 62 entries, but the TIT for TAT implementation still won.

The Stanford Encyclopedia of Philosophy defines a Prisoner's Dilemma case for multiple moves [4]. It incorporates the options to Cooperate, Defect, or Neither (C, D, or N). For this test 'neither' represents the uncertainty selection. The matrix lay out is shown in Table 3.

Table 3. Prisoner's Dilemma matrix

	C	D	N
C	R,R	S,T	T,S
D	T,S	P,P	R,S
N	S,T	S,R	S,S

The T, R, S, and P values are selected based on the criteria that $S < P < R < T$. The values used for this analysis are similar to the original values present by Alexrod for his tournament in the 1980s. The matrix representation used is shown in Table 4,

Table 4. Prisoner's Dilemma possibility variables

Support Variable N_{ij}	P2 Cooperate	P2 Defect	P2 Defect
P1 Cooperate	R,R N_{11}	S,T N_{12}	T,S N_{13}
P1 Defect	T,S N_{21}	P,P N_{22}	R,S N_{23}
P1 Defect	S,T N_{31}	S,R N_{32}	S,S N_{33}

where $R = 3$ years, $S = 1$ years, $T = 5$ years, and $P = 2$ years.

The way the *SnA*, *SpA*, *SnB*, and *SpB* values are determined is different from the original layout of the possibility reasoning system. Instead of N_{11} having only one possible value in this situation each matrix position can have multiple values depending on whether you are looking at the result for Prisoner 1 (P1) or the result for Prisoner 2 (P2). As you can see in cell N_{12} there are two possible values, T for Prisoner 1 and a S for Prisoner 2. So, if A represents Prisoner 1, B represents Prisoner 2, and

$$\text{Prisoner 1 Cooperates implies: } SnA = N_{11} + N_{12} + N_{13}$$
$$\text{Prisoner 2 Cooperates implies: } SnB = N_{11} + N_{21} + N_{31}$$

and

$$\text{Prisoner 1 Cooperate implies: } SnA = R + T + S$$
$$\text{Prisoner 2 Cooperates implies: } SnB = R + T + S.$$

4.1. *TIT for TAT Approach*

In order to test the Prisoner's Dilemma, the TIT for TAT case was simulated by optimizing the variables with respect to '*A AND B*' and '*NOT A AND NOT B*'. The TIT for TAT strategy is a simple decision process that is solely based on the previous decision of the other prisoner. If Prisoner 1 defects and Prisoner 2 cooperates then on the next play Prisoner 1 will cooperate and Prisoner 2 will defect. This first example has Prisoner 1 and Prisoner 2 with equal possibilities and necessities, and an uncertainty of 0.20.

```
Case: Prisoner's Dilemma


Input Data
Proposition A: Prisoner 1 with Possibility 0.60 and Necessity 0.20
Proposition B: Prisoner 2 with Possibility 0.60 and Necessity 0.20
Solution for 25 equations 43 variables and iterations 24
Maximize → Prisoner 1 AND Prisoner 2
C(1) = 1.00 − X(1) = 0.000 - Prisoner 1 AND Prisoner 2
C(2) = 0.00 − X(2) = 0.000 - NOT (Prisoner 1) AND NOT (Prisoner 2)
C(3) = 0.00 − X(3) = 0.400 - UNCERTAIN (Prisoner 1 AND Prisoner 2)
C(4) = 0.00 − X(4) = 0.200 - NOT (Prisoner 1) AND Prisoner 2
C(5) = 0.00 − X(5) = 0.000 - UNCERTAIN (Prisoner 1) AND Prisoner 2
C(6) = 0.00 − X(6) = 0.200 - Prisoner 1 AND NOT (Prisoner 2)
C(7) = 0.00 − X(7) = 0.000 - Prisoner 1 AND UNCERTAIN (Prisoner 2)
C(8) = 0.00 − X(8) = 0.000 - UNCERTAIN (Prisoner 1) AND NOT (Prisoner 2)
C(9) = 0.00 − X(9) = 0.000 - NOT (Prisoner 1) AND UNCERTAIN (Prisoner 2)

The highest possibility is UNCERTAIN (Prisoner 1 AND Prisoner 2) with value
0.400

Inferences from the possibilities
Possibility that Prisoner 1 is true has support of 0.20
Possibility that Prisoner 2 is true has support of 0.20
Possibility that Prisoner 1 is false has support of 0.20
Possibility that Prisoner 2 is false has support of 0.20
Possibility that Prisoner 1 is uncertain has support of 0.40
Possibility that Prisoner 2 is uncertain has support of 0.40
Max inference that Prisoner 1 is uncertain has support of 0.40
```

With the above possibility and necessity, the possibility that Prisoner 1 and Prisoner 2 will defect is 0.40 and the possibility of both cooperating is 0.20, if both of them are using the TIT for TAT method to make their decisions.

If the OR conjunction is used the Prisoner's Dilemma will result in the same possibilities. The possibility that Prisoner 1 or Prisoner 2 will defect is again 0.20.

```
Case: Prisoner's Dilemma


Input Data
Proposition A: Prisoner 1 with Possibility 0.60 and Necessity 0.20
Proposition B: Prisoner 2 with Possibility 0.60 and Necessity 0.20
Solution for 25 equations 43 variables and iterations 17
Maximize → Prisoner 1 OR Prisoner 2
```

$C(1) = 1.00 - X(1) = 0.000$ - Prisoner 1 OR Prisoner 2
$C(2) = 1.00 - X(2) = 0.000$ - NOT (Prisoner 1) OR NOT (Prisoner 2)
$C(3) = 0.00 - X(3) = 0.400$ - UNCERTAIN (Prisoner 1 OR Prisoner 2)
$C(4) = 0.00 - X(4) = 0.200$ - NOT (Prisoner 1) OR Prisoner 2
$C(5) = 0.00 - X(5) = 0.000$ - UNCERTAIN (Prisoner 1) OR Prisoner 2
$C(6) = 0.00 - X(6) = 0.200$ - Prisoner 1 OR NOT (Prisoner 2)
$C(7) = 0.00 - X(7) = 0.000$ - Prisoner 1 OR UNCERTAIN (Prisoner 2)
$C(8) = 0.00 - X(8) = 0.000$ - UNCERTAIN (Prisoner 1) OR NOT (Prisoner 2)
$C(9) = 0.00 - X(9) = 0.000$ - NOT (Prisoner 1) OR UNCERTAIN (Prisoner 2)

```
The highest possibility is UNCERTAIN (Prisoner 1 OR Prisoner 2) with value
0.400


Inferences from the possibilities
Possibility that Prisoner 1 is true has support of 0.20
Possibility that Prisoner 2 is true has support of 0.20
Possibility that Prisoner 1 is false has support of 0.20
Possibility that Prisoner 2 is false has support of 0.20
Possibility that Prisoner 1 is uncertain has support of 0.40
Possibility that Prisoner 2 is uncertain has support of 0.40
Max inference that Prisoner 1 is uncertain has support of 0.40
```

Both the AND and OR examples have a probability of uncertainty higher than the possibility of cooperation, but the possibility that Prisoner 1 and Prisoner 2 will cooperate and defect is 0.2.

The next example shows a case where Prisoner 2 has a higher necessity than Prisoner 1. In this case *Sn* for Prisoner 1 is 0.2 while the *Sn* for Prisoner 2 is 0.30.

```
Case: Prisoner's Dilemma

Input Data
Proposition A: Prisoner 1 with Possibility 0.60 and Necessity 0.20
Proposition B: Prisoner 2 with Possibility 0.60 and Necessity 0.30
Solution for 25 equations 43 variables and iterations 27
Maximize -> Prisoner 1 AND Prisoner 2
```

$C(1) = 1.00 - X(1) = 0.033$ - Prisoner 1 AND Prisoner 2
$C(2) = 1.00 - X(2) = 0.000$ - NOT (Prisoner 1) AND NOT (Prisoner 2)
$C(3) = 0.00 - X(3) = 0.286$ - UNCERTAIN (Prisoner 1 AND Prisoner 2) $C(4) = 0.00 - X(4) = 0.200$ - NOT (Prisoner 1) AND Prisoner 2
$C(5) = 0.00 - X(5) = 0.000$ - UNCERTAIN (Prisoner 1) AND Prisoner 2
$C(6) = 0.00 - X(6) = 0.029$ - Prisoner 1 AND NOT (Prisoner 2)
$C(7) = 0.00 - X(7) = 0.014$ - Prisoner 1 AND UNCERTAIN (Prisoner 2)
$C(8) = 0.00 - X(8) = 0.114$ - UNCERTAIN (Prisoner 1) AND NOT (Prisoner 2)
$C(9) = 0.00 - X(9) = 0.000$ - NOT (Prisoner 1) AND UNCERTAIN (Prisoner 2)

```
The highest possibility is UNCERTAIN (Prisoner 1 AND Prisoner 2) with value
0.286

Inferences from the possibilities
Possibility that Prisoner 1 is true has support of 0.08
Possibility that Prisoner 2 is true has support of 0.23
Possibility that Prisoner 1 is false has support of 0.20
Possibility that Prisoner 2 is false has support of 0.14
Possibility that Prisoner 1 is uncertain has support of 0.40
Possibility that Prisoner 2 is uncertain has support of 0.30
Max inference that Prisoner 1 is uncertain has support of 0.40
```

When the Prisoner's Dilemma is implemented in the Reason and Possibilities example using the TIT for TAT decision base, the Max inference that Prisoner 1 and Prisoner 2 are uncertain has support of 0.40. This shows that the Prisoner 1 and Prisoner 2 are '*uncertain*' more often than they will '*cooperate*' or '*defect*'. Prisoner 1 and Prisoner 2 also '*cooperate*' and '*defect*' with a support of 0.20. This shows that when both prisoners are using the same method, probability, and necessity against each other they both do the exact same thing that their accomplice does. When all the variables are constant the prisoners have an equal level of trust for each other. If you decrease the necessity of Prisoner 1 then Prisoner 2 will take advantage of that and will then cooperate more than Prisoner 1 does. The lower the necessity of each prisoner, the more they trust the other. If Prisoner 2 cooperates more than Prisoner 1, Prisoner 2 will yield a higher score.

5. Epiminides Paradox

The paradox of Epiminides the Cretan, who said, "All Cretans are liars," can be modeled by possibility reasoning with uncertainty. If we assert both propositions independently that Epiminides is a Cretan and all Cretans are liars with possibility 1.0 and necessity 0.0 we get:

`Case: Epiminides Paradox`

`Input Data`

`Proposition A: Epiminides is Cretan with Possibility 1.00 and Necessity 0.00`

`Proposition B: Cretans are liars with Possibility 1.00 and Necessity 0.00`

`Solution for 25 equations 43 variables and iterations 18`

`Maximize` $\rightarrow$ `Epiminides is Cretan AND Cretans are liars`

$C(1) = 1.00 - X(1) = 0.00$ `- Epiminides is Cretan AND Cretans are liars`

$C(2) = 1.00 - X(2) = 0.00$ `- NOT (Epiminides is Cretan) AND NOT (Cretans are liars)`

$C(3) = 1.00 - X(3) = 1.00$ `- UNCERTAIN (Epiminides is Cretan AND Cretans are liars)`

$C(4) = 1.00 - X(4) = 0.00$ `- NOT (Epiminides is Cretan) AND Cretans are liars`

$C(5) = 1.00 - X(5) = 0.00$ `- UNCERTAIN (Epiminides is Cretan) AND Cretans are liars`

$C(6) = 1.00 - X(6) = 0.00$ `- Epiminides is Cretan AND NOT (Cretans are liars)`

$C(7) = 1.00 - X(7) = 0.00$ `- Epiminides is Cretan AND UNCERTAIN (Cretans are liars)`

$C(8) = 1.00 - X(8) = 0.00$ `- UNCERTAIN (Epiminides is Cretan) AND NOT (Cretans are liars)`

$C(9) = 1.00 - X(9) = 0.00$ `- NOT (Epiminides is Cretan) AND UNCERTAIN (Cretans are liars)`

```
The highest possibility is UNCERTAIN (Epiminides is Cretan AND Cretans are
liars) with value 1.00

Inferences from the possibilities
Possibility that Epiminides is Cretan is true has support of 0.00
Possibility that Cretans are liars is true has support of 0.00
Possibility that Epiminides is Cretan is false has support of 0.00
Possibility that Cretans are liars is false has support of 0.00
Possibility that Epiminides is Cretan is uncertain has support of 1.00
Possibility that Cretans are liars is uncertain has support of 1.00
Max inference that Epiminides is Cretan is uncertain has support of 1.00
```

This indicates that the conclusion that both propositions are uncertain has the possibility 1.0 for both statements. Changing the propositions possibilities to 0.5 yields the following solution:

Case: Epiminides Paradox

Input Data

Proposition A: Epiminides is Cretan with Possibility 0.50 and Necessity 0.00

Proposition B: Cretans are liars with Possibility 0.50 and Necessity 0.00

Solution for 25 equations 43 variables and iterations 22

Maximize $\rightarrow$ Epiminides is Cretan AND Cretans are liars

$C(1) = 1.00 - X(1) = 0.00$ - Epiminides is Cretan AND Cretans are liars

$C(2) = 1.00 - X(2) = 0.50$ - NOT (Epiminides is Cretan) AND NOT (Cretans are liars)

$C(3) = 1.00 - X(3) = 0.50$ - UNCERTAIN (Epiminides is Cretan AND Cretans are liars)

$C(4) = 1.00 - X(4) = 0.00$ - NOT (Epiminides is Cretan) AND Cretans are liars

$C(5) = 1.00 - X(5) = 0.00$ - UNCERTAIN (Epiminides is Cretan) AND Cretans are liars

$C(6) = 1.00 - X(6) = 0.00$ - Epiminides is Cretan AND NOT (Cretans are liars)

$C(7) = 1.00 - X(7) = 0.00$ - Epiminides is Cretan AND UNCERTAIN (Cretans are liars)

$C(8) = 1.00 - X(8) = 0.00$ - UNCERTAIN (Epiminides is Cretan) AND NOT (Cretans are liars)

$C(9) = 1.00 - X(9) = 0.00$ - NOT (Epiminides is Cretan) AND UNCERTAIN (Cretans are liars)

```
The highest possibility is NOT (Epiminides is Cretan) AND NOT (Cretans are
liars) with value 0.50

Inferences from the possibilities
Possibility that Epiminides is Cretan is true has support of 0.00
Possibility that Cretans are liars is true has support of 0.00
Possibility that Epiminides is Cretan is false has support of 0.50
Possibility that Cretans are liars is false has support of 0.50
Possibility that Epiminides is Cretan is uncertain has support of 0.50
Possibility that Cretans are liars is uncertain has support of 0.50
Max inference that Epiminides is Cretan is false has support of 0.50
```

This still has the solution that it is uncertain with possibility 0.5 for both propositions along with the possibility 0.5 that each proposition is false. By increasing the necessity from 0.0 to 0.2 for each proposition the solution is:

Case: Epiminides

Paradox Input

Data Proposition A: Epiminides is Cretan with Possibility 0.50 and Necessity 0.20

Proposition B: Cretans are liars with Possibility 0.50 and Necessity 0.20

Solution for 25 equations 43 variables and iterations 22

Maximize → Epiminides is Cretan AND Cretans are liars

$C(1) = 1.00 - X(1) = 0.20$ - Epiminides is Cretan AND Cretans are liars

$C(2) = 1.00 - X(2) = 0.50$ - NOT (Epiminides is Cretan) AND NOT (Cretans are liars)

$C(3) = 1.00 - X(3) = 0.30$ - UNCERTAIN (Epiminides is Cretan AND Cretans are liars)

$C(4) = 1.00 - X(4) = 0.00$ - NOT (Epiminides is Cretan) AND Cretans are liars

$C(5) = 1.00 - X(5) = 0.00$ - UNCERTAIN (Epiminides is Cretan) AND Cretans are liars

$C(6) = 1.00 - X(6) = 0.00$ - Epiminides is Cretan AND NOT (Cretans are liars)

$C(7) = 1.00 - X(7) = 0.00$ - Epiminides is Cretan AND UNCERTAIN (Cretans are liars)

$C(8) = 1.00 - X(8) = 0.00$ - UNCERTAIN (Epiminides is Cretan) AND NOT (Cretans are liars)

$C(9) = 1.00 - X(9) = 0.00$ - NOT (Epiminides is Cretan) AND UNCERTAIN (Cretans

```
are liars)

The highest possibility is NOT (Epiminides is Cretan) AND NOT (Cretans are
liars) with value 0.50

Inferences from the possibilities
Possibility that Epiminides is Cretan is true has support of 0.20
Possibility that Cretans are liars is true has support of 0.20
Possibility that Epiminides is Cretan is false has support of 0.50
Possibility that Cretans are liars is false has support of 0.50
Possibility that Epiminides is Cretan is uncertain has support of 0.30
Possibility that Cretans are liars is uncertain has support of 0.30
Max inference that Epiminides is Cretan is false has support of 0.50
```

The possibility that each proposition is true rises to 0.2. The solution that each is uncertain has possibility 0.3 along with the possibility 0.5 that each proposition is false.

These three different versions of the Epiminides Paradox give different results, but the two with a probability of 0.5 are similar compared to the paradox with a probability of 1.0 (see Figure 6).

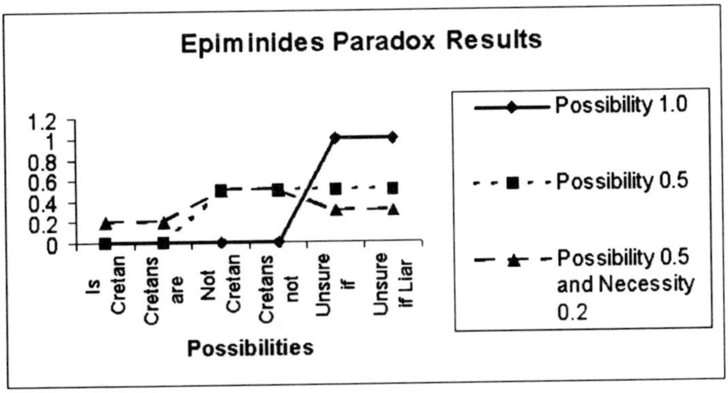

Fig. 6. Results from studies of Epiminides Paradox.

Many other solutions can be constructed to illustrate how this model incorporates simultaneous conflicting logical inferences. This is a feature not supported by classical reasoning using Boolean algebra with or without the use of probability. Of course there is a danger of committing serious reasoning errors by constructing illogical propositions, assigning unsupported possibilities, and drawing suspect inferences. However, the ability to provide a solution to this 2000-year-old logic paradox is a unique feature of possibility reasoning with uncertainty.

6. Conclusion

A method for possibility reasoning with uncertainty was developed for evaluating logical proposition inferences. The approach was to include "uncertain" as a logic state along with "true" and "false". This leads to a model for possibility variable computation that can be solved as a linear programming problem. The logical inferences of the proposition states can then be computed in terms of the possibility variables. This reasoning was used to find possibility of cooperation from Prisoner 1 and Prisoner 2 in the Prisoner's Dilemma. This also shows how different the optimized probabilities are using different conjunctions between the variables. For this research the AND and OR conjunction are incorporated, with plans to incorporate the XOR conjunction later. This capability for uncertain reasoning was also used to solve the classic Epiminides Paradox and illustrates the unique capability of evaluating simultaneous conflicting logical statements. The flexibility and complex reasoning capability of this model indicates a wide range of future application possibilities.

References

[1] Zimmerman H., Fuzzy Set Theory and It's Applications - Second Edition, Kluwer, 1991.

[2] Zadeh L., "Fuzzy Sets," Information and Control 8, 1965.

[3] Baldwin J., "Support Logic Programming," International Journal of IntelligentSystems 1, 1986.

[4] Kuhn Steven, "Stanford Encyclopedia of Philosophy" Copyright 1997, Georgetown University, `http://setis.library.usyd.edu.au/stanford/archives/win1997/entries/prisoner-dilemma/#Sym`

[5] Axelrod, R., The evolution of strategies in the iterated Prisoner's Dilemma. In L. D. Davis, Ed., Genetic Algorithms and Simulated Annealing, New York: Morgan Kaufmann, pages 32-41, 1987.

[6] Axelrod, R., The Complexity of Cooperation: Agent-Based Models of Competition and Collaboration Princeton University Press, 1998.

[7] Pfister H., "Possibility Reasoning With Uncertainty," Artificial Neural Networks in Engineering Conference, St Louis - ANNIE 2003, November 2003.

[8] Pfister H., "Uncertain Reasoning with Linear Programming," Institute for Operations Research and Management Science, INFORMS 2003, Atlanta, October 2003

CHAPTER 19

THE GROUP ASSIGNMENT PROBLEM ARISING IN MULTIPLE TARGET TRACKING

Aubrey B. Poore

Department of Mathematics
Colorado State University
Fort Collins, 80523
and
Numerica
PO Box 271246
Fort Collins, CO 80527-1246
aubrey.poore@colostate.edu, abpoore@numerica.us

Sabino M. Gadaleta

Numerica
PO Box 271246
Fort Collins, CO 80527-1246
smgadaleta@numerica.us

The central problem in multiple target tracking is the data association problem of partitioning sensor reports into tracks and false alarms. This problem occurs at all levels of tracking involving a single sensor, multiple sensors on a single platform, and multiple sensors on multiple platforms and multiple networks. Multiple frame data association, whether it is based on multiple hypothesis tracking (MHT) or multiple frame assignments (MFA), has established itself as the method of choice for difficult tracking problems, principally due to the ability to hold difficult data association decisions in abeyance until additional information is available. Over the last twenty years, these methods have focused on one-to-one assignments and occasionally on many-to-one or many-to-many assignments. Recent re-emphasis on closely spaced objects and track-to-track multiple hypothesis correlation over time have clearly demonstrated the need for a new class of data association problems and algorithms. The goal then for this work is the formulation of some of these group assignment problems, which represent a generalized data association problem in the sense that it reduces to the classical assignment problems when there are no overlapping groups.

Keywords:
Multidimensional assignment problem, group assignment problem, cluster tracking, merged measurement problem, multiple hypothesis correlation

1. Introduction

The central problem in multiple target tracking is the data association problem of partitioning sensor reports into tracks and false alarms. This problem occurs at all levels of tracking: single sensor, multiple sensors on a single platform, and multiple sensors on multiple platforms and multiple networks. There are two basic association and fusion problems, namely measurements (e.g., sensor observations such as range, azimuth, elevation, range rate, or some subset thereof) and track states (e.g., position and velocity).

For measurement-to-measurement or measurement-to-track fusion, multiple frame data association based on multidimensional assignments problems (often called multiple frame assignments (MFA) or multiple hypothesis tracking (MHT)) has established itself as the method of choice for difficult tracking problems, principally due to the ability to hold difficult data association decisions in abeyance until additional information is available. Over the last twenty or thirty years, these methods have focused mostly on individual object tracking using one-to-one assignments with an occasional use of many-to-one and many-to-many assignments. In the last four or five years, renewed interests in tracking closely spaced objects have produced two primary classes of problems that do not fit within this framework.

The first is that of grouping (or clustering) many closely spaced observations or tracks together and tracking the group. Examples include group formation tracking for ground targets and clustering radar or IR measurements and tracking the centroids. The second is that of breaking groups or clusters apart and tracking the subgroups (or subclusters) or individual objects. Pixel-cluster tracking for IR sensors and the merged measurement problem in (narrow band) radar are examples of the second.

A third broad class of problems is that in which tracks from multiple sources must be associated and fused. While measurement fusion generally yields superior tracks, many systems (sensors, platforms, and networks) produce only track states without any information regarding which measurements are associated with the track. In this case, the central problem is to correlate and fuse tracks to produce "composite" tracks superior to any of the individual tracks so combined or to correlate tracks to produce a consistent air picture from platform to platform. A multidimensional assignment approach properly expresses this track-to-track association problem when the tracks are pairwise time-aligned. (A key problem with which one must deal is the statistical cross-correlation between tracks due, e.g., to common process noise.) A second aspect of this problem is that of maintaining a consistent set of track numbers over time to preserve track continuity and ID at the system level. It is this latter problem to which group assignments is applicable.

One of the requirements in the development of the group tracking concepts is that it must fit within the traditional two and multidimensional assignment problems so that both individual and group tracking can occur within one framework, which additionally must allow transitions between the two types of tracking. In this sense this new class of data association formulation must accommodate both types of tracking. Thus, the goal of this work is the formulation of the assignment problems representing these *generalized data association problems.* These same group-assignment problems also appear to have much broader application to new problems arising in auctions, network management, procurement, and resource scheduling. Section 2 illustrates this relationship.

Although many tracking applications could be used as motivation, group-cluster tracking may be one of the easier ones on which to base a rigorous formulation of the group assignment problem. Thus, Section 3 reviews cluster tracking and clustering methods. The general cluster assignment problem is formulated in Section 4, the merged measurement problem is briefly discussed in Section 5, and Section 6 contains a brief summary.

2. Combinatorial Auctions, Coalitions, and Their Relationship to Target Tracking

The group assignment problem that will be presented in this chapter covers a broad range of important problems that reach far beyond the area of target tracking. This is *motivated* in this section by illustrating the relationship between a special set of auctions and the similar problem in target tracking. The auction setting also serves as an easily accessible framework to introduce the different classes of assignment problems that the group assignment problem encompasses. Figure 1 illustrates the relationship between a set of auction problems and a set of assignment problems that arise in target tracking.

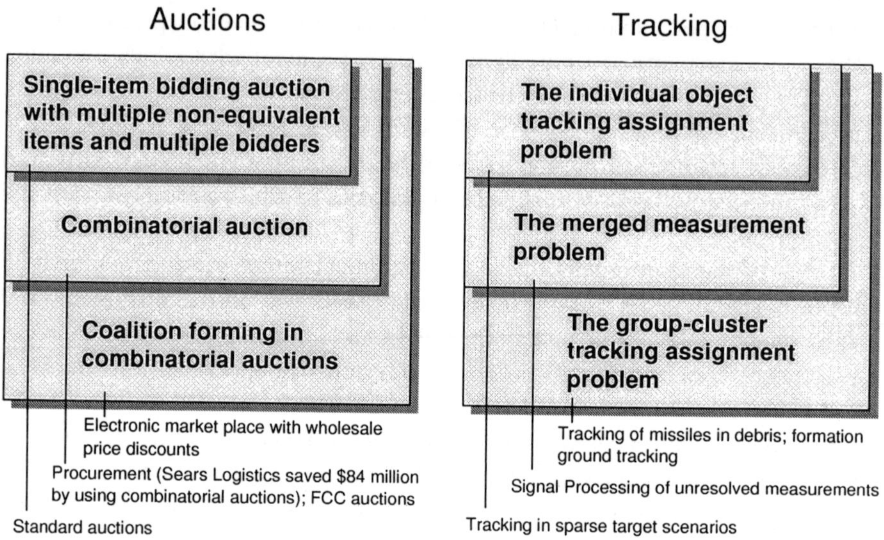

Fig. 1. A set of auctions and a set of tracking assignment problems that are similar in scope.

The group assignment problem includes the regular one-to-one or multi-assignment problem and the merged measurement problem. The regular one-to-one or multi-assignment problem is similar to the *standard single-unit auction problem* where a number of bidders may acquire a single unit. The merged measurement assignment problem is similar to a class of auction problems referred to as *combinatorial auctions* where bidders can bid on bundles of items. The general group assignment problem is similar to a new class of auctions referred to as *coalition forming auctions* where multiple bidders can form groups to bid on items or bundles.

Single-item Auctions and the One-to-One Assignment Problem of Tracking In single-item auctions a group of bidders makes bids on a list of non-identical sale-items. We may assume that all bidders bid simultaneously on all sale items and offer a price (assignment cost) for a bid on individual items. Figure 2 illustrates such an auction that considers three sale items and three bidders. The resulting assignment problem needs to consider two constraints: (1) every item can be sold to at most one buyer, and (2) every bidder j can bid on at most n_j items. This problem can be described through an assignment problem that is equivalent to the two-dimensional assignment problem for individual object tracking. This is discussed in Section 4.1 and given through Eqn. (1) (setting $m_i = 1$ in Eqn. (1)).

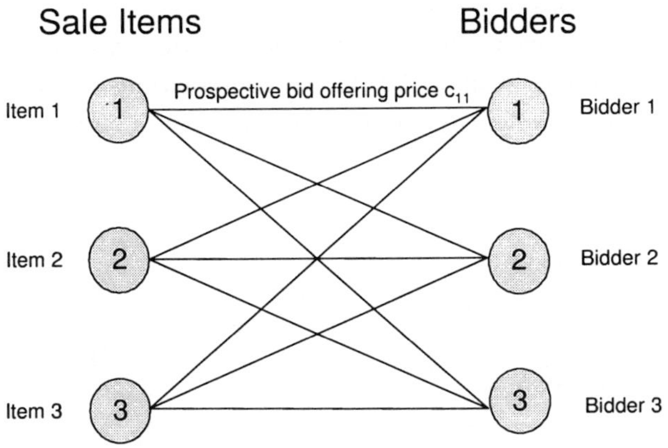

Fig. 2. Single item bidding auction.

Figure 2 shows connections between all sale items and all bidders. In practice a bidder will only bid on a selected set of items, i.e., only a subset of all feasible assignment arcs will need to be considered in the auction. In target tracking, where the lists of items and bidders may represent measurements or tracks, the assignment problem is reduced through *gating* methods that identify dynamically infeasible assignment arcs.

Combinatorial Auctions and the Merged Measurement Assignment Problem More recently a more sophisticated form of *combinatorial auctions* is considered where a bidder is allowed to bid on bundles or groups of sale items [8]. These problems are instances of the set packing problem and similar in scope to the merged measurement problem discussed in Section 5. The importance for combinatorial auctions arises from the fact that a bundle of sale items may be more valuable than the sum of their individual values. A recent example is the FCC spectrum auction, where bidders, comprised of US telecommunications companies, cellular telephone companies, and cable-television companies, competed to win various spectrum licenses for different geographical areas. The synergies arising from owning licenses in adjoining geographical areas create dependencies in (some) bidders' valuations for individual licenses [10]. Other examples include manufacturing, networking, or logistics. Sears Logistics recently saved over $84 million running six combinatorial auctions [22]. The combinatorial auction problem is becoming more mature but the need for development of (near) optimal and fast solution methods still exists.

Figure 3 illustrates a combinatorial auction with three items and three bidders where a buyer is allowed to bid on combinations, bundles, or groups of items. Note that a single item is also interpreted as a bundle for notational convenience.

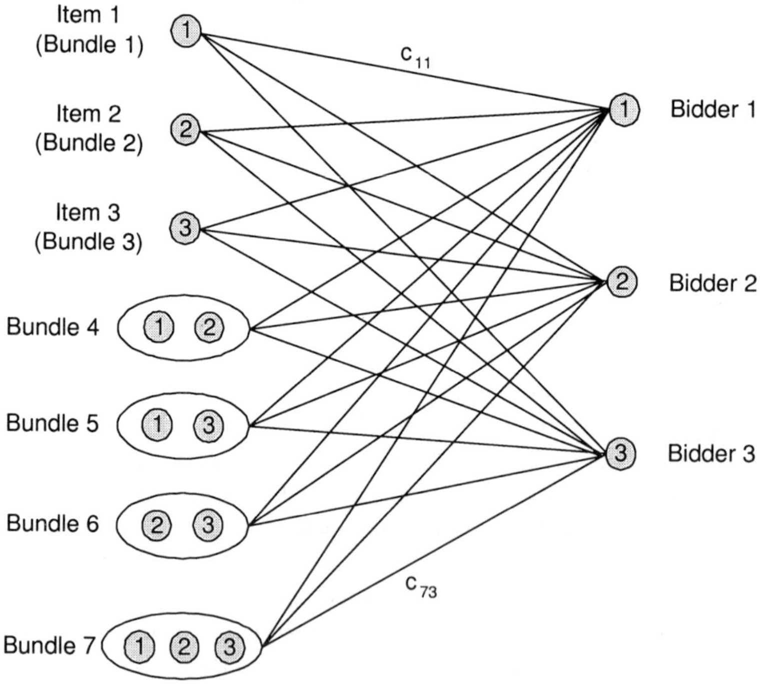

Fig. 3. Combinatorial auction.

An important constraint in this assignment problem is the *set packing constraint*: in the final assignment, an item may only be assigned to a single bundle (or not assigned at all). Otherwise, a single item could be sold to two different buyers which is not feasible. (Unless one may sell fractions or shares of an item which motivates a "soft set packing constraint".) Figure 4 illustrates the set packing constraint and the possible final bundles that would satisfy a set packing constraint for this example.

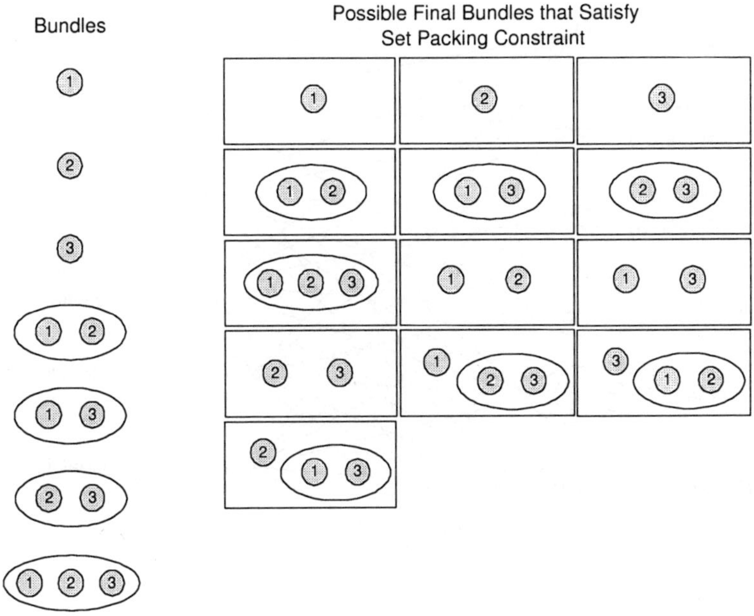

Fig. 4. The set packing constraint.

The resulting assignment problem is equivalent to the merged measurement assignment problem of target tracking, Eqn. (9), discussed in Section 5. This assignment problem is an instance of the more general group assignment problem.

Cooperative Bidding and the Group Assignment Problem With the emergence of the electronic market place an even more general form of auctions has appeared that combines *coalition forming of bidders* with combinatorial auction [14]. In this problem, several bidders may form a coalition to bid on bundles of items. For example, when items are offered in bundles at wholesale prices, several bidders may improve their payoff by buying bundles in a coalition compared to buying the items of interest separately. Figure 5 illustrates such an auction problem. In the example we assume that a coalition between bidder 1 and 3 or between all bidders is not desired. Furthermore, a single bidder will be interpreted as a (trivial) coalition.

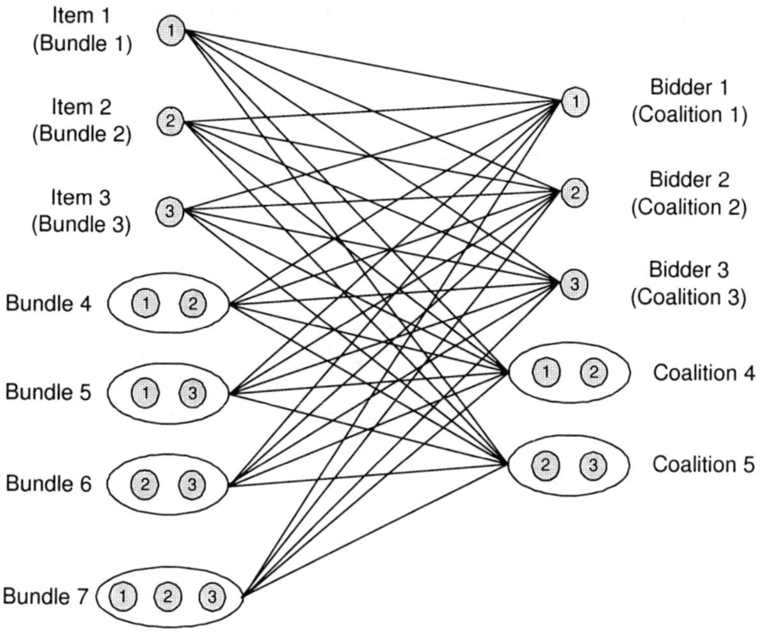

Fig. 5. Cooperative bidding in combinatorial auctions.

In the final assignment of such an auction both the items and the bidders need to satisfy a set packing constraint. This coalition auction problem is similar to the problem addressed by the general group assignment problem Eqn. (5) that represents a novel formulation for this new problem.

Fractional or Soft Constraints in Bidding To date all auction problems enforce (hard) set packing constraints. The group-cluster tracking problem motivates an extension of the group assignment problem that may also be of value for more general auction problems. When using soft-clustering approaches, an item in the final assignment may belong to more than one group. This most general group assignment problem is obtained from the assignment problem Eqn. (5) by replacing the hard set packing constraint with a soft set packing constraint. For auctions this soft assignment implies that bidders can bid on fractions, or shares, of products, where the actual percentage is to be assigned through the optimization algorithm. In the future this *soft group assignment problem* may very well provide for the optimal framework for closely spaced object target tracking *and* auctions.

3. Cluster Tracking Background and Motivation

The purpose of this section is to give a brief background on group-cluster tracking and clustering techniques. We motivate the potential benefits of multiple frame cluster tracking which bases clustering decisions on the information from multiple consecutive frames of data. An important aspect in the formulation is to allow for multi-assignments between the clusters of consecutive frames.

3.1. *Brief Review of Cluster Tracking*

One of the major challenges in modern tracking applications is the tracking of large numbers of closely spaced objects. A typical example is the flight of aircraft in formation or ground vehicles traveling in formation [1]. In these problems, objects can be so close that almost all measurements on one frame of data can be associated with any measurement on subsequent frames of data even when the best preprocessing techniques are used. (A "frame of data" as used here refers to a collection of sensor returns in which an object is seen at most once. Examples include a radar sweep of a region and a sensor dwell.) Thus, one needs to give up the goal of tracking individual objects and track clusters (or groups) of objects, at least until the objects begin to separate.

Early work on cluster or group tracking is discussed by Blackman [1]. He distinguishes centroid group tracking and formation group tracking. In centroid group tracking, group track centroids are correlated with and updated by the measurement centroids. Formation group tracking preserves individual target information within a group. Formation group tracking can provide more stable tracking solutions but is computationally more involved [1]. Drummond *et al.* [9] developed a cluster tracking algorithm for multiple passive sensors. In their approach, termed *cluster ellipsoid tracking*, cluster centroids are represented by a six-state position and velocity vector and a covariance estimating the size of the cluster. Centroids are propagated over time through a single Kalman filter. In the above investigations cluster tracking was not considered in a multiple frame association tracking environment.

Multiple target tracking methods divide into two broad classes, namely single frame and multiple frame methods. The single frame methods include nearest neighbor, global nearest neighbor and JPDA (joint probabilistic data association). The most successful of the multiple frame association methods are multiple hypothesis tracking (MHT) [2] and multiple frame assignment (MFA) [20, 21]. The performance advantage of the multiple frame methods over the single frame methods for tracking individual objects follows from the ability to hold difficult decisions in abeyance until more information is available and the opportunity to change past decisions to improve current decisions, thereby making it the preferred solution for modern tracking applications.

One approach to cluster tracking is to use clustering methods to group the data on each frame of data and to match the clusters over multiple frames of data just as in MFA/MHT applications. A key problem here is that of determining the number of clusters into which to group the data as well as the correct clustering of the data. Rather than making a firm (or hard) decision on each frame, a soft decision approach is to form multiple clustering hypotheses on each frame and make decisions on a single frame by considering the clusterings over multiple frames. The goal here is to formulate this "generalized data association" problem in which both individual objects and clusters are present. The association problem is called "generalized" because it reduces to the classical one when clusters (or groups) do not overlap.

In recent works [5, 23], unrelated to target tracking, space-time clustering techniques have been suggested. These algorithms consider the clustering of sequences, or frames of data, which are related over time. Carlotto [5] develops a space-time clustering method for moving target indicator radars. Scheirer [23] develops a dynamic auditory cluster algorithm that correlates multiple frames of data. For each frame several clustering hypotheses are formed using a Bayesian clustering technique. The optimal evolution of frames is obtained from the solution of a suitable assignment problem by means of dynamics programming.

Finally, one should observe that the term "clustering" has previously been used in MHT/MFA applications to mean "partitioning." The goal was to partition the association problem into a list of independent association problems to reduce the size of the measurement-to-track association problem [13, 18, 19, 16, 3]. This partitioning of the problem is distinct from the "clustering" considered here.

In previous work [11] we introduced a class of group-cluster assignment problems. In this chapter we will show that group-cluster tracking and the merged measurement or pixel-cluster tracking problem can be formulated within this general class of group-cluster tracking assignment problems.

One of the most important aspects to any group-cluster tracking system is the ability to correctly partition the data points (i.e., measurements, tracks, target features) into common groups. To this end, we review in the next subsection different clustering methods of special interest to cluster tracking. To fully exploit the spatio-temporal nature of the data, i.e., the frame-to-frame dependence of the data, it is important to base single-frame clustering decisions on the information of multiple frames of data. We term this approach multiple frame cluster tracking.

3.2. *Review of Clustering Techniques*

The main use of clustering is data compression. Given a data set (e.g. a set of measurements) $\mathcal{Z} = \{\mathbf{z}_1, \cdots, \mathbf{z}_N\}$ from an input space $Z \subset \mathbb{R}^d$ a clustering algorithm attempts to partition Z into natural groups or clusters based on some measure of similarity. The clustering result depends on the specific cluster algorithm and the similarity criteria used. One can distinguish between sequential, hierarchical, and cost function optimizing clustering algorithms. We will only discuss algorithms from the latter two classes here. The cost function optimizing algorithms can also be separated into hard and soft algorithms [25].

Definition 1: (Hard M-Clustering) A hard M-clustering of a data set $\mathcal{Z} \subset Z$ denotes the partitioning of $\mathcal{Z}$ into M sets (clusters, groups) $\{\mathcal{C}_1, \cdots \mathcal{C}_M\}$, such that a) $\mathcal{C}_i \neq \emptyset$, i = 1,...,M, b) $\mathcal{C}_i \cap \mathcal{C}_j = \emptyset, i \neq j, i, j = 1, \ldots, M$, and c) $\cup_{i=1}^{M} \mathcal{C}_i = \mathcal{Z}$.

In this chapter we do not make a formal distinction between groups and clusters. A hard clustering assigns a data point to exactly one cluster. A famous hard clustering algorithm is the k-means or Isodata algorithm [15]. Soft (or fuzzy) clustering, on the other hand, allows the assignment of a data point to multiple classes through a membership function which, for the Bayesian approach, represents a probability that the data vector belongs to a given class.

Definition 2: (Soft M-Clustering) A soft M-clustering of $\mathcal{Z}$ is characterized by M membership functions $u_i : \mathcal{Z} \to [0,1]$ $(i = 1, \cdots, M)$ such that $\sum_{i=1}^{M} u_i(\mathbf{z}) = 1$ for all $\mathbf{z} \in \mathcal{Z}$ and $0 < \sum_{j=1}^{N} u_i(\mathbf{z}_j) < N$ $(i = 1, \cdots, M)$.

The last requirement assures that the soft clustering does not produce a hard clustering. Given any soft clustering, a hard partitioning can be obtained by assigning a data point only to its most likely group. A widely used soft-clustering algorithm is the Expectation-Maximization (EM) algorithm [7].

Hierarchical cluster algorithms produce a hierarchy of nested clusterings [25]. Given a data set $\mathcal{Z}$ we denote by $\mathcal{H}^k(\mathcal{Z})$ a clustering (either hard or soft) of $\mathcal{Z}$ containing k clusters.

Definition 3: A clustering $\mathcal{H}^i$ is *nested* in a clustering $\mathcal{H}^j$, denoted by $\mathcal{H}^i \sqsubset \mathcal{H}^j$, if $j < i$ and each cluster in $\mathcal{H}^i$ is a subset of a set in $\mathcal{H}^j$ and at least one cluster of $\mathcal{H}^i$ is a proper subset of $\mathcal{H}^j$ [25].

Agglomerative hierarchical algorithms start from a clustering $\mathcal{H}^i$ and produce a clustering $\mathcal{H}^{i-1}$ such that $\mathcal{H}^i \sqsubset \mathcal{H}^{i-1}$, while divisive hierarchical algorithms start from a clustering $\mathcal{H}^i$ and produce a clustering $\mathcal{H}^{i+1}$ such that $\mathcal{H}^{i+1} \sqsubset \mathcal{H}^i$ [25]. While a number of specific hard hierarchical clustering methods exist one can in principle produce hierarchical clusterings with most soft or hard clustering methods. To this end, given an initial clustering $\mathcal{H}^i$, one obtains a divisive clustering $\mathcal{H}^j$ $(j > i)$ by dividing selected clusters. Similarly one can produce an agglomerative clustering $\mathcal{H}^k, k < i$ by combining selected clusters. Following such an approach one can produce a set of clustering hypotheses $\{\mathcal{H}^{i_{\min}}, \cdots, \mathcal{H}^{i-1}, \mathcal{H}^i, \mathcal{H}^{i+1}, \cdots, \mathcal{H}^{i_{\max}}\}$ based on an initial clustering $\mathcal{H}^i$. This represents one feasible approach to produce multiple clustering hypotheses of a data set.

One of the most difficult problems in clustering is to determine the correct number of clusters M to partition the data set. Many variants of algorithms exist to solve this problem. Typical algorithms start with a minimal set and increase the number of clusters until a certain stopping criterion is reached [17]. One useful criterion for density based clustering is the Bayesian Information Criterion (BIC) [24]. Given a model Θ, its BIC value is given by $\mathrm{BIC}(\Theta) = -2\log\mathcal{L}(\Theta) + d\log N$, with the model likelihood $\mathcal{L}(\Theta)$, the total number of parameters in the model d, and the number of samples N.

Given a single frame of data a model selection method allows one to choose the correct number of clusters for an optimal clustering of the data.

The final clustering will however only be based on the information from a single frame of data. If the frame of data is generated by a time-evolving system then it is possible that a different clustering is better suited to describe the evolving system. In other words, *the best model to fit a single frame of data might not be the best model to fit multiple dependent frames of data.* Thus, in order to find the best clustering for a single frame-of-data we can produce a number of candidate clustering hypotheses and select the best one based on information from multiple frames of data. This idea is motivated in the following subsection.

If it would be feasible to consider all possible clustering hypotheses for the frame of data then it would be guaranteed that the optimal one is contained in the set of clustering hypotheses; however, this is not possible in general. Given a set of clustering hypotheses we can in principle form new hypotheses by combining clusters from the different hypotheses. This suggest another approach to group-cluster tracking. After forming a set of clustering hypotheses we will collect all unique clusters from all the hypotheses and form a "best" clustering from the set of clusters based on information from multiple frames of data.

3.3. *Benefits of Multiple Frame Cluster Tracking*

While typical clustering techniques consider stationary systems, many realistic systems are in fact non-stationary. Indeed, the sensor measurements are precisely of this nature with the data changing its characteristics from frame to frame in such a way that one must consider multiple frames of data (past, present, and future) to decide on the correct clustering of a single frame of data. Thus, the objective in this section is to illustrate this phenomena in Figure 6 through the three panels (a) through (c). Panels (a) and (b) of Figure 6 show two different sets of four consecutive frames of data of a time-evolving system. A single frame clustering algorithm required to find a partitioning of Frame 1 into at most three clusters, can produce any (and more) of the partitions illustrated in panel (c) of Figure 6. It is impossible for the algorithm to decide which of these four partitions will fit the evolving system best. However, considering additional frames of data, it becomes clear which partitioning is most suitable to describe the system. In the time evolution of panel (a) the partitioning 3) of panel (c) would have best described the system, while in the time evolution depicted in panel (b) the partitioning 2) of panel (c) would have been best. It is clear from this example that even if the additional knowledge would have been available that a two-cluster partitioning would fit best, the algorithm would have not been able to produce the most suitable clustering from the information of a single frame of data. This type of example is typical for scenarios involving spawning missiles and countermeasures. The correct clustering of the data at the earliest instant, only possible through multi-frame clustering, will allow an accurate initiation of the spawning object.

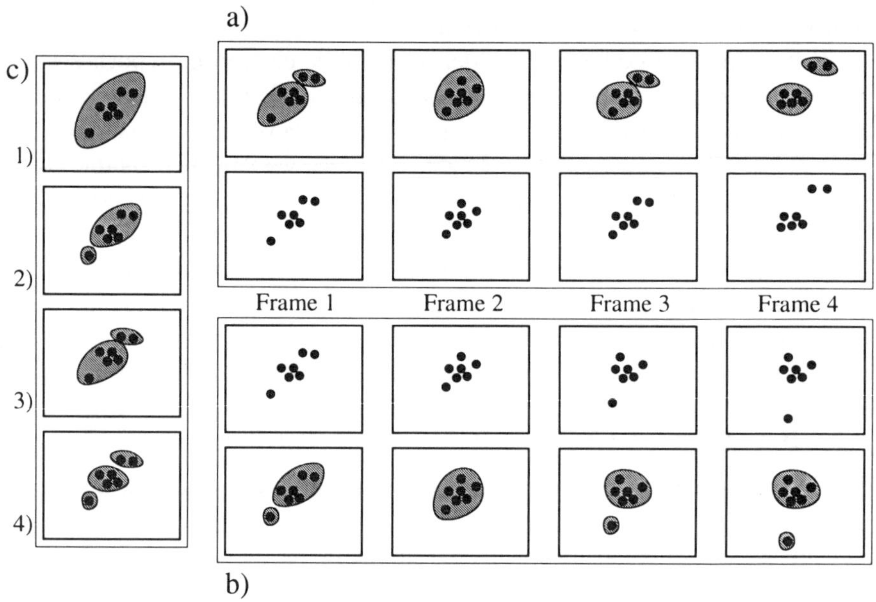

Fig. 6. The benefits of multi-frame clustering (see text for details.)

The goal of the next section is to formulate the group-cluster assignment problem for cluster tracking which incorporates multiple clustering hypotheses on each frame of data, one-to-one, many-to-one, and many-to-many assignments between frames of data, and imposes the set packing property on each frame of data. This generalized data association problem also governs the assignment problem for merged measurements in radar.

4. The Two-Dimensional Cluster Assignment Problem

The goal of this section is to give a formulation of the group-cluster assignment problem for group-cluster tracking association and for the merged measurement problem. While the ideas apply equally well to single and multiple frame association, the technical development will be restricted to clustering and matchings between two frames of objects, e.g., tracks and measurements, for the sake of brevity. The multiple frame analogue is reasonably straight forward and the three dimensional version is given as an example to illustrate the generalization.

The idea in the formulation of the problem is to consider several clustering (or grouping) hypotheses for each frame of data. Then, the distinct subclusters from all the cluster hypotheses are listed on each frame. The subclusters are then assigned across multiple frames of data subject to the set packing constraint on each frame. When the subclusters are each composed of a single individual object, then the resulting cluster assignment problem reduces to the usual multi-assignment or one-to-one assignment classically used in tracking.

4.1. *Individual Object Tracking*

As background for the cluster tracking problem, the two dimensional assignment problem most often used to track individual objects is briefly reviewed in this section. Since multi-assignment is part of the cluster tracking problem, it is included here in the individual object tracking.

The formulation can be expressed in either the dense or sparse form; however, the sparse form is used here. Starting with objects enumerated by $I = \{1, \ldots, m\}$ and $J = \{1, \ldots, n\}$, one first decides which objects in I can be associated with which objects in J (e.g., by using gating procedures) and denotes the feasible parings by $\mathcal{A} \subseteq \{(i,j) : i \in I,\ j \in J\}$. Further we denote the objects to which an object $i \in I$ can be assigned to in J by the set $A(i) = \{j : (i,j) \in \mathcal{A}\}$, and the objects in I to which $j \in J$ can be assigned to by the set $B(j) = \{i : (i,j) \in \mathcal{A}\}$. In addition, one must develop a cost function. (Although this development is not addressed specifically in this work, the cost coefficients can be based on the negative of the log of a likelihood ratio [2] c_{ij}). The resulting problem then is

$$\begin{aligned} \text{Minimize} \quad & \sum_{(i,j)\in\mathcal{A}} c_{ij}x_{ij}, \\ \text{Subject To:} \quad & \sum_{j\in A(i)} x_{ij} \le m_i \quad (i = 1, ..., m), \\ & \sum_{i\in B(j)} x_{ij} \le n_j \quad (j = 1, ..., n), \\ & x_{ij} \in \{0,1\}, \end{aligned} \tag{1}$$

where each $m_i \ge 1$ and $n_j \ge 1$.

The usual assignment problem used for data association in single frame processing in tracking is the *one-to-one assignment* obtained from (1) by using $m_i = 1$ and $n_j = 1$ for all i and j. The case $m_i \ge 1$ and $n_j \ge 1$ allows for the multi-assignment of tracks to measurements or vice-versa.

If, for example, the first index i denotes track numbers and j refers to a measurement number, then, in one-to-one assignments, each track can be assigned to at most one measurement and vice-versa. Also, the inequalities are present as opposed to equalities because a measurement may or may not be assigned to a track (e.g., it may be a false alarm) or a track may or may not be assigned (e.g., a target may not be detected).

While the problem of *one-to-one assignments* is genuinely an assignment problem, the *multi-assignment* problem is more appropriately identified with the classical (Hitchcock) transportation problem with integer capacity constraints on the assignments.

4.2. *Multiple Clustering Hypotheses*

We assume that we start with two lists of objects (e.g., measurements, features, or tracks). In a first step we hypothesize a set of complete candidate clusterings of the two data lists. Here is a formal definition.

Definition 4: Let P and Q denote two lists of objects and let $\mathcal{H}(P) = \{H_i(P)\}_{i\in I_H}$ and $\mathcal{H}(Q) = \{H_j(Q)\}_{j\in J_H}$ denote collections of complete clusterings of P and Q, respectively. In addition, let $\mathcal{P} = \{P_i\}_{i\in I}$, and $\mathcal{Q} = \{Q_j\}_{j\in J}$ denote the collection of all distinct clusters from the hypotheses $\mathcal{H}(P)$ and $\mathcal{H}(Q)$, respectively.

The first formulation of the cluster assignment problem will be based on explicit enumeration while the second and third formulation formulate the problem as a single assignment problem in which the distinct subclusters in $\mathcal{P}$ are matched to subclusters in $\mathcal{Q}$ in such a way that (1) the set packing property is maintained for both sets and (2) multiple assignments between the subclusters in the different frames are allowed. We distinguish between a *hard set packing property* and a *soft set packing property.*

Definition 5: (Hard Clusterings Set Packing Property) Find a subcollection $\{P_{i_1}, \ldots, P_{i_M}\}$ $(M \leq I)$ of $\mathcal{P}$ that is matched to a subcollection $\{Q_{j_1}, \ldots, Q_{j_N}\}$ $(N \leq J)$ of $\mathcal{Q}$ with the requirements that $\{P_{i_p}\}_{p=1}^{M}$ and $\{Q_{j_q}\}_{q=1}^{N}$ are set packings of P and Q, respectively, i.e.,

$$(a) P_{i_p} \neq \emptyset; \quad (b) \cup_{p=1}^{M} P_{i_p} \subseteq P; \quad (c) P_{i_p} \cap P_{i_q} = \emptyset$$

and similarly for $\mathcal{Q}$. In addition, each P_i should be allowed to be multiply assigned to a Q_j, and vice-versa.

Soft clusterings that allow overlap between the clusters cannot satisfy the partitioning property $P_{i_p} \cap P_{i_q} = \emptyset$ of the hard cluster. A number of schemes attempt to approximate this requirement and a discussion can be found in the book *Pattern Recognition* by Theodoridis and Koutroumbas [25], which is also an excellent reference for clustering methods. The soft clustering conditions similar to (a), (b), and (c) of the hard set packing property are as follows.

Definition 6: (Soft Clusterings Set Packing Property) Out of all the class memberships functions, find $u_{i_1}, \ldots, u_{i_M}$ satisfying

$$(a) \sum_{z \in \mathcal{Z}} u_{i_p}(z) > 0 \ (p = 1, \ldots, M), (b) \sum_{p=1}^{M} u_{i_p}(z) \leq 1, \quad \text{for all} \quad z \in \mathcal{Z}, \text{ and}$$

$$(c) \frac{1}{N} \leq PC = \frac{1}{M} \sum_{p=1}^{M} \sum_{j=1}^{N} u_{i_p}^2(z_j) \leq 1.$$

If the partition coefficient $PC \sim 1$, then the cluster is almost hard. For the remainder of the work we will restrict development to use of hard clusterings.

Figure 7 illustrates the two classes of group-cluster tracking assignment problems that are formulated in this chapter. The illustration shows two frames of data where Frame 1 consists of 11 observations, and Frame 2 of 10 observations. Figure 7(a) illustrates the approach that matches complete clustering hypotheses between frames and shows three candidate

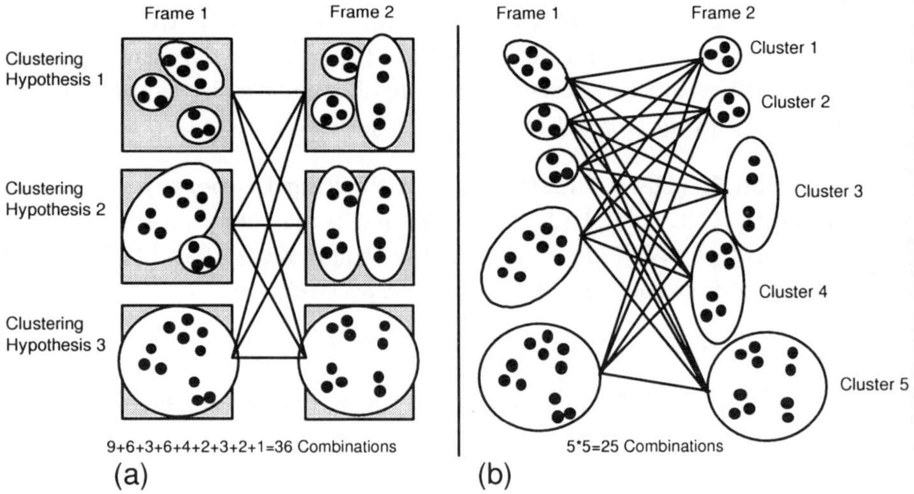

Fig. 7. Illustration of two formulations of the group-cluster assignment problem. (a) Matching complete clusterings between frames, and (b) matching clusters between frames.

group-clusterings of the respective frames. The resulting assignment problem would require to compute a total of 36 cost coefficients (however, since some of the clusters are equivalent, 11 of the computations are redundant). The following paragraph discusses the solution of this assignment problem through explicit enumeration which is adequate for "small" problems. Figure 7(b) shows the unique clusters from all clustering hypotheses and illustrates the resulting assignment problem that matches group-clusters between frames. This second and general class of the group assignment problem is discussed in Section 4.3.

Solution Through Explicit Enumeration of the Assignments One possible solution of the cluster tracking assignment problem is to determine the best score among the different complete clusterings in $\mathcal{H}(P)$ and a complete clustering in $\mathcal{H}(Q)$. Let $H_i(P) = \{P_{ik}\}_{k=1}^{P_i}$ and $H_j(Q) = \{Q_{jl}\}_{l=1}^{Q_j}$ denote the ith and j^{th} complete clusterings of P and Q respectively. If the subclusters P_{ik} and Q_{jl} to be assigned m_k^{ij} and n_l^{ij} times, respectively, then the assignment problem between $\mathcal{H}_i(P)$ and $\mathcal{H}_j(Q)$ is

$$\begin{aligned} &\text{Minimize} \quad \sum_{(k,l)\in\mathcal{A}} c_{kl}^{ij} x_{kl}, \\ &\text{Subject To:} \sum_{l\in A(k)} x_{kl} \le m_k^{ij} \quad (k=1,\dots,P_i), \\ &\qquad\qquad \sum_{k\in B(l)} x_{kl} \le n_l^{ij} \quad (l=1,\dots,Q_j), \\ &\qquad\qquad x_{ij} \in \{0,1\}, \end{aligned} \tag{2}$$

where each $m_k^{ij} \ge 1$ and $n_l^{ij} \ge 1$. Having computed the optimal score for each pairing (i,j), one chooses the one with the best score from this list. Note that if the number of complete clusterings in $\mathcal{H}(P)$ is M and the number in $\mathcal{H}(Q)$ is N, then one must solve $M \times N$ assignment problems. This number grows substantially over multiple frames of data.

4.3. *The Group-Cluster Assignment Problem*

In the previous section, we considered a formulation of the cluster tracking problem wherein the objective was to find the best matching between a complete clustering on one frame to one on the next, chosen from multiple possible complete clusterings on each frame. This approach essentially enumerates the assignment problems to find the best matching of a complete clustering on two distinct frames of data. In this section, the collection of all of these problems is collapsed into a single assignment problem. While this formulation is not guaranteed to solve the same problem as in the previous section, it is guaranteed to produce a matching whose overall score is at least as good as, if not better than, that found by matching complete clusterings to complete clusterings. As before, let $\mathcal{H}(P)$ and $\mathcal{H}(Q)$ denote collections of complete clusterings of P and Q, respectively. Next, let $\mathcal{P} = \{P_i\}_{i\in I}$ and $\mathcal{Q} = \{Q_j\}_{j\in J}$ denote the collection of all unique clusters from the hypotheses $\mathcal{H}(P) = \{H_i(P)\}_{i\in I_H}$ and $\mathcal{H}(Q) = \{H_j(Q)\}_{j\in J_H}$, respectively. The second formulation of the cluster tracking assignment problem attempts to match the clusters in $\mathcal{P}$ to clusters in $\mathcal{Q}$ while maintaining the set packing property for each. Note that the set packing property discussed in the previous subsection does not require that all the data be used. If not, then the remaining objects in P can be put into an additional set and combined with those actually assigned to form a set partitioning as used in the definition in the clustering. Thus, the problem formulated in this section is in a sense more general than that formulated in the previous section. We next present several formulations of the group-cluster assignment problem.

4.3.1. *First Formulation: Constraints Enumerated by Individual Objects in P and Q*

The case in which each group in one list is assigned to at most one group in the other list has a particularly attractive form. While this appears to restrict this approach to one-to-one assignments, the use of *subgroups* within a particular group adds additional flexibility for multi-assignment as explained later. To preserve the selection of subsets of $\mathcal{P}$ and $\mathcal{Q}$, we introduce the following definitions.

Definition 7: Let P and Q denote two lists of objects and let $\mathcal{P} = \{P_i\}_{i \in I}$ and $\mathcal{Q} = \{Q_j\}_{j \in J}$ denote collections of subsets of P and Q, respectively. Define the indicator functions

$$m_{ki} = \begin{cases} 1 & \text{if object } k \in P \text{ is in } P_i, \\ 0 & \text{otherwise} \end{cases}, \text{ and } n_{lj} = \begin{cases} 1 & \text{if object } l \in Q \text{ is in } Q_j, \\ 0 & \text{otherwise.} \end{cases}$$

Given this definition, the problem formulation is

$$\begin{aligned} \text{Minimize } & \sum_{(i,j) \in \mathcal{A}} c_{ij} x_{ij}, \\ \text{Subject To: } & \sum_{(i,j) \in \mathcal{A}} m_{ki} x_{ij} \leq 1 \quad (k \in P), \\ & \sum_{(i,j) \in \mathcal{A}} n_{lj} x_{ij} \leq 1 \quad (l \in Q), \\ & x_{ij} \in \{0, 1\}. \end{aligned} \tag{3}$$

The key new component of this formulation (3) is the use of the constraints $\sum_{(i,j) \in \mathcal{A}} m_{ki} x_{ij} \leq 1 \quad (k \in P)$ which says that an object $k \in P$ can be present in at most one pairing $(i, j) \in \mathcal{A}$ and that any group i can be assigned to at most one group j. A similar statement holds for objects $l \in Q$. Thus the groups that end up actually being assigned have the properties of a set packing. This particular formulation incorporates a set packing formulation for a single data set commonly used, e.g., in auctions. Also, the constraints in this formulation are posed in terms of the individual objects themselves rather than groups and thus may contain many redundant ones.

Multiple Assignments Via Subgroups The formulation (3) admits multi-assignment in a very structured manner if one allows subgroups within a group. Here is an example of how this might be used. Suppose a group (cluster) P_i on the first frame is to be allowed to be assigned to two groups Q_r, Q_s on the second frame. One way to accomplish this within the current formulation is to form another group, say $Q_{n+1} = \{Q_r, Q_s\}$ composed of subgroups Q_r and Q_s, and add this group to $\mathcal{Q}$. In fact, this formulation may be the preferred one for controlling multiple assignments between groups in one data set to those in another, especially for many-to-one assignments.

The Case of Singleton Groups The usual one-to-one assignment problem can be seen as a special case of (3) with the following identification. Let $P = \{1,\ldots,m\}$, $Q = \{1,\ldots,n\}$, $P_i = \{i\}$ for $i = 1,\ldots,m$, $(I = \{1,\ldots,m\})$, $Q_j = \{j\}$ for $j = 1,\ldots,n$, $(J = \{1,\ldots,n\})$, $m_i = 1$, and $n_j = 1$. Then the $m_{ik} = \delta_{ik}$ and $n_{lj} = \delta_{lj}$, so that the above assignment problem (3) reduces to the usual *one-to-one assignment* problem. The δ_{ik} are defined to be one if $i = k$ and zero otherwise.

$$
\begin{aligned}
\text{Minimize} \quad & \sum_{(i,j)\in\mathcal{A}} c_{ij}x_{ij}, \\
\text{Subject To:} \quad & \sum_{(i,j)\in\mathcal{A}} \delta_{ki}x_{ij} \le 1 \quad (k \in P) \Rightarrow \sum_{(j\in A(i)} x_{ij} \le 1 \text{ for } i = 1,\ldots,m \\
& \sum_{(i,j)\in\mathcal{A}} \delta_{lj}x_{ij} \le 1 \quad (l \in Q) \Rightarrow \sum_{i\in B(j)} x_{ij} \le 1 \text{ for } j = 1,\ldots,n \\
& x_{ij} \in \{0,1\}.
\end{aligned} \tag{4}
$$

4.3.2. *Second Formulation: Constraints Enumerated by Subclusters in $\mathcal{P}$ and $\mathcal{Q}$*

A more general formulation of the cluster tracking multi-assignment problem allows the multi-assignment between groups P_i of $\mathcal{P}$ and Q_j of $\mathcal{Q}$ directly and then adds the set packing as additional constraints. Using the hard set packing constraint this problem can be expressed as:

$$
\begin{aligned}
\text{Minimize} \quad & \sum_{(i,j)\in\mathcal{A}} c_{ij}x_{ij}, \\
\text{Subject To} \quad & \sum_{j\in\mathcal{A}(i)} x_{ij} \le m_i \quad (i \in I), \\
& \sum_{i\in\mathcal{B}(j)} x_{ij} \le n_j \quad (j \in J), \\
\text{(HSP)} \quad & x_{i_1 j_1} + x_{i_2 j_2} \le 1 \text{ for all } (i_1, j_1) \text{ and } (i_2, j_2) \in \mathcal{A} \\
& \qquad \text{for which } i_1 \ne i_2 \text{ and } P_{i_1} \cap P_{i_2} \ne \emptyset \\
& \qquad \text{or } j_1 \ne j_2 \text{ and } Q_{j_1} \cap Q_{j_2} \ne \emptyset, \\
& x_{ij} \in \{0, 1\}.
\end{aligned} \tag{5}
$$

The constraint (HSP) of Eqn. (5) is the aforementioned constraint on the (hard) set packing requirement for the final assignment.

A Special Case for One-to-One Assignments In the above formulation when $m_i = 1$ and $n_j = 1$ for all (i, j), the constraints $x_{i_1 j_1} + x_{i_2 j_2} \le 1$ for all (i_1, j_1) and $(i_2, j_2) \in \mathcal{A}$ for which $i_1 \ne i_2$ and $P_{i_1} \cap P_{i_2} \ne \emptyset$ or $j_1 \ne j_2$ and $Q_{j_1} \cap Q_{j_2} \ne \emptyset$ can be replaced by sums. The resulting problem for this special case is

$$
\begin{aligned}
\text{Minimize} \quad & \sum_{(i,j)\in\mathcal{A}} c_{ij}x_{ij}, \\
\text{Subject To} \quad & \sum_{j\in\mathcal{A}(i)} x_{ij} \le 1 \quad (i \in I), \\
& \sum_{i\in\mathcal{B}(j)} x_{ij} \le 1 \quad (j \in J), \\
& \sum_{j\in A(i_1)} x_{i_1 j} + \sum_{j\in A(i_2)} x_{i_2 j} \le 1 \text{ for which } i_1 \ne i_2 \text{ and } P_{i_1} \cap P_{i_2} \ne \emptyset, \\
& \sum_{i\in B(j_1)} x_{i j_1} + \sum_{i\in B(j_2)} x_{i j_2} \le 1 \text{ for which } j_1 \ne j_2 \text{ and } Q_{j_1} \cap Q_{j_2} \ne \emptyset, \\
& x_{ij} \in \{0, 1\}.
\end{aligned} \tag{6}
$$

A Lagrangian Relaxation Algorithm This second formulation of the cluster assignment problem is particularly well-suited to a Lagrangian relaxation algorithm in that the set packing constraint can be Lagrangian relaxed to the base problem consisting of either the usual one-to-one assignment problem or the multi-assignment problem. The nonsmooth optimization of the resulting problem is relatively straight forward. The final step in restoring the set packing constraint is that which remains.

4.4. *The Three-Dimensional Problem*

The multidimensional assignment versions of the above problems have been presented elsewhere [12]. Here is a brief summary. As before, let $\mathcal{H}(P)$, $\mathcal{H}(Q)$, and $\mathcal{H}(R)$ denote collections of complete clusterings of P, Q, and R, respectively. Next, let $\mathcal{P} = \{P_i\}_{i \in I}$, $\mathcal{Q} = \{Q_j\}_{j \in J}$, and $\mathcal{R} = \{R_k\}_{k \in K_c}$ denote the collection of all unique clusters from the hypotheses $\mathcal{H}(P) = \{H_i(P)\}_{i \in I_H}$, $\mathcal{H}(Q) = \{H_j(Q)\}_{j \in J_H}$, and $\mathcal{H}(R) = \{H_k(R)\}_{k \in K_H}$, respectively.

Definition 8: Let P, Q, and R denote three lists of objects and let $\mathcal{P} = \{P_i\}_{i \in I}$, $\mathcal{Q} = \{Q_j\}_{j \in J}$, and $\mathcal{R} = \{R_k\}_{k \in K}$ denote collections of subsets of P, Q, and R, respectively. Define the indicator functions

$$m_{pi} = \begin{cases} 1 & \text{if object } p \in P \text{ is in } P_i, \\ 0 & \text{otherwise} \end{cases}, \quad n_{qj} = \begin{cases} 1 & \text{if object } q \in Q \text{ is in } Q_j, \\ 0 & \text{otherwise} \end{cases},$$

$$o_{rk} = \begin{cases} 1 & \text{if object } r \in R \text{ is in } R_k, \\ 0 & \text{otherwise.} \end{cases}$$

Given this definition, the problem formulation is

$$\begin{aligned} \text{Minimize} \quad & \sum_{(i,j,k) \in \mathcal{A}} c_{ijk} x_{ijk}, \\ \text{Subject To:} \quad & \sum_{(i,j,k) \in \mathcal{A}} m_{pi} x_{ijk} \le 1 \quad (p \in P), \\ & \sum_{(i,j,k) \in \mathcal{A}} n_{qj} x_{ijk} \le 1 \quad (q \in Q), \\ & \sum_{(i,j,k) \in \mathcal{A}} o_{rk} x_{ijk} \le 1 \quad (r \in R), \\ & x_{ijk} \in \{0, 1\}. \end{aligned} \tag{7}$$

The constraints are enumerated based on the objects in P, Q, and R. Analogous to the second formulation of the two-dimensional cluster assignment problem, we have

$$
\begin{aligned}
\text{Minimize} \quad & \sum_{(i,j,k)\in\mathcal{A}} c_{ijk}x_{ijk}, \\
\text{Subject To} \quad & \sum_{(j,k)\in\mathcal{A}(i)} x_{ijk} \le m_i \qquad (i \in I), \\
& \sum_{(i,k)\in\mathcal{B}(j)} x_{ijk} \le n_j \qquad (j \in J), \\
& \sum_{(i,j)\in\mathcal{C}(k)} x_{ijk} \le o_k \qquad (k \in K), \\
& x_{i_1j_1k_1} + x_{i_2j_2k_2} \le 1 \text{ for all } (i_1,j_1,k_1) \text{ and } (i_2,j_2,k_2) \in \mathcal{A} \\
& \text{for which } i_1 \ne i_2 \text{ and } P_{i_1} \cap P_{i_2} \ne \emptyset \text{ or} \\
& \qquad\qquad j_1 \ne j_2 \text{ and } Q_{j_1} \cap Q_{j_2} \ne \emptyset \text{ or} \\
& \qquad\qquad k_1 \ne k_2 \text{ and } R_{k_1} \cap R_{k_2} \ne \emptyset, \\
& x_{ijk} \in \{0,1\},
\end{aligned}
\tag{8}
$$

where $m_i \ge 1$, $n_j \ge 1$, and $o_k \ge 1$. If $m_i = 1$, $n_j = 1$, and $o_k = 1$, then the set packing constraints can be replaced by sums similar to that discussed above for the two-dimensional problem.

5. The Merged Measurement Assignment Problem

Much of the motivation for the formulation of the cluster assignment problem presented in the previous sections has been based on forming multiple clustering hypotheses on each frame of data and deciding which clustering hypothesis is correct based on viewing a window of frames of data. The objective in this section is to explain how the merged measurement problem, originally formulated by Blair, Slocumb, Brown, and Register [4] follows exactly this same approach.

We assume that we have a set of tracks P and a set of measurements Q and let $\mathcal{H}(P)$ denote a collection of hypotheses that two or more tracks can be associated with a merged measurement and let $\mathcal{H}(Q)$ denote collections of complete clusterings of P and Q, respectively. Next, let $\mathcal{P} = \{P_i\}_{i\in I}$ and $\mathcal{Q} = \{Q_j\}_{j\in J}$ where $Q_j = \{j\}$ denotes the jth measurement. In this notation, we hypothesize that $P_i = \{i\}$ denotes the individual tracks for $i = 1,\ldots,M$, and P_i $(i = M+1,\ldots,M+U)$ denotes combinations of these M tracks that might be associated with the unresolved measurements. Then, the problem

$$
\begin{aligned}
&\text{Minimize} \sum_{(i,j)\in\mathcal{A}} c_{ij}x_{ij}, \\
&\text{Subject To:} \sum_{(i,j)\in\mathcal{A}} m_{ki}x_{ij} \le 1 \quad (k \in P), \\
&\qquad\qquad \sum_{i\in B(j)} x_{ij} \le 1 \quad (j = 1, \dots, N) \\
&\qquad\qquad x_{ij} \in \{0,1\}.
\end{aligned}
\tag{9}
$$

where

$$
m_{ki} = \begin{cases} 1 & \text{if object } k \in P \text{ is in } P_i, \\ 0 & \text{otherwise,} \end{cases}
$$

is equivalent to the formulation presented in the work of Blair, et al, but perhaps a little different notation in that we have used the indicator function m_{ki} instead of the double sum found in that paper. Thus, this formulation fits within the second cluster assignment formulation.

A third formulation of the merged measurement assignment problem is given by H Chen, T. Kirubarajan, and Y. Bar-Shalom [6], but this formulation is equivalent to the second cluster assignment formulation above wherein only one-to-one assignments are allowed in the association between groups.

6. Summary

In cluster tracking the fundamental problem is to partition data (either tracks or measurements) into groups of data points that can be represented by the parameters, which in turn describe the cluster to which the group of data points belong. Thus, finding an optimal clustering for a given set of data is a critical issue in cluster tracking. Through a simple example we have illustrated that it can be suboptimal to base clustering decisions on the information from a single frame of data, i.e., a single look at the data. Basing clustering decisions on multiple looks (or frames of data) at the data representing time dynamic objects shows considerable promise in improving these decisions in much the same way MHT/MFA tracking does when compared to single frame processing. The proposed approach requires the formation of multiple clustering hypotheses for each given frame of data using either hard or soft clustering techniques. The optimal clustering for a frame of data can then be obtained from the solution of the group assignment problem which minimizes the cost of assigning clusters between frames of data. The formulated group assignment problem is of sufficient generality to deal with three major classes of problems, namely the (a) group-cluster tracking problem, (b) pixel-cluster tracking problem, and (c) merged measurement problem. In addition, the formulation accommodates one-to-one, many-to-one, and many-to-many assignments. Most importantly, these formulations represent generalized data association in the sense that the assignment problem reduces to the classical one if the groups do not overlap.

Acknowledgments

This work was supported in part by the Air Force Office of Scientific Research under Grant Number F49620-00-1-0108.

References

[1] S. Blackman, *Multiple-Target tracking with radar applications,* Artech House, Norwood, MA, 1986.
[2] S. Blackman and R. Popoli, *Design and analysis of modern tracking systems,* Artech House, Boston, London, 1999.
[3] M. Chummun, T. Kirubarajan, K. Pattipati, and Y. Bar-Shalom, *Fast data association using multidimensional assignment with clustering,* IEEE Transactions on Aerospace and Electronic Systems, Vol. 37, pages 898–913, 2001.
[4] W. D. Blair, W. D., B. J. Slocumb, G. C. Brown, and A. H. Register, *2-D Measurement-to-Track Association for Tracking Closely Spaced, Possibly Unresolved, Rayleigh Targets: Idealized Resolution,* Aerospace Conference Proceedings, Vol. 4, pages 4.1543-4.1550, 2002.
[5] M. Carlotto, *MTI data clustering and formation recognition,* IEEE Transactions on Aerospace and Electronic Systems, Vol. 37, pages 524–536, 2001.
[6] H. Chen, T. Kirubarajan, and Y. Bar-Shalom, *Multiple Target Finite Resolution Sensors,* preprint, 2002.
[7] A. Dempster, N. Laird, and D. Rubin, *Maximum-likelihood from incomplete data via the EM algorithm,* Journal of the Royal Statistical Society, B, Vol. 39, page 39, 1977.
[8] S. DeVries and R. Vohra, *Combinatorial auctions: A survey,* Technical Report, http://citeseer.nj.nec.com/devries01combinatorial.html, 2000.
[9] O. Drummond, S. Blackman, and G. Petrisor, *Tracking clusters and extended objects with multiple sensors,* SPIE Vol. 1305, Signal and Data Processing of Small Targets, pages 362-375, 1990.
[10] W. Elmaghraby and P. Keskinocak, *Combinatorial Auctions in Procurement,* Technical Report, School of Industrial and Systems Engineering, Georgia Institute of Technology, 2002.
[11] S. Gadaleta, M. Klusman, A. B. Poore, and B. J. Slocumb, *Multiple Frame*

Cluster Tracking, SPIE Vol. 4728, Signal and Data Processing of Small Targets, pages 275–289, 2002.

[12] S. Gadaleta, A. B. Poore, and B. J. Slocumb, *Some Assignment Problems Arising From Cluster Tracking*, ORNL Workshop on Signal Processing, Communications and Chaotic Systems: A Tribute to Rabinder N. Madan, Harbor Island Conference Center, 2002.

[13] M. Kovacich, *An application of MHT to group to object tracking*, Proceedings SPIE Vol. 1481, Signal and Data Processing of Small Targets, pages 357–370, 1991.

[14] C. Li and K. Sycara (2001), *Algorithms for combinatorial coalition forming and payoff division in an electronic marketplace*, Technical Report, Carnegie Mellon University, 2001.

[15] J. MacQueen, *Some methods for classification and analysis of multivariate observations*, Proceedings of the Fifth Berkeley Symposium on Mathematical statistics and probability, University of California Press, pages 281–297, 1967.

[16] N. Nabaa and R. Bishop, *Clustering approach to the multitarget multisensor tracking problem*, SPIE Vol. 3163, Signal and Data Processing of Small Targets, pages 226–237, 1997.

[17] Nasa, Bayesian Learning Group, http://ic.arc.nasa.gov/projects/bayes-group/autoclass/autoclass-refs.html.

[18] A. B. Poore and N. Rijavec, *A Numerical Study of Some Data Association Problems Arising in Multitarget Tracking*, in W. W. Hager, D. W. Hearn, and P. M Pardalos, editors, Large Scale Optimization: State of the Art, Kluwer Academic Publishers B. V., Boston, MA, pages 339–361, 1994.

[19] A. B. Poore, N. Rijavec, T. Barker, and M. Munger, *Data association problems posed as multidimensional assignment problems: numerical simulations*, in Oliver E. Drummond, editor, Proceedings of SPIE, pages 564–573, 1993.

[20] A. B. Poore and A. J. Robertson, III, *A new class of Lagrangian relaxation based algorithms for a class of multidimensional assignment problems*, Computational Optimization and Applications, Vol. 8, No. 2, pages 129-150, 1997.

[21] A. B. Poore and X. Yan, *Some Algorithmic Improvements in Multi-frame Most Probable Hypothesis Tracking*, Signal and Data Processing of Small Targets, Oliver E. Drummond, editor, SPIE, 1999.

[22] D. Porter, D. Torma, J. Ledyard, J. Swanson, and M. Olson, *The First Use of a Combined Value Auction for Transportation Services*, Interfaces, Vol. 32, pages 4–12, 2002.

[23] E. Scheirer, *Music-listening systems*, Massachusetts Institute of Technology, S.M. Media Arts and Sciences, PhD Thesis, 2000.

[24] G. Schwarz, *Estimating the dimension of a model*, Annals of Statistics, Vol. 6, pages 461–464,

[25] S. Theodoridis and K. Koutroumbas, *Pattern Recognition*, Academic Press, 1999.

CHAPTER 20

COORDINATING VERY LARGE GROUPS OF WIDE AREA SEARCH MUNITIONS

Paul Scerri, Elizabeth Liao, Justin Lai, Katia Sycara
Carnegie Mellon University
pscerri@cs.cmu.edu, eliao@andrew.cmu.edu, guomingl@andrew.cmu.edu, katia@cs.cmu.edu

Yang Xu, Mike Lewis
University of Pittsburgh
xuy3@pitt.edu, ml@sis.pitt.edu

Coordinating hundreds or thousands of Unmanned Aerial Vehicles (UAVs), presents a variety of new exciting challenges, over and above the challenges of building single UAVs and small teams of UAVs. We are specifically interested in coordinating large groups of Wide Area Search Munitions (WASMs), which are part UAV and part munition. We are developing a "flat", distributed organization to provide the robustness and flexibility required by a group where team members will frequently leave. Building on established *teamwork* theory and infrastructure we are able to build large teams that can achieve complex goals using completely distributed intelligence. However, as the size of the team is increased, new issues arise that require novel algorithms. Specifically, key algorithms that work well for relatively small teams, fail to scale up to very large teams. We have developed novel algorithms meeting the requirements of large teams for the tasks of instantiating plans, sharing information and allocating roles. We have implemented these algorithms in reusable software proxies using the novel design abstraction of a *coordination agent* that encapsulates a piece of coordination protocol. We illustrate the effectiveness of the approach with 200 WASMs coordinating to find and destroy ground based targets in support of a manned aircraft.

1. Introduction

Wide Area Search Munitions (WASMs) are a cross between an unmanned aerial vehicle and a munition. With an impressive array of onboard sensors and autonomous flight capabilities WASMs can play a variety of roles in a modern battle field including reconnaissance, search, battle damage assessment, communications relays and decoys. Also being able to play the role of munition makes WASMs a very valuable asset for battlefield commanders. In the foreseeable future, it is envisioned that groups of the order of 100 WASMs will support and protect troops in a battlespace.

Getting large groups of WASMs to cooperate in dynamic and hostile environments is an exciting though difficult challenge. There have been significant successes in automated coordination [5, 8, 18, 31], but the number of entities involved has been severely limited due to the failure of key algorithms to scale to the challenges of large groups. When coordinating small groups of WASMs there are a variety of challenges such as formation flying and avoiding mid-air collisions. However, when we scale up the number of WASMs in the group, a new set of challenges, attributable to the scale of the team, come to the fore. For example, communication bandwidth becomes a valuable commodity that must be carefully managed. This is not to say that the challenges of small teams disappear, only that there are additional challenges. The focus of this chapter, is on those challenges that occur only when the size of the group is scaled up.

Given the nature of the domain, we are pursuing a completely distributed organization that does not rely on any specific entity, either WASM or human, for continued operation. This makes the overall system more robust to enemy activity. Our flat organization builds on well understood theories of *teamwork* [7, 13, 19, 16, 35]. Teamwork has the desirable properties of flexibility and robustness we require. Coordination based on ideas of teamwork requires that a number of algorithms work effectively together. We encapsulate our teamwork algorithms in a domain independent, reusable software proxy [27, 18]. A proxy works in close cooperation with a domain level agent to control a single team member. Specifically, the proxy works in close cooperation with an autopilot to control a single WASM. The proxies communicate among themselves and with their domain agent to achieve coordination.

The proxies execute Team Oriented Plans (TOPs) that break team activities down into individual activities called *roles*. TOPs are specified a priori, typically by a human designer, and specify the means by which the team will achieve its joint goals. Typically, TOPs are parameterized in templates and can be instantiated at runtime with specific details of the environment. For example, a TOP for destroying a target might have the specific target as a parameter. Importantly, the TOP does not specify who performs which role, nor does the TOP specify low level coordination details. Instead, these generic coordination "details" are handled by the proxies at runtime, allowing the team to leverage available resources and overcome failures.

The proxies must implement a range of algorithms to facilitate the execution of a TOP, including algorithms for instantiating TOPs, allocating roles and sharing relevant information. To build large teams, novel approaches to key algorithms are required. Specifically, we have developed novel approaches to creating and managing team plans, to allocating roles and to sharing information between team members. Our approach to plan instantiation allows any proxy to instantiate a TOP. The team member can then initiate coordination for, and execution of, that TOP and then the whole team (or just a part) can be involved in the coordination and execution.

We are also developing new communication reasoning that works by simply passing pieces of information to group members more likely to know who needs that information. Previous algorithms for reasoning about communication have made assumptions that do not hold in very large groups of WASMs. Specifically, previous algorithms have either assumed that centralization is possible or have assumed that agents have accurate models of other members of the group. Because of a phenomenon called "small world networks" [38] (in human groups this phenomenon is captured informally by the notion of "six degrees of separation") the result of our simple communication technique is targeted information delivery in an efficient manner. Our algorithm avoids the need for accurate information about group members and functions well even when group members have only very vague information about other group members.

Our implementation of the proxies is based on the abstraction of a *coordination agent*. Each coordination agent is responsible for a "chunk" of the overall coordination and encapsulates a protocol for one aspect of the coordination. We use a separate coordination agent for each plan or subplan, role and piece of information that needs to be shared. Specifically, instead of distributed protocols, which provide no single agent a cohesive view of the state of coordination, that state is encapsulated by the coordination agent and moves with that agent. Thus, the proxies can be viewed as a mobile agent platform upon which the coordination agents execute the TOPs. A desirable side effect of this design abstraction is that it is easier to build and extend complex "protocols" since the complexity of the protocol is hidden in the coordination reasoning, rather than being spread out over many agents.

We are evaluating our approach in a WASM simulation environment that emphasizes the coordination issues, without requiring too much attention to aerodynamic or low-level control issues. We have implemented two different forms of control, centralized and distributed, to allow us to quickly test ideas then perform more detailed validation. Our initial experiments have revealed some interesting phenomena including that very simple target allocation algorithms can perform surprisingly well under some circumstances.

2. Wide Area Search Munitions

Wide Area Search Munitions (WASMs) are a cross between an unmanned aerial vehicle and a standard munition. The WASM has fuel for about 30 minutes of flight, after being launched from an aircraft. The WASM cannot land, hence it will either end up hitting a target or self destructing. The sensors on the WASM are focused on the ground and include video with automatic target recognition, ladar and GPS. It is not currently envisioned that WASMs will have an ability to sense other objects in the air. WASMs will have reliable high bandwidth communication with other WASMs and with manned aircraft in the environment. These communication channels will be required to transmit data, including video streams, to human controllers, as well as for the WASM coordination.

The concept of operations for WASMs are still under development, however, a wide range of potential missions are emerging as interesting. A driving example for our work is for a team of WASMs to be launched from an AC-130 aircraft supporting special operations forces on the ground. The AC-130 is a large, lumbering aircraft, vulnerable to attack from the ground. While it has an impressive array of sensors, those sensors are focused directly on the small area of ground where the special operations forces are operating. The WASMs will be launched as the AC-130 enters the battlespace. The WASMs will protect the flight path of the AC-130 into the area of operations of the special forces, destroying ground based threats as required. Once the AC-130 enters a circling pattern around the special forces operation, the WASMs will set up a perimeter defense, destroying targets of opportunity both to protect the AC-130 and to support the soldiers on the ground. Even under ideal conditions there will be only one human operator on board the AC-130 responsible for monitoring and controlling the group of WASMs. Hence, high levels of autonomous operation and coordination are required of the WASMs themselves.

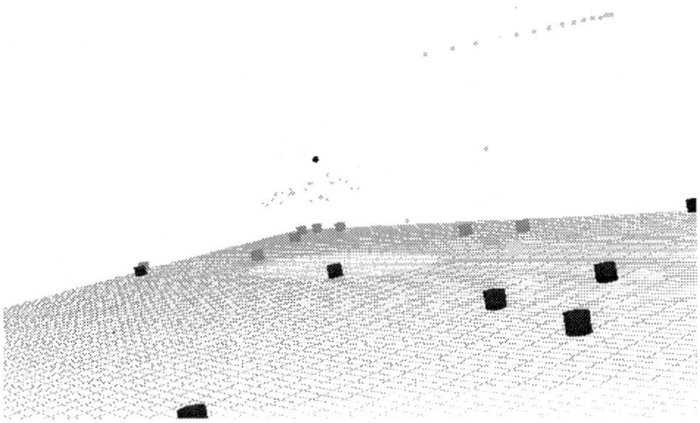

Fig. 1. A screenshot of the simulation environment. A large group of WASMS (small spheres) are flying in protection of a single aircraft (large sphere). Various SAM sites are scattered around the environment. Terrain type is indicated by the color of the ground.

Many other operations are possible for WASMs. Given their relatively low cost compared to Surface-to-Air Missiles (SAMs), WASMs can be used simply as decoys, finding SAMs and drawing fire. WASMs can be used as communication relays for forward operations, forming an adhoc network to provide robust, high bandwidth communications for ground forces in a battle zone. Since a WASM is "expendable", it can be used for reconnaissance in dangerous areas, providing real-time video for forward operating forces. Many other operations could be imagined in support of both manned air and ground vehicles, if issues related to coordinating large groups can be adequately resolved.

While our domain of interest is teams of WASMs, the issues that need to be addressed have close analogies in a variety of other domains. For example, coordinating resources for disaster response involves many of the same issues [23], as does intelligent manufacturing [29] and business processes. These central issues of distributed coordination in a dynamic environment are beginning to be addressed, but in all these domains current solutions do not efficiently scale to large numbers of group members.

3. Large Scale Teamwork

The job for the proxies is to take the TOP templates, instantiate TOPs as events occur in the environment then manage the execution of the instantiated TOPs. To achieve this a number of algorithms must work effectively together. Events occurring in the environment will only be detected by some agents (depending on sensing abilities). The occurrance of these events may need to be shared with other proxies so that a single proxy has all the information required to instantiate a plan. Care must be taken to ensure that there are not duplicate or conflicting team plans instantiated. Events occurring in the environment need to be shared with agents performing roles that are impacted by those events. Once the plans are instantiated, roles need to be allocated to best leverage the team capabilities. Plans also need to be terminated when they are completed, irrelevant or unachievable. Other algorithms, such as ones for allocating resources, may also be required but are not considered here. All the algorithms must work together efficiently and robustly in order for the team to achieve its goals.

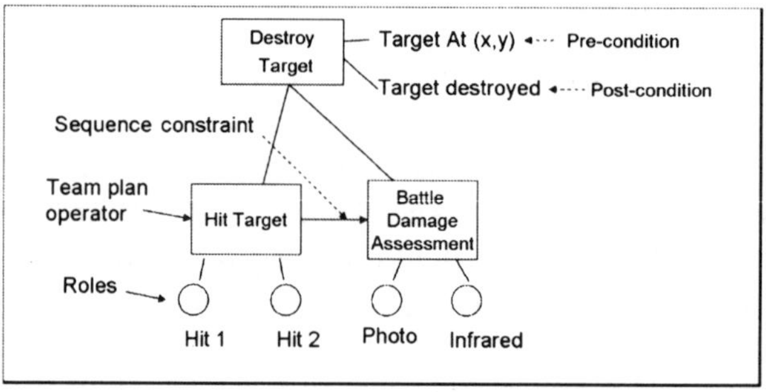

Fig. 2. An example team plan for destroying a ground based target. There are four roles, that will be instantiated in two stages, destroying the target (which requires that two WASMs hit the target) and the subsequent battle damage assessment (which requires both a photo and infrared imagining).

Viewed abstractly, the reasoning of the team can be seen as a type of hierarchical reasoning. At the top of the hierarchy are the plans that will be executed by the team. Those plans get broken down into more detailed plans, until the pieces, which we call roles, can be performed by a single team member. The next layers of the hierarchy deal with allocating those roles and finding coalitions for sets of roles that must be performed together. Finally, at the bottom of the hierarchy, is the detailed reasoning that allows team members performing as a part of a coalition to work together effectively. In these small coalitions we can apply standard teamwork coordination techniques such as STEAM. The basic idea is shown in Figure

3. The important caveat is that there is no hierarchical reasoning imposed on the team, the hierarchical view is simply a way of understanding what is happening. In the remainder of this section, we describe the proxies, the coordination agents and some of the key algorithms.

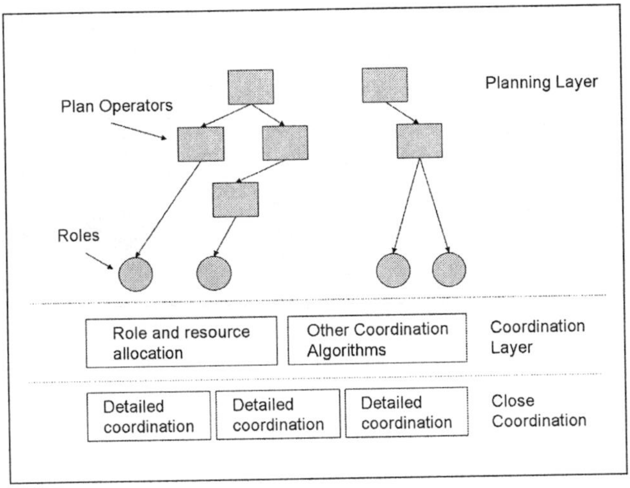

Fig. 3. Conceptual view of teamwork reasoning hierarchy. At the top, boxes represent team plans which are eventually broken down into individual roles. The roles are sent to the coordination layer which allocates the roles and resources to execute the plans. Finally, at the detailed level, specific sub-teams must closely coordinate to execute detailed plans.

3.1. *Machinetta Proxies*

To enable transitioning our coordination techniques to higher fidelity simulation environments or other domains, we separate the low level dynamic control of the WASM from the high level coordination code. The general coordination code is encapsulated in a *proxy* [18, 36, 26, 32]. There is one proxy for each WASM. The basic architecture is shown in Figure 4. The proxy communicates via a high level, domain specific protocol with an intelligent agent that encapsulates the detailed control algorithms of the WASM. Most of the proxy code is domain independent and can be readily used in other domains requiring distributed control. The proxy code, known as Machinetta, is a substantially extended and updated version of the TEAMCORE proxy code [36]. TEAMCORE proxies implement *teamwork* as described by the STEAM algorithms [35], which are in turn based on the theory of *joint intentions* [19, 7].

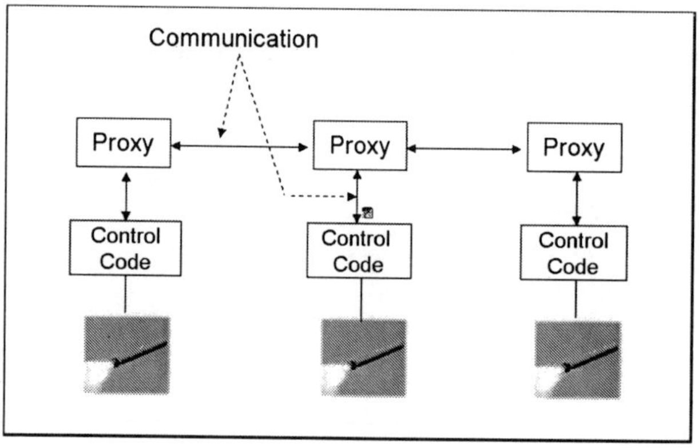

Fig. 4. The basic system architecture showing proxies, control code and WASMs being controlled.

3.1.1. *Coordination Agents*

In a dynamic, distributed system protocols for performing coordination need to be extremely robust. When we scale the size of a team to hundreds of agents, this becomes more of an issue than simply writing bug-free code. Instead we need abstractions and designs that promote robustness. Towards this end, we are encapsulating "chunks" of coordination in *coordination agents*. Each coordination agent manages one specific piece of the overall coordination. When control over that piece of coordination moves from one proxy to another proxy, the coordination agent moves from proxy to proxy, taking with it any relevant state information. We have coordination agents for each plan or subplan (PlanAgents), each role (RoleAgents) and each piece of information that needs to be shared (InformationAgents). For example, a RoleAgent looks after everything to do with a specific role. This encapsulation makes it far easier to build robust coordination.

Coordination agents manage the coordination in the network of proxies. Thus, the proxy can be viewed simply as a mobile agent platform that facilitates the functioning of the coordination agents. However, the proxies play the additional important role of providing and storing local information. We divide the information stored by the proxies into two categories, domain specific knowledge, K, and the coordination knowledge of the proxy,

CK. K is the information this proxy knows about the state of the environment. For example, the proxy for a WASM knows its own location and fuel level as well as the location of some targets. This information comes both from local sensors, reported via the domain agent, and from coordination agents (specifically InformationAgents, see below) that arrive at the proxy. CK is what the proxy knows about the state of the team and the coordination the team is involved in. For example, CK includes the known team plans, some knowledge about which team member is performing which role and the TOP templates. At the most abstract level, the activities of the coordination agents involve moving around the proxy network adding and changing information in C and CK for each agent. The content of K as it pertains to the local proxy, e.g., roles for the local proxy, govern the behavior of that team member. The details of how a role is executed by the control agent, i.e., the WASM, are domain (and even team member) dependant.

A *Factory* at each proxy is responsible for creating coordination agents as required.[a] It creates a *PlanAgent* when the pre-conditions of a plan template are met and an *InformationAgent* when a new piece of domain information is sensed locally by the proxy allowing the team to share information sensed locally by a proxy. The algorithm is shown in Figure 5.

Factory
loop
 Wait for state change
 foreach template ∈ TOP Templates
  if matches (*template*, K)
   Create PlanAgent(*template*, K)
 end for
 if new locally sensed information in K
  Create InformationAgent (new information)
end loop

Fig. 5. Algorithm for a proxy's factory.

[a] *Factory* is a software engineering term for, typically, an object that creates other objects

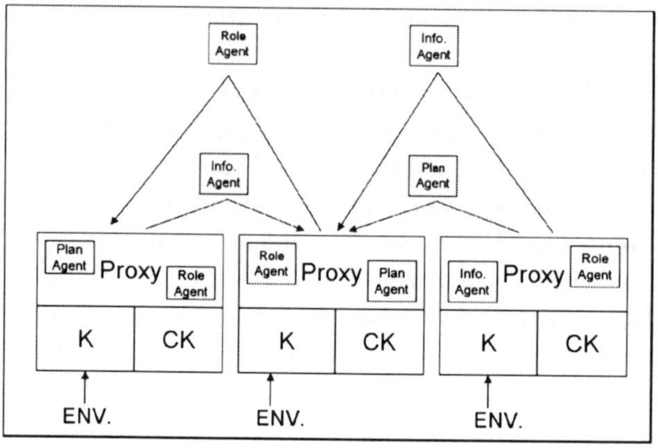

Fig. 6. High level view of the implementation, with coordination agents moving around a network of proxies.

3.2. *Team Oriented Plans*

The basis of coordination in the Machinetta proxies are a *Team Oriented Plans* (TOP) [28]. A TOP describes the joint activities that must take place for the team to achieve its goals. At any point in time, the team may be executing a number of TOPs simultaneously. TOPs are instantiated from TOP *templates*. These templates are designed before the team begins operation, typically by humans to ensure compliance with established doctrine or best practices. A TOP is a tree structure, where leaf nodes are called *roles* and are intended to be performed by a single team member. For example, a typical TOP for the WASM domain is to destroy a ground based target as shown in Figure 2. Such a plan is instantiated when a ground based target is detected. The plan is terminated when the target is confirmed as destroyed or the target becomes irrelevant. The plan specifies that the roles are to actually hit the target and to perform battle damage assessment. The battle damage assessment must be performed after the target has been hit. The coordination algorithms built into the proxies handle the execution of the TOP, hence the plan does not describe the required coordination nor how the coordination needs to be performed. Instead the TOP describes the high level activities and the relationship and constraints between those activities.

3.2.1. *Plan Monitoring with PlanAgents*

A *PlanAgent* is responsible for "managing" a plan. This involves instantiating and terminating roles as required and stopping execution of the plan when the plan either succeeds, becomes irrelevant or is no longer achievable. These conditions are observed from K in the proxy state. Currently, the PlanAgent must simply match conditions using string matching against post-conditions in the template, but we can envision more sophisticated reasoning in the future.

Because plans are instantiated in a distributed manner, the PlanAgents need to ensure that there are not other plans that are attempting to achieve the same goal (e.g., hit the same target) or other plans that may conflict. We discuss the mechanisms by which a PlanAgent can avoid these conflicts below. To facilitate the conflict avoidance (and detection) process, as well as keeping the team appraised of ongoing activities, the first thing a PlanAgent does is create an InformationAgent to inform the other proxies (who will update CK.)

If the PlanAgent does not detect any conflicts, it executes its main control loop until the plan becomes either irrelevant, unachievable or is completed. For each role in the plan, a *RoleAgent* is created. RoleAgents are coordination agents that are responsible for a specific role. We do not describe the RoleAgent algorithms in detail here, see [12] for details. Suffice it to say that the RoleAgent is responsible for finding a team member to execute that role. As the plan progresses, the required roles may change, in which case the PlanAgent must terminate the current RoleAgents and create new RoleAgents for the new roles. It is also possible that a previously undetected plan conflict is found and one plan needs to be terminated. The PlanAgents responsible for the conflicting plans jointly determine which plan to terminate (not shown for clarity). When the plan is completed, the PlanAgent terminates any remaining RoleAgents and finishes. The overall algorithm is shown in Figure 7.

3.2.2. *Instantiating Team Oriented Plan Templates*

The TOP templates typically have open parameters which are instantiated with specific domain level information at run time. Specifically, the *Factory* uses K to match against open parameters in plan templates. The matching

```
PlanAgent
  Wait to detect conflicts between plans
  if conflict detected then end
  else
   Create InformationAgent to inform others of plan
   Instantiate initial RoleAgents
   while (¬irrelevant & ¬complete & ¬unachievable)
    Wait for change in K or CK
    Check if RoleAgents need to be terminated
    Instantiate new RoleAgents if required
    if newly detected plan conflicts then
     Terminate this plan or conflicting
    end if
   end while
  end if
  Terminate all RoleAgents
```

Fig. 7. Algorithm for a PlanAgent

process is straightforward and currently involves simple string matching.[b] The *Factory* must also check CK to ensure that the same TOP has not been previously instantiated. When the team is very large, it is infeasible to have all team members agree on which plan to instantiate or even for all team members to know that a particular plan has been instantiated. For example, in a team with 100 members, it may take on the order of minutes to contact all members, significantly delaying execution of the plan. However, this is what is typically required by teamwork models. Instead, we allow any proxy that detects all the preconditions of a plan to instantiate that plan. Hence, notice that when a factory at any proxy notices that preconditions are met, the TOP is initiated immediately and a PlanAgent is created (see below).

[b]We can envision more sophisticated matching algorithms and even runtime planning, however to date this has not been required.

3.2.3. *Avoiding Conflicting Plans*

While the distributed plan instantiation process allows the team to instantiate plans efficiently and robustly, two possible problems can occur. First, the team could instantiate different plans for the same goal, based on different preconditions detected by different members of the team. For example, two different plans could be instantiated by different factories for hitting the same target depending on what particular team members know or sense. Second, the team may initiate multiple copies of the same plan. For example, two WASMs may detect the same target and different factories instantiate identical plans to destroy the same target. While our algorithms handle conflict recognition and resolution (see PlanAgent algorithm), minimizing conflicts to start with minimizes excess communication and wasted activity. When a PlanAgent is created for a specific plan, the first thing it does is "wait to detect conflict". This involves checking CK to determine whether there are conflicting plans, since CK contains coordination knowledge and will contain information about the conflicting plans,. Clearly, there may be conflicting plans the proxy does not know about, because they are not in CK, and thus there may be a conflict, not immediately apparent to the PlanAgent.

We are currently experimenting with a spectrum of algorithms for minimizing instantiations of conflicting plans. Each of the algorithms implements the "Wait to detect conflict" part of the PlanAgent algorithm in a different way. At one end of the spectrum we have a specific, deterministic rule based on specific information about the state of the team. We refer to this instantiation rule as the *team status instantiation rule.* When using this rule, we attached a mathematical function to each TOP. The value of that function can be computed from information in K. For example, the function attached to the TOP for destroying a target is based on distance to the target. Unless the PlanAgent computes that the local proxy has the highest possible value for that function, it should not proceed. The advantage of this rule is that there will be no conflicts, provided that K is accurate. The disadvantage of the rule is that many InformationAgents must move around the proxies often to keep K up-to-date.

At the other end of the spectrum, we have a probabilistic rule that requires no information about other team members. This rule, which we refer to as the *probabilistic instantiation rule*, requires that the PlanAgent wait a random amount of time, to see whether another team member instantiates that plan (or a conflicting plan.) Thus, InformationAgents for newly instantiated TOPs at other proxies have some time to reach the proxy, update CK and avoid a costly conflict. The advantage of this rule, is that no information is required about other team members to use this rule, thus reducing the volume of InformationAgents required. There are two disadvantages. First, there may be conflicting plans instantiated. Second, there may be a significant delay between detection of pre-conditions and the instantiation of the plan depending on how long the PlanAgents wait.

In between these two extremes, we define another rule, which we refer to as the *local information rule*, that requires that a proxy must detect some of the TOP's preconditions locally, in order to instantiate the plan. Specifically, at least one of the TOPs preconditions must have come into K directly from the environment, rather than via an InformationAgent. Although this will lead to conflicting plans when multiple proxies locally sense preconditions, it is easier to determine where the conflicts might occur and resolve them quickly. Specifically we can look for proxies with the ability to locally sense information, e.g., those in a specific part of the environment. The major disadvantage of this rule is that when a TOP has many preconditions the team members that locally detect specific preconditions may never get to know all the preconditions and thus not instantiate the plan.

Figure 8(a) shows the result of a simple simulation of the three instantiation rules. We used simple models of the environment to work out how often InformationAgents must move around in order to implement the three rules. This "cost" is indicated by the left-hand column and uses a logarithmic scale. The right hand column shows the number of plan conflicts that result. A conflict occurs when two or more PlanAgents proceed before they have been informed that the other has proceeded. Clearly, the *team status* rule gives a different tradeoff between conflicts and cost than the other rules. Notice that the precise behavior of the *probabilistic rule* depends on the specific parameter settings. Figure 8(b) shows how many conflicts result from this approach as we increase the number of PlanAgents. The precise slope of the line depends on the amount of time the PlanAgent is willing to wait and the length of time it takes to communicate that the PlanAgent has been instantiated.

3.3. *Information Sharing*

Information or events sensed locally by an agent will often not be sensed by other agents in the team. In some cases, however, that information will

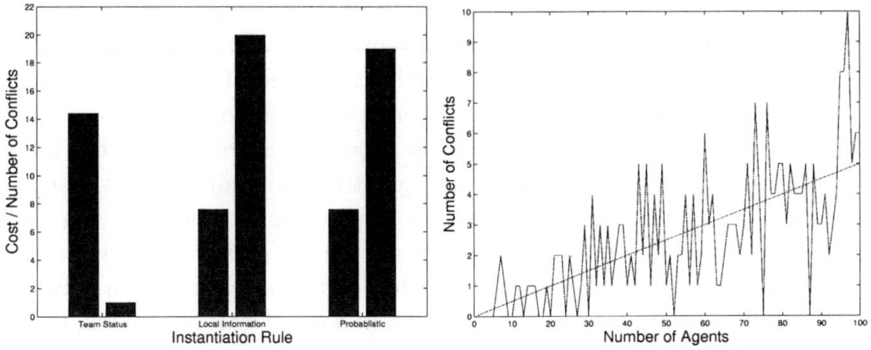

(a) (b)

Fig. 8. (a) The number of plan instantiations as we increase the number of agents using the probabilistic instantiation rule. The straight line represents the average of a large number of runs. The jagged line shows output from specific runs, highlighting the high variance. (b) The number of plan instantiations using the three different rules. In this simulation, there were 200 agents and a message took 600ms to be transmitted. For the probabilistic instantiation rule, the PlanAgent would wait upto 10s.

be critical to other members of the team, hence should be communicated to them. For example, consider the case where one agent detects that a ground target has moved into some trees. It needs to inform the WASM that is tasked with destroying that target, but will typically not know which WASM that is or whether any WASM is or whether the WASM has already been informed of the move (perhaps many times). A successful information sharing algorithm needs to deliver information where it is required without over loading the communication network.

Previous algorithms for sharing information in a multiagent system have made assumptions that do not hold in very large groups of WASMs (or large teams in general). Specifically, algorithms either assume that centralization is possible [33] or assume that agents have accurate models of other members of the group [35]. Often techniques for communication assume that an agent with some potentially relevant information will have an accurate model of the rest of the group. The model of the group is used to reason about which agents to communicate the information to (and whether there is utility in communicating at all [35, 26]). However, in large groups, individual agents will have very incomplete information about the rest of the group, making the decision about to whom to communicate some information much more difficult. Moreover, both as a design decision and for practical reasons, communication in a centralized way is not appropriate.

We are developing new communication reasoning that reduces the need to know details about other team members by exploiting the fact that, even in very large groups, there is a low *degree of separation* between group members. We assume the agents have point-to-point communication channels with a small percentage ($< 1\%$) of other group members. Having a low degree of separation means that a message can be passed between any two agents via a small number of the point-to-point connections. Such networks are known as *small worlds networks* [38]. In a small worlds network, agents are separated from any other agent by a small number of links. Such networks exist among people and are popularized by the notion of "six degrees of separation" [1]. When agents are arranged in a network, having a small number of neighbors relative to the number of members in the team, the number of agents through which a message must pass to get from any agent to any other, going only from neighbor to neighbor, is typically very small.

The intuition behind our approach is that agents can rapidly get information to those requiring it simply by "guessing" which acquaintance to send the information to. The agent attempts to guess which of its neighbors either require the information or are in the best position to get the information to the agent that requires it. In a small worlds network, an agent only needs to guess correctly slightly more often than it guesses wrong and information is rapidly delivered. Moreover, due to the low degree of separation, there only needs to be a small number of correct "guesses" to get information to its destination. Since the agents are working in a team, they can use information about the current state of the coordination to inform their guesses. While members of large teams will not have accurate, up-to-date models of the team, our hypothesis is that they will have sufficiently accurate models to "guess" correctly often enough to make the algorithm work.

InformationAgents are responsible for delivering information in our proxy architecture. Thus, these "guesses" about where to move next are made by the InformationAgents as they move around the network. The basic algorithm is shown in Figure 9. The InformationAgent guesses where to move next, moves there, updates the proxy state and moves on. This process continues until the information is likely to be out of date or the InformationAgent has visited enough proxies that it believes there are unlikely to be more proxies requiring the information. In practice, we typically stop an InformationAgent after it has visited a fixed percentage of the proxies, but we are investigating more optimal algorithms.

InformationAgent
while Worth Continuing
Guess which link leads closer to proxy requiring information
Move to that proxy
Add information to proxy state (either K or CK)
end while

Fig. 9. Algorithm for an InformationAgent

To test the potential of the approach we ran an experiment where proxies are organized in a three dimensional grid. One proxy is randomly chosen as the source of some information and another is randomly picked as the sink for that information. For testing, a probability is attached to each link, indicating the chance that passing information down that link will get the InformationAgent a smaller number of links from the sink. (These probabilities need to be inferred in the real proxies, see below for details.) In the experiment shown in Figure 10(a) we adjust the probability on links that actually lead to an agent requiring the information. For example, for the "59%" setting, links that lead closer to the sink agent have a probability of 0.59 attached, while those that lead further away have a 0.41 probability attached. The InformationAgent follows links according to their probability, e.g., in the "59%" setting, it will take links that lead it closer to the sink 59% of the time. Figure 10(a) shows that the information only needs to move closer to the target slightly more than 50% of the time to dramatically reduce the number of messages required to deliver information efficiently to the sink. To test the robustness of the approach, we altered the probability on some links so that the probability of moving further from the sink was actually higher than moving toward it. Figure 10(b) shows that even when a quite large percentage of the links had these "erroneous" probabilities, information delivery was quite efficient. While this experiment does not show that the approach works, it does show that if the InformationAgents can guess correctly only slightly more than 50% of the time, we can get targeted, efficient information delivery.

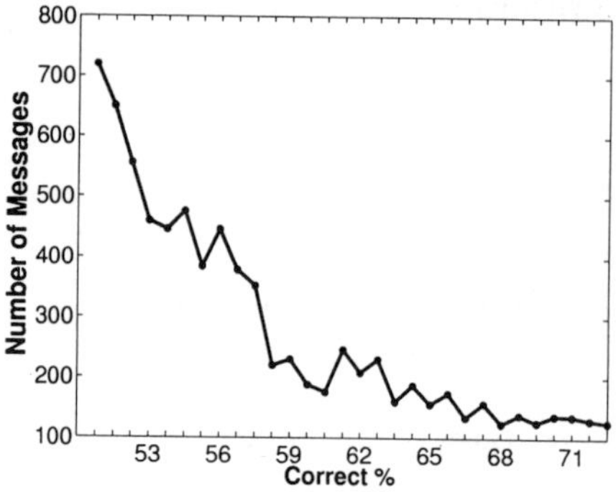

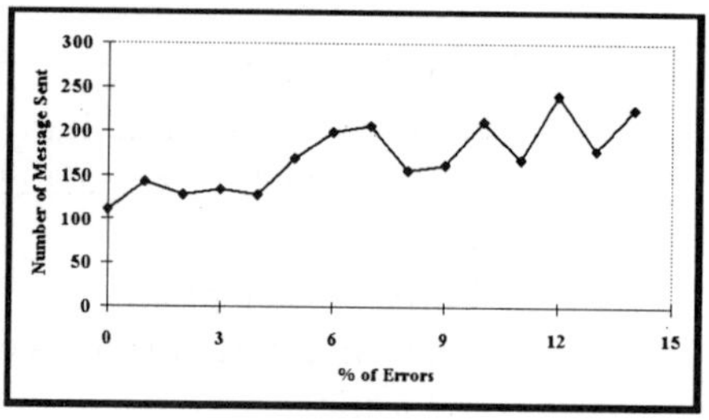

(a) (b)

Fig. 10. (a) The number of messages required to get a piece of information from one point in a network to another as we increase the likelihood that agents pass information closer to the target. There were 800000 agents arranged in a three dimensional grid. (b) The total number of messages required as the percentage of agents with probabilities indicating the wrong direction to send the information.

3.3.1. *Sharing Information with InformationAgents*

An initial approach to determining where InformationAgents should travel relies on inferring the need for one piece of information from the receipt of another piece. To understand the motivation for the idea, consider the following example. When a proxy receives a message about a role that is being performed at coordinates (1,1) from neighbor a, it can infer that if it found out about a SAM site at coordinates (1,2), passing that information to neighbor a is likely to get the information to a proxy that needs it. Notice, that it need not be the neighbor a that actually needs the information, but it will at least likely be in a good (or better) position to know who does. These inferences can be inferred using Bayes' Rule. In the following, we present a model of the small worlds network and an algorithm, based on Bayes' Rule, for updating where an InformationAgent should move next.

3.3.2. *Proxy Network Models*

Our proxy network model is composed of three elements, A, N and I, where A are the proxies, N is the network between the agent and I is the information to be shared. The team consists of a large number of proxies, $A(t) = \{a_1, a_2, ..., a_n\}$.

N denotes the communication of network among proxy team. A proxy a

can only communicate directly with a very small subset of its team mates. The acquaintances, or neighbors, of a at time t are written $n(a,t)$ and the whole network as $N(t) = \bigcup_{a \in A(t)} n(a,t)$. A message can be transferred from proxies that are not neighbors by passing through intermediate proxies but proxies will not necessarily know that path. We define the minimum number of proxies a message must pass through to get from one agent to another as the *distance* between those agents. The maximum distance between any two proxies is the network's "degree of separation". For example, if proxies a_1 and a_2 are not neighbors, but share a neighbor $distance(a_1, a_2) = 1$. We require the network, N, to be a small worlds network, which imposes two constraints. First, $|n(a,t)| < K$, where K is a small integer, typically less than 10. Second, $\forall a_i, a_j \in A, distance(a_i, a_j) < D$ where D is a small integer, typically less than 10. While N is a function of time, we assume that it typically changes slowly relative to the rate messages are sent around the network.

I is the alphabet of information that the team knows, $I = CK \cup K$. $i \in I$ denotes a specific piece of information, such as "There is a tank at coordinates (12, 12)".

The internal state of the team member a is represented by $S_a =< H_a, P_a, K_a >$. H_a is the history of messages received by the proxy. In practice, this history may be truncated to leave out old messages for spaces reasons. $K_a \subseteq I$ is the local knowledge of the proxy (it can be derived from H_a). If $i \in K_a$ at time t we say $knows(a,i,t)$.

The matrix P is the key to our information sharing algorithm.

$$P : I \times N(a) \rightarrow [0,1]$$

P maps a proxy and piece of information to a probability that that proxy is the best to pass that piece of information to. To be "best" means that passing the information to that proxy will most likely get the information to a sink. For example, if $P[i_1, a_2] = 0.9$, then given the current state of a_1 suggests that passing information i_1 to proxy a_2 is the best proxy to pass that information to. To obey the rules of probability, we require:

$$\forall i \in I, \sum_{b \in N(a)} P[i,b] = 1$$

Using P, when the proxy has a piece of information to send, it chooses a proxy to send the message to according to the likelihood sending to that proxy is the best. Notice, that it will not always send to the best proxy, but will choose a proxy relative to its probability of being the best.

The state of a proxy, S_a, gets updated in one of three ways. First, local sensing by the proxy can add information to K_a. Second, over time the information in K_a changes as information becomes old. For example, information regarding the location of an enemy tank becomes more uncertain over time. Maintaining K_a over time is an interesting and difficult challenge, but not the focus of this chapter, hence we ignore any such effects. Finally, and most importantly, K_a changes when a message m is sent to the proxy a from another proxy b at time t, $sent(m, a, b, t)$. In the case that m contains a piece of information i, we define a transition function, δ, that specifies the change to the proxy state. Two parts of the δ function, namely the update to the history part of the state, $H_a(t+1) = H_a(t) \cup m$, and the knowledge part of the state, $K_a(t+1) = K(t) \cup i$, are trivial. The other part of the transition function, the update to P_a due to message m, is written δ_P. This is the most difficult part of the transition function, and is the key to the success of the algorithm. The function is discussed in detail in later sections.

The reason for sharing information between team mates is to improve the individual performance and hence the overall performance. To quantify the importance of a piece of information i to a proxy a at time t we use the function $R : I \times A \times t \rightarrow \mathcal{R}$. The importance of the information i is calculated by determining the expected increase in utility of the proxy with the information versus without it. That is, $R(a, i, t) = EU(a, K + i) - EU(a, K - i)$, where $EU(a, K)$ is the expected utility of the proxy a with knowledge K. When $R(a, i, t) > 0$, it means that the specific information i supports a's decision making. The larger the value of $R(a, i, t)$ the more a needs the information.

$O(A, I, N)$ is the objective function:

$$reward(A, t) = \frac{\sum_{a \in A(t)} r(a, i, t)}{\sum_{a \in A(t)} knows(a, i, t)}$$

The numerator sums the reward received for getting the information to proxies that need it, while the denominator gives the total number of agents to whom the information was given. Intuitively, the objective function is maximized when information is transferred to as many as possible proxies that need that information and as few as possible of those that do not.

3.3.3. *Updating Proxy Network Models*

The key question for the algorithm is how we define δ_P, i.e., how we update the matrix P when a new message arrives. To update where to send a piece of information j based on a message containing information i, we need to know the relationship, if any between those pieces of information. Such relationships are domain dependant, hence we assume that a relationship function, $rel(i,j) \to [0,1]$, is given. The intuition captured by rel is that if $rel(i,j) > 0.5$ then an agent interested in i will also be interested in j, while if $rel(i,j) < 0.5$ then an agent interested in i is unlikely to be interested in j. For example, if i corresponds to a particular event in the environment, if j corresponds to an event near the event at i, we can expect $rel(i,j) > 0.5$, otherwise we expect a smaller value. If there is no relationship between i and j, then $rel(i,j) = 0.5$.

Utilizing Bayes' rule, we interpret a message containing information i arriving from a proxy b as *evidence* that proxy b is the best associate to pass information j to. Specifically, we can define define δ_P as follows:

$$\delta_P(P, recv(i,a)) = Pr(P[j,b] | recv(i,a)) \times P[j,b]$$

where

$$Pr(P[j,b] | recv(i,a)) = \begin{cases} rel(i,j) \times \frac{2}{|N|} & \text{if } a = b \\ \frac{1}{|N|} & \text{otherwise} \end{cases}$$

After δ_P has been applied, P must again be normalized:

$$P'[i,a] = \frac{P[i,a]}{\sum_{b \in N(a)} P[i,b]}$$

4. Results

The most important aspect of our results is that we have run a team of 200 simulated WASMs, controlled by proxies in a simulation of a mission to protect a manned aircraft. Such a team is an order of magnitude bigger than previously published teams. The proxies are implemented in Java and 200 ran on two 2GHz linux machines with 1Gb of RAM on each machine. In the following, we present selected runs from experiments with this scenario, plus the results of experiments using a simpler centralized controller that mimics the coordination, but is more lightweight.

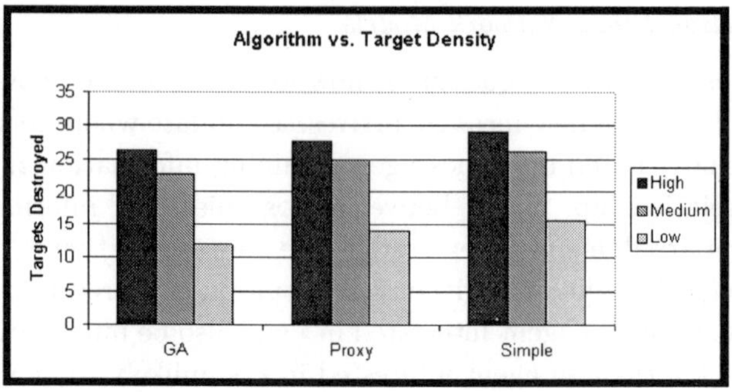

Fig. 11. Comparing the number of targets hit by three different role allocation algorithms under three different target densities.

The first experiment compared three different algorithms for allocating WASMs to targets. We compared two centralized algorithms with our distributed allocation. The first centralized algorithm was very simple, allocating the closest available WASM to every newly discovered target. The second centralized algorithm was a genetic algorithm based approach. Figure 11 shows the number of randomly distributed targets destroyed by each of the algorithms in a fixed amount of time. For each algorithm we tried three different levels of target density, few targets spread out to many targets in a small area. Somewhat surprisingly, the simple algorithm performed best, followed by our distributed algorithm, finally followed by the genetic algorithm. It appears that the random distribution of targets is especially amenable to simple allocation algorithms. However, notice that the performance of the distributed algorithm is almost as good as the simple algorithm, despite having far lower communication overheads. We then performed more detailed experiments with the distributed algorithm, varying the threshold for accepting a role to destroy a target. The threshold is inversely proportional to the distance of the WASM to the target. A team member will not accept a role unless its capability is above the threshold and it has available resources. Figure 12(a) shows that unless the threshold is very high and WASMs will not go to some targets, the number of targets hit does not vary. Even the rate of targets hit over time does not change much as we vary the thresholds, see Figure 12(b).

In our second experiment, we used the centralized version of our teamwork algorithms to run a very large number of experiments to understand how WASMs should coordinate. The mission was to protect a manned aircraft and the output measure was the closest distance an undestroyed target

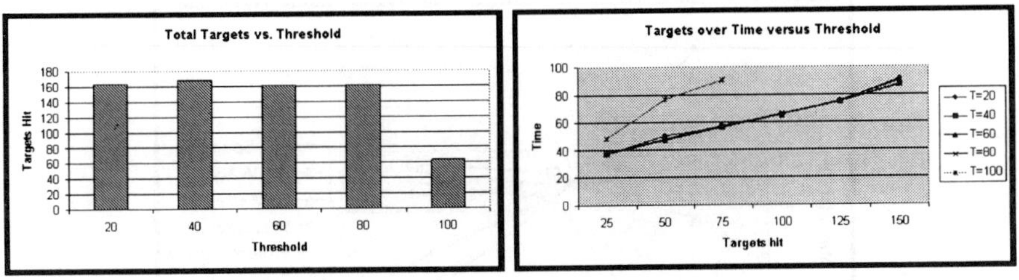

(a) (b)

Fig. 12. (a) The number of targets hit as the threhold is varied. Threshold is the minimum capability of a WASM assigned a target and is inversely proportional to the WASMs distance from the target. (b) The time taken to hit a specific number of targets as the threshold is varied.

got to the manned aircraft (higher is better) which followed a random path. The WASMs had two options, stay with the aircraft or spread out ahead of the aircraft path. We varied six parameters, giving them low, medium and high values and performed over 8000 runs. The first parameter was the speed of the aircraft relative to the WASM (A/C Speed). The second parameter was the number of WASMs (No. WASM). The third parameter was the number of targets (SAM sites). The fourth parameter was the percentage of WASMs that stayed with the aircraft versus the percentage that spread out looking for targets. The fifth parameter is the distance that the WASMs which stayed with the aircraft flew from it (Protect Spread). Finally, we varied the WASM sensor range. Figure 13 shows the results. Notice the speed of the aircraft relative to the WASMs is one of the most critical factors, alongside the less surprising Number of WASMs.

Finally, we ran two experiments to evaluate the information sharing algorithm. In the first experiment, we arranged around 20000 agents in a small worlds network. Then we passed 150 pieces of information from a particular source randomly around the network. After these 150 pieces of information had been sent, we created a new piece of information randomly and applied our algorithms to get it to a specific sink agent. In Figure 14(a) we show the average number of steps taken to deliver the message from the source to the sink as we varied the strength of the relationship between the information originally sent out and the new piece of information. As expected, the stronger the relationship between the originally sent information and the new information the better the information delivery. In the second experiment, we started information from various sources, moving

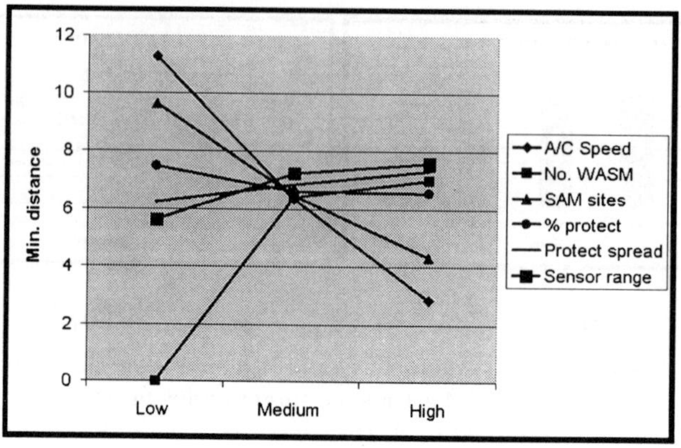

Fig. 13. Effects of a variety of parameters on the minimum distance a SAM site gets to a manned aircraft the WASMs are protecting.

them 150 steps, as in the first experiment. In this case, there were multiple "sinks" for the piece of information that we randomly added. The reward received, based on the objective function above, is proportional to the ratio of the number of agents receiving the information that wanted it and the number that did not need it. Figure 14(b) shows that our algorithm dramatically outperforms random information passing. While important work remains, the initial information sharing experiments show the promise of our approach.

5. Related Work

Coordination of distributed entities is an extensively studied problem [7, 6, 21, 25, 34]. A key design decision is how the control is distributed among the group members. Solutions range from completely centralized [11], to hierarchical [10, 17] to completely decentralized [39]. While there is not yet definitive, empirical evidence of the strengths and weaknesses of each type of architecture, it is generally considered that centralized coordination can lead to behavior that is closer to optimal, but more distributed coordination is more robust to failures of communications and individual nodes [2]. Creating distributed groups of cooperative autonomous agents and robots that must cooperate in dynamic and hostile environments is a huge challenge that has attracted much attention from the research community [22, 24]. Using a wide range of ideas, researchers have had moderate success in building and understanding flexible and robust teams that can effectively

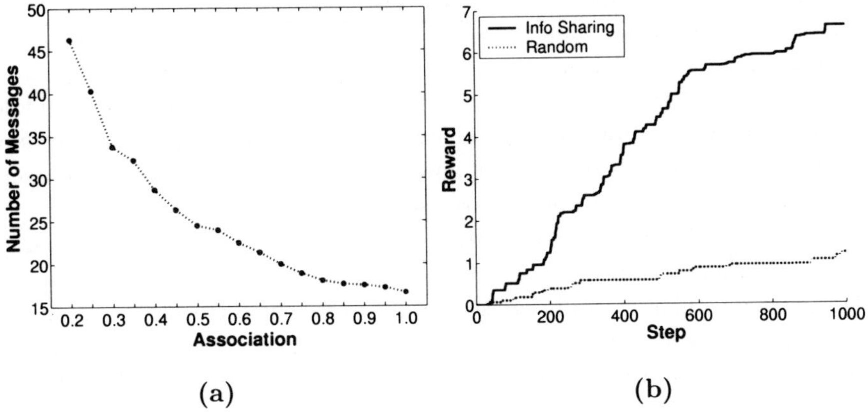

Fig. 14. (a) The reduction in the number of messages as the association between information received and information to be sent increases. (b) The reward received over time, based on our information sharing algorithm and on a random information passing algorithm.

act towards their joint goals [5, 8, 18, 31].

Tidhar [37] used the term "team-oriented programming" to describe a conceptual framework for specifying team behaviors based on mutual beliefs and joint plans, coupled with organizational structures. His framework also addressed the issue of team selection [37] — team selection matches the "skills" required for executing a team plan against agents that have those skills. Jennings's GRATE* [18] uses a teamwork module, implementing a model of cooperation based on the joint intentions framework. Each agent has its own *cooperation level* module that negotiates involvement in a joint task and maintains information about its own and other agents' involvement in joint goals. The Electric Elves project was the first human-agent collaboration architecture to include both proxies and humans in a complex environment [5]. COLLAGEN [30] uses a proxy architecture for collaboration between a single agent and user. While these teams have been successful, they have consisted of at most 20 team members and will not easily scale to larger teams.

Jim and Giles [20] have show that communication can greatly improve multiagent system performance greatly by analyzing a general model of multi-agent communication. However, these techniques rely on a central message board. Burstein implemented a dynamic information flow framework and proposed an information delivery algorithm based on two kinds of information communication: Information Provision advertisements and

Information Requirements advertisements [4]. But its realization was based on broadcast or using middle agents as brokers who respond to all the information disseminated. Similar research can be found in Decker and Sycara's RETSINA multiagent system [9, 13] which defined information and middle agents who were supposed to be able to freely deliver information with any of the others without delay. Such approaches, while clearly useful for some domains, are not applicable to large scale teams.

Yen's CAST proposed a module that expedites information exchange between team members based on a shared mental model, but almost the same shortcoming exists because in a huge team who is working in an ad-hoc environment, any team member can only sense a very limited number of teammates' status as well as their mental [42]. Xuan [41] and Goldman [15] proposed a decentralized communication decision model in multi-agent cooperation based on Markov decision processes (MDP). Their basic idea is that an explicit communication action will incur a cost and they assume the global reward function of the agent team and the communication cost and reward are known. Xuan used heuristic approaches and Goldman used a greed meta-level approaches to optimize the global team function. Moveover, Goldman [14] put forward a decentralized collaborative multiagents communication model and mechanism design based on MDP, which assumed that agents are fully-synchronized when they start operating, but no specific optimal algorithm was presented. Furthermore, there are no experiment result was shown that their algorithm can work on huge team very well.

Bui [3] and Wie [40] solved the information sharing problems in novel ways. In Bui's work, he presented a framework for team coordination under incomplete information based on the theory of incomplete information game that agents can learn and share their estimates with each other. Wie's RHINO used a probability method to coordinate agent team without explicit communication by observing teammates' action and coordinating their activities via individual and group plan inference. The computational complexity of these approaches makes them inapplicable to large teams.

6. Conclusions and Future Work

In this Chapter we have presented a novel approach and initial results to the challenges presented by coordination of very large groups of WASMs. Specifically, we presented Machinetta proxies as the basic architecture for flexible, robust distributed coordination. Key coordination algorithms encapsulated by the proxies were presented. These algorithms, including plan instantiation and information sharing, address new challenges that arise when a large group is required to coordinate. These novel algorithms replace existing algorithms that fail to scale when the group involves a large number of entities. We implemented the proxies using the novel abstraction of coordination agents, which gave us high levels of robustness. With the novel algorithms and architecture we were able to execute scenarios involving 200 simulated WASMs flying coordinated search and destroy missions.

Our initial experiments reveal that while our algorithms are capable of dealing with some of the challenges of the domain, many challenges remain. Perhaps more interestingly, new unexpected phenomena are observed. Understanding and dealing with these phenomena will be a central focus of future efforts. Further down the track, the coordinated behavior must be able to adapt strategically in response to the tactics of the hostile forces. Specifically, it should not be possible for enemy forces to exploit specific phenomena of the coordination, the coordination must react to such attempts by changing their coordination. Such reasoning is currently far beyond the capabilities of large teams.

Acknowledgments

This research has been supported by AFRL/MNK grant F08630-03-1-0005.

References

[1] Albert-Laszla Barabasi and Eric Bonabeau. Scale free networks. *Scientific American*, pages 60–69, May 2003.

[2] Johanna Bryson. Hierarchy and sequence vs. full parallelism in action selection. In *Intelligent Virtual Agents 2*, pages 113–125, 1999.

[3] H. H. Bui, S. Venkatesh, and D. Kieronska. A framework for coordination and learning among team members. In *Proceedings of the Third Australian Workshop on Distributed AI*, 1997.

[4] Mark H. Burstein and David E. Diller. A framework for dynamic information flow in mixed-initiative human/agent organizations. *Applied Intelligence on Agents and Process Management*, 2004. Forthcoming.

[5] Hans Chalupsky, Yolanda Gil, Craig A. Knoblock, Kristina Lerman, Jean Oh, David V. Pynadath, Thomas A. Russ, and Milind Tambe. Electric Elves: Agent technology for supporting human organizations. *AI Magazine*, 23(2):11–24, 2002.

[6] D. Cockburn and N. Jennings. *Foundations of Distributed Artificial Intelligence*, chapter ARCHON: A Distributed Artificial Intelligence System For Industrial Applications, pages 319–344. Wiley, 1996.

[7] Philip R. Cohen and Hector J. Levesque. Teamwork. *Nous*, 25(4):487–512, 1991.

[8] K. Decker and J. Li. Coordinated hospital patient scheduling. In *Proceedings of the 1998 International Conference on Multi-Agent Systems (ICMAS'98)*, pages 104–111, Paris, July 1998.

[9] K. Decker, K. Sycara, A. Pannu, and M. Williamson. Designing behaviors for information agents. In *Procs. of the First International Conference on Autonomous Agents*, 1997.

[10] Vincent Decugis and Jacques Ferber. Action selection in an autonomous agent with a hierarchical distributed reactive planning architecture. In *Proceedings of the Second International Conference on Autonomous Agents*, 1998.

[11] T. Estlin, T. Mann, A. Gray, G. Rapideau, R. Castano, S. Chein, and E. Mjolsness. An integrated system for multi-rover scientific exploration. In *Proceedings of AAAI'99*, 1999.

[12] Alessandro Farinelli, Paul Scerri, and Milind Tambe. Building large-scale robot systems: Distributed role assignment in dynamic, uncertain domains. In *Proceedings of Workshop on Representations and Approaches for Time-Critical Decentralized Resource, Role and Task Allocation*, 2003.

[13] Joseph Giampapa and Katia Sycara. Team oriented agent coordination in the RETSINA multi-agent system. In *Proceedings of Agents02*, 2002.

[14] C. V. Goldman and S. Zilberstein. Mechanism design for communication in cooperative systems. In *Fifth Workshop on Game Theoretic and Decision Theoretic Agents*, 2003.

[15] C. V. Goldman and S. Zilberstein. Optimizing information exchange in cooperative multi-agent systems. In *Proceedings of the Second International Conference on Autonomous Agents and Multi-agent Systems*, 2003.

[16] Barbara Grosz and Sarit Kraus. Collaborative plans for complex group actions. *Artificial Intelligence*, 86:269–358, 1996".

[17] Bryan Horling, Roger Mailler, Mark Sims, and Victor Lesser. Using and maintaining organization in a large-scale distributed sensor network. In *In Proceedings of the Workshop on Autonomy, Delegation, and Control (AAMAS03)*, 2003.

[18] N. Jennings. The archon systems and its applications. Project Report, 1995.

[19] N. R. Jennings. Specification and implementation of a belief-desire-joint-intention architecture for collaborative problem solving. *Intl. Journal of Intelligent and Cooperative Information Systems*, 2(3):289–318, 1993.

[20] Kam-Chuen Jim and C. Lee Giles. How communication can improve the performance of multi-agent systems. In *Proceedings of the fifth international conference on Autonomous agents*, 2001.

[21] David Kinny. The distributed multi-agent reasoning system architecture and language specification. Technical report, Australian Artificial intelligence institute, Melbourne, Australia, 1993.

[22] Hiraoki Kitano, Minoru Asada, Yasuo Kuniyoshi, Itsuki Noda, Eiichi Osawa, , and Hitoshi Matsubara. RoboCup: A challenge problem for AI. *AI Magazine*, 18(1):73–85, Spring 1997.

[23] Hiroaki Kitano, Satoshi Tadokoro, Itsuki Noda, Hitoshi Matsubara, Tomoichi Takahashi, Atsushi Shinjoh, and Susumu Shimada. Robocup rescue: Searh and rescue in large-scale disasters as a domain for autonomous agents research. In *Proc. 1999 IEEE Intl. Conf. on Systems, Man and Cybernetics*, volume VI, pages 739–743, Tokyo, October 1999.

[24] John Laird, Randolph Jones, and Paul Nielsen. Coordinated behavior of computer generated forces in TacAir-Soar. In *Proceedings of the fourth conference on computer generated forces and behavioral representation*, pages 325–332, Orlando, Florida, 1994.

[25] V. Lesser, M. Atighetchi, B. Benyo, B. Horling, A. Raja, R. Vincent, T. Wagner, P. Xuan, and S. Zhang. The UMASS intelligent home project. In *Proceedings of the Third Annual Conference on Autonomous Agents*, pages 291–298, Seattle, USA, 1999.

[26] David Pynadath and Milind Tambe. Multiagent teamwork: Analyzing the optimality and complexity of key theories and models. In *First International Joint Conference on Autonomous Agents and Multi-Agent Systems (AAMAS'02)*, 2002.

[27] David V. Pynadath and Milind Tambe. An automated teamwork infrastructure for heterogeneous software agents and humans. *Journal of Autonomous Agents and Multi-Agent Systems, Special Issue on Infrastructure and Requirements for Building Research Grade Multi-Agent Systems*, page to appear, 2002.

[28] D.V. Pynadath, M. Tambe, N. Chauvat, and L. Cavedon. Toward team-oriented programming. In *Intelligent Agents VI: Agent Theories, Architectures, and Languages*, pages 233–247, 1999.

[29] Paul Ranky. *An Introduction to Flexible Automation, Manufacturing and Assembly Cells and Systems in CIM (Computer Integrated Manufacturing), Methods, Tools and Case Studies.* CIMware, 1997.

[30] C. Rich and C. Sidner. COLLAGEN: When agents collaborate with people. In *Proceedings of the International Conference on Autonomous Agents (Agents'97)"*, 1997.

[31] P. Rybski, S. Stoeter, M. Erickson, M. Gini, D. Hougen, and N. Papanikolopoulos. A team of robotic agents for surveillance. In *Proceedings of the fourth international conference on autonomous agents*, pages 9–16, 2000.

[32] P. Scerri, D. V. Pynadath, L. Johnson, Rosenbloom P., N. Schurr, M Si, and M. Tambe. A prototype infrastructure for distributed robot-agent-person teams. In *The Second International Joint Conference on Autonomous Agents and Multiagent Systems*, 2003.

[33] Daniel Schrage and George Vachtsevanos. Software enabled control for intelligent uavs. In *Proceedings of the 1999 IEEE International Symposium on Computer Aided Control System Design*, Hawaii, August 1999.

[34] Munindar Singh. Developing formal specifications to coordinate hetrogeneous agents. In *Proceedings of third international conference on multiagent systems*, pages 261–268, 1998.

[35] Milind Tambe. Agent architectures for flexible, practical teamwork. *National*

Conference on AI (AAAI97), pages 22–28, 1997.

[36] Milind Tambe, Wei-Min Shen, Maja Mataric, David Pynadath, Dani Goldberg, Pragnesh Jay Modi, Zhun Qiu, and Behnam Salemi. Teamwork in cyberspace: using TEAMCORE to make agents team-ready. In *AAAI Spring Symposium on agents in cyberspace*, 1999.

[37] G. Tidhar, A.S. Rao, and E.A. Sonenberg. Guided team selection. In *Proceedings of the Second International Conference on Multi-Agent Systems*, 1996.

[38] Duncan Watts and Steven Strogatz. Collective dynamics of small world networks. *Nature*, 393:440–442, 1998.

[39] Tony White and Bernard Pagurek. Towards multi swarm problem solving in networks. In *Proceedings of the International conference on multi-agent systems*, pages 333–340, Paris, July 1998.

[40] Michael Van Wie. A probabilistic method for team plan formation without communication. In *Proceedings of the fourth international conference on Autonomous agents*, 2000.

[41] P. Xuan, V. Lesser, and S. Zilberstein. Communication decisions in multiagent cooperation: Model and experiments. In *Proceedings of the Fifth International Conference on Autonomous Agents*, 2001.

[42] J. Yen, J. Yin, T. R. Ioerger, M. S. Miller, D. Xu, and R. A. Volz. Cast: Collaborative agents for simulating teamwork. In *Proceedings of the International Joint Conference on Artificial Intelligence*, pages 1135–1142, 2001.

CHAPTER 21

COOPERATIVE CONTROL SIMULATION VALIDATION USING APPLIED PROBABILITY THEORY

Capt. Chris S. Schulz, [a]
LtCol David R. Jacques, [b]
and Dr. Meir Pachter [c]

Air Force Institute of Technology, Wright-Patterson AFB, OH
chris.schulz@afit.edu

Several research simulations have been created to support development and refinement of teamed autonomous agents using decentralized cooperative control algorithms. Simulation is the necessary tool to evaluate the performance of decentralized cooperative control algorithms, however these simulations lack a method to validate their output. This work presents a method to validate the performance of a decentralized cooperative control simulation environment for an autonomous Wide Area Search Munition (WASM). Rigorous analytical methods for six wide area search and engagement scenarios involving Uniform, Normal, and Poisson distributions of N real targets and M false target objects are formulated to generate expected numbers of target attacks and kills for a searching WASM. The mean value based on the number of target attack and kills from Monte Carlo simulations representative of the individual scenarios are compared to the analytically derived expected values. Emphasis is placed on Wide Area Search Munitions operating in a multiple target environment where a percentage of the total targets are either false targets or may be misconstrued as false by varying the capability of the WASM's Automatic Target Recognition (ATR) capability.[d]

[a] Dept. of Aeronautics and Astronautics
[b] Asst. Prof., Dept. of Aeronautics and Astronautics
[c] Prof., Dept. of Electrical and Computer Engineering
[d] The views expressed in this article are those of the authors and do not reflect the official policy of the U.S. Air Force, Department of Defense, or the U.S. Government.

Nomenclature

$\alpha \equiv$ False target density parameter $[1/km^2]$
$A \equiv$ Area $[km^2]$
$A_S \equiv$ Area of battle space $[km^2]$
$\beta \equiv$ Target density parameter $[1/km^2]$
$\lambda \equiv$ Poisson probability law parameter
$P_A \equiv$ Probability of attack
$P_E \equiv$ Probability of encounter given target in search area
$P_K \equiv$ Probability of kill given attack
$P_{TR} \equiv$ Probability of correct target report
$P_{FTR} \equiv$ Probability of false target report
$r \equiv$ Radial distance $[km]$
$s \equiv$ Time $[sec]$
$t \equiv$ Time $[sec]$
$\tau \equiv$ Time $[sec]$
$T \equiv$ Time duration of mission $[sec]$

1. Introduction

The United States Department of Defense (DoD) is investigating opportunities to expand the future battlefield capabilities of multiple Wide Area Search Munitions (WASMs) through cooperative control. Current emphasis is placed on exploiting a WASMs' ability to search, detect, identify, and attack a host of targets autonomously. Ultimately, this research will pave the way to sophisticated unmanned weapon systems capable of efficiently performing high risk / high payoff tasks such as Suppression of Enemy Air Defenses (SEAD), Persistent Area Denial (PAD), and Combat Intelligence, Surveillance, and Reconnaissance (Combat- ISR). Research to improve multiple WASM mission efficiency is exploring the use of cooperative teams rather than individual autonomous WASMs. The need for this tactical capability is driving basic research in cooperative behavior for teamed agents. These works include [1], [2], and [3], which address hierarchal decomposition, decentralized execution, task coupling, and task timing for a team of WASMs. Due to the complexities of decentralized cooperative controller development for WASM teams, simulation remains the most viable methods for analysis. Applied research has relied on empirical results from simulations such as MultiUAV [4] to analyze controller performance. As to date, a method to independently validate the performance of the

MultiUAV environment has yet to be developed. The work presented here is concerned with the development of a method to validate the performance of a decentralized cooperative control simulation for autonomous wide area search munitions. A rigorous analytical treatment of six persistent area denial scenarios involving N+M targets and ψ WASMs are used to validate the results of identical simulation runs. Emphasis is placed on WASMs operating in a multiple target environment where a percentage of the total targets are either decoys or targets that may be misconstrued as false targets by the WASM's Automatic Target Recognition (ATR) software. The chapter is organized as follows. Section 2 introduces the MultiUAV simulation environment by providing an overview of its components and operation methodology. Emphasis is placed on the analytical expressions for six basic scenarios, which are introduced and explained in detail beginning in section 3.1. This is followed by an explanation of the simulation configuration used to match the six analytical scenarios in section 4. Finally, results and conclusions of the comparative evaluation are made in section 5 and section 6, respectfully.

2. MultiUAV Simulation Environment

MultiUAV is a Matlab/Simulink simulation designed to enable algorithm development for research in cooperative WASM control. It is built around a discrete time state engine that progresses the event flow for the WASMs as they proceed in their task to search and attack targets of opportunity. The simulation environment allows researchers to use a maximum of 8 WASMs searching for a user specified number of targets and non-targets. The simulation allows for five target types, including decoys and false targets. This provides the ability for investigating the effects of dissimilar target priorities on cooperative behavior. Furthermore, MultiUAV permits users to vary the detection capability of the ATR function. This, combined with an environment of multiple target types permits researches to explore the ill effects of false targets on cooperative control. Cooperative behavior in terms of target identification, target classification, and task allocation in order to improve mission effectiveness is investigated. While MultiUAV is a basic research tool, it includes a continuous time vehicle dynamics model for the search munitions, in addition to the discrete state engine. This allows researchers to exercise the control algorithms in hybrid discreet/continuous time environment. Finally, MultiUAV is designed in a modular fashion, which allows the user to modify or replace any of the simulation functions quickly and easily to accommodate the needs of their research.

2.1. *Simulation Operation*

The simulation models the general characteristics of wide area search munitions performing search, classify, and attack functions. Searching WASMs perform actions based on rules that control the event flow of a generic search, classify, and attack mission, as seen in Figure 1

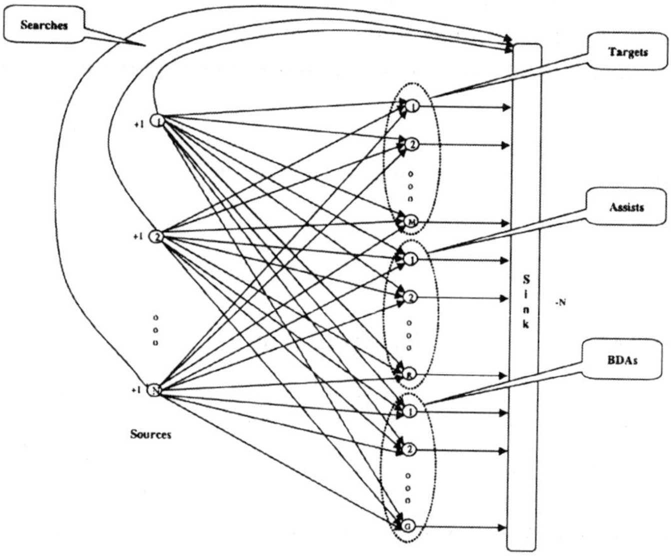

Fig. 1. MultiUAV State Engine

The orders of operation for the rules that govern the chain of events, or 'Kill Chain', are as follows;

- Detected
- Classified
- Attacked
- Verified Destroyed

The MultiUAV environment performs all operations based on the flow of these events. A typical simulation begins with the vehicles starting from pre-determined positions and flying pre-determined routes. When an object enters a vehicle's field of regard, the vehicle classifies the object as a target or non-target and assigns a probability of correct classification based on the angle from which the vehicle viewed the object. Each vehicle then calculates the benefits of performing certain tasks. Possible tasks are

- Continue searching
- Reclassify a previously classified target
- Perform target attack
- Perform battle damage assessment on an attacked target

Vehicle tasks are assigned such that the overall benefit is maximized. This task allocation occurs each time the state of a target changes until the maximum simulation time is reached.

While the MultiUAV environment relies on several functions to perform the entire kill chain, Target Classification and Task Allocation have the greatest effect on the results of the simulation and thus will be explained in further detail.

2.2. *Target Classification via ATR Operation*

When a WASM classifies an object, the ATR function calculates a confidence level for that classification based on the angle from which the vehicle viewed the object. If the confidence is below a user-defined threshold, a second WASM may be assigned to assist in classifying the object if the user specifies cooperation of more than one WASM. The second WASM flies to the object and assigns its own confidence of correct classification. The individual confidences are combined into a single confidence level that is compared to the threshold value. Once the confidence of correct classification is greater than the threshold, the object is deemed classified. In order to provide realistic modelling to the ATR function a method to model error associated with it is required. This error is represented by a method referred to as a confusion matrix [5], and is described in section 2.2.1.

2.2.1. *Confusion Matrix Definition*

When a WASM encounters a target, error associated with ATR has the possibility to cause false target detections. The confusion matrix method models the ATR function based on probability of target report, P_{TR} , and probability of false target report, P_{FTR}. An example of the single target type case is shown below in Table 2.2.1.

Table 1. 2x2 Confusion Matrix

True/Rpt	T	FT
T	P_{TR}	$1 - P_{FTR}$
FT	$1 - P_{TR}$	P_{FTR}

The confusion matrix provides a method to determine the probability of a falsely declared target, as represented by the rows of the matrix, based on the actual target encountered by the vehicle, as represented by the columns of the matrix. The confusion matrix is expandable to accommodate several target types, thus providing a realistic event generator for scenarios involving a more complex battlespace.

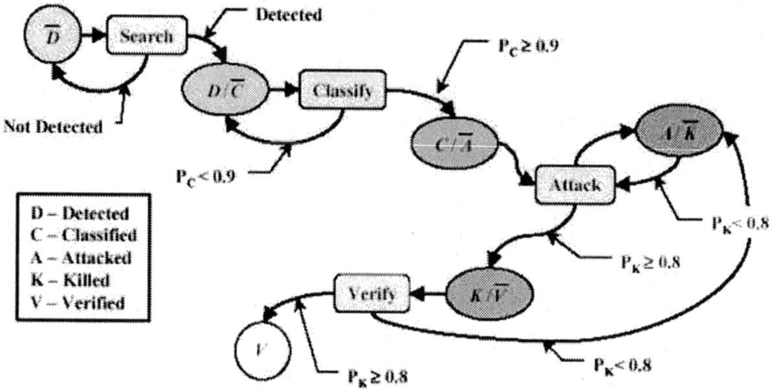

Fig. 2. Capacitated Transshipment Network

2.3. *Task Allocation*

The capacitated transshipment network, as used in [3], provides the method for task allocation generation for the WASMs modelled in the MultiUAV environment. A graphical representation of the network is shown in Figure 2. Capacitated transshipment is based on optimal routing of resources to meet demand in a network of defined capacity. At the other end of the network is a demand of φ, where φ is the number of WASMs. The targets are transshipment nodes with supplies and demands of zero. Each WASM must travel through the network to satisfy the end demand where as the vehicles travel through the network along arcs that represent specific tasks and have certain benefits associated with them. The flow through the network is determined in part by the benefits associated with the various arcs. The values of these benefits are crucial to solving the network flow problem. By analyzing all arcs, the optimal task allocation provides the greatest overall benefit to the system. The capacitated transshipment network is modified linear programming problem that yields an integer solution. MultiUAV employs a heuristic approach for benefit assignment, wherein emphasis is placed on killing high value targets over performing other tasks such as reclassification or battle damage assessment.

3. Analytical Theory for Cooperative Search, Classification, and Attack

Simulation tools such as MultiUAV provide a tool for cooperative control development. However, before any controller development can take place, an independent baseline performance comparison is necessary to ensure the proper operation of the simulation environment. [6] provide a method of system analysis based on applied probability theory for vehicles performing search, classification, and attack on encountered targets within a battle space. For ψ WASMs armed with ζ munitions, a progression of analytical expressions for six scenarios is provided that consider both real and false target distributions in the defined search area. This work represents the foundation for the baseline comparison used for the MultiUAV environment performance validation. The baseline comparison outlined in this work focuses on a single WASM ($\psi = 1$), armed with a single munition ($\zeta = 1$). The scenarios considered allow for various amounts of real and false targets, and in addition vary the type of target distribution by either a Uniform or Poisson field. Uniform distributions provide for a known quantity of targets or false targets in a given area, and thus are used ensure that number of targets are encountered in the search. Poisson fields, however, do not guarantee an absolute quantity over a specified range as they are defined by a target density parameter, $\alpha\ [\frac{1}{km^2}]$. As an area A is searched, the Poisson probability law parameter, λ, is defined, $\lambda = \alpha A$. Therefore, the Poisson probability function $P(\cdot)$ is specified by

$$P(\{k\}) = e^{-\lambda}\ \frac{\lambda^k}{k!}\ , \quad k = 0, 1, 2, ... \tag{1}$$

which specifies the total number of targets. Poisson fields are used in the scenarios so that while a density of targets/false targets may be specified it is not guaranteed that you will encounter one.

The baseline comparisons for all four scenarios will focus on four parameters.

- Probability of real target attack, P_{A_T}
- Probability of false target attack $P_{A_{FT}}$
- Probability of successful target kill P_{TK}
- Probability of successful false target kill P_{FTK}
- Longevity of WASM given attack occurs $\frac{\bar{s}}{T}$

where P_{TK} and P_{FTK} are calculated by

$$P_{TK}, P_{FTK} = P_{A_T}, P_{A_{FT}} \cdot P_k \tag{2}$$

given P_k, the probability of kill. P_k is a function of the warhead lethality, and in this case was selected as either 50% or 80%. As a note, T is defined as the total time required performing a search of the battle space, t is time target is attacked, and where s is time of target attack.

Scenarios 1-4 assume the WASM performs searches over a linearly defined area, as represented in Figure 3. Here, the WASM has the parameters of forward velocity V, and sensor swath width W to create the total search area. Scenarios 5 and 6 are similar to 1-4, with the exception that they search over a circular area. Below is a brief description of the six individual scenarios used in the MultiUAV baseline comparison.

3.1. *Scenario 1 (Single Uniform T, Poisson FTs)*

Scenario 1 presents a single target (T) uniformly distributed amongst a Poisson field of false targets (FTs) in a battlespace of area A_s. For the Poisson field of FTs assuming a non-zero P_{FTR} , α is modified as follows: $\alpha \equiv (1 - P_{FTR})\alpha$ With this, the probability of attack, P_{A_T} , is defined as

$$P_{A_T} = P_{TR}\frac{1 - e^{-(1-P_{FTR})\lambda}}{(1 - P_{FTR})\lambda} \tag{3}$$

Additionally, the probability of false target attack, P_{A_FT} , is defined as

$$P_{A_{FT}} = [1 - \frac{P_{TR}}{(1 - P_{FTR})\lambda}][1 - e^{-(1-P_{FTR})\lambda}] + P_{TR}\ e^{-(1-P_{FTR})\lambda} \tag{4}$$

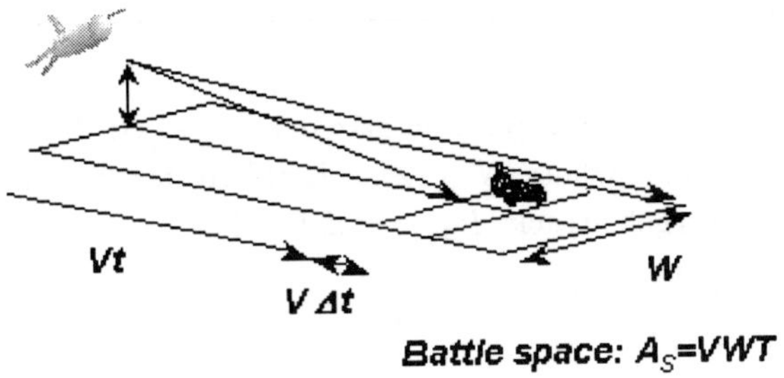

Fig. 3. Linear Search

And finally, the longevity of the WASM, assuming a performed attack on target is defined as

$$\frac{\bar{s}}{T} = \frac{(1-P_{FTR})\lambda - P_{TR}}{[(1-P_{FTR})\lambda]^2[1-(1-P_{TR})e^{-(1-P_{FTR})\lambda}]} + \frac{[P_{TR}-(1-P_{TR})(1-P_{FTR})\lambda(1+\lambda-P_{FTR}\lambda)]e^{-(1-P_{FTR})\lambda}}{[(1-P_{FTR})\lambda]^2[1-(1-P_{TR})e^{-(1-P_{FTR})\lambda}]} \tag{5}$$

3.2. *Scenario 2 (Poisson T, Poisson FT)*

For the second scenario considered a search environment consisting of both a Poisson field of targets, Ts, and false targets, FTs, is considered. For the Poisson field of real targets, the Poisson probability law parameter describing real targets, λ_T, is defined as $\lambda_T = \beta A_S$. Here, the Poisson field of real targets is parameterized by β $[\frac{1}{km^2}]$ and false targets by α $[\frac{1}{km^2}]$. So, the probability of real target attack, P_{A_T}, is defined as

$$P_{A_T} = \frac{P_{TR}\lambda_T}{(1-P_{FTR})\lambda_{FT} + P_{TR}\lambda_T}\{1 - e^{-[(1-P_{FTR})\lambda_{FT}+P_{TR}\lambda_T]}\} \tag{6}$$

Additionally, the probability of false target attack, $P_{A_{FT}}$, is defined as

$$P_{A_{FT}} = \frac{(-P_{FTR})\lambda_{FT}}{(1-P_{FTR})\lambda_{FT} + P_{TR}\lambda_T}\{1 - e^{-[(1-P_{FTR})\lambda_{FT}+P_{TR}\lambda_T]}\} \tag{7}$$

And finally, for scenario 2, the longevity of the WASM, assuming a performed attack on target is defined as

$$\frac{\bar{s}}{T} = \frac{1-[1+(1-P_{FTR})\lambda_{FT}+P_{TR}\lambda_T]\,e^{-[(1-P_{FTR})\lambda_{FT}+P_{TR}\lambda_T]}}{[(1-P_{FTR})\lambda_{FT}+P_{TR}\lambda_T]\{1-e^{-[(1-P_{FTR})\lambda_{FT}+P_{TR}\lambda_T]}\}}\,. \tag{8}$$

3.3. *Scenario 3 (N Uniform Ts, Poisson FT)*

Scenario 3 presents a search environment represented by a Poisson field of FTs, with a uniform distribution of N real targets, Ts. As in scenario 1 and 2, the Poisson field of FTs is parameterized by $\alpha[\frac{1}{km^2}]$. A recursive form is used to present cases where $N \geq 2$. Therefore, for N real targets the probability of real target attack, P_{A_T} , is defined as

$$P_{A_T}^{(N)} = \frac{P_{TR}N}{(1-P_{FTR})\lambda_{FT}}\,[1-(1-P_{TR})^{N-1}e^{-(1-P_{FTR})\lambda_{FT}} - P_{A_T}^{(N-1)}]$$

$$N = 2,... \tag{9}$$

The initial probability $P_{A_T}^{(1)}$ was calculated for Scenario 1.
Also, for false targets the probability of false target attack, $P_{A_{FT}}$,

$$P_{A_{FT}}^{(N)} = 1-(1-P_{TR})^N\, e^{-(1-P_{FTR})\lambda_{FT}} - P_{A_{FT}}^{(N-1)}\ , \quad N = 2,... \tag{10}$$

with the initial probability, $P_{A_{FT}}^{(1)}$, calculated for Scenario 1. And finally, to calculate the longevity of the munition, $\frac{\bar{s}}{T}$, given the munition has attacked a target or false target is expressed as

$$1 - H^{(N)}(T) = 1 - e^{-(1-P_{FTR})\lambda_{FT}} \tag{11}$$

where

$$H^{(N)}(s) = (1-P_{TR}\frac{s}{T})^N\, e^{-(1-P_{FTR})\lambda_{FT}\frac{s}{T}} \tag{12}$$

3.4. *Scenario 4(N Uniform Ts, M Uniform FTs)*

In scenario 4, a uniform distribution environment is used to ensure real and false target encounters. The search environment consists of N uniformly distributed targets and M uniformly distributed false targets. Scenario 4 is unique in that the analytical solution for the probability of false target attack, $P_{A_{FT}}$, and the probability of real target attack, P_{A_T}, is represented by a system of partial differential equations with given boundary conditions. This system is represented by

$$P_{A_T}^{(M,N)} = 1-(1-P_{TR})^N P_{FTR}^M - P_{A_{FT}}^{(M,N)}\ ,$$
$$M = 2,3,...;\quad N = 2,3,... \tag{13}$$

Also, for false targets the probability of false target attack, $P_{A_{FT}}$,

$$P_{A_{FT}}^{(M+1,N-1)} = \frac{M+1}{N}\,\frac{1-P_{FTR}}{P_{TR}}P_{A_T}^{(M,N)}\ ,$$
$$M = 1,2,...;\quad N = 2,3,... \tag{14}$$

with boundary conditions

$$P_{A_T}^{(M,1)} = \frac{1}{M+1}\frac{P_{TR}}{1-P_{FTR}}(1-P_{FTR}^{M+1}) \tag{15}$$

$$P_{A_{FT}}^{(1,N)} = \frac{1}{N+1}\frac{1-P_{FTR}}{P_{TR}}[1-(1-P_{TR})^{N+1}] \tag{16}$$

And finally, for scenario 4, the longevity of the WASM, assuming a performed attack on target, is calculated by the following probability distribution function, $g^{(M,N)}(\tau)$,

$$g^{(M,N)}(\tau) = \tfrac{1}{T}M(1-P_{FTR})(1-P_{TR}\tfrac{\tau}{T})^N\ [1-(1-P_{FTR})\tfrac{\tau}{T}]^{M-1} \tag{17}$$

and the probability

$$1-H^{(M,N)}(T) = 1-(1-P_{TR})^N\ P_{FTR}^M \tag{18}$$

3.5. *Scenario 5(N Normal Ts, Poisson FTs)*

In scenario 5, a circular battlespace of radius r centered at the origin is considered. The search area contains N Normally distributed targets with variance σ and M Poisson distributed false targets parameterized by $\alpha[\frac{1}{km^2}]$. Scenario 5 and later 6 are different from the previous 4 scenarios in that they search in a spiral pattern from the outside of the circle inward. [6] presents the probability of attack and false target attack for this case as

$$P_{A_T}^{(N)}(r) = P_{TR}\,N\int_0^{\frac{r^2}{2\sigma^2}} x\,e^{-[1+2(1-P_{FTR})\alpha\pi\sigma^2]}\,(1-P_{TR}+P_{TR}\,e^{-x})^{N-1}dx \tag{19}$$

$$P_{A_{FT}}^{(N)}(r) = \int_0^r (1-P_{FTR})2\pi\alpha\rho e^{-(1-P_{FTR})\alpha\pi p^2}[1-P_{TR}+P_{TR}e^{-\frac{p^2}{2\sigma^2}}]^N\ dp \tag{20}$$

Similarly, the probability of an attack occurring is characterized as

$$1-H^{(N)}(r) = 1-[1-P_{TR}+P_{TR}\ e^{-\frac{r^2}{2\sigma^2}}]^N\ e^{-\alpha\pi(1-P_{FTR})r^2} \tag{21}$$

where

$$H^{(N)}(r) = e^{-(1-P_{FTR})\alpha\pi r^2}\ [1-P_{TR}+P_{TR}e^{-\frac{r^2}{2\sigma^2}}]^N \tag{22}$$

3.6. *Scenario 6(N Normal Ts, M Circular FTs)*

Scenario 6 consists of a similar battlespace configuration as scenario 5, with the exception that false targets are distributed according to a Normal distribution with variance σ_{FT}. As with scenario 5, the munition search path starts from the outer rim of the circular battlespace and searches inward. The p.d.f.s of interest are

$$f^{(M,N)}(r) = P_{TR}\, N\, \frac{1}{\sigma_T^2} r\, e^{-\frac{r^2}{2\sigma_T^2}} \left(1 - P_{TR} + P_{TR} e^{-\frac{r^2}{2\sigma_T^2}}\right)^{N-1} \cdot \left[P_{FTR} + (1 - P_{FTR}) e^{-\frac{r^2}{2\sigma_{FT}^2}}\right]^{M} \tag{23}$$

$$g^{(M,N)}(r) = (1 - P_{FTR})\, M\, \frac{1}{\sigma_{FT}^2} r\, e^{-\frac{r^2}{2\sigma_{FT}^2}} \left(1 - P_{TR} + P_{TR} e^{-\frac{r^2}{2\sigma_T^2}}\right)^{N} \cdot \left[P_{FTR} + (1 - P_{FTR}) e^{-\frac{r^2}{2\sigma_{FT}^2}}\right]^{M-1} \tag{24}$$

and

$$H^{(M,N)}(r) = \left(1 - P_{TR} + P_{TR} e^{-\frac{r^2}{2\sigma_T^2}}\right)^{N} \left[P_{FTR} + (1 - P_{FTR}) e^{-\frac{r^2}{2\sigma_{FT}^2}}\right]^{M} \tag{25}$$

4. Simulation Configuration

In order to mimic the environments detailed in scenarios 1-6, identical environmental parameters including the general characteristics of the searching WASM were established as to ensure evaluation validity. First, along with their respective distributions all targets were considered non-mobile. Secondly, the WASM swath width for the WASM is modelled at 600m wide by 15m in length. In addition to the ATR parameters, the WASM moves at a fixed velocity of $140\frac{m}{s}$.

The battlespace for scenarios 1-4 considers search strip 600 meters wide by 270,000 meters in length. This provides a search area with equal width of the WASM's primary target acquisition sensor, and with length that can be traversed by the WASM in a 30-minute time of flight. This search area was selected as it modelled as closely as possible the battlespace considered in scenarios 1-4. For this simulation evaluation only two target types are considered, representing a real and false target, respectively. In order for the analytic models for P_{A_T} and $P_{A_{FT}}$ to be valid, the ATR sensor is not assumed to have a fixed view in order to perform all searches without overlapping any previously searched area. Finally, every scenario is evaluated with a single, non-cooperating, searching WASM having a predetermined search path. An example of this search is seen in Figure 3.

Scenarios 5 and 6 require special attention in the construction of their respective search areas. The analytic models depicted in section 3.5 and 3.6 are constructed based on a circular search area. The search area used in the simulation, however, was of identical configurations as those of scenarios 1-4, with the exception of the overall length of the search area. This was necessary due to modelling limitations in the simulation. As a note, the analytic models for scenarios 5,6 were modified to reflect the change from a circular to linear search area.

5. Results

In order to validate the MultiUAV simulation, the analytic results of the six scenarios developed in section 3 were compared to empirical results from Monte Carlo simulations. Each scenario was configured in the simulation to match the false and true target density, the distribution type. Additionally, the WASM lethality and the ability of the ATR algorithm to correctly classify the target type were set to match those values used in the scenario analytical formulation. These parameters can be sorted into two categories.

- WASM Parameters
  - ATR capabilities modelled in the confusion matrix as P_{TR} and P_{FTR}
  - Warhead lethality, P_k
- Battlespace Characteristics
  - Uniform Target density, N
  - Uniform False Target density, M
  - Real Target Poisson Probability Law Parameter, λ_T
  - False Target Poisson Probability Parameter,λ_{FT}
  - Standard Deviation of Target location, σ_{loc}
  - Standard Deviation of False Target location, σ_{locFT}

These results represent several combinations varying P_{TR} and P_{FTR} over a range of realistic values. These test metrics introduced in section 3 represent the expectations that both real and false targets are attacked, destroyed, and on average how long the WASM searched the battlespace before engaging either a T or FT.

The simulation model for the validity investigation of scenario 1 was set up using a single uniformly distributed T, and Poisson distribution of FTs. Hence, $\lambda_{FT} = 10$, for the expectation of 10 false targets over the battlespace, and T =1. The results are tabulated in Table 2.

Table 2. Scenario 1 Results

P_{TR}	P_K	Metric	Simulation Value	Analytic Value	Difference
.85	.5	P_{AT}	46.0	44.0	2.0
		$P_{A_{FT}}$	48.0	45.1	2.9
		P_{TK}	22.0	22.0	0.0
		P_{FTK}	24.0	24.0	0.0
		$\frac{\bar{s}}{T}$	0.3	0.3	0.0
	.8	P_{AT}	46.0	44.0	2.0
		$P_{A_{FT}}$	48.0	52.6	4.6
		P_{TK}	37.0	35.2	1.8
		P_{FTK}	41.0	42.1	1.1
		$\frac{\bar{s}}{T}$	0.3	0.3	0.0
.95	.5	P_{AT}	74.0	74.8	0.8
		$P_{A_{FT}}$	20.0	22.2	2.2
		P_{TK}	37.0	37.4	0.4
		P_{FTK}	11.0	11.1	0.1
		$\frac{\bar{s}}{T}$	0.4	0.4	0.0
	.8	P_{AT}	74.0	74.8	0.8
		$P_{A_{FT}}$	20.0	22.2	2.2
		P_{TK}	61.0	59.8	1.2
		P_{FTK}	19.0	17.7	1.3
		$\frac{\bar{s}}{T}$	0.4	0.4	0.0

Table 3. Scenario 2 Results

P_{TR}	P_K	Metric	Simulation Value	Analytic Value	Difference
.85	.5	P_{AT}	31.0	32.7	1.7
		$P_{A_{FT}}$	59.0	57.7	1.3
		P_{TK}	15.0	16.4	1.4
		P_{FTK}	30.0	28.9	1.1
		$\frac{\bar{s}}{T}$	29.4	32.0	2.6
	.8	P_{AT}	31.0	32.7	1.7
		$P_{A_{FT}}$	59.0	57.7	1.3
		P_{TK}	24.0	26.2	2.2
		P_{FTK}	48.0	46.2	1.8
		$\frac{\bar{s}}{T}$	29.4	32.0	2.6
.95	.5	P_{AT}	50.0	50.1	0.1
		$P_{A_{FT}}$	27.0	26.4	0.6
		P_{TK}	25.0	25.1	0.1
		P_{FTK}	12.0	13.2	1.2
		$\frac{\bar{s}}{T}$	33.2	38.3	5.1
	.8	P_{AT}	50.0	50.1	0.1
		$P_{A_{FT}}$	27.0	26.4	0.6
		P_{TK}	37.0	40.1	3.1
		P_{FTK}	22.0	21.1	0.9
		$\frac{\bar{s}}{T}$	33.2	38.3	5.1

The simulation model for the validity investigation of scenario 2 was set up using a Poisson distribution of Ts, and Poisson distribution of FTs. Hence, $\lambda_{FT} = 10$ for the expectation of 10 false targets over the battlespace, and $\lambda_T = 1$. This setup resembles that of Scenario 1, with the exception that the targets are all modelled via Poisson distributions. The results are tabulated in Table 3.

Table 4. Scenario 3 Results

P_{TR}	P_K	Metric	Simulation Value	Analytic Value	Difference
.85	.5	P_{AT}	77.0	77.0	0.0
		$P_{A_{FT}}$	22.0	26.5	4.5
		P_{TK}	39.0	38.5	0.5
		P_{FTK}	8.0	13.3	5.3
		$\frac{\bar{s}}{T}$	16.1	15.6	0.5
	.8	P_{AT}	77.0	77.0	0.0
		$P_{A_{FT}}$	22.0	26.5	4.5
		P_{TK}	64.0	61.6	2.4
		P_{FTK}	20.0	21.2	1.2
		$\frac{\bar{s}}{T}$	16.1	15.6	0.5
.95	.5	P_{AT}	92.0	91.2	0.8
		$P_{A_{FT}}$	8.0	11.2	3.2
		P_{TK}	46.0	45.6	0.4
		P_{FTK}	3.0	5.6	2.6
		$\frac{\bar{s}}{T}$	17.2	16.3	0.9
	.8	P_{AT}	92.0	91.2	0.8
		$P_{A_{FT}}$	8.0	11.2	3.2
		P_{TK}	77.0	73.0	4.0
		P_{FTK}	8.0	8.9	0.9
		$\frac{\bar{s}}{T}$	17.2	16.3	0.9

The simulation model for the validity investigation of scenario 3 was set up using N uniformly distributed Ts, and Poisson distribution of FTs. Hence, $\lambda_{FT} = 10$ for the expectation of 10 false targets over the battlespace, and N = 5 for Ts. The results are tabulated in Table 4.

The simulation model for the validity investigation of scenario 4 was set up using a N uniformly distributed T, and M uniformly distributed FTs. Hence, M = 10, for the expectation of 10 false targets over the battlespace, and N=1 for Ts. The results are tabulated in Table 5.

The simulation model for the validity investigation of scenario 5 was set up using a N=1 Normally distributed T having σ_T of 98.46 and Poisson distributed FTs. This was realized using $\lambda_F T = 10$ for the expectation of 10 false targets over the battlespace, and N = 1 for Ts. The results are tabulated in Table 6.

Table 5. Scenario 4 Results

P_{TR}	P_K	Metric	Simulation Value	Analytic Value	Difference
.85	.5	P_{AT}	41.0	42.9	1.9
		$P_{A_{FT}}$	57.0	54.1	2.9
		P_{TK}	22.0	21.5	0.6
		P_{FTK}	35.0	27.1	8.0
		$\frac{\bar{s}}{T}$	35.6	34.5	1.1
	.8	P_{AT}	41.0	42.9	1.9
		$P_{A_{FT}}$	57.0	54.1	2.9
		P_{TK}	35.0	34.3	0.7
		P_{FTK}	49.0	43.3	5.7
		$\frac{\bar{s}}{T}$	35.6	34.5	1.1
.95	.5	P_{AT}	72.0	74.5	2.5
		$P_{A_{FT}}$	24.0	22.5	1.5
		P_{TK}	37.0	37.3	0.3
		P_{FTK}	18.0	11.3	6.7
		$\frac{\bar{s}}{T}$	45.4	44.3	1.1
	.8	P_{AT}	72.0	74.5	2.5
		$P_{A_{FT}}$	24.0	22.5	1.5
		P_{TK}	59.0	59.6	0.6
		P_{FTK}	23.0	18.0	5.0
		$\frac{\bar{s}}{T}$	45.4	44.3	1.1

Table 6. Scenario 5 Results

P_{TR}	P_K	Metric	Simulation Value	Analytic Value	Difference
.85	.5	P_{AT}	43.0	40.0	3.0
		$P_{A_{FT}}$	32.0	32.0	0.0
		P_{TK}	21.0	20.4	0.6
		P_{FTK}	16.0	15.8	0.2
		$\frac{\bar{s}}{T}$	27.1	24.2	2.9
	.8	P_{AT}	43.0	40.0	0.3
		$P_{A_{FT}}$	39.0	32.0	7.0
		P_{TK}	36.0	34.3	1.7
		P_{FTK}	26.0	25.4	0.6
		$\frac{\bar{s}}{T}$	29.3	24.2	5.1
.95	.5	P_{AT}	67.0	64.4	2.6
		$P_{A_{FT}}$	10.0	10.5	0.5
		P_{TK}	31.0	32.2	1.2
		P_{FTK}	6.0	5.2	1.2
		$\frac{\bar{s}}{T}$	35.6	33.0	2.6
	.8	P_{AT}	67.0	64.4	2.6
		$P_{A_{FT}}$	15.0	10.5	4.5
		P_{TK}	55.0	51.5	3.5
		P_{FTK}	9.0	8.4	0.6
		$\frac{\bar{s}}{T}$	35.3	33.0	2.3

Table 7. Scenario 6 Results

P_{TR}	P_K	Metric	Simulation Value	Analytic Value	Difference
.85	.5	P_{AT}	37.0	30.2	6.8
		$P_{A_{FT}}$	54.0	43.0	11.0
		P_{TK}	18.0	15.0	3.0
		P_{FTK}	30.0	21.1	8.9
		$\frac{\bar{s}}{T}$	26.1	25.0	1.1
	.8	P_{AT}	32.0	30.2	1.8
		$P_{A_{FT}}$	52.0	43.0	9.0
		P_{TK}	29.0	24.0	5.0
		P_{FTK}	44.0	34.4	9.6
		$\frac{\bar{s}}{T}$	24.5	25.0	0.5
.95	.5	P_{AT}	70.0	71.6	1.6
		$P_{A_{FT}}$	20.0	16.1	3.9
		P_{TK}	36.0	35.8	0.2
		P_{FTK}	11.0	8.0	3.0
		$\frac{\bar{s}}{T}$	38.3	29.0	9.3
	.8	P_{AT}	69.0	71.6	2.6
		$P_{A_{FT}}$	16.0	16.1	0.1
		P_{TK}	56.0	57.2	1.2
		P_{FTK}	14.0	12.8	1.2
		$\frac{\bar{s}}{T}$	32.1	29.0	3.1

The simulation model for the validity investigation of scenario 6 was set up using a N Normally distributed T having σ_T of 98.46 and M Normally distributed FT having $\sigma_F T$ of 98.46. This was realized using M = 10, for the expectation of 10 false targets over the battlespace, and N = 1 for Ts. The results are tabulated in Table 7.

Tables 2 through 7 represent the comparative results of the simulation vs. analytical formulations of the six scenarios outlined in sections 3.1 through 3.6. The results are presented in tabular form, with the analytical solution to the expected probabilities in the Analytical Calculation column, and the results of simulation in the Simulation Result column. Each Monte Carlo simulation of a scenario was run 400 times, 100 per variation of P_{TR} and P_k.

In review one can see the analytic predictions for all scenarios closely align with the simulation results. This is evident as the percent differences between the analytical and empirical data fall well within a 9.6% error bound defined by the confidence interval based on 100 samples per simulation run [7]. This is true for all cases except scenario 6 where P_{FTA} and P_{FTk} for $P_{TR} = .85$ exceed the error bound by 2%. This is the result of data generated from several machines that have dissimilar random number generators. The results indicate strong correlation between the analytical models for all scenarios, as the errors in all cases have fallen within the statistical confidence interval calculated for 100 simulation runs per scenario configuration.

6. Conclusions

An evaluation methodology to provide a baseline performance validation of the MultiUAV simulation tool has been proposed. This evaluation compares the simulation vs. analytical results for scenarios comprised of both real and false target attacks, in addition to the lifetime of a single WASM over a range of vehicle performance parameters. Six analytical scenarios provide the necessary variations in the type of multi-target distributions in order to evaluate the simulation performance parameters for varying battlespace conditions.

Comparative results presented in the previous section indicate the use of the MultiUAV simulation can provide valid target classification and kill information. The validation methodology presented here is crucial for further research involving MultiUAV for use in the study of cooperative WASMs. This allows future decentralized cooperative control research to focus on control algorithms, as the results of each target attack and kill are now deemed valid.

References

[1] Robert E. Dunkel, "Investigation of cooperative behavior in autonomous wide area search munitions," M.S. thesis, Air Force Institute of Technology, Wright-Patterson AFB OH, March 2002.

[2] Daniel P. Gillen, "Cooperative behavior schemes for improving the effectiveness of autonomous wide area search munitions," M.S. thesis, Air Force Institute of Technology, Wright-Patterson AFB OH, March 2001.

[3] Phillip R. Chandler, Corey Schumacher and Steven R. Rasmussen, "Task allocation for wide area search munitions via network flow optimization," Guidance, Navigation and Control Conference, Aug 2001.

[4] P. R. Chandler and S. J. Rasmussen, "MultiUAV: A multiple UAV simulation for investigation of cooperative control," in *Winter Simulation Conference*, San Diego, CA, November 2002.

[5] David R. Jacques, *Search, Classification and Attack Decisions for Cooperative Wide Area Search Munitions*, Work in Progress, 2002.

[6] Meir Pachter and David R. Jacques, *Theory of Cooperative Search, Classification, and Target Attack*, Work in Progress, 2002.

[7] G. M. Bragg, *Principles of Experimentation and Measurement*, New Jersey: Prentice-Hall, 1974.

CHAPTER 22

COOPERATIVE CONTROL OF MULTIPLE UAV'S IN CLOSE FORMATION FLIGHT VIA NONLINEAR ADAPTIVE APPROACH

Y. D. Song,[a] Y. Li, M. Bikdash and T. Dong

Department of Electrical Engineering
North Carolina A&T State University
Greensboro, NC
songyd@ncat.edu

Close formation control of multi-UAVs is addressed in this chapter. Nonlinear dynamic model reflecting the aerodynamic coupling effects introduced by close formation flight (such as vortex of the adjacent lead aircraft) is considered. Adaptive control algorithms for asymptotic lateral, longitudinal, and vertical separation tracking are developed. Simulation on three F16 class aircrafts performing Δ-shaped formation was conducted. Both theoretical studies and simulation results demonstrate the effectiveness of the proposed control method.

Keywords:
Close formation, adaptive control, multi-UAVs, tracking stability

1. Introduction

Unmanned Aerial Vehicles (UAVs) are remotely piloted or self-piloted aircrafts that can carry cameras, sensors, communications equipment or other payloads. They have been used in a reconnaissance and intelligence-gathering role since the 1950s, and more challenging roles are envisioned, including combat missions. In fact, UAVs and UCAVs (Unmanned Combat Aerial Vehicles) will be used increasingly to counter threats from mobile targets and high-value targets of opportunity in battlefield, as conceptually illustrated in Figure 1.

For this reason, the problem of close formation flying control of UAVs in the patterns (i.e. V-shaped, Δ-shaped) similar to those flown by flocks

[a]Corresponding author

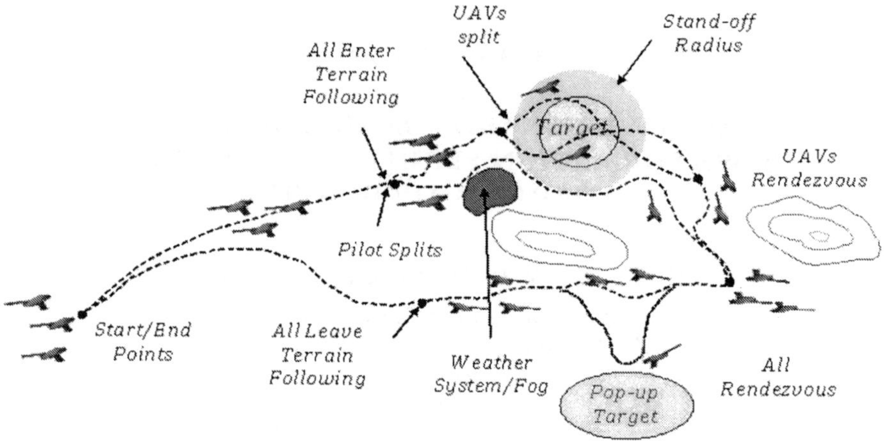

Fig. 1. Typical Formations in Battlefield.

of birds has been an interesting yet challenging topic of research for many years.

A number of studies on modeling and control of multiple UAVs in close formation flight have been carried out lately. Through aerodynamic calculations Blake and Multhopp [1] investigated the effect of vortexes created by the leader aircraft on the follower aircraft. Such effect on the wingman's flight dynamics was studied by D'Azzo and his coworkers [7]. Early contributors on close formation control include Buzogany and Pachter [2], Reyna and Pachter [8], Proud et al. [7], Fierro et al. [3], Giulitti et al. [4], Jongusuk et al. [5], Richards et al. [9], and Wolfe et al. [5]. Most of the results are based on linear flight models that either linearize or ignore the effect of vortex. Singh [10] considered the nonlinear property of vortex and studied the control problem using backstepping design method.

In this chapter, we present a control scheme based on nonlinear flight model in which the effect of vortex introduced by the adjacent leader UAV is addressed. By using orthogonal coordinate transformation, we develop a set of control algorithms capable of maintaining the desirable separation of wingman with the leading UAV. The design procedure presented here is simple and the results are global. Simulation tests on three F16 class aircrafts performing Δ-shaped formation are conducted and satisfactory results are achieved.

2. Flight Model

The formation geometry is determined by the relative position between the Leader and Wingman as shown in Figure 2. The formation control objective is to steer the Wingman (follower) to maintain certain separation distance in longitudinal, lateral and vertical directions.

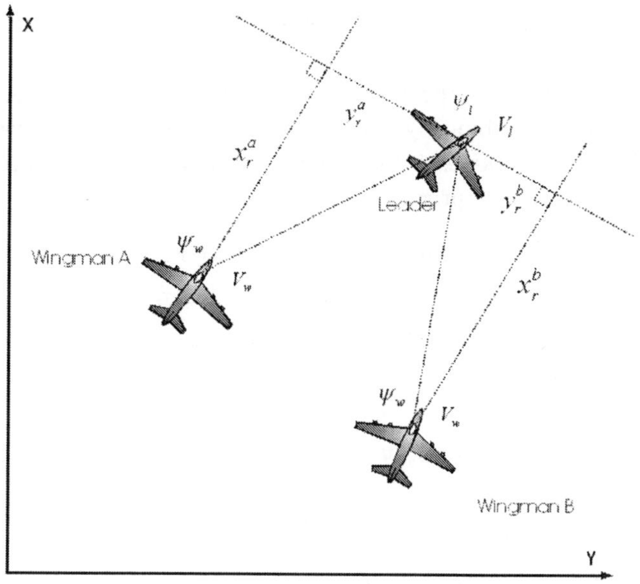

Fig. 2. Multi-UAVs in close formation flight.

Table 1. Nomenclature of the flight

Parameter	Variable
Aircraft Mass	m
Dynamic Pressure	$\overline{q}$
Wing Area	s
Distance of X Coordinate	x
Distance of Y Coordinate	y
Altitude	h
Heading Velocity	V
Heading Angle	ψ
Autopilot Time Constant	τ

Table 2. Subscripts of the variables

Parameter	Subscript
Desired Value	d
Separation/Difference	e
Leader Aircraft	l
Wingman Aircraft	w
Drag Coefficient	D
Side-wash Coefficient	I
Lift Coefficient	L
Relative Measurement	r

By properly defining the body frame and inertia frame, the following three equations describing dynamic behavior of wingman aircraft can be established (refer to Tables 1 and 2 for nomenclature and definition of the variables used in the chapter),

$$\dot{V}_w = f_v(V_w) + g_v u_v + \Delta f_v(\cdot) \tag{1}$$

$$\ddot{\psi}_w = f_\psi(\psi_w, \dot{\psi}_w) + g_\psi u_\psi + \Delta f_\psi(\cdot) \tag{2}$$

$$\ddot{h}_w = f_h(h_h, \dot{h}_w) + g_h u_h + \Delta f_h(\cdot) \tag{3}$$

with

$$\Delta f_v(\cdot) = \frac{\overline{q}s}{m}\Delta c_{Dw_y}(y - y_d) + \delta_1(\cdot) \tag{4}$$

$$\Delta f_\psi(\cdot) = \frac{\overline{q}s}{mV_w}[\Delta c_{Iw_y}(y - y_d) + c_{Iw_y}(h - h_d)] + \delta_2(\cdot) \tag{5}$$

$$\Delta f_h(\cdot) = \frac{\overline{q}s}{m}\Delta c_{Lw_y}(y - y_d) + \delta_3(\cdot) \tag{6}$$

where V_w, ψ_w and h denote the wingman (follower's) heading velocity, heading angle and altitude, g_v, g_ψ and g_h are the system constant, u_v, u_ψ and u_h are the control inputs. The effect of vortex and external disturbances are represented by $\Delta f_v(\cdot)$, $\Delta f_\psi(\cdot)$ and $\Delta f_h(\cdot)$, in which s is the wing area, m is gross mass, $\overline{q}$ is dynamic pressure, Δc_{Dw} is drag coefficient, Δc_{Iw} is side-wash coefficient, Δc_{Lw} is lift coefficient, and $\delta_i(\cdot)$ are lumped disturbances. The flight model presented here includes most existing UAV models as a special case.

For close formation, it is important to precisely keep (separate) the wingman certain distance away from the leader UAV in lateral (x-axis), longitudinal (y-axis) and vertical (h-axis) directions to prevent possible collision. For this reason, we define a relative frame as shown in Figure 2 and introduce the following relative coordinates (for simplicity, leader with only one wingman is considered hereafter, while the development applies to multi-UAVS):

$$x_r = x_l - x_w, \quad y_r = y_l - y_w, \quad h_r = h_l - h_w$$

Then we have the relative kinematics equations,

$$\dot{x}_r = V_l \cos(\psi_l - \psi_w) + \dot{\psi}_w y_r - V_w \tag{7}$$

$$\dot{y}_r = V_l \sin(\psi_l - \psi_w) - \dot{\psi}_w x_r \tag{8}$$

$$\dot{h}_r = \dot{h}_l - \dot{h}_w \tag{9}$$

The formation control problem can be stated as follows: design control algorithms so that wingman's relative position in terms of x_r, y_r, h_r coordinates is kept at the desired value with respect to the leader aircraft. Namely, the heading velocity control u_v, heading angle control u_ψ, and altitude control u_h are to be designed so that

$$x_r = x_l - x_w \rightarrow x_d \tag{10}$$

$$y_r = y_l - y_w \rightarrow y_d \tag{11}$$

$$h_r = h_l - h_w \rightarrow h_d \tag{12}$$

where x_d, y_d and h_d are the desired formation distance in x, y and h coordinates. Jongusak et al. [5] addressed this problem by ignoring $\Delta f_v(\cdot)$, $\Delta f_\psi(\cdot)$ and $\Delta f_h(\cdot)$, Proud et al. [7] studied the control problem with the assumption that $\Delta f_v(\cdot)$, $\Delta f_\psi(\cdot)$ and $\Delta f_h(\cdot)$ are known and linear. In this chapter, we explicitly consider the effect of $\Delta f_v(\cdot)$, $\Delta f_\psi(\cdot)$ and $\Delta f_h(\cdot)$ on the system dynamics using robust and adaptive methods. Since the flight conditions and vortex effects are not known precisely in general, then $\Delta f_v(\cdot)$, $\Delta f_\psi(\cdot)$ and $\Delta f_h(\cdot)$ will be treated totally unavailable and will not be used directly.

3. Control Algorithms Design

As the first step, we introduce the separation error

$$e_x = x_r - x_d \tag{13}$$

$$e_y = y_r - y_d \tag{14}$$

$$e_h = h_r - h_d \tag{15}$$

The control objective is to design the control inputs u_v, u_ψ and u_h to steer the separation error to zero. Using the relative kinematics equations (7)-(8) we obtain the following lateral, longitudinal and vertical formation error dynamics equation

$$\begin{bmatrix} \dot{e}_x \\ \dot{e}_y \\ \dot{e}_h \end{bmatrix} = A(\psi_e) + R(x_r, y_r) \begin{bmatrix} V_w \\ \dot{\psi}_w \\ \dot{h}_w \end{bmatrix} \tag{16}$$

where

$$A(\psi_e) = \begin{bmatrix} V_l \cos\psi_e \\ V_l \sin\psi_e \\ \dot{h}_l \end{bmatrix}, \; R = \begin{bmatrix} -1 & y_r & 0 \\ 0 & -x_r & 0 \\ 0 & 0 & -1 \end{bmatrix}, \; \psi_e = \psi_l - \psi_w$$

noting that heading speed V_w and heading angle ψ_w are controlled by u_v and u_ψ via

$$\begin{bmatrix} \dot{V}_w \\ \ddot{\psi}_w \\ \ddot{h}_w \end{bmatrix} = \begin{bmatrix} f_v \\ f_\psi \\ f_h \end{bmatrix} + G \begin{bmatrix} u_h \\ u_\psi \\ u_h \end{bmatrix} + \begin{bmatrix} \Delta f_v \\ \Delta f_\psi \\ \Delta f_h \end{bmatrix} \tag{17}$$

where $G = diag(g_v, g_\psi, g_h)$, one may attempt to combine (16) and (17) to design control for stabilizing e_x and e_y directly. However, since the inverse of R is not defined at $x = 0$ (this physically corresponds to the situation that the relative distance in x-axis from Wingman toward Leader becomes zero, which could happen anytime), direct stabilization of e_x, e_y and e_h is infeasible. To circumvent this problem, we introduce the following coordinate transformation

$$E = \begin{bmatrix} E_x \\ E_y \\ E_h \end{bmatrix} = B_i(\psi_w) \begin{bmatrix} e_x \\ e_y \\ e_h \end{bmatrix} \tag{18}$$

where $B_i(\psi_w)$ is an orthogonal matrix satisfying

$$B_i^T(\psi_w) B_i(\psi_w) = \begin{bmatrix} 1 & 0 & 0 \\ 0 & 1 & 0 \\ 0 & 0 & 1 \end{bmatrix} \; \forall \; \psi_w \tag{19}$$

It is interesting to note that there exist many such matrices, i.e.,

$$B_1 = \begin{bmatrix} \sin\psi_w & \cos\psi_w & 0 \\ \cos\psi_w & -\sin\psi_w & 0 \\ 0 & 0 & 1 \end{bmatrix}, \; B_2 = \begin{bmatrix} \cos\psi_w & -\sin\psi_w & 0 \\ \sin\psi_w & \cos\psi_w & 0 \\ 0 & 0 & 1 \end{bmatrix}$$

$$B_3 = \begin{bmatrix} -\cos\psi_w & \sin\psi_w & 0 \\ \sin\psi_w & \cos\psi_w & 0 \\ 0 & 0 & 1 \end{bmatrix}, \; B_4 = \begin{bmatrix} -\sin\psi_w & -\cos\psi_w & 0 \\ \cos\psi_w & -\sin\psi_w & 0 \\ 0 & 0 & 1 \end{bmatrix}$$

Also note that with (18) we have

$$\|E\|^2 = \begin{bmatrix} e_x \ e_y \ e_h \end{bmatrix} B_i^T B_i \begin{bmatrix} e_x \\ e_y \\ e_h \end{bmatrix} = \left\| \begin{matrix} e_x \\ e_y \\ e_h \end{matrix} \right\|^2 \tag{20}$$

which implies that $\|E\| \to 0$ as $t \to \infty$ leads to

$$\left\| \begin{matrix} e_x \\ e_y \\ e_h \end{matrix} \right\| \to 0 \text{ as } t \to \infty.$$

Therefore, it is sufficient to design control law to stabilize E. From (16) and (18) (where B_1 is used), it can be shown that

$$\dot{E} = \begin{bmatrix} V_l \sin \psi_l \\ V_l \cos \psi_l \\ \dot{h}_l \end{bmatrix} + C \begin{bmatrix} V_w \\ \dot{\psi}_w \\ \dot{h}_w \end{bmatrix} \tag{21}$$

where

$$C = \begin{bmatrix} -\sin \psi_w & y_d \sin \psi_w - x_d \cos \psi_w & 0 \\ -\cos \psi_w & y_d \cos \psi_w + x_d \sin \psi_w & 0 \\ 0 & 0 & -1 \end{bmatrix} \tag{22}$$

It can be verified that the matrix C is always invertible. In order to design control law to derive E toward zero asymptotically, we use (17) and (21) to get

$$\ddot{E} = \begin{bmatrix} \dot{V}_l \sin \psi_l + \dot{\psi}_l V_l \cos \psi_l \\ \dot{V}_l \cos \psi_l - \dot{\psi}_l V_l \sin \psi_l \\ \ddot{h}_l \end{bmatrix} + \dot{C} \begin{bmatrix} V_w \\ \dot{\psi}_w \\ \dot{h}_w \end{bmatrix} + C \begin{bmatrix} \dot{V}_w \\ \ddot{\psi}_w \\ \ddot{h}_w \end{bmatrix} \tag{23}$$

which can be further expressed as

$$\ddot{E} = D + C \left\{ G \begin{bmatrix} u_v \\ u_\psi \\ u_h \end{bmatrix} + \begin{bmatrix} \Delta f_v \\ \Delta f_\psi \\ \Delta f_h \end{bmatrix} \right\} \tag{24}$$

where

$$D = \begin{bmatrix} \dot{V}_l \sin\psi_l + \dot{\psi}_l V_l \cos\psi_l \\ \dot{V}_l \cos\psi_l - \dot{\psi}_l V_l \sin\psi_l \\ \ddot{h}_l \end{bmatrix}$$

$$+ \dot{\psi}_w \begin{bmatrix} -\cos\psi_w & x_d \sin\psi_w + y_d \cos\psi_w & 0 \\ \sin\psi_w & -y_d \sin\psi_w + x_d \cos\psi_w & 0 \\ 0 & 0 & 0 \end{bmatrix} \begin{bmatrix} V_w \\ \dot{\psi}_w \\ \dot{h}_w \end{bmatrix}$$

$$+ C \begin{bmatrix} f_v \\ f_w \\ f_h \end{bmatrix} \tag{25}$$

Since D is computable and $\det(C) = x_d$, the matrix CG is invertible, therefore the transformation as introduced in (18) makes the following control law well defined,

$$\begin{bmatrix} u_v \\ u_w \\ u_h \end{bmatrix} = G^{-1}C^{-1}\left\{ -D - 2(\alpha+\beta)\dot{E} - \alpha\beta E + \begin{bmatrix} u_1 \\ u_2 \\ u_3 \end{bmatrix} \right\} \tag{26}$$

where $\alpha > 0$ and $\beta > 0$ are design parameters chosen arbitrarily and u_1, u_2 and u_3 are the compensating signals to be determined based on the following conditions on $\Delta f_v(\cdot)$, $\Delta f_\psi(\cdot)$ and $\Delta f_h(\cdot)$.
Case 1: If $\Delta f_v(\cdot)$, $\Delta f_\psi(\cdot)$ and $\Delta f_h(\cdot)$ are negligible, then $u_1 = u_2 = u_3 = 0$.
Case 2: If $\Delta f_v(\cdot)$, $\Delta f_\psi(\cdot)$ and $\Delta f_h(\cdot)$ are available precisely, then

$$\begin{bmatrix} u_1 \\ u_2 \\ u_3 \end{bmatrix} = -C \begin{bmatrix} \Delta f_v(\cdot) \\ \Delta f_\psi(\cdot) \\ \Delta f_h(\cdot) \end{bmatrix}.$$

Case 3: If $\begin{bmatrix} \Delta f_v(\cdot) \\ \Delta f_\psi(\cdot) \\ \Delta f_h(\cdot) \end{bmatrix} = \Psi a$ where $\Psi \in R^{3\times q}$ is a known and bounded regressor matrix, and $a \in R^q$ is an unknown parameter vector, then

$$\begin{bmatrix} u_1 \\ u_2 \\ u_3 \end{bmatrix} = -C\Psi\hat{a} \ \text{ and } \ \dot{\hat{a}} = (C\Psi)^T(\dot{E} + \alpha E). \tag{27}$$

Case 4: If $\left\| C \begin{bmatrix} \Delta f_v(\cdot) \\ \Delta f_\psi(\cdot) \\ \Delta f_h(\cdot) \end{bmatrix} \right\| \leq c < \infty$, then $\begin{bmatrix} u_1 \\ u_2 \\ u_3 \end{bmatrix} = -c\,sign(\dot{E} + \alpha E)$.

Proof: Case 1 and Case 2 can be easily shown. For Case 3, we have

$$\ddot{E} = -(\alpha + \beta)\dot{E} - \alpha\beta E + C\Psi\tilde{a} \tag{28}$$

Consider the Lyapunov function candidate

$$V = \frac{1}{2}(\dot{E} + \alpha E)^T(\dot{E} + \alpha E) + \frac{1}{2}\tilde{a}^T\tilde{a} \tag{29}$$

The control scheme (26)-(27) will lead to

$$\dot{V} = -\beta(\dot{E} + \alpha E)^T(\dot{E} + \alpha E) \leq 0 \tag{30}$$

Therefore, it is readily shown that $\dot{E} + \alpha E \in L_\infty \cap L_2$ and $\tilde{a} \in L_\infty$. Therefore we have $E \in L_\infty \cap L_2$, $\dot{E} \in L_\infty \cap L_2$ and $E \in L_\infty$. Since Ψ is bounded, therefore $\ddot{E} \in L_\infty$ which can be proved from (23). By Barbalat lemma [11], we conclude that E, e_x, e_y and e_h tend to zero as time increases (similarly for Case 4). □

The overall control scheme is illustrated in Figure 3

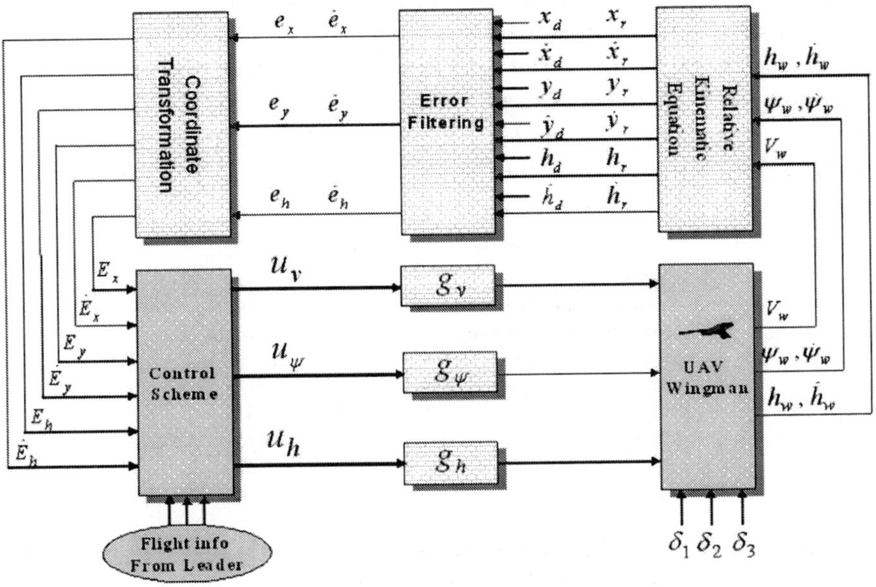

Fig. 3. Control Scheme Diagram.

4. Simulation

To verify the effectiveness of the developed control law, we conduct computer simulation on three F-16 class aircrafts in close formation. It has been determined through aerodynamic calculations that the optimal spacing between the wingman and leader aircraft is $\pi\ell/4$, where ℓ is wingspan of the lead aircraft. The vortex effect is considered as given in Proud et al. [7]. Note that during varying flight conditions it is hard to posses the precise value of dynamic pressure $\bar{q}$ and the lift, drag and side-wash coefficients. In this work, all these parameters are treated as completely unknown. Namely,

$$a = \begin{bmatrix} \bar{q}\Delta C_{Dw_y}, \\ \bar{q}\Delta C_{Iw_y}, \\ \bar{q}\Delta C_{Iw_h}, \\ \bar{q}\Delta C_{Lw_y}, \end{bmatrix}, \quad \Psi = \begin{bmatrix} \frac{se_y}{m} & 0 & 0 & 0 \\ 0 & \frac{se_y}{mV_w} & \frac{se_h}{mV_w} & 0 \\ 0 & 0 & 0 & \frac{se_y}{m} \end{bmatrix}$$

The flight characteristics and simulation parameters for Leader UAV, Wingman A and B are shown in Table 3. The initial relative flying positions for Wingman A and Wingman B are $[100\ \ 60\ \ 5,000](ft)$ and $[90\ \ -50\ \ -29,000](ft)$, respectively. The simulation of Δ-shaped (triangle) formation was conducted, which requires the final formation positions for Wingman A is maintained at $[60\ \ 30\pi/4\ \ 0](ft)$ and Wingman B is $[60\ \ -30\pi/4\ \ 0](ft)$. Once the formation is established, three UAVs will maintain the flight dynamics as heading speed $V_l = 825 ft/s$, heading angle $\psi_l = \frac{\pi}{8}(rad)$ and altitude $h_l = 30,000(ft)$ (Note that three UAVs are finally flight on the same altitude).

Table 3. Aircraft Characteristic Values

Parameter	*Value*	*Unit*
Velocity time constant	5	Seconds
Heading time constant	0.75	Seconds
Dynamic Pressure	155.8	lb/ft^2
Formation Heading Velocity	825	ft/s
Wing Area	300	ft^2
Wing Span	30	ft
Mass	776.4	lb

The simulation results are presented in Figures 4 through 7, where Figure 4 illustrates the 3D tracking process of the three UAVs performing Δ-shaped (triangle) formation. Figure 5 is the separation distance trajectory tracking on lateral (x-axis), longitudinal (y-axis) and vertical (h-axis) direction, respectively. The heading velocity and heading angle trajectory tracking are shown in Figure 6. The control signal for u_v, u_ψ and u_h are depicted in Figure 7 As can be seen, the proposed control scheme works very well in maintaining the desired formation under the effect of vortex. The control action is bounded and smooth.

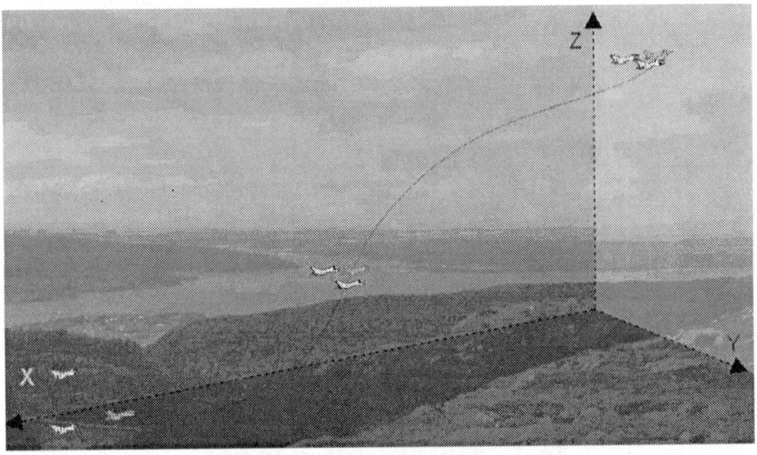

Fig. 4. Triangle formation flight simulation result.

5. Conclusions

This chapter proposed a new method to design control law for close formation tracking control of multi-UAVs. The developed strategy is based on highly nonlinear flight model and the effect of vortex is considered. It is shown that the adaptive control scheme is able to deal with the system nonlinearities and external disturbances due to the close formation. Simulation of one-leader-two-wingman formation pattern was conducted. Both theoretical studies and simulation results demonstrate that this developed adaptive control scheme is effective and robust for multi-UAVs formation flight under varying flight conditions.

Acknowledgments

This project was supported in part by ONR (Office of Naval Research) through the grant N00014-03-1-0462.

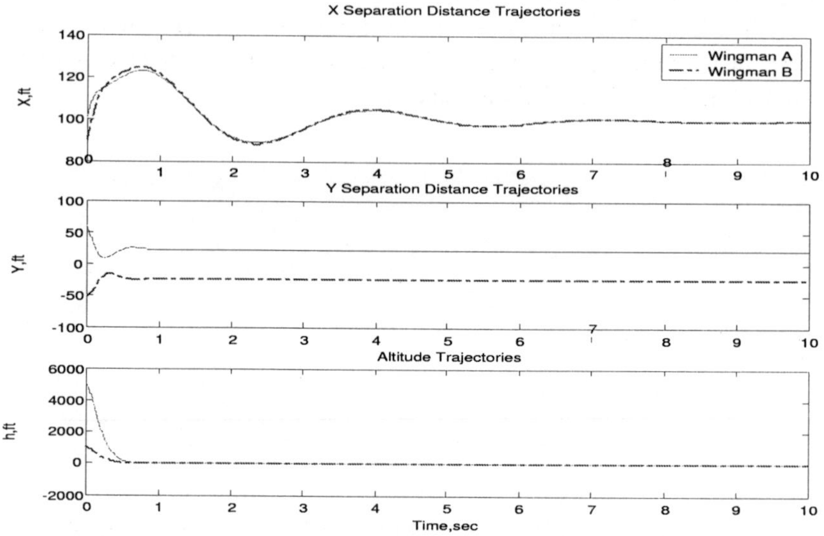

Fig. 5. Separation distance tracking.

References

[1] W. Blake and D. Multhopp, Design, Performance and Modeling Consideration for Close Formation Flight. AIAA Guidance, Navigation and Control Conference, Boston, MA, July 1998.

[2] L. E. Buzogany and M. Pachter. Automated Control of Aircraft in Formation Flight. AIAA Guidance, Navigation, and Control Conference, Part. 3, pages 1349-1370 , 1993.

[3] R. Fierro, C.Belta, and J.P. Desai. On Controlling Aircraft Formation. In Proc. 40th IEEE Conf on Decision and Control., pp. 1065-1070, Orlando, FL, December 2001.

[4] F. Giulitti, L. Pollini, and M. Innocenti. Autonomous formation flight. IEEE Control System, vol 20, no.6, pages 34-44, 2000.

[5] J. Jongusuk, T. Mita, Y. Masuko. Tracking Control of UAV in 3D Space - Toward Formation Control. The 30th Symposium on System Theory, Oita, Japan, 2001.

[6] M. Kristic, I. Kanellakopoulos and P. Kokotovic. Nonlinear and Adaptive Control Design. Wiley-Inerscience, 1995.

[7] A. W. Proud, M. Pachter, and J. J. D'Azzo. Close Formation Flight Control. AIAA Guidance, Navigation, and Control Conference, Vol. 2, pages 1231-1246, 1999.

[8] V. P. Reyna and M. Pachter. Formation Flight Control Automation. AIAA Guidance, Navigation, and Control Conference, Part. 3, pages 1379-1404 ,

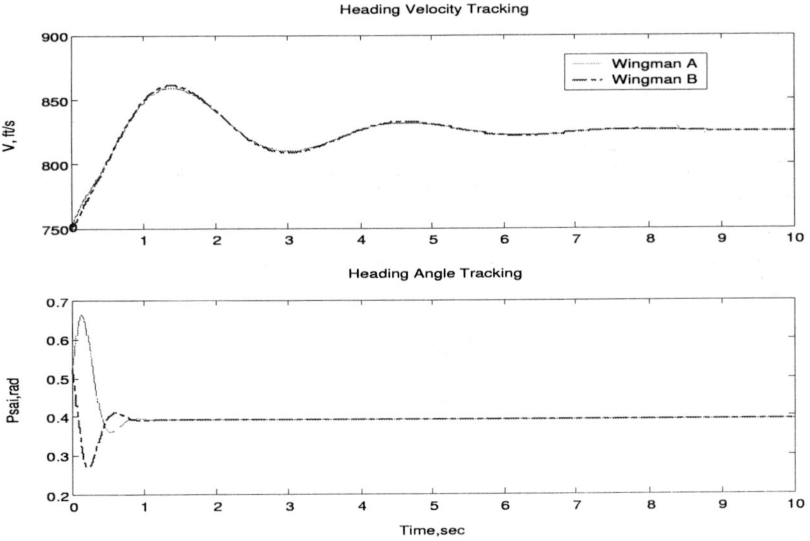

Fig. 6. Heading speed and heading angle tracking.

1994.

[9] A. Richards, J. Bellingham, M. Tillerson, and J. P. How. Coordination and Control of Multiple UAVs, AIAA Guidance, Navigation, and Control Conference, Monterey, CA, 2002.

[10] S. N. Singh. Adaptive Feedback Linearization Nonlinear Close Formation Control of UAVs. Proceedings of American Control Conference, pages 854-858, 2000.

[11] J. J. Slotine and W. Li. Applied Nonlinear Control. Prentice-Hall, 1991.

[12] J. D. Wolfe, D. F. Chichka, and J. L. Speyer. Decentralized controllers for unmanned aerial vehicle formation flight. AIAA Guidance Navigation and Control Conference, San Diego, CA, July, 1996.

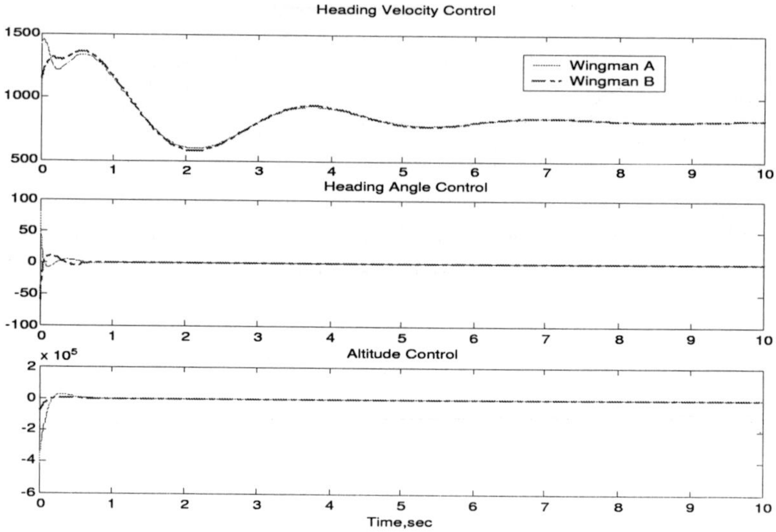

Fig. 7. Control signal for heading speed, heading angle and altitude channels.

CHAPTER 23

A VEHICLE FOLLOWING METHODOLOGY FOR UAV FORMATIONS

Stephen Spry
Andy Vaughn
Xiao Xiao
and J. Karl Hedrick
University of California, Berkeley

This chapter develops a control methodology which allows a group of Unmanned Aerial Vehicles (UAVs) to follow a ground vehicle, or, more generally, a moving or stationary point, while maintaining a desired formation pattern. This capability could be used in a number of applications, including surveillance missions such as convoy protection or search and rescue operations.

Assuming that the point of interest is moving at a speed less than the maximum flight speed of the aircraft, the point is used to define the location and orientation of a moving orbital trajectory. This trajectory is designed to satisfy aircraft speed and turn rate constraints, and is developed such that an aircraft which tracks the trajectory will cross over the point periodically, with a specified time interval. As the ratio of point speed to aircraft speed varies from zero to one, the path traced by the aircraft changes smoothly from a figure-eight to a periodic curve to a straight line. A tracking law is developed which steers the aircraft along the trajectory using heading and airspeed commands.

In order to apply this approach to a formation of UAVs, we use a formation controller which is based on the use of generalized coordinates. These coordinates characterize the location (L), orientation (O), and shape (S) of the formation. This provides a natural and convenient way of specifying configuration and makes it possible to control a group of aircraft as a single entity. This controller is used as an intermediate layer between the orbit tracking control and the individual aircraft. It accepts orbit tracking commands as group motion commands, and produces heading and airspeed commands for the individual aircraft in the formation. These individual commands are designed to move the group along a desired LO trajectory while maintaining desired relative positioning of the aircraft.

The methodology is illustrated through several hardware-in-the-loop simulations, in which two aircraft follow a truck moving at different speeds and headings. In addition, experimental results from a two-aircraft flight test are presented.

1. Introduction

A current area of research for military and civilian applications is the use of small, inexpensive unmanned aerial vehicles (UAVs) to provide useful services to personnel. Some of the services that UAVs may assist in are: surveillance, convoy protection, border patrol, search and rescue, and weather monitoring [4, 7]. UAVs are particularly suited to these tasks because they are economical, they minimize the risk of loss of human life, they are able to perform monotonous duties for long periods of time, and they may be operated by a limited number of personnel (hopefully, many UAVs operated by one person).

A core capability which may be useful for convoy protection, search and rescue, and border patrol applications is the ability of either a single UAV or a group of UAVs to track a point which may move arbitrarily. In convoy protection, for example, we may wish to maintain video coverage of a region surrounding the convoy, where the center of the region moves with the convoy. In a search and rescue mission, we would like to keep a group of UAVs moving with a human-piloted helicopter. In border patrol, we might like to track a group of intruders.

For a number of reasons, it is desirable to use fixed-wing UAVs if possible, as they are simpler, less expensive, and have greater maximum flight times than rotary-wing aircraft. The main difficulty with fixed-wing aircraft is that they are subject to constraints on airspeed and turn rate. Because of this, special tracking algorithms must be used to allow fixed-wing aircraft to track points of interest which can move arbitrarily. If we want to track a moving point with a group of UAVs, then the tracking algorithm should also be able to maintain a particular group shape and orientation (to allow optimal spacing of multiple cameras, for example).

In this chapter, we focus on the convoy protection problem. The objective is to have a group of fixed-wing UAVs perform a surveillance routine while tracking a ground vehicle that is moving unpredictably, but at a speed less than the maximum flight speed of the UAVs. The aircraft are subject to both airspeed and turn-rate constraints. We assume that information on the position and heading of the vehicle is obtained from vision, radar, or GPS sensors. If desired, the UAVs can travel at a specified offset distance ahead of the ground vehicle; this scheme was addressed in [1].

Our previous work on convoy protection featured a single UAV tracking a ground vehicle while flying in a sinusoidal path [1]. This approach to flying was based on allowing the UAV to fly at a constant speed while the ground vehicle was able to travel at any speed from a standstill up to the velocity of the UAV. The trajectory changed amplitude based on the speed of the ground vehicle relative to the UAV. Although this approach worked well, the sinusoidal path was only applicable to ground vehicle speeds above a certain value. For slower ground vehicle speeds, the UAV had to change its desired trajectory and switch into a different mode.

This chapter expands upon the approach developed in [1] on several fronts. First, the need for mode switching has been eliminated in our new approach. We replace the path generation by an orbit trajectory generation method. Second, we advance the methodology by developing an algorithm that can be tuned for the flight parameters of a given aircraft. Third, we employ a formation control algorithm that allows multiple aircraft to track the ground vehicle as a group, performing convoy protection and surveillance while maintaining safe, collision-free flight.

The outline of the chapter is as follows. In section 2, we formulate the orbit trajectory and establish some of its key properties. We also discuss the control strategy which is used to track this trajectory. In section 3, we present the formation control which allows multiple aircraft to track the orbit as a group. Section 4 discusses implementation issues. Section 5 includes results of some hardware-in-the-loop (HIL) simulations, and section 6 highlights the results from a recent two-aircraft flight test. Conclusions are given in section 7.

2. Orbital Trajectory

In this section, we describe a trajectory generation algorithm that allows a UAV with a limited range of flight speeds and limited turn rate to track a point which moves arbitrarily. This algorithm will generate a feasible path for the UAV that will "slow it down" and allow it to track the point, in the sense that it stays within a certain distance of the point and passes over it periodically. In the convoy protection application, the point could be a specified ground vehicle, the centroid of an entire convoy, or a point that stays some distance ahead of the convoy.

The trajectory is based on a parameterized family of figure eight orbits which are defined in a coordinate frame that moves with the point of interest and has its positive y-axis aligned with the velocity of the point. Figures 1 and 2 illustrate the trajectory in point-fixed coordinates and ground-fixed coordinates respectively, for a point speed of 4 m/s and a UAV speed of 20 m/s. It will be shown later in this section how the orbit parameters are chosen for different point and aircraft speeds.

In contrast to the approach presented in [1], that used two different trajectory modes (sinusoidal and loitering) for different ground vehicle speeds, this approach uses a single mode for all ground vehicle speeds. As the ratio of ground vehicle speed to aircraft speed varies from zero to one, the path traced by the aircraft changes smoothly from a figure-eight to a periodic curve to a straight line.

The trajectories may be used with either a single UAV or with a group of UAVs having compatible flight characteristics. When used with a group of UAVs, the trajectory tracking commands are sent to a formation controller, which is explained in the next section of this chapter.

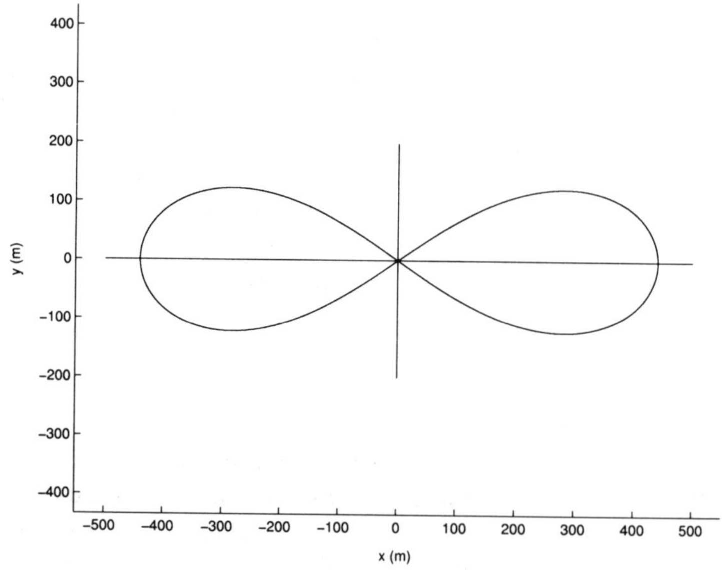

Fig. 1. Lemniscate trajectory in point-fixed coordinates.

Assuming steady (constant-velocity) motion of the ground vehicle, we define two reference frames, A, and B. Frame B is a right-handed frame, fixed in the ground vehicle, with its y-axis aligned with the vehicle heading

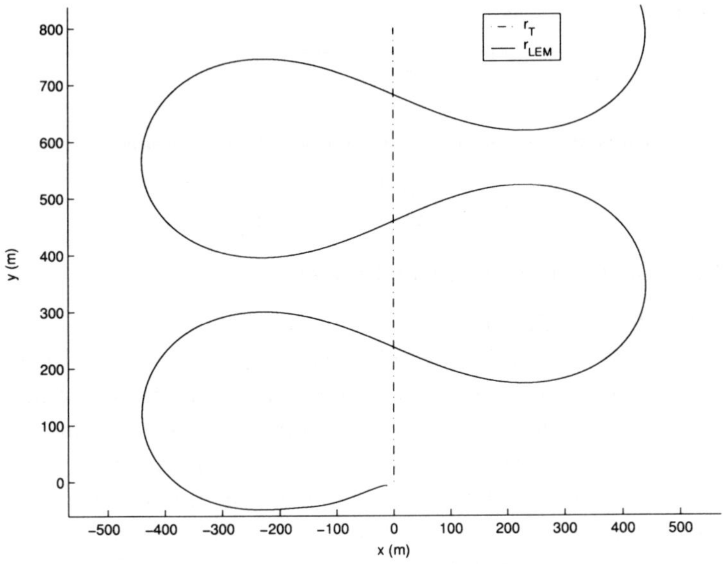

Fig. 2. Lemniscate trajectory in ground-fixed coordinates.

and its z-axis pointing up. Frame A is an earth-fixed frame with its axes parallel to those of B. As it is assumed that the ground vehicle does not accelerate, both A and B are inertial frames.

Orbital trajectories are defined in terms of frame B using the equation for a lemniscate curve, which is:

$$r = A\sqrt{\cos p\theta} \tag{1}$$

In this equation, r and θ are cylindrical coordinates in frame B, with θ being the angle from the local x-axis. The constant parameters A and p determine the amplitude and shape of the curve and are to be chosen based on desired trajectory properties.

For $\theta \in [0, \frac{\pi}{2p}]$, the position of a point L on the lemniscate curve is given by

$$\mathbf{r}_L = \begin{bmatrix} r\cos\theta \\ r\sin\theta \end{bmatrix} = A\sqrt{\cos p\theta}\begin{bmatrix} \cos\theta \\ \sin\theta \end{bmatrix} \tag{2}$$

where $\mathbf{r}_L$ is the position vector in frame B. This is a curve in the first quadrant. Symmetry is used to reflect the lemniscate curve into all quadrants to produce a figure-eight orbital trajectory.

The velocity of a point P relative to the earth-fixed frame A, may be written in terms of its velocity relative to B and the velocity of the ground vehicle T relative to A as

$$\mathbf{v}_A := \begin{bmatrix} v_x \\ v_y \end{bmatrix} := ({}^A\mathbf{v}^P)_B = ({}^A\mathbf{v}^T)_B + ({}^B\mathbf{v}^P)_B \tag{3}$$

where $(\cdot)_B$ indicates components relative to B. This leads to:

$$\mathbf{v}_A = \begin{bmatrix} 0 \\ V_T \end{bmatrix} + \begin{bmatrix} v_x \\ v_y - V_T \end{bmatrix} \tag{4}$$

where V_T is the speed of the ground vehicle.

Parameter Determination

Now that we have chosen the governing shape of the trajectory, we will show how the trajectory parameters are chosen. A key limitation of fixed-wing aircraft is the maximum turn rate achievable while maintaining relatively stable flight. The UAVs that were used in the experiments, for example, had a maximum turn rate of 10 deg/sec. Therefore, the first constraint that will govern the choice of trajectory parameters is the maximum turn rate.

The turn rate of a point P, moving at constant speed V_P in frame A, satisfies the equation:

$$|\dot{\psi}| = |{}^A\mathbf{a}^P|/V_P \tag{5}$$

where $V_P := |{}^A\mathbf{v}^P|$. If we assume that the UAV moves at constant speed V_P and tracks the lemniscate perfectly, then ${}^A\mathbf{a}^P$ becomes a function of V_P, V_T, p, A, and θ. Therefore, for given constants V_P, V_T, p, and A, we can find the maximum turn rate magnitude on the trajectory as

$$\max_\theta |\dot{\psi}(\theta)| = \max_\theta |\mathbf{a}_A(\theta)|/V_P \tag{6}$$

where $\mathbf{a}_A := ({}^A\mathbf{a}^P)_B$. The details of this are outlined below.

With the assumption of perfect tracking,

$$({}^B\mathbf{v}^L)_B = \dot{\mathbf{r}}_L \tag{7}$$

Combining (2),(3), and (7) yields

$$\mathbf{v}_A := \begin{bmatrix} 0 \\ V_T \end{bmatrix} + \frac{d}{dt}\left(A\sqrt{\cos p\theta}\begin{bmatrix} \cos\theta \\ \sin\theta \end{bmatrix}\right) \tag{8}$$

Carrying out the differentiation gives,

$$\mathbf{v}_A := \begin{bmatrix} 0 \\ V_T \end{bmatrix} + \frac{Ap\dot{\theta}}{2\sqrt{\cos p\theta}} \begin{bmatrix} -\sin p\theta\cos\theta - \frac{2}{p}\cos p\theta\sin\theta \\ -\sin p\theta\sin\theta + \frac{2}{p}\cos p\theta\cos\theta \end{bmatrix} \tag{9}$$

This expression gives velocity as a function of θ and $\dot{\theta}$.

Since we are given the UAV speed, V_P, we can derive an alternate expression which gives $\mathbf{v}_A$ as a function of θ. First, we define the velocity ratio in frame B as

$$m := \frac{v_y - V_T}{v_x} = \frac{\sin p\theta \sin\theta - \frac{2}{p}\cos p\theta \cos\theta}{\sin p\theta \cos\theta + \frac{2}{p}\cos p\theta \sin\theta} \tag{10}$$

where we have used (4) and (9). We will also define $\sigma := V_P/V_T$, to simplify the final result. Note that m is undefined for $\theta = v_x = 0$, which occurs at the outer edges of the trajectory, and σ is undefined for $V_T = 0$. These special cases are handled below.

Continuing, we use (10) to write $v_x = (v_y - V_T)/m$ and combine it with the aircraft velocity squared, $V_P^2 = v_x^2 + v_y^2$. Using the above relations, we can find v_y to be

$$v_y = V_T\left(\frac{1 \pm m\sqrt{\sigma^2(1+m^2)-1}}{1+m^2}\right) =: V_T h^{\pm}(m,\sigma) \tag{11}$$

where h^- is used when $v_x < 0$ and h^+ is used when $v_x > 0$. Note that the sign of v_x is easily determined based on quadrant and the direction of travel around the orbit.

In the special case $v_x = 0$, (11) is replaced by $v_y = \pm V_P - V_T$. In the special case $V_T = 0, v_x \neq 0$, (11) is replaced by

$$v_y = \pm V_P \frac{m}{\sqrt{1+m^2}} \tag{12}$$

With v_y available, we can find v_x using (10), to get

$$\mathbf{v}_A = \begin{bmatrix} v_x(\theta) \\ v_y(\theta) \end{bmatrix} \tag{13}$$

Equating (9) and (13) allows us to solve for $\dot{\theta}$ as a function of θ. With $\dot{\theta}$ at hand, the acceleration as a function of θ can be found as:

$$\mathbf{a}_A := \dot{\theta}\frac{d\mathbf{v}_A}{d\theta} \tag{14}$$

As indicated in (6), we calculate the maximum turn rate by finding the magnitude of the trajectory's acceleration over all values of θ and dividing by the UAV's speed. It is evident that the maximum turn rate will be different for every value of V_T and V_P.

The turn rate is a decisive part of choosing the parameters p and A, but we also require the UAVs to travel over the target point periodically in order to fulfill the mission of tracking the point and performing surveillance. Therefore, the orbit parameters must also be chosen such that return time requirements are satisfied, where return time is defined as the time between each pass over the point. The return time, T, is calculated using

$$T = \int dt = \int \frac{dt}{ds} ds \tag{15}$$

where s is the arclength parameter. The time derivative of s is

$$\frac{ds}{dt} = |\frac{d\mathbf{r}_L}{dt}| = |({}^B\mathbf{v}^L)_B| =: |\mathbf{v}_L| \tag{16}$$

We can find ds by taking the derivative of $\mathbf{r}_L$ with respect to θ:

$$ds^2 = |d\mathbf{r}_L|^2 = |\frac{d\mathbf{r}_L}{d\theta}|^2 d\theta^2 = A^2(\frac{p^2}{4} \tan p\theta \sin p\theta + \cos p\theta) d\theta^2 \tag{17}$$

$$ds = A\sqrt{\frac{p^2}{4} \tan p\theta \sin p\theta + \cos p\theta d\theta} \tag{18}$$

Combining (15) and (18) produces

$$T = \int \frac{dt}{ds} ds = 2A \int_0^{\frac{\pi}{2p}} \frac{1}{|\mathbf{v}_L|} \sqrt{\frac{p^2}{4} \tan p\theta \sin p\theta + \cos p\theta d\theta} \tag{19}$$

where the return time is computed as twice the time to traverse a single quadrant. We can integrate (19) numerically, using (13), (10), and (11) to assist in the solution of $|\mathbf{v}_L|$.

Using the expressions above, for each V_P and V_T, parameters p and A are determined such that the turn rate and return time constraints are satisfied. The algorithm to choose p and A is as follows:

i.) Choose A very small
ii.) Calculate p to minimize the maximum turn rate over the entire orbit
iii.) If turn rate and return time constraints are satisfied, stop. Otherwise, increase A and go to step ii).

Note that the existence of p and A are dependent on the choice of a return time which is achievable for a given aircraft.

A sample of the resulting orbital trajectories that were chosen for our application are shown in Figure 3. The plots are shown in the point coordinate frame. V_T is varied while V_P is held constant at 20 m/s. It is evident that the amplitude decreases as the speed of the point increases. Also, p increases as V_T increases, which is apparent through the narrowing of the trajectory. At $V_T = 20$ m/s, the trajectory becomes a point.

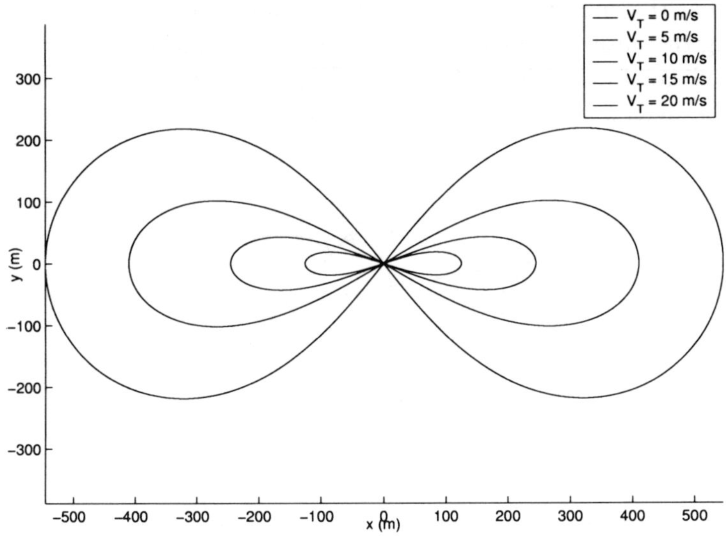

Fig. 3. Trajectories in the point coordinate frame, $V_P = 20$ m/s.

Control

The tracking control law that is developed for the UAV to follow the trajectory is defined in terms of the trajectory's tangent line. Given a point on the trajectory, the unit tangent vector, $\mathbf{t}_L$, is calculated as

$$\mathbf{t}_L = \frac{\frac{d\mathbf{r}_L}{d\theta}}{|\frac{d\mathbf{r}_L}{d\theta}|} \tag{20}$$

where the derivative of the position vector is given by:

$$\frac{d\mathbf{r}_L}{d\theta} = \frac{Ap}{2\sqrt{\cos p\theta}} \begin{bmatrix} -\sin p\theta \cos\theta - \frac{2}{p}\cos p\theta \sin\theta \\ -\sin p\theta \sin\theta + \frac{2}{p}\cos p\theta \cos\theta \end{bmatrix} \tag{21}$$

This leads to

$$\mathbf{t}_L = \frac{\frac{p}{2}(\cos p\theta)^{-1/2}}{\sqrt{\frac{p^2}{4}\tan p\theta \sin p\theta + \cos p\theta}} \begin{bmatrix} -\sin p\theta \cos\theta - \frac{2}{p}\cos p\theta \sin\theta \\ -\sin p\theta \sin\theta + \frac{2}{p}\cos p\theta \cos\theta \end{bmatrix} \tag{22}$$

Given the current aircraft position, we seek a point on the trajectory such that the vector from that point to the aircraft is orthogonal to the trajectory tangent vector at that point. Depending on the aircraft position, the trajectory may have several such points. The point that will be used is chosen by a routine that predicts which quadrant in the point coordinate frame that the aircraft should be in or be heading towards, combined with choosing the position closest to the origin if two solutions are found in the predicted quadrant. This point is called p_L. We define $\mathbf{n}$ as the normal vector between p_L and $\mathbf{r}_P$, as shown in Figure 4.

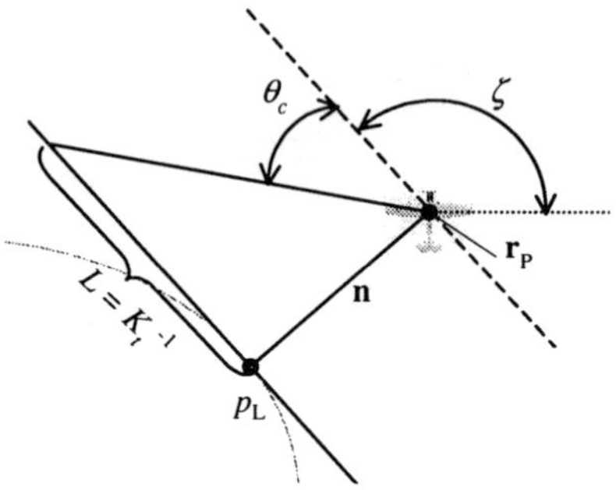

Fig. 4. Control law development for a single UAV following an orbit trajectory.

Once p_L is found, the tangent vector angle, ζ, and the control angle, θ_c, are calculated, with

$$\theta_c = \arctan(K_t|\mathbf{n}|) \tag{23}$$

where K_t is a controller gain. Intuitively, $K_t = 1/L$, where L is the distance from p_L, along the tangent line, of a point that we steer the plane towards. Based on simulations, L was chosen to be 125m, or $K_t = 0.008$.

The desired velocity vector of the UAV in frame B becomes

$$({}^B\mathbf{v}^P)_B = v \begin{bmatrix} \cos(\theta_c + \zeta) \\ \sin(\theta_c + \zeta) \end{bmatrix} \tag{24}$$

Both v and ${}^A\mathbf{v}^P$ are then determined using (3),(4), V_T, and V_P.

3. Formation Control

With an orbit developed as in the previous section, we now develop a control law which will allow a group of aircraft to track that orbit in a coordinated fashion. The control law is a modified version of that presented in [6], which offers a general approach to modeling and control of vehicle formations. This approach accommodates either two or three-dimensional motion of formations consisting of particles and/or bodies, and allows for connections between elements. In addition, it provides for simplified trajectory planning and allows the system controller to be formulated in terms of quantities which are closely related to performance objectives. This method is an alternative to other methods such as leader-follower [3], and artificial potential [2]. Here, it is modified to work in terms of desired aircraft speeds and headings.

Kinematics

We begin by deriving expressions for aircraft velocities in terms of a set of generalized coordinates and speeds which characterize the motion of the formation. These coordinates and speeds represent the location (L), orientation (O), and shape (S) of the formation, which we will define as the position of a formation reference point (FRP), the orientation of a formation reference frame (FRF), and the set of aircraft positions relative to the FRF, respectively. The location of the FRP and the orientation of the FRF are defined in terms of the formation configuration.

The FRF F is defined by a right-handed set of orthogonal unit vectors $\mathbf{f}_1$, $\mathbf{f}_2$ and $\mathbf{f}_3$. Similarly, the inertial frame A is defined by $\mathbf{a}_1$, $\mathbf{a}_2$ and $\mathbf{a}_3$. The position of a point i is defined in terms of components relative to A as

$$\mathbf{R}_i = \mathbf{R}_0 + Q\mathbf{r}_i \tag{25}$$

where $\mathbf{R}_i = (\mathbf{R}_i)_A$ is the position vector of the point relative to the origin of A, $\mathbf{R}_O = (\mathbf{R}_O)_A$ is the position vector of the FRF origin relative to the origin of A, $\mathbf{r}_i = (\mathbf{r}_i)_F$ is the position vector from the FRF origin to i, and Q is the rotation matrix of frame F relative to frame A. The position of the FRP is denoted by $\mathbf{R}_p = (\mathbf{R}_p)_A$. $\mathbf{R}_p$, Q, and $\mathbf{r}_i$ are parameterized by the coordinate vectors $\mathbf{q}_L$, $\mathbf{q}_O$, and $\mathbf{q}_S$ respectively.

The velocity of point i can be written as

$$\dot{\mathbf{R}}_i = \dot{\mathbf{R}}_0 + \dot{Q}\mathbf{r}_i + Q\dot{\mathbf{r}}_i = \dot{\mathbf{R}}_0 + Q\Omega\mathbf{r}_i + Q\dot{\mathbf{r}}_i \tag{26}$$

As $\mathbf{r}_i = \mathbf{r}_i(\mathbf{q}_S)$, and assuming that $\mathbf{v}_S$ satisfies $\dot{\mathbf{q}}_S = \beta_S(\mathbf{q}_S)\mathbf{v}_S$,

$$\dot{\mathbf{r}}_i = \frac{\partial \mathbf{r}_i}{\partial \mathbf{q}_S}\dot{\mathbf{q}}_S = \frac{\partial \mathbf{r}_i}{\partial \mathbf{q}_S}\beta_S\mathbf{v}_S := D_i(\mathbf{q}_S)\mathbf{v}_S \tag{27}$$

Defining the 3 by 3 matrix C_i as:

$$C_i(\mathbf{q}_S) = -\widehat{\mathbf{r}}_i \tag{28}$$

where $\widehat{\mathbf{r}}_i$ is the skew form of $\mathbf{r}_i$, we can write

$$\dot{\mathbf{R}}_i = \dot{\mathbf{R}}_0 + QC_i\boldsymbol{\omega} + QD_i\mathbf{v}_S \tag{29}$$

where $\boldsymbol{\omega} = ({}^A\boldsymbol{\omega}^F)_F$ is the angular velocity of F relative to A.

Similarly, for a point p,

$$\dot{\mathbf{R}}_p = \dot{\mathbf{R}}_0 + QC_p\boldsymbol{\omega} + QD_p\mathbf{v}_S \tag{30}$$

Now, defining $\mathbf{v}_L = \dot{\mathbf{R}}_p$ and $\mathbf{v}_O = \boldsymbol{\omega}$, we can write

$$^A\mathbf{v}^i = \dot{\mathbf{R}}_i = V_i \begin{bmatrix} \mathbf{v}_L \\ \mathbf{v}_O \\ \mathbf{v}_S \end{bmatrix} := V_i\mathbf{v} \tag{31}$$

where

$$V_i = \begin{bmatrix} I_3 \; Q(C_i - C_p) \; Q(D_i - D_p) \end{bmatrix} := \begin{bmatrix} I_3 \; QC_{ip} \; QD_{ip} \end{bmatrix} \tag{32}$$

Note that, for a formation of N unconnected elements, where $\mathbf{R}_i$ is the position of the ith element, the matrix V, defined by

$$V := \begin{bmatrix} V_1 \\ \vdots \\ V_N \end{bmatrix} \tag{33}$$

is invertible. Note also that $\dot{\mathbf{q}}$ and $\mathbf{v}$ are related by a block diagonal matrix β:

$$\dot{\mathbf{q}} = \begin{bmatrix} \dot{\mathbf{q}}_L \\ \dot{\mathbf{q}}_O \\ \dot{\mathbf{q}}_S \end{bmatrix} = \begin{bmatrix} \beta_L(\mathbf{q}_L) & 0 & 0 \\ 0 & \beta_O(\mathbf{q}_O) & 0 \\ 0 & 0 & \beta_S(\mathbf{q}_S) \end{bmatrix} \begin{bmatrix} \mathbf{v}_L \\ \mathbf{v}_O \\ \mathbf{v}_S \end{bmatrix} = \beta\mathbf{v} \tag{34}$$

If the reference point is defined as a weighted sum of other points:

$$\mathbf{R}_p = \frac{\sum_i a_i\mathbf{R}_i}{\sum_i a_i} \tag{35}$$

then

$$C_p = \frac{\sum_i a_i C_i}{\sum_i a_i} \quad \text{and} \quad D_p = \frac{\sum_i a_i D_i}{\sum_i a_i} \tag{36}$$

If a mass m_i is associated with each point i, choosing $a_i = m_i$ places the reference point at the center of mass. Choosing $a_i = 1$ places the reference point at the geometric center of the points.

Velocity-Based Formation Control

Using results of the previous section, we now develop a velocity-based formation control law. From this point on, we will assume that the β matrix is invertible. We first define the tracking error

$$\mathbf{e} := \mathbf{q} - \mathbf{q}_d \tag{37}$$

and the vector:

$$\begin{aligned}\mathbf{s} &:= \beta^{-1}(\dot{\mathbf{e}} + \Lambda\mathbf{e}) \\ &= \beta^{-1}[\dot{\mathbf{q}} - (\dot{\mathbf{q}}_d - \Lambda\mathbf{e})]\end{aligned}$$

Note that with this definition,

$$\mathbf{s} = 0 \Rightarrow \dot{\mathbf{e}} + \Lambda\mathbf{e} = 0 \tag{38}$$

By defining the 'reference velocity' vector [5] as:

$$\mathbf{v}_r := \beta^{-1}(\dot{\mathbf{q}}_d - \Lambda\mathbf{e}) \tag{39}$$

$\mathbf{s}$ can be written as:

$$\mathbf{s} = \mathbf{v} - \mathbf{v}_r \tag{40}$$

so that

$$\mathbf{v} \to \mathbf{v}_r \Rightarrow \mathbf{s} \to 0 \Rightarrow \mathbf{e} \to 0 \tag{41}$$

Now, define the desired velocity for aircraft i as

$${}^A\mathbf{v}_d^i = V_i\mathbf{v}_r \tag{42}$$

Due to the fact that V is invertible,

$${}^A\mathbf{v}^i \to {}^A\mathbf{v}_d^i \quad i = 1, \ldots, N \Leftrightarrow \mathbf{v} \to \mathbf{v}_r \tag{43}$$

so that

$${}^A\mathbf{v}^i \to {}^A\mathbf{v}_d^i \quad i = 1, \ldots, N \Rightarrow \mathbf{e} \to 0 \tag{44}$$

Furthermore, if the tracking errors

$${}^A\mathbf{v}^i - {}^A\mathbf{v}_d^i \quad i = 1, \ldots, N \tag{45}$$

remain bounded, then so do $\mathbf{s}$ and $\mathbf{e}$.

For vehicle following, an alternate form of the reference velocity:

$$\mathbf{v}_r := \begin{bmatrix} \mathbf{v}_{rL,com} \\ \mathbf{v}_{rO,com} \\ \beta_S^{-1}(\dot{\mathbf{q}}_{Sd} - \Lambda_S\mathbf{e}_S) \end{bmatrix} \tag{46}$$

is used, to directly steer and rotate the group while maintaining its shape. The velocity command $\mathbf{v}_{rL,com}$ is obtained from the orbit tracking controller. The rotation command $\mathbf{v}_{rO,com}$ rotates the group to its desired orientation.

Communication Requirements

We now consider the element-element communication requirements for implementation of the control outlined above. We will assume that the gain matrix Λ is diagonal. When this is true, the controls can be implemented in a semi-decentralized fashion which does not require extensive two-way communication between elements. We define two sets of elements, S_F and S_P, which contain those elements which define the FRF and FRP respectively.

With the reference velocity partitioned into L, O, and S components, the desired velocity for the ith element is computed as:

$$^{A}\mathbf{v}_d^i = V_i \mathbf{v}_r = \mathbf{v}_{rL} - Q[C_p \mathbf{v}_{rO} + D_p \mathbf{v}_{rS}] + Q[C_i \mathbf{v}_{rO} + D_i \mathbf{v}_{rS}] \tag{47}$$

In general, computation of the terms involving D_i and D_p could require information from all elements of the formation. If, however, we assume that the shape coordinate vector is chosen as $\mathbf{q}_S = \begin{bmatrix} \mathbf{q}_{S1} \cdots \mathbf{q}_{SN} \end{bmatrix}$, where $\mathbf{q}_{Si}$ contains the nonzero cartesian components of the position vector $\mathbf{r}_i$, and $\mathbf{v}_S = \dot{\mathbf{q}}_S$, then

$$D_i \mathbf{v}_{rS} = D_i \mathbf{v}_{rSi} \tag{48}$$

and

$$^{A}\mathbf{v}_d^i = V_i \mathbf{v}_r = \mathbf{v}_{rL} - Q[C_p \mathbf{v}_{rO} + D_p \mathbf{v}_{rSP}] + Q[C_i \mathbf{v}_{rO} + D_i \mathbf{v}_{rSi}] \tag{49}$$

where $\mathbf{v}_{rSi}$ is the S component of the reference velocity vector with all entries except those associated with element i set to zero, and $\mathbf{v}_{rSP} := \sum_{i \in S_P} \mathbf{v}_{rSi}$.

Computation of the desired velocity requires knowledge of the position and orientation of the FRF, the position of the FRP, and the positions of the elements of S_P. All of these can be obtained from a broadcast of position data from the elements of S_F and S_P. This information, combined with formation configuration data, desired trajectory data, and control parameters is sufficient for computation of the first two terms on the right-hand side of the expression. These terms are independent of i, and can be thought of as a common coordination signal which is applied to all elements of the group. Adding knowledge of $\mathbf{R}_i$ allows computation of the remaining term, which is specific to element i.

Motion Constraints

In this section, we look at the constraints which are imposed on the motion of the formation due to flight speed and turn rate limitations of the individual aircraft. These constraint conditions can be used as feasibility criteria for group trajectory generation.

Letting v_i denote the speed of the ith aircraft, the speed constraints

$$0 \le v_{i,min} \le v_i \le v_{i,max} \tag{50}$$

lead to the constraint conditions:

$$v_{i,min}^2 \le ({}^A\mathbf{v}^i)^2 \le v_{i,max}^2 \tag{51}$$

Defining $\mathbf{t}$ and $\mathbf{k}$ as the tangent and curvature vectors of the aircraft path, the turn rate is given by

$$\frac{d\mathbf{t}}{dt} = \mathbf{k}v_i \tag{52}$$

This leads to the turn rate constraint

$$\mathbf{k}^2 v_i^2 \le \dot{\phi}_{i,max}^2 \tag{53}$$

In terms of aircraft velocity and acceleration, $\mathbf{k}^2 v_i^2$ can be written as:

$$\mathbf{k}^2 v_i^2 = \mathbf{k}^2 ({}^A\mathbf{v}^i)^2 = \frac{({}^A\mathbf{v}^i)^2({}^A\dot{\mathbf{v}}^i)^2 - ({}^A\mathbf{v}^i \cdot {}^A\dot{\mathbf{v}}^i)^2}{({}^A\mathbf{v}^i)^4} \tag{54}$$

By using the equations ${}^A\mathbf{v}^i = V_i\mathbf{v}$ and ${}^A\dot{\mathbf{v}}^i = \dot{V}_i\mathbf{v} + V_i\dot{\mathbf{v}}$, the motion constraints can be expressed in terms of $\mathbf{q}$ and $\mathbf{v}$.

4. Implementation

For control implementation, we are using the Piccolo system by CloudCap Technology [8]. This system consists of a miniature autopilot unit (Piccolo) mounted in each UAV, and a Piccolo ground station (GS), which is connected to a ground-based PC running command and control software. Communications between the GS and the Piccolos are via a 900MHz radio link.

The Piccolo system also provides a convenient hardware-in-the-loop (HIL) simulation capability. In HIL mode, each Piccolo unit is attached to a PC which runs an aircraft simulator. The simulator is driven by commands from the Piccolo, and the Piccolo receives simulated sensor data from the simulator. The only change which occurs when going from HIL simulation to actual flight mode is that the Piccolos are removed from the simulator PC's and installed in aircraft.

For initial testing, we have used the native Piccolo architecture, in which aircraft communicate only with the GS. The data flow is as follows: 1. All aircraft send telemetry to a single GS. 2. The data is passed from the GS to a PC which performs the control calculations. 3. The resulting commands are then sent out to each aircraft via the GS.

This architecture has the advantage of being relatively simple to implement, as no plane-plane communication is required, and all higher-level software is located on a single ground-based computer. Aside from being convenient for testing, the simplicity of this arrangement makes it an attractive choice for many applications. The major limitation is that the system does not provide high communication rates. Also, the aircraft must stay close enough to the GS that communication dropouts do not become a problem.

As the Piccolo is a waypoint-based system, velocity control is implemented by setting a distant waypoint which is in the desired direction and a desired waypoint approach speed.

Due to the low communications rate, it may be desirable to operate the system in true waypoint mode. By closing an additional loop on the aircraft itself (at much higher bandwidth than is possible from the ground), this approach may improve performance considerably in the presence of significant disturbances. To do this, the formation control described above is applied to a set of virtual reference aircraft; the actual aircraft then track these reference models. The model equations are:

$$\dot{x} = v\cos(\theta) \tag{55}$$

$$\dot{y} = v\sin(\theta) \tag{56}$$

$$\dot{\theta} = sat[-(\theta - \theta_d)/\tau_\theta] \tag{57}$$

$$\dot{v} = sat[-(v - v_{d,lim})/\tau_v] \tag{58}$$

$$v_{d,lim} = sat[v_d] \tag{59}$$

The model provides a first-order rate-limited response to desired heading and airspeed inputs.

The x,y positions of the virtual aircraft are sent to the actual aircraft as desired waypoint values. With appropriate parameter choices, this provides a reasonable tracking target for the actual aircraft. In this mode, the higher-level control acts as a trajectory generator. Note that as waypoints are not valid setpoints for a fixed-wing aircraft, suitable logic must be in place for when an aircraft reaches its waypoint.

5. HIL Simulations

Figure 5 shows a three-aircraft formation following a ground vehicle which moves at varying speed. From a stationary start at the origin, the ground vehicle heads north, accelerating uniformly to a speed of 20 m/s at the top of the figure. As the ground vehicle begins to move, the aircraft trajectories transition smoothly from a figure-eight to a periodic wave. As the speed increases, the wave amplitude decreases, going to zero at 20 m/s. The formation control maintains spacing between aircraft and keeps the formation perpendicular to the vehicle path during the motion.

In Figure 6, a two aircraft formation follows a ground vehicle which moves arbitrarily. The ground vehicle is initially at a standstill. It is then accelerated to a speed of 26 mph and driven in a variety of directions before slowing to a stationary final position. The orbit point and orientation were chosen to be the ground vehicle location and heading respectively. As the vehicle heading varies, the formation is rotated to remain perpendicular to it.

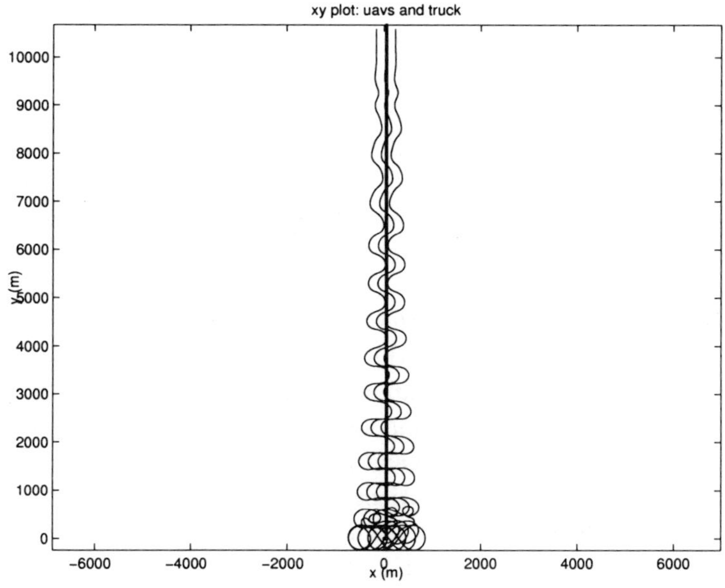

Fig. 5. HIL Simulation: Varying Truck Speed.

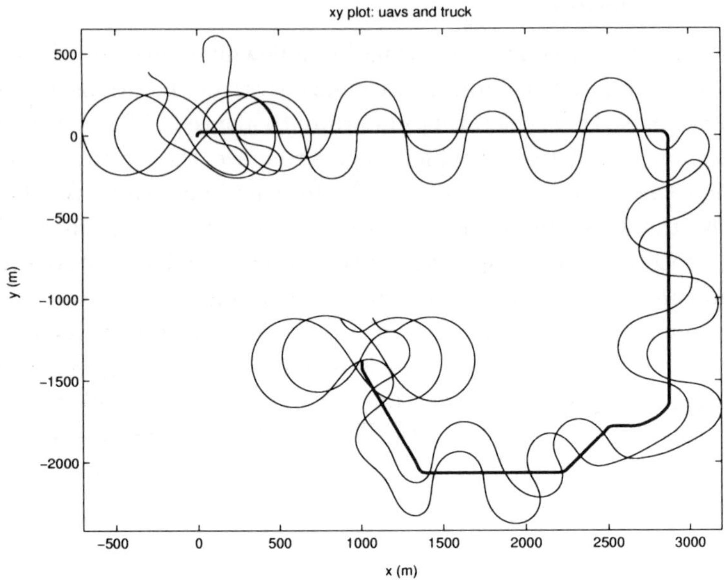

Fig. 6. HIL Simulation: Varying Truck Direction.

6. Flight Tests

A number of flight tests with two aircraft tracking a moving truck were performed in August, 2003 at an R/C airfield near Tucson, Az. The aircraft were modified SIG Rascal 110's, each fitted with a Piccolo unit and supporting hardware. The Piccolo ground station was mounted in the moving truck.

Figure 7 shows the paths of the two aircraft and the truck during the test. From a starting point near the runway, the truck drove to a nearby north-south road, then approximately two miles north along the road to an intersection, where it turned around and returned to the starting point. The truck speed was kept between 15 and 25 mph. The orbit center was chosen to be the truck location and the orientation was chosen to be perpendicular to the truck heading, with these values determined from GPS data. The region shown in the figure is approximately 4km x 4km.

Figure 8 shows the ground video coverage which would be achieved with these flight paths, assuming a flight altitude of 250m, and a gimballed, downward-pointing camera with a 50 degree field of view mounted on each aircraft. For clarity, only the data up to the turnaround point is shown.

Under these conditions, almost complete coverage of the region surrounding the road is achieved.

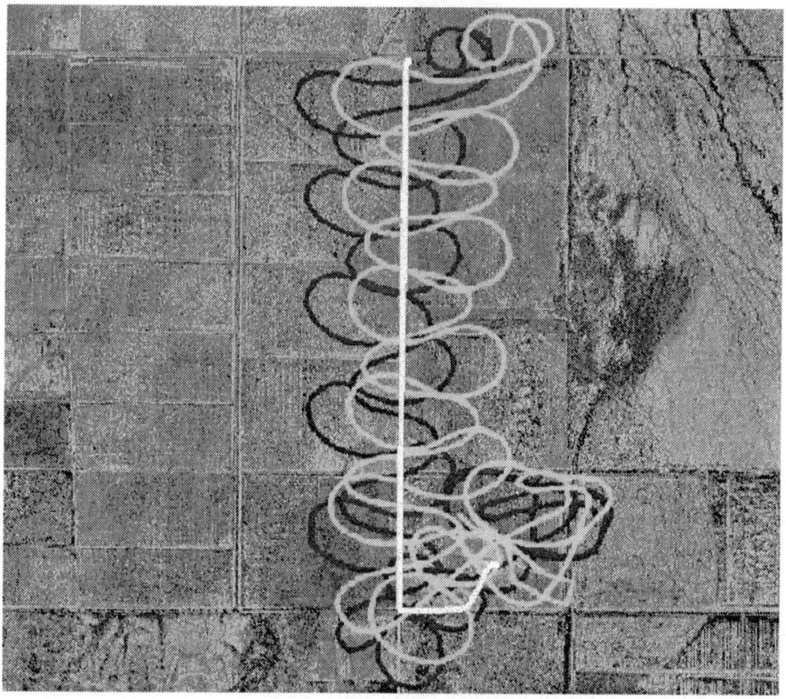

Fig. 7. Flight Test: Aircraft and Truck Paths.

7. Conclusions

In this chapter, we have presented an approach for using fixed-wing aircraft to track objects having unconstrained, arbitrary motion. This could facilitate the use of fixed-wing UAVs in a number of applications, including convoy protection, border patrol, and search-and-rescue.

The approach consists of a time-varying orbital trajectory combined with a formation control law. The orbital trajectory moves with the target and allows the motion-constrained aircraft to track the target in the sense that it crosses over the target point at a specified frequency. Trajectory parameters may be chosen to accommodate the motion constraints of a particular aircraft type. The formation controller allows this approach to

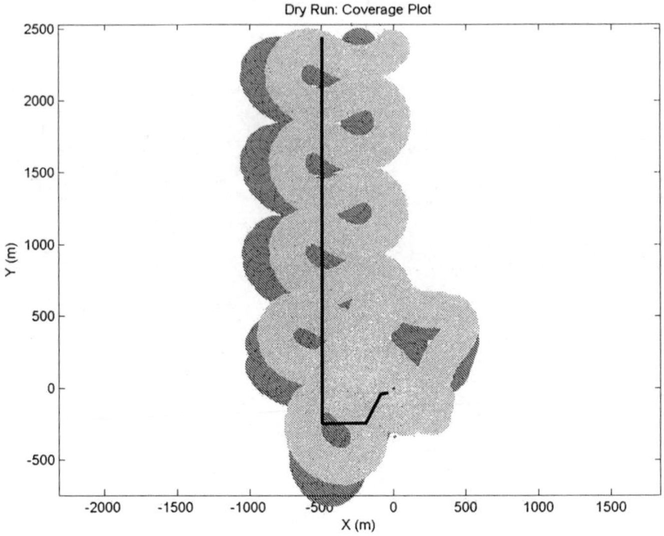

Fig. 8. Flight Test: Camera Coverage.

be extended to multiple aircraft. It produces desired group motion while maintaining the relative positions of the aircraft. One use of flying a group of aircraft in this way is to produce a large multi-camera sensor platform providing a wide band of gapless coverage.

Simulations and flight tests were presented to demonstrate the approach.

Acknowledgments

This research was supported by the Office of Naval Research, AINS program, led by Dr. Allen Moshfegh (contract number N00014-03-C-0187). We would like to express our appreciation for this support. We would also like to thank Advanced Ceramics Research of Tucson, Az. for providing hardware and technical assistance during flight testing.

References

[1] Lee, J., Huang, R., Vaughn, A., Xiao, X., Hedrick, J.K., Zennaro, M., and Sengupta, R., "Strategies of Path Planning for a UAV to Track a Ground Vehicle," *Second Annual Symposium on Autonomous Intelligent Networks and Systems*, Palo Alto, CA., June 2003.

[2] Leonard, N.E., and Fiorelli, E., "Virtual Leaders, Artificial Potentials, and

Coordinated Control of Groups" *Proceedings of the 40th IEEE Conf. on Decision and Control* Orlando, FL., Dec. 2001, pages 2968-2973.

[3] Pant, A., Seiler P., Koo, T. J., and Hedrick, J. K., "Mesh Stability of Unmanned Aerial Vehicle Clusters," *Proceedings of the American Control Conference*, Arlington, VA., June 2001, pages 62-68.

[4] Schoenwald, D.A., "AUVs: In Space, Air, Water, and on the Ground," *IEEE Control Systems Magazine*, Vol. 20, No. 6, Dec. 2000, pages 15-18.

[5] Slotine, J.J., and Li, W., *Applied Nonlinear Control*, Prentice Hall, 1991.

[6] Spry, S.C., "Modeling and Control of Vehicle Formations," PhD Thesis, University of California, Berkeley, 2002.

[7] *Unmanned Aerial Vehicles Roadmap: 2002-2027*, Office of the Secretary of Defense, December 2002.

[8] www.cloudcaptech.com

CHAPTER 24

COORDINATED UAV TARGET ASSIGNMENT USING DISTRIBUTED TOUR CALCULATION

David H. Walker
Department of Mechanical Engineering
Brigham Young University
dhw9@email.byu.edu

Timothy W. McLain
Department of Mechanical Engineering
Brigham Young University
mclain@byu.edu

Jason K. Howlett
San Jose State University Foundation
NASA Ames Research Center
jhowlett@mail.arc.nasa.gov

In this chapter a method for assigning unmanned aerial vehicle agents to targets through the use of preplanned vehicle tours is presented. Assignments are based on multi-target tours that consider the spread of the targets and the sensor capabilities of the vehicles. In this way, the individual agents and the team as a whole make better use of team resources and improve team cooperation. Planning and assignments are accomplished in reasonable computational time through the use of heuristics to reduce the problem size.

Keywords:
UnManned aerial vehicles, task allocation, path planning, cooperative control

1. Introduction

A growing number of applications require the coordination and cooperation of multiple autonomous agents to accomplish a team goal. Many of these efforts utilize Unmanned Aerial Vehicles (UAVs) due to the unique capabilities they provide. In a growing number of these applications, agents must make both tactical and practical decisions autonomously. This is particularly true of systems involving teams of agents which are too complicated to be controlled or efficiently monitored by a human operator. This work applies to the coordination and cooperation of multiple autonomous fixed-wing UAVs that are subject to dynamic and sensory constraints. The vehicles cooperate in an effort to visit a number of targets and to perform a number of different tasks on those targets.

This work is relevant to the implementation of autonomous Wide Area Search Munitions (WASM). A common scenario for a WASM team is for the team to visit multiple potential targets in order to properly classify them, attack classified targets (that prove not to be decoys), and then to revisit the attacked targets to perform Battle Damage Assessment (BDA) [20, 18]. An example of this scenario is depicted in Figure 1.

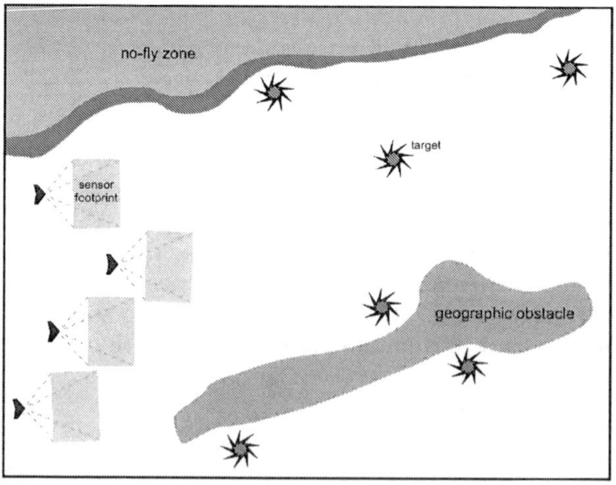

Fig. 1. Example scenario for cooperative assignment.

The general problem is to resolve who goes where and to determine how they are going to get there. These questions are subject to vehicle and problem constraints, as well as computational and timing limitations.

Challenging aspects of the problem include the dynamic constraints on the individual vehicles and the overall rate of problem growth associated with multi-vehicle, multi-target assignment problems. Dynamic vehicle limitations make it difficult to plan flyable paths that make effective use of UAV sensory capabilities. Problem growth is a complication because the number of possible individual UAV tour paths and team assignments grows rapidly with increasing numbers of vehicles, targets, and tasks [5]. This growth makes global path planning and assignment evaluation computationally intractable for problems of even modest size.

The principal issues that need to be addressed are optimal (or at least effective) path planning, target assignment, and the coupled relationship of these two tasks. In order for a team to effectively coordinate the mission plan between vehicles, it requires the management of these two coupled decision tasks. The execution order of these tasks is not obvious due to the coupled relationship between them [7, 11]. In order for an effective assignment to be selected it must be known how and when a vehicle will arrive at the specified target — the path or tour must be known. However, for the vehicle to plan a path it must know where it is expected to visit — the assignment must first be known.

Path planning is the process of generating a flyable trajectory that the vehicle follows in accomplishing all of its desired tasks. Planning optimal, dynamically constrained paths is a complicated nonlinear optimization problem of high degree [15]. There has been significant work exploring methods for effective path planning including: the use of piecewise optimal, geometrically constructed path segments and iterative assignments [18, 6]; the use of mixed-integer linear programming [17]; the use of probabilistic and random search methods [8]; the construction of Voronoi diagrams [13]; the assembly of paths from preconstructed automaton path segments [16]; and the implementation of an A* path tree search [12].

The majority of methods plan paths between two fixed and known points. When paths that pass through multiple points are required, paths are generated by assembling multiple point-to-point path segments end to end. These methods guarantee path-length optimality only for a given order of waypoint visits. Using conventional methods, optimal multiple-point tour paths can only be generated when the required waypoints and the order in which the waypoints will be visited has been previously determined.

The one exception to these path planning requirements is the method described by Howlett [12]. This planner finds the optimal path through a series of targets while also finding the best order for visiting those targets through a Learning Real-Time A* (LRTA*) tree search. The search method also takes advantage of the full sensing capabilities of the vehicle. By utilizing the full area of the sensor footprint, this planner produces shorter, more efficient paths. It is hypothesized that when individual UAVs plan better paths and make better use of individual UAV resources, assignments constructed from those paths will also result in improved use of team resources and increased cooperation. It is on this path planning method that this coordinated assignment work is primarily based.

The coupled problem of allocating vehicles to tasks has also received considerable attention in the literature. One method that has been used is a market driven approach in which the vehicles bid for tasks based on flight costs related to accomplishing the task [4]. Another method used to iteratively assign tasks to vehicles is accomplished through a network flow optimization model [18, 19]. Others have formulated the vehicle routing problem, with various constraints and degrees of freedom, as a Mixed Integer Linear Program (MILP) [1, 2]. The problem has also been studied using gaming theory [9]. Still other methods have been applied to ground-based robots that have relevance to the task allocation problem in UAVs [3]. The allocation methods described in these papers address some of the coupled problems of path planning and task allocation, but also often prove to be optimal only for restricted problems. These paths are often only piecewise optimal, used in situations where path planning is performed one step at a time without regard for future possible vehicle actions.

The work herein represents an alternative method for task allocation that is enabled by the use of an improved path planner. The concept is summarized in this statement: when each vehicle makes better use of individual resources through planning efficient tour paths, the team is able to improve the overall use of resources and the coordination between agents. The computationally intense path planning and combinatorially large number of assignments are managed through heuristics and estimates so that the system can produce near real-time assignments and path plans. A method using path planning developed from geometric constructions described in [6] and a iterative greedy assignment method are developed and used as benchmarks for comparison.

2. Problem Statement

The problems to which this work applies involve systems of agents that must cooperate to accomplish a team goal. The specific problem addressed involves multiple vehicles that must cooperatively visit multiple targets. Further, each target must be visited multiple distinct times by a vehicle. The need for repeated visits to the targets arises from the distinct tasks that must be performed on the targets. Multiple visits may be required in order to properly classify a target. After classification, the target may need to be attacked and then receive a Battle Damage Assessment (BDA) sensory pass to verify that the target has been destroyed. We refer to this type of problem as a Multiple Vehicle, multiple Target, multiple Visit (MVTV) problem.

The MVTV problem described here applies to WASM which are typically fixed-wing aircraft with limited sensors that must accomplish each of the tasks mentioned above. The munitions have dynamic limitations associated with fixed-wing aircraft. The vehicles must maintain a minimum speed to prevent stalling, and they have a limited turning radius or maximum turning rate. For simplicity, the vehicles are assumed to fly at their maximum velocity, at a constant altitude, and are assumed to make all turns at their constant minimum turning radius.

There are a number of sensory simplifications made in this work. Each vehicle is equipped with a sensor that views the ground in a fixed position relative to the vehicle. The sensor footprint is large relative to the size of the vehicle and is placed so that it views the ground directly below the vehicle. Any target on the ground inside the sensor footprint of the vehicle is considered detected. The sensor is gimballed so that it views the ground below the vehicle whether the vehicle is in level flight or is banked in a turn. Another simplification is that the vehicles are assumed to be equally capable of accomplishing all task types. This implies that all requirements for task completion are equal to the path planner and the assignment manager, reducing the different tasks to a sequence of visits by the vehicles. A final simplification is that target positions in the area of interest are already known. This can be accomplished by a preliminary sensory pass through the area of interest by the agents resulting in a clear picture of potential targets to be visited.

A vehicle tour is a set of targets that the vehicle must visit. Problems such as the MVTV problem, in which the vehicles are subject to dynamic limitations, have the added complication of targets that are spatially coupled. The coupling is most severe when the spacing of the targets is on the order of the turning radius of the vehicles. Coupling between path segments is apparent whenever a path segment concludes in a heading that prevents the vehicle from readily accomplishing a subsequent visit.

Many path planning methods are based on point-to-point optimal planning. The benchmark path planning method that is used for comparison of results is such a planner and is used in [18, 4, 19]. This planner is based on the mathematical work of L.E. Dubins [6]. In a point-to-point planner the initial and final positions and headings of a given flight segment influence the optimal path for the segment. When a path is required to pass through multiple points, the points to be visited and the order in which they are to be visited must be specified to the planner. The point-to-point path planner designates the position and heading of the vehicle at the completion of a path segment, and thereby also fixes the initial position and heading of the vehicle for the subsequent path segment. Spatial coupling occurs because the route to a subsequent target depends heavily on how previous visits were completed. Path planners that find an optimal path for a given sequence of positions and headings may not obtain the optimal trajectory simply because the sequence of waypoints was not optimal. Even when the sequence is optimal, and each of the point-to-point segments are optimal, the resulting multi-target path may be significantly longer than necessary due to this spatial coupling and incorrect selection of vehicle headings at the completion of each task. A case illustrating how this can happen is shown in Figure 2.

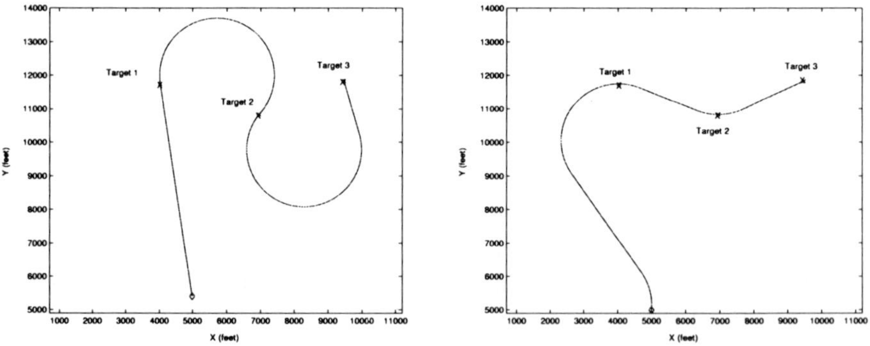

(a) Correct sequence may yield sub-optimal paths.

(b) Optimal multi-target path plan needs correct sequence and headings.

Fig. 2. Coupling between path segments.

An effective solution for MVTV problems requires an improved trajectory planner and an efficient method of assignment selection that is capable of managing problem growth and meeting computational speed requirements. The planner should

- plan optimal or near-optimal tour paths for closely spaced targets
- make full use of the entire sensor footprint
- plan complete tours over multiple targets, some requiring multiple visits
- find the best tour without specification of tour visit order.

The planner utilized here determines the best path through a given set of targets, including the tour order and the optimal multi-target path. This trajectory planning method will be described in greater detail in Section 3.1.

As discussed earlier, the coupling between path planning and target allocation is a significant issue in MVTV problems. The dilemma is that an assignment is needed for the vehicle to plan a path, but the details of the path are required for effective team assignments to be made. A common approach to overcome this dilemma is to plan path segments and make single assignments iteratively. The vehicles plan optimal path segments from their current location to the various targets that need immediate attention. Greedy assignments are then made based on the costs for the vehicles to accomplish the immediate tasks. The problem that arises is that the assignments and planned paths make no consideration for the state of the system at the conclusion of the various tasks. The vehicles often complete the present task in an optimal manner, but are in poor condition to address subsequent unfinished tasks. Furthermore, iterative methods may lead to "churning" in the assignment. Churning occurs when a vehicle is assigned to a task, but is later unassigned on a subsequent assignment iteration because it is determined that another vehicle will be able to accomplish the task first. Iterative assignment methods, although fast, often lend themselves to overall system inefficiencies, lengthy paths, and poor cooperation among the agents because the assignment is myopic with no concern for future actions.

An improved planner that results in better tour paths can be used to improve assignments. In selecting assignments the managing algorithm must take a number of factors into consideration. The assignment algorithm should

- efficiently setup the problem — find complete assignments and possible UAV tours

- utilize paths planned by the individual vehicles' tour planners
- effectively manage problem growth issues
- efficiently evaluate assignment costs, returning good assignments in reasonable time.

For the MVTV problem, increases in the number of cooperating vehicles, the number of targets, and the number of required visits to each target result in explosive growth in the number of possible tours and team assignments. This growth in problem size affects the computational requirements for both the tour planner and the algorithms used in assignment setup and selection. As a result of this explosive growth, viable methods must focus on the development of fast algorithms and methods for reducing the problem size.

The objective of this work is to improve team cooperation through improved tour paths. A tour planner creates optimal tour options for each UAV without a priori knowledge of tour order. Assignments are then selected by combining appropriate tours from the separate UAVs. These assignments and paths fulfill the global team goal rather than looking only one step ahead, and improve use of team resources and overall cooperation between the agents.

3. Technical Approach

The approach presented here achieves the goals set forth through the use of an improved path planner for individual flight tours coupled with an efficient approach for task management. The calculation of a tour path allows the consideration of the overall benefit of an entire team assignment, rather than iteratively evaluating the immediate gain of individual vehicle subassignments.

The path planner uses a learning algorithm that makes it capable of accomplishing the various required tasks. The planner is described in Section 3.1, defining how it works and its limitations. The assignment algorithm is presented in Section 3.2. Various aspects of the assignment process are described. First, the problem setup and the utilization of the tour path planner are explained. Methods for controlling problem growth in the assignment process are then discussed. The overall algorithm is presented in Section 3.3, illustrating how the computation can be distributed across multiple computers to further manage the computational load.

3.1. *Tour Path Planner*

The tour path planner developed in [12] implements a discrete-step path planner to search a tree of possible paths. The goal of this path-planning approach is to find the branches of the tree that result in the agent meeting the objectives set forth. Once a set of branches have proven to meet the objectives of the planner, the shortest branch is selected as the planned path. Due to the well defined nature of the discrete tree, it lends itself to a Learning Real-Time A* (or LRTA*) search to explore the tree for branches that meet the desired objectives. The specific implementation of the LRTA* algorithm developed by Howlett is unique because there is no set goal node. The objective is met as the path weaves its way through the spatially close targets and is able to sense each of them. In the original work [12], the objective was to sense the multiple targets only a single time each. The algorithms, heuristics, and path goals have been modified in this work to allow multiple repeated visits to individual targets as required by the MVTV problem definition.

An example of an LRTA* path tree is depicted in Figure 3. The tree is constructed of left-turn, straight, and right-turn segments of discrete length. The root of the tree is at the initial vehicle location. The tree is constructed so that the branches span the area of interest.

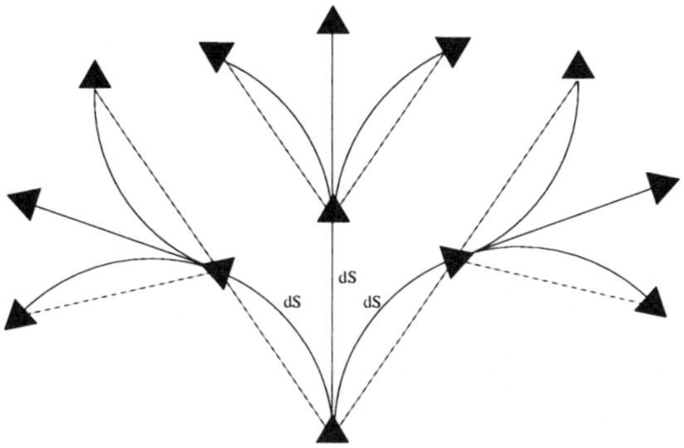

Fig. 3. Primitive turn and straight path segments of equal length, dS, are assembled to form a tree of flyable paths.

The LRTA* algorithm is actually quite simple and proceeds in the following manner [21]. Each discrete-step node, i, has a heuristic, h_i, which estimates the path length to be travelled by the vehicle before the multiple-target sensing objective is accomplished. Every node has a set of m neighbor nodes, which are the discrete-step nodes that the vehicle can proceed to next. At each step of the search, the current node, i, calculates

$$f_j = k_{ij} + h_j \quad \forall j = 1, \ldots, m. \tag{1}$$

The value of k_{ij} is the cost for the vehicle to travel from node i to node j. The value of f_j is the estimated path length before the objective is met if the vehicle at node i proceeds to neighbor j. The node i heuristic is updated so that $h_i = \min_j f_j$, and then the algorithm proceeds to the corresponding minimum cost neighbor. The algorithm proceeds from node to node in this manner, updating the heuristics as it goes, until the objective is reached (all targets sensed), and the search is begun again at the initial node. For each step of the search, the heuristic for the current node is updated with a better estimate until the updates converge to the actual minimum path.

There are two major issues for consideration when initializing heuristics in the LRTA* planner. The fundamental requirement of the LRTA* path search method is that the individual node path heuristics must always initially underestimate the true path length. This heuristic admissibility restriction is required by the algorithm because if it initially overestimates the path length then the algorithm may never explore branches of the discrete tree that actually lead to the optimal solution. The second issue that is pertinent to the effectiveness of LRTA* search is the initial value of the heuristic. The closer the initial heuristics are to the actual path-length value, the faster the algorithm will converge to the optimal path.

The learning algorithm that is used is actually a non-improving version of the LRTA* algorithm. The Non-Improving LRTA* (or NILRTA*) is identical to the general LRTA* algorithm except that it has an additional search terminating condition. The LRTA* algorithm only terminates when the heuristics along the optimal path have converged to the actual path-length value. The LRTA* algorithm quickly finds optimal or near-optimal paths, but spends most of the computation time either tweaking the path for minor improvement or simply verifying that the path found is optimal. The NILRTA* algorithm, described in [12] and used in this work, uses a search terminating condition in addition to the heuristic convergence used in LRTA*. When the algorithm has gone through a given number of iterations without finding a better path, the algorithm terminates and returns the best current path. In this way the algorithm is able to trade off minor improvements in path planning performance for major gains in speed of the computation. Two sample paths for the same multi-target tour are shown in Figures 4(a) and 4(b). The tour in Figure 4(a) represents a sample path from a point-to-point planner. Figure 4(b) illustrates a tour planned using the NILRTA* tour planner. In the case shown, the tour-planned path is only 41 percent as long as the point-to-point path. The tour-planned path is capable of completing the identical tour in significantly less time due to the effective use of the entire sensor footprint enabled by the NILRTA* planner.

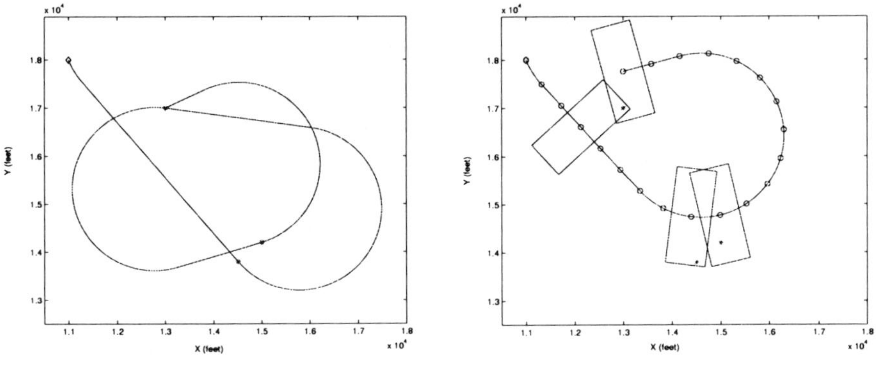

(a) Optimal point-to-point segments. (b) NILRTA* tour plan.

Fig. 4. Sample paths generated using a point-to-point planner and NILRTA* tour path planner.

3.2. *Team Assignment Strategy*

A large portion of the assignment problem is tied up in generating assignments that are both complete and not redundant. An assignment is complete when every target is fully serviced by the UAV team. A redundant assignment is one in which more visits are made to a given target than are required. Assignments cannot make effective use of team resources if they either fail to service the targets or if they are assigned to over-service certain targets. An iterative approach that has been used in [18, 19] guarantees complete assignments that are not redundant. However, the iterative approach can result in assignments and paths that are shortsighted in scope and objective, and can often result in a less effective use of team resources. When the vehicles plan paths through an entire tour they make better use of resources that result in better team assignments. This is the objective of the method presented here.

The problem is setup in a manner that produces only complete and non-redundant assignments for the vehicles on the team. The first step taken in generating a complete assignment is to make a list of all possible ways that each target can be visited. For instance, a target that must be visited three times by a team of three vehicles can be visited in the combinations shown in Table 1. The way the data is presented, the assignments (1 2 1)

Table 1. The ten possible combinations of three UAVs that can be assigned to visit a three-visit target. Assignment 2 results in vehicle 1 visiting the target twice and vehicle 2 visiting the target once

assignment number	assigned vehicles		
1	1	1	1
2	1	1	2
3	1	1	3
4	1	2	2
5	1	2	3
6	1	3	3
7	2	2	2
8	2	2	3
9	2	3	3
10	3	3	3

and (2 1 1) are identical to the shown assignment (1 1 2), and therefore are not listed. This is because the planner finds the best order to accomplish the three tasks and does not need to be told explicitly.

The number of possible vehicle combinations for servicing the i^{th}target, T_i (ten in the case illustrated in Table 1), is a function of the number of visits the target requires, n_i, and the number of vehicles on the team that are used in the assignment, m, and is given by the relationship

$$T_i = \frac{((m-1)+n_i)!}{(m-1)! \cdot n_i!}. \tag{2}$$

The complete and non-redundant assignments are obtained from all possible combinations of the individual target service combinations. When multiple targets are involved, the total number of possible assignments, A, is obtained from the product of all the T_i's from the individual vehicle visit combinations for each target

$$A = \Pi_{i=1,2,\ldots,I}(T_i). \tag{3}$$

Making assignments in this manner will always result in a complete assignment that will service all targets without redundancies.

Figure 5 illustrates the combinatorial growth that occurs in MVTV problems. The growth data presented involves targets that must each be visited three distinct times. The total number of possible assignments make it computationally intractable to perform exhaustive searches to find global solutions in near real-time applications.

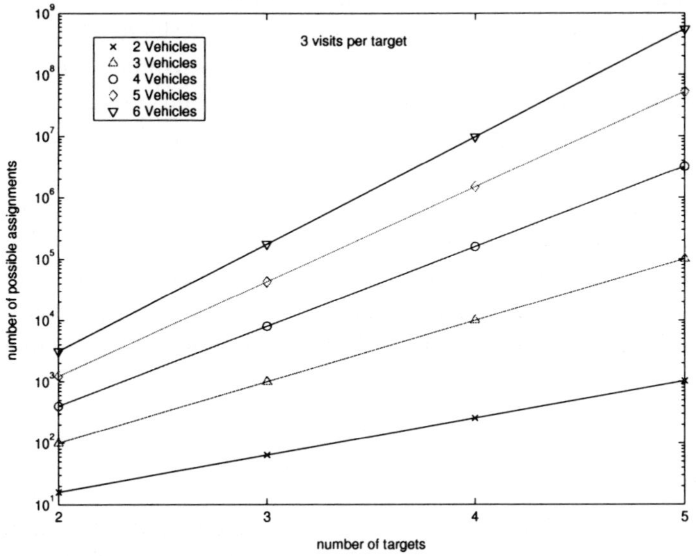

Fig. 5. The number of possible assignments grows exponentially with the number of vehicles and the number of targets.

Path-length heuristics and team cost estimates are used to quickly approximate the cost or value of a given assignment without actually planning the paths. The large number of tours in MVTV problems makes it impractical to plan all paths with a NILRTA* path planner for global solutions within reasonable time constraints. As a result, the assignment algorithm requires more simplistic approximations of path length in order to get preliminary estimates of assignment costs and benefits. These initial approximations are used to prune obviously poor vehicle tour paths and team assignments from consideration so that computational time and effort is not further wasted planning or evaluating them. This is a necessary step to get the near real-time response that is desired.

The length of each individual tour path for each of the vehicles is approximated using a functional relationship rather than a learning search. In estimating the length of a path, the function considers the spread of the targets (the distance between the two targets furthest from each other), the number of visits required by each target, the spatial position of each target with respect to other targets in the group, and the size and orientation of the UAV sensor footprint relative to the vehicle flight path.

The individual path heuristic costs are combined to get estimates for entire team assignments. The cost of an assignment is estimated by combining tour heuristics from several vehicles as though the heuristics were the actual path lengths of the complete tours. The assignment cost estimates allow the assignments to be ordered according to their approximate relative value. The ordered list gives the priority for planning and evaluating the actual paths and assignments.

The ordered list is also used to reduce the number of assignments and paths under consideration. After the assignments have been ordered, only the N best assignments are kept for actual evaluation. The value of N is determined by the problem size and is used to control problem growth. Effective control of problem growth through the choice of N is demonstrated in Section 4.

3.3. *Algorithm*

Computations associated with path planning and assignment can be broken into portions that are either centralized or distributed. MVTV problems, by definition, are composed of multiple distinct agents that work together. The ability to manage problem growth can be improved by distributing the computational burden. The computational load is distributed to each of the individual UAV agents for path planning, and to a managing agent for problem setup, information management, and assignment evaluation. The assignment manager can be an additional computer agent in the lead UAV, or it can be in a separate agent in a command center location — possibly in a nearby ground station or in a high flying UAV.

The calculation of the individual vehicle path-length heuristics is initially performed by both the assignment manager and the individual UAV agents. The heuristic calculations execute fast enough that it is simpler, more robust, and requires less communication to have every agent perform this initial estimation independently. The individual agents calculate the path-length heuristics for all the tours they can conceivably be asked to perform. As the heuristics are calculated, each UAV does a preliminary ordering of tour paths based on their potential benefit. The individual UAV agents do not have the benefit of knowing how their tour will fit in with the rest of the team, but they are able to determine whether or not the tour effectively uses their individual resources. While the UAV agents are awaiting further instructions from the assignment manager, they continually calculate actual tour paths in the order of this initial ordering. In this way, the agents waste no time waiting, and instead perform calculations that they deem most useful to the team.

The managing agent is responsible for initial problem setup as well as the preliminary estimation and ordering of path heuristics and team assignments. The assignment manager calculates tour path heuristics for every tour of every vehicle in the team and then assembles team cost estimates by combining tour heuristics from the several vehicles. As the estimates are calculated, they are also ordered by estimated cost. The manager uses the ordered assignment estimates to initially reduce the size of the problem under consideration by keeping only the N best assignments based on estimated costs. The ordered list of team assignment estimates and the associated tours of each vehicle are then communicated to the individual UAVs for calculation.

Upon receiving a list of tour paths from the manager, each vehicle will have a limited number of potential tour paths present in the top N ordered team assignments. It is only these tours that the individual UAVs must calculate with the NILRTA* tour planning method. The UAVs plan their own individual tours in the order they appear in the ordered list of team assignment cost estimates obtained from the manager. Once a vehicle has planned a NILRTA* discrete path, the resulting path is immediately communicated to the managing agent for evaluation.

As new tour path data comes into the manager, the tour costs are combined and actual assignment costs are determined. A team assignment is then ordered on a separate list based on the actual cost of the assignment. The best assignment yet evaluated will always be at the beginning of the ordered list, ready for execution should a valid assignment be immediately required. This method can return a valid, executable solution at any time. In this way, the algorithm lends itself to situations where the planning times out, requiring a ready solution to be executed immediately.

Figure 6 gives an overview of the algorithm and shows the separate distributed and centralized aspects of the computation. First, the managing agent is responsible for problem setup and initialization. Similarly, the central manager is responsible for prioritizing the calculation of team assignments and individual vehicle tours. In a fully distributed manner, the UAV agents are then responsible for calculating their own individual NILRTA* discrete-step tour paths. After the tours have been calculated, the results are communicated to the managing agent for centralized evaluation and team assignment selection.

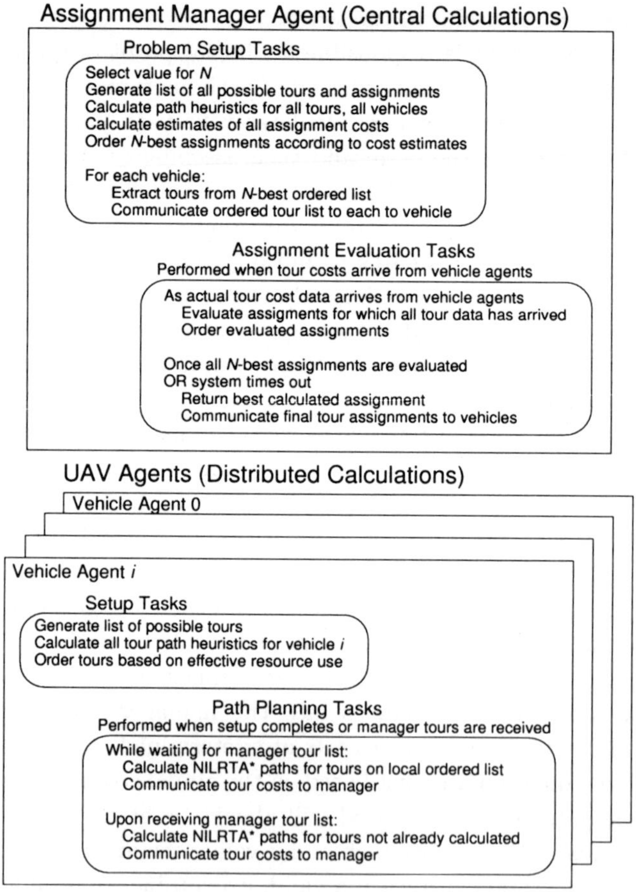

Fig. 6. Tour planning and assignment selection algorithms.

There are a number of factors that contribute to the speed of the algorithms and the overall method, and each requires individual tuning to maximize speed without reducing the quality of the result. The factors listed below have been tuned for best results in speed and quality:

- number of nodes in the discrete planner — determined by the world dimensions and the size of the discrete step;
- limits on the number of tasks a UAV can perform in a tour and on the length of the tour path;
- number of iterations of the Non-Improving planner before the search times out;
- limits on the total problem size, in number of assignments that must be setup — a function of the number of vehicles, targets, and the number of visits needed to each target;
- number of assignments kept by manager in the N-best assumption.

The values to which these factors are tuned depend on computer resources that are available to both the assignment manager and the individual UAVs. It is particularly important that the individual UAV agents have sufficient computer memory for calculation of the NILRTA* discrete paths. The world dimensions and planner step size are limited by the memory available to the UAV agent. Though memory is an issue for the assignment manager in extremely large problems, the speed of the processor is of much greater importance for this agent.

4. Results and Discussion

In this section the results of using the tour path planning method and a team assignment methodology are compared to established methods. The baseline method that is used for comparison uses a point-to-point path planner similar to the planner developed by the AFRL/VACA [18, 4, 19], which is based on the geometric study of L.E. Dubins [6]. The baseline method also uses an assignment method that is iterative and greedy. The greedy method is used to compare myopic, iterative results with those obtained using tour paths and overall team assignments.

The greedy and myopic methods used here are straightforward implementations similar to existing iterative assignment and segment-optimal point-to-point path planning methods. The results and successes of the method are similar to those reported in previous works [18, 19]. These types of methods are useful in dealing with large problems since only a small portion of the problem is considered at any one time. The result is that large problems are automatically reduced in size and are evaluated in computationally tractable pieces.

Although computationally efficient, applications of iterative methods often result in team assignment inefficiencies and lengthy vehicle paths. A typical assignment that demonstrates this can be seen in Figure 7. The iterative assignment is created by determining which vehicles can visit a target most immediately. A target visit is accomplished by flying directly over the target and makes no additional use of the sensor footprint. The example shows that vehicles passed close enough to targets to have them within their sensor footprint, but the vehicle was not expecting it and had to return and fly directly over it. The result of many assignments such as this is that multiple vehicles are used to accomplish what a single UAV could do. Churning in the assignment is evident in this example. The vehicle represented by the star waypoint path made its final turn to return to a target that was reassigned to another vehicle just before the diamond UAV could complete the task. Even though one vehicle was able to visit two targets in quick succession, it still took longer than would have been necessary if the vehicle had been able to utilize the full sensor footprint.

The same scenario as was used in Figure 7 was run using the tour plan assignment method to compare resulting assignments. In contrast to the inefficient assignment and lengthy tours obtained with greedy, iterative methods, the method presented here results in shorter individual tours, better team cooperation, and as a result, faster overall completion of the team goal. The assignment obtained from the tour plan assignment method is presented in Figure 8. The use of planned tour paths results in tours that accomplish more in less time through the effective use of the entire sensor footprint and better overall cooperation. The tour-planned paths in this case result in an assignment that is completed in approximately half the time required to complete the iterative greedy tour.

4.1. *Method Comparisons*

Iterative assignments can lead to poor use of vehicle and team resources. The proposed method overcomes these weaknesses through better individual UAV tour planning and overall team assignments. The approach used here plans for both immediate and future target visits. When cooperating UAVs plan multi-target tours and make assignments based on these tours, the team can better utilize the mission capabilities of the individual UAVs.

Over 215 randomized tests were performed in an effort to quantify the difference between the benchmark method and the approach discussed here.

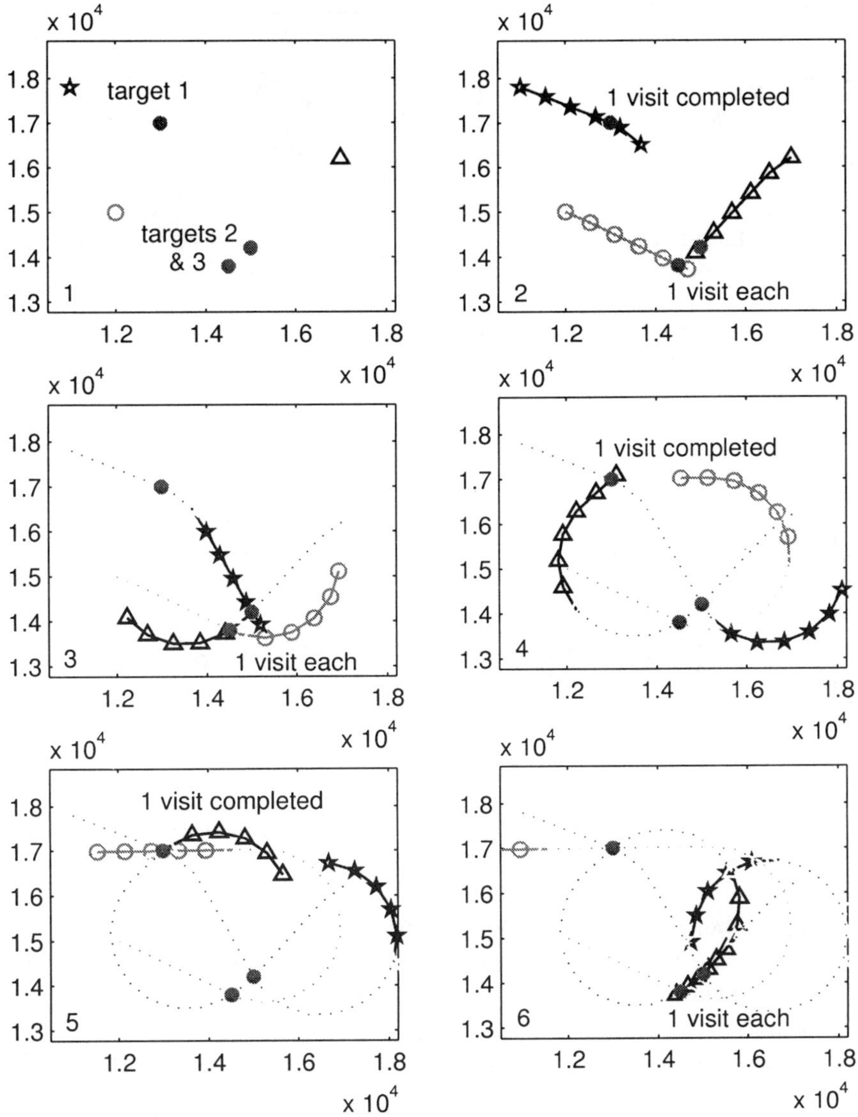

Fig. 7. A sample assignment reached through execution of a greedy and iterative assignment method that employs segment-optimal path planning. Dimensions in feet.

Each test involved randomizing the following parameters:

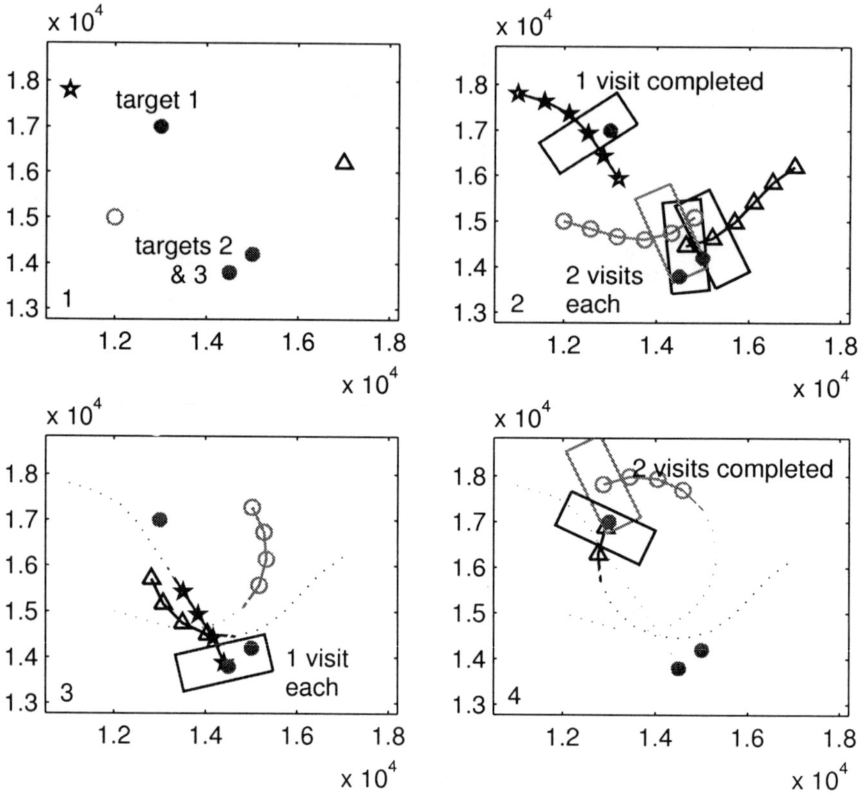

Fig. 8. Team assignment generated through the use of individual UAV tour paths and an overall assignment selection. Dimensions in feet.

- number of vehicles in the scenario — between 2 and 5
- starting UAV positions and headings — anywhere within 9000 ft. by 9000 ft. area
- number of targets to be visited — between 2 and 4
- target positions — anywhere within an 8000 ft. by 8000 ft. area.

In the randomized tests, the approach of planning tours and using those tours in making assignments proved to produce significantly better tour paths and team assignments than did the iterative and piecewise benchmark method. On average, the iterative assignment method produces tours that are 89 percent longer in time to completion than the tour-planning method proposed here. In multiple cases, the iterative assignment produced an assignment that was over six times longer than the tour-based solution. At the other extreme, the minimum benefit of the tour-based approach was an assignment completion time that was 8 percent shorter. These results are summarized in Table 2.

Table 2. Iterative assignment costs compared with tour-based approach costs

Average:	iterative cost = 1.89 × tour-based cost
Maximum:	iterative cost = 6.56 × tour-based cost
Minimum:	iterative cost = 1.08 × tour-based cost

Situations in which our method obtains the greatest improvement in overall team cooperation are exactly where previous methods have obtained the most undesirable results. Point-to-point planners are weakest when the Euclidean distance between any two targets in the tour is less than twice the turning radius of the UAVs. The complication from target spread proximity is compounded in MVTV problems when multiple visits are required by each target. In these cases the vehicles stand to gain the most from the effective use of the full sensor footprint, something that point-to-point planners are not capable of providing.

4.2. *Reducing Problem Size*

The average improvement of a team assignment using the proposed approach is considerable, but it does not give a complete picture of the value or cost of the approach. Assignment benefits include faster completion time of the team assignment, improved UAV cooperation, better use of vehicle sensors and resources, and an improved ability to visit and service spatially close targets. However, even with these gains, if the approach is to be useful, the results need to be obtained within reasonable time limits and with reasonable computational resources.

A necessary part of ensuring that the problem remains computationally tractable is reducing the size of the MVTV problem space that is explored for the selection of a final assignment. MVTV problem reduction is possible through the use of tour path-length heuristics and estimations of team assignment costs. In this way, the team can weed out obviously poor paths and assignments so they will not need to be fully planned and evaluated. Iterative methods control problem size by only considering a portion of the total assignment at a time, while the tour planning assignment method controls problem size through efficient elimination of tours and assignments that are unlikely to produce good results.

MVTV problems can be effectively reduced due to the nature of the assignment cost estimates generated from the heuristics. Each of the 215+ scenarios tested were solved globally while also maintaining a record of the ordered heuristics. In this way, reduced problem solutions and ordered heuristics can be compared directly to the global solution and actual ordered costs, and used to determine the effect of maintaining only a fraction of the potential assignments in the N-best assignments assumption. Figure 9 illustrates the average position of the actual global solution on the list of assignments ordered by the heuristic cost estimate for problems of various sizes. The chart suggests an average value for N to be used in problems of different sizes when a high probability of finding the optimal or global solution is desired. As can be seen, the percentage of assignments improperly ordered above the global optimum decreases as the problem size increases.

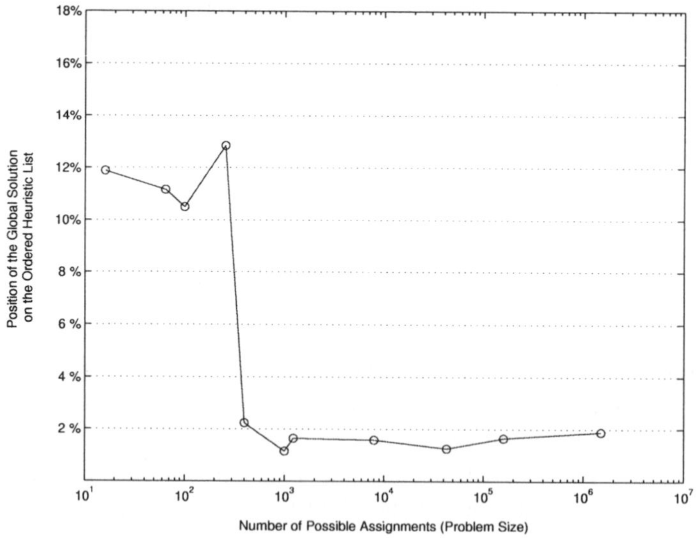

Fig. 9. Average position of the globally optimal solution on the list of assignments ordered by cost estimate.

Although the percentage of team assignments that must be maintained in the N-best assumption decreases as the problem gets larger, the total number of assignments and tour paths calculated still increases. The result is an upper limit on problem size that is governed by computer speed and by the desired quality of the result. The value of N depends on the problem size. For problems with fewer than 1000 possible assignments, globally optimal solutions could be reliably found by fully computing only the top 20 percent of those assignments. For problems with more than 1000 possible assignments only the top ten percent would need to be computed to reliably find the global optimum. Improved accuracy of tour path-length heuristics and team assignment cost estimates would result in a better initial ordering of assignments and a reduction in the percentage of the total assignments that would need to be included in the N-best path list. However, if the value of N is reduced beyond what the accuracy of the path heuristics and assignment cost estimates can effectively predict, the assignments and tours included in the N-best ordered estimates may not reflect the best actual paths and assignments, jeopardizing the quality of the final assignment.

The pruning of poor tour paths and team assignments can only be as good as the path heuristics and team assignment estimates that are used in the pruning. Effective path pruning comes when the tour heuristics properly represent the actual length of the path, and more importantly, when they properly represent the order of the tours from shortest to longest. The tour path heuristics used in pruning are calculated in nearly the same manner as the NILRTA* heuristics. The only difference is that the pruning heuristics include additional factors in calculating the path-length heuristic that are intentionally left out of the NILRTA* path planning heuristics to satisfy the heuristic admissability requirement of the A* algorithm. The additional factors are necessary because they prevent the heuristics from "breaking down" on smaller problems. In Figure 9, it can be seen that the heuristics begin to break down for the larger problems considered, resulting in the optimal assignment being found further down the ordered list of cost estimates.

Using the N-best assignments reduction method effectively reduces problem size while still producing improved team assignment results. Figure 10 shows that keeping only the N-best assignments reduces the number of individual tour paths needed for each individual vehicle to fully plan and calculate, in addition to reducing the number of assignments evaluated by the manager. The data shows that only a fraction of the possible individual vehicle tours are represented in the top N ordered assignments. Therefore, the N-best assignments assumption reduces the problem size and computational load for both the assignment manager and the individual UAV agents. By reducing the problem in this manner, improved assignments can be determined for near real-time applications.

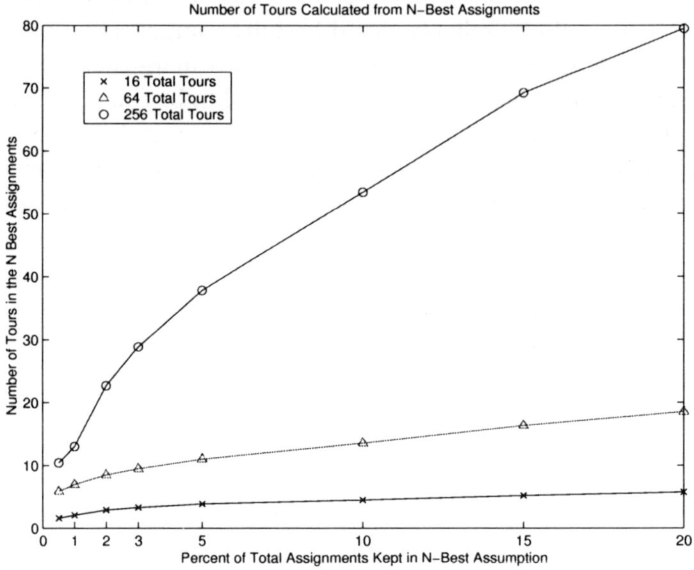

Fig. 10. Number of tours calculated versus fraction of ordered team assignments kept.

The nature of the MVTV problem as outlined is similar to the multiple Travelling Salesman Problem (or mTSP), with the added complication that each salesman is a dynamically constrained vehicle. The TSP and mTSP have been shown to be NP-complete problems [14, 10], and by extension, so too is the MVTV problem. The implication is that no algorithm other than an exhaustive search can guarantee the optimal or global solution. Maintaining a limited number of assignment estimates in the N-best assumption removes any guarantees that the solution will be optimal or even improved, but the accurate development and effective use of path-length heuristics and assignment cost estimates has been shown to reduce the problem size to a manageable level while still statistically improving the assignments that are returned. The motivation for such tradeoffs is the need for speed, which is discussed in Section 4.3.

At times the need for speed requires an even further reduction in the number of assignments kept (the value of N) than can be justified by the statistics shown in Figure 9. The ordering of assignment cost estimates allows for this additional reduction. When the optimal assignment is not found, near-optimal assignments usually result. Figure 11 shows the average length of an assignment returned when compared to the length of the global solution. When keeping only 0.5 percent of the possible assignments for larger problems, the resulting path is only 10 percent longer than the global solution. This is still significantly better than the iterative assignments which are 89 percent longer, on average, than the overall assignment obtained with tour-planned paths. It is noteworthy that a smaller percentage of assignments is needed for large problems for effective solutions. This is significant because it demonstrates the feasibility of the proposed method for solving large problems in near real time.

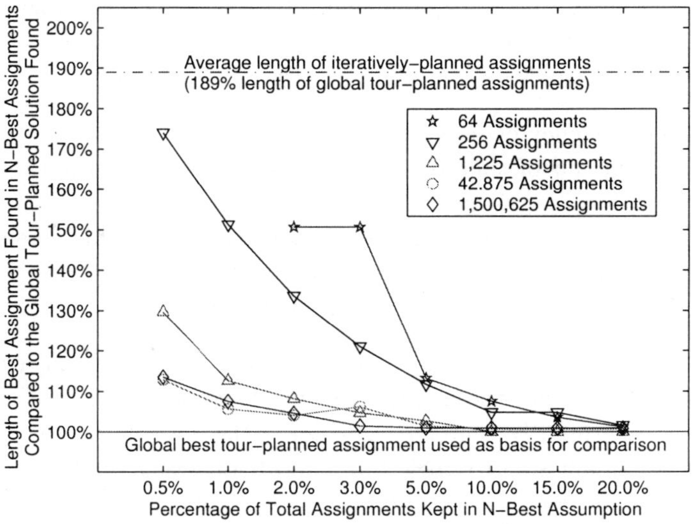

Fig. 11. Assignment costs compared to the globally best tour assignment solution comparing the effectiveness of problem reduction methods and sizes.

4.3. *Speed of Calculations*

The size and complexity of the MVTV problem require certain tradeoffs to be made between the optimality of the solution and the speed with which the result is returned for execution. The N-best assignment assumption increases the speed but also reduces the probability of obtaining the optimal assignment. The non-improving modification to the LRTA* planner has a similar result. By timing out of a non-improving tour path search, the planner increases the speed with which a path is planned but also decreases the probability that the path is truly optimal. The data shown in preceding sections demonstrate that these tradeoffs have not significantly compromised the ability to obtain better results through this method. The question that remains is whether the quality has been obtained with adequately low computational burden.

Assignments are obtained from the tour plan assignment method in sufficient time for execution in near real-time situations. The speed of the method is much slower than the Dubins paths/iterative assignment method used as a benchmark for comparison purposes, but it is not intended to be run as frequently. The assignment process only needs to be run a single time for an entire team assignment to be reached. By contrast, the iterative method runs every time the system state changes and a new subassignment needs to be made. Deciding whether or not the proposed assignment method is fast enough depends on a number of variables including

- the frequency of assignment calculations
- the amount of time in advance that agents know the target positions before assignment execution is required
- the quantity of previous calculations still applicable when the assignment needs to be recalculated
- the level of confidence required in the solution

The speed of the algorithm depends on the computational capability of both the UAVs and the manager agent. The computation of the manager agent is primarily centered on three tasks: generating the complete and non-redundant set of vehicle tours and team assignments; calculating, evaluating and ordering of team cost estimates; and finally, the evaluation and ordering of actual assignment costs when vehicles report tour lengths and costs. For the manager, the amount of time required depends mostly on the number of total team assignments that are being kept and ordered (the value of N). Problem setup involves the first two steps mentioned. The assignment manager can entirely setup most problems, which would be considered small, in less than four seconds. Setup for global solutions (ordering all assignment cost estimates) for larger problems takes much longer, as can be seen in Table 3. In the table all targets are assumed to be visited three times each. In practice, the limit on the value of N has been set at 80,000 assignments that are explicitly kept and ordered so that setup can be fully executed on the order of seconds rather than minutes or hours.

The calculation of the individual tour path trajectories can be fully dis-

Table 3. Setup times for problems of various sizes. All targets are assumed to be visited three distinct times

Number Vehicles	Number Targets	# Assignments Kept & Ordered	Avg Time to Setup
3	3	1,000	0.4 sec
3	4	10,000	2.8 sec
4	3	8,000	2.1 sec
4	4	160,000	5 min
5	4	1,500,625	2 hrs
5	4	75,000	6.6 sec
5	4	20,000	4.7 sec

tributed to the several UAV agents. Complete NILRTA* paths involving multiple targets and tasks are calculated in 1.1 seconds[a] on average. Actual times range between 0.2 and 1.8 seconds depending on the size of the world, the length of the path, the number of targets and the spread of their positions, and the number of tasks assigned in the tour. Problems solved in this work ranged from 16 to 512 tour paths per vehicle. Global solutions require each vehicle to calculate all tours, but as Figure 10 shows, the individual UAVs are generally asked to plan only a fraction of the total possible tours when using the N-best assignments assumption.

5. Conclusions

The MVTV problem poses significant challenges for both path planning and task assignment. Path planning challenges include dynamical vehicle limitations and spatial coupling of targets and tasks. Task assignment is made more difficult by the need to prepare for both immediate needs and for future tasks. Path planning and task assignment are also coupled, leading to complications in determining effective path plans and assignments.

MVTV problems can be successfully addressed through the use of an improved tour planner that plans near-optimal paths through a sequence of multiple targets. Tour trajectory planning is accomplished through a Non-Improving LRTA* search. The NILRTA* search is effective at planning flyable paths for dynamically constrained vehicles. Through the search process, vehicles learn the best trajectory through a set of targets by taking advantage of the full sensor footprint to help overcome the spatial coupling of targets and individual tour segments.

[a]Computations were performed on a desktop computer with an AMD Athlon 2700 chip and 1024 MB RAM.

Finally, improved assignments are made that specifically take advantage of tour-planned paths. When assignments are made using tour-planned paths the cooperative team can accomplish tasks in less time. Exponential growth in problem size can be controlled sufficiently through initial ordering of paths based on heuristics and team assignment estimates. Ordering by estimated cost leads to effective assignments, improved cooperation, and better use of team and individual resources. The resulting paths and assignments can be computed in near real time.

References

[1] Alighanbari, M., Kuwata, Y., and How, J. P., Coordination and control of multiple uavs with timing constraints and loitering. In *Proceedings of the American Control Conference*, pages 5311–5316, Denver, CO, 2003.

[2] Bellingham, J., Tillerson, M., Richards, A., and How, J., Multi-task allocation and trajectory design for cooperating UAVs. In Butenko, S., Murphey, R., and Pardalos, P. M., editors, *Cooperative Control: Models, Applications and Algorithms*. Kluwer Academic Publishers, 2003.

[3] Brummit, B. L. and Stentz, A., Dynamic mission planning for multiple mobile robots. In *Proceedings of the IEEE International Conference on Robotics and Automation*, volume 3, pages 2396–2401, Minneapolis, MN, 1996.

[4] Chandler, P. R. and Pachter, M., Hierarchical control for autonomous teams. In *Proceedings of AIAA Guidance, Navigation, and Control Conference*, Montreal, Canada. AIAA paper 2001-4149, 2001.

[5] Chandler, P. R., Pachter, M., Swaroop, D., Fowler, J. M., Howlett, J. K., Rasmussen, S., Schumacher, C., and Nygard, K., Complexity in UAV cooperative control. In *Proceedings of American Control Conference*, pages 1831–1836, 2002.

[6] Dubins, L., On curves of minimal length with a constraint on average curvature and with prescribed initial and terminal positions and tangents. *American Journal of Math*, 79:497–516, 1957.

[7] Fowler, J. M., Coupled task planning for multiple unmanned air vehicles. Technical report, AFRL/VACA WPAFB, Dayton, OH, 2001.

[8] Frazzoli, E., Dahleh, M. A., and Feron, E., Real-time motion planning for agile autonomous vehicles. *AIAA Journal of Guidance, Control, and Dynamics*, 25(1):116–129, 2002.

[9] Ganapathy, S. and Passino, K. M., Agreement strategies for cooperative control of uninhabited autonomous vehicles. In *Proceedings of the American Control Conference*, pages 1026–1031, 2003.

[10] Goldberg, A. V., Combinatorial optimization. Lecture Notes for CS363/OR349, Department of Computer Science, Stanford University, Stanford, CA, 1993.

[11] Howlett, J. K., Path planning and cooperative assignment. Technical report, AFRL/VACA WPAFB, Dayton, OH, 2001.

[12] Howlett, J. K., Path planning for sensing multiple targets from an aircraft. Master's thesis, Brigham Young University, Provo, UT, 2002.

[13] McLain, T., Chandler, P., Rasmussen, S., and Pachter, M., Cooperative control of uav rendezvous. In *Proceedings of the American Control Conference*, pages 2309–2314, Arlington, VA, 2001.

[14] Motwani, R., Lecture notes on approximation algorithms. Lecture Notes for CS351, Department of Computer Science, Stanford University, Stanford, CA 94305–2140, 1991-1992.

[15] Reif, J., Complexity of the mover's problem and generalizations. In *Proceedings of the 20th IEEE Symposium on the Foundations of Computer Science*, pages 421–427, Washington DC. IEEE, 1979.

[16] Schouwenaars, T., Mettler, B., Feron, E., and How, J. P., Robust motion planning using a maneuver automaton with built-in uncertainties. In *Proceedings of the American Control Conference*, volume 3, pages 2211–2216, Denver, CO, 2003.

[17] Schouwenaars, T., Moor, B. D., Feron, E., and How, J., Mixed integer programming for multi-vehicle path planning. In *Proceedings of the European Control Conference*, pages 2603–2608, 2001.

[18] Schumacher, C., Chandler, P. R., and Rasmussen, S. J., Task allocation for wide area search munitions via iterative network flow. In *Proceedings of AIAA Guidance, Navigation, and Control Conference.* AIAA paper 2001-4586, 2002.

[19] Schumacher, C., Chandler, P. R., Rasmussen, S. J., and Walker, D., Task allocation for wide area search munitions with variable path length. In *Proceedings of the American Control Conference*, pages 3472–3477, Denver, CO, 2003.

[20] Swaroop, D., A method of cooperative classification and attack for LOCAAS vehicles. Technical report, AFRL/VACA WPAFB, Dayton, OH, 2000.

[21] Weiss, G., editor, *Multiagent Systems: A Modern Approach to Distributed Artificial Intelligence*, pages 182–185. The MIT Press, Cambridge, MA, 2000.

CHAPTER 25

DECENTRALIZED OPTIMIZATION VIA NASH BARGAINING

Steven L. Waslander, Gokhan Inalhan and Claire J. Tomlin
Stanford University, Stanford CA

We present a new method for solving multi-player coordination problems using decentralized optimization. The algorithm utilizes the Nash Bargaining solution as the preferable outcome for all players among the set of Pareto optimal points, under assumptions of convexity. We demonstrate the concept on a multi-agent kinematic trajectory planning problem with collision avoidance. An analysis and numeric comparison of complexity is performed between centralized and decentralized penalty method based optimization. The analysis and the simulations suggest operation regimes where the decentralized method incurs no increase in complexity and even improvement in computation time proportional to the number of players over the centralized method. Experimental results from the MIT rover testbed are presented as well, showing very good correlation between the planned and executed trajectories.

Keywords:
Decentralized optimization, Nash Bargaining solution, multi-agent control

1. Introduction

Multi-agent systems, such as collections of vehicles, autonomous robots and supply chain networks, can often benefit from coordination between agents in achieving system level goals and satisfying inter-agent constraints. In the case of aircraft traffic flow through a constricted airspace, coordination of aircraft trajectories can improve fuel consumption or reduce flight duration while maintaining a minimum safe distance between vehicles at all times. There are many approaches to multi-agent coordination which broadly fall into three categories. Centralized approaches require information about each agent's goals and constraints to be available to a central planner which makes decisions for all agents. Distributed approaches allow individual agents to make decisions, but require some central coordination of the decision process to maintain a complete mathematical model. Finally, decentralized approaches remove the requirements of central coordination and allow individual agents to determine their own actions based on only locally available information. This chapter focuses on a decentralized approach to coordination that can provide methods for systems where central coordination is undesirable due to the structure of the problem (ie. competing businesses in a supply chain network) or due to a large number of agents (ie. automobile collision avoidance).

There are many areas of research that touch on aspects of this problem. Decomposition and distributed optimization dates back to results by Benders [2], and were extended to a general class of convex optimization problems by [9], and to non-convex problems by [25, 18]. For multiple decision makers, distributed computation of Pareto-optimal solutions has been studied by [27]. However, the result is limited to quasi-concave cost functions and problems with no constraints. [13] provides a decentralized method for calculating Pareto-optimal solutions in multi-party negotiations, using a structure similar to distributed optimization methods. The notion of decentralization optimization for stochastic discrete-event systems has been studied by [26]. In addition, team algorithms [1] have been developed to solve nonlinear systems of equations in a parallel distributed fashion. We utilize ideas from multi-objective optimization covered by [19, 14]. Refer to [7] for an extensive review on this topic. Additionally, we use concepts of decomposition and overlapping given by [29] that aid in analyzing large-scale interconnected systems.

Our recent work, [15] and [16], formulated the multi-agent coordination problem as a cooperative decentralized optimization, and guaranteed that the solution satisfies necessary conditions for Pareto optimality of a centralized formulation. Furthermore, sufficiency conditions for Pareto optimality are met for convex optimization problems, and hence the algorithm is guaranteed to converge to within ϵ of a Pareto optimal solution. In this chapter, we select a mutually agreeable solution to convex decentralized optimization problems by constructing an algorithm to search for a specific Pareto optimal point, the Nash Bargaining Solution, as first proposed by John Nash, [20].

The Nash Bargaining Solution was extended to multi-player games with coalitions by Harsani, [12], and modified for non-convex problems by Conley and Wilkie, [8]. Objections have been raised to one of the axioms needed to define the Nash Bargaining Solution by Kalai and Smorodinsky, who proposed an alternate solution which focuses on global information, [17]. In the decentralized framework, the Nash Bargaining solution remains of interest, however, due to its differentiability and focus on local information.

To the best of our knowledge, this chapter presents the following novel results. With the addition of requirements of convexity and communication between all agents, we modify our previous algorithm for decentralized optimization to seek the Nash Bargaining Solution (NBS). We compare, through analysis and simulation, the computational complexity of centralized and decentralized penalty method optimization for non-convex problems. Finally, we demonstrate real-time operation of the decentralized non-convex optimization algorithm on the MIT rover testbed, courtesy of the Aerospace Controls Laboratory under the supervision of Professor Jonathan How.

2. Problem Formulation

Consider a system of p agents, where each agent $i, \in \mathbb{P} = [1 \ldots p]$ has associated with it a vector of optimization variables, $x_i \in \mathbb{R}^{n_i}$ with $x = [x_1, \ldots, x_p] \in \mathbb{R}^n$. For each agent, we define an independent cost function, $f_i(x_i)$, where $f_i : \mathbb{R}^{n_i} \to \mathbb{R}_+$. The *centralized optimization problem* can be defined as,

Definition 1: *[Centralized Optimization Problem]*

$$\begin{aligned} \min_{x \in \mathbb{R}^n} \quad & [f_1(x_1), \ldots, f_p(x_p)] \\ \text{subject to} \quad & g(x) \leq 0 \\ & h(x) = 0 \end{aligned} \tag{1}$$

where $g : \mathbb{R}^n \to \mathbb{R}^q$, $h : \mathbb{R}^n \to \mathbb{R}^r$ are lists of inequality and equality constraints which can include both local and global requirements. The notation $g^k(\cdot)$ refers to the kth constraint in $g(\cdot)$. In the example of agents as vehicles, the local cost function can be constructed to penalize, for example, deviations from a desired trajectory or fuel consumption. Local constraints can include vehicle dynamics, minimum and maximum control limits, and obstacle avoidance constraints. Global requirements can account for collision avoidance between vehicles, coordinated search requirements or resource allocation among agents. We assume that f_i, g, h are continuously differentiable functions of continuous variables, and that the complete set of constraints is regular [4].

Optimality for the centralized optimization problem is defined using Pareto optimality.

Definition 2: *[Pareto Optimal Solution]* The vector $x^{*_P} \in \mathbb{F} = \{x \in \mathbb{R}^n | g(x) \leq 0, h(x) = 0\}$ is a Pareto optimal (minimal) solution of the centralized optimization problem if there exists no $x \in \mathbb{F}$ and $j \in \mathbb{P}$ such that $f_i(x_i) \leq f_i(x_i^{*_P}) \quad \forall i \in \mathbb{P}$ and $f_j(x_j) < f_j(x_j^{*_P})$.

The multi-agent coordination problem can also be posed in a decentralized manner. Let us first define an agent i's *Neighborhood* as the set $\mathbb{P}_i \subseteq \mathbb{P}$ of agents j for which there exists a constraint that involves both agents i and j. Intuitively, the notion of neighborhood bounds the scope of interest for an agent to those members of the system that may have an impact on its optimization process. We use the notation $\{x_j\}_i = \{x_j \in \mathbb{R}^{n_j} | j \in \mathbb{P}_i\}$ to refer to the set of optimization variables of all agents j in the neighborhood of agent i. The decentralized framework requires that each agent solve a local optimization based exclusively on information concerning other agents in its neighborhood. The *decentralized optimization problem* can be written as,

Definition 3: *[Decentralized Optimization Problem]*

$$\begin{aligned} \min_{x_i \in \mathbb{R}^{n_i}} \quad & f_i(x_i) \\ \text{subject to } & g_i(x_i | \{x_j\}_i) \leq 0 \\ & h_i(x_i | \{x_j\}_i) = 0 \end{aligned} \tag{2}$$

Here $g_i(x_i|\{x_j\}_i)$, $h_i(x_i|\{x_j\}_i)$ are lists of inequality constraints and equality constraints on x_i, given that the states of all agents j in the neighborhood of i are held constant. g_i and h_i can be further subdivided into local constraints, $(g_{l_i}(x_i), h_{l_i}(x_i))$, involving only local optimization variables, and interconnected or global constraints, $(g_{g_i}(x_i|\{x_j\}_i),\ h_{g_i}(x_i|\{x_j\}_i))$ involving the optimization variables of at least one other agent in the neighborhood, $\mathbb{P}_i$. We include similar assumptions as in the centralized formulation, namely that f_i, g_i, h_i are continuously differentiable functions of continuous variables, and that the complete set of constraints is regular. Furthermore, we assume that all interconnected constraints enter each associated local optimization identically.

For the decentralized optimization problem of Eq. 3, we define optimality using the Nash equilibrium.

Definition 4: *[Nash Equilibrium]* Let $\mathbb{F}_i = \{x_i \in \mathbb{R}^{n_i} | g_i(x_i|\{x_j\}_i) \leq 0, h_i(x_i|\{x_j\}_i) = 0\}$. Then $x^{*_{ne}} \in \mathbb{F}$ is a Nash equilibrium of the decentralized optimization problem if, $\forall i \in \mathbb{P}$, given $\{x_j\}_i$, $f_i(x_i^{*_{ne}}) \leq f_i(x_i)$, $\forall x_i \in \mathbb{F}_i$.

3. Solution Algorithms

Centralized Algorithms Two well known techniques for solving the centralized formulation are the Lagrange Multiplier and Penalty methods. In both cases, the local cost function is augmented to include costs which penalize the violation of constraints. If a solution can be found, the Lagrange multiplier method guarantees that constraints will be satisfied, but the method requires a new optimization variable for each constraint. In order to solve the vector optimization defined in Eq. 2, we introduce $\omega \in \mathbb{R}^p$ as a weighting vector of agents' costs. The Lagrange multiplier method can then be written as follows (see [4] for a more general formulation).

$$\min_{x \in \mathbb{R}^n} \max_{\lambda \in \mathbb{R}^q_+, \mu \in \mathbb{R}^r} [f_1(x_1) \dots f_p(x_p)] \cdot \omega + \lambda^T g(x) + \mu^T h(x) \tag{3}$$

where the Lagrange multiplier vectors for equality constraints are defined as $\mu \in \mathbb{R}^r$. For inequality constraints, the Lagrange multiplier vector is

$$\lambda \in \mathbb{R}^q_+ \quad \text{where} \quad \begin{cases} \lambda_k \geq 0 \\ \lambda_k = 0 \text{ if } g^k \text{ is inactive} \end{cases} \quad \forall k \in \{1 \dots q\} \tag{4}$$

With a linear combination of local cost functions, it is not necessarily possible to achieve all Pareto optimal solutions, however, this simplification is required in order to pose an optimization that can be solved using standard non-linear programming methods.

Formulation of the centralized optimization problem via penalty methods allows for a separate treatment of constraints that does not increase the dimension of the optimization problem, but requires iteration of the entire optimization process until convergence. The penalty method assigns costs to the violation of constraints by including a penalty function in the minimization. For comparison to the decentralized penalty method, we use the penalty method formulation only for interconnected constraints. Let each agent's locally feasible region,

$$\mathbb{X}_i = \{x_i \in \mathbb{R}^{n_i} | g_{l_i}(x_i) \leq 0, h_{l_i}(x_i) = 0\} \quad \forall i \in \mathbb{P}$$

and let $\mathbb{X} = \{x \in \mathbb{R}^n | x_i \in \mathbb{X}_i, \forall i \in \mathbb{P}\}$. With equality constraints recast as inequality constraints using slack variables [6], let us define a class of inexact differentiable penalty functions, $P : \mathbb{R}^n \rightarrow \mathbb{R}_+$, that penalize all interconnected constraints of a system by,

$$P(x) = \sum_{k=1}^{q_g} \max\left(0, g^k(x)\right)^\gamma \tag{5}$$

where q_g now defines the total number of interconnected constraints in the system and $\gamma \in \mathbb{R}$, $\gamma \geq 2$ defines the order of the penalty function. The centralized optimization problem in penalty method form solves multiple iterations of the following optimization as the penalty parameter, $\beta \in \mathbb{R}_+$, tends to 0.

$$\lim_{\beta \rightarrow 0} \left(\min_{x \in \mathbb{X}} \left[f_1(x_1), \ldots, f_p(x_p)\right] \cdot \omega + \tfrac{1}{\beta} P(x) \right) \tag{6}$$

The centralized method, using inexact penalty functions, is guaranteed to converge to a solution, given a feasible solution exists and assuming the penalty parameter is selected such that it converges to some value. If the parameter converges to zero, the solution found meets necessary conditions for Pareto optimality. Furthermore, since the optimal solution is feasible and results in $P(x) = 0$, each intermediate solution of the optimization is bounded above by the optimal cost, and thus the optimization cannot become ill-conditioned at any stage of the process.

Decentralized Algorithm The decentralized algorithm first defined in [16] ties a localized penalty method formulation to a bargaining process between agents. A distinction is made between local and interconnected constraints, for in the decentralized approach, interconnected constraints require special treatment. Local penalty functions are defined analogously to Eq. 5, $P_i : \mathbb{R}_i^p \rightarrow \mathbb{R}_+$,

$$P_i(x_i | \{x_j\}_i) = \sum_{k=1}^{q_{g_i}} \max\left(0, g_{g_i}^k(x_i | \{x_j\}_i)\right)^\gamma \tag{7}$$

where, for each agent, i, q_{g_i} now defines the number of interconnected constraints, respectively. In order to convert from a centralized to a decentralized formulation of the penalty method, a $\beta_i \in \mathbb{R}_+$ pre-multiplier is included in the penalty augmented cost function, $F_i : \mathbb{X}_i \rightarrow \mathbb{R}_+$, defined as,

$$F_i = \beta_i \left[f_i(x_i) + \frac{1}{\beta_i} \Big(P_i(x_i | \{x_j\}_i) \Big) \right] \tag{8}$$

This modification is required since local cost functions are no longer bounded above by the optimal solution; only the local optimization variables x_i can be modified by a local optimization. To ensure convergence, the decentralized approach reduces the weight on the local cost at each iteration, instead of increasing the weight on the violation of constraints. This approach, as first defined in [16], ensures that the augmented cost functions converge as long as the local penalty parameter, β_i, tends toward 0. Then, for each agent i, the local penalty method formulation for decentralized optimization is,

$$\lim_{\beta_i \to 0} \left(\min_{x_i \in \mathbb{X}_i} \beta_i \left[f_i(x_i) + \tfrac{1}{\beta_i} \left(P_i(x_i | \{x_j\}_i) \right) \right] \right) \tag{9}$$

The decentralized algorithm can proceed in a number of fashions. In sequential form, all agents calculate a desired trajectory in the absence of interconnected constraints. Agent 1 receives the desired solutions from all other agents in its neighborhood and then solves a local optimization problem with the other agents' solutions fixed, to form a new solution set for all p vehicles. This set is passed along to agent 2 who also performs the local optimization and passes on the updated solution set to agent 3, etc. This method causes a bias in the solution against lower numbered agents in favor of higher numbered agents.

In "multi-threaded" form, all agents initially optimize based on the complete set of desired solutions, then pass out solution sets to each other and re-optimize for each solution set received. At each step, an agent could receive up to $p-1$ solution sets for agents in its neighborhood and must select a preferred solutions to ensure the number of solution threads does not expand exponentially. Trimming of solutions threads can be done based exclusively on local information or by considering global preferences defined in terms of other agents' local cost information which can be included in each solution set.

As presented in detail in [16], the above algorithm has been shown to converge to a decentralized Nash Equilibrium solution, which is also a Nash Equilibrium of the centralized problem and comes within ϵ of a solution that satisfies the necessary conditions for Pareto optimality. The proof of this assertion hinges on the fact that the bargaining parameters $\beta_i \to 0, \quad \forall i$, which ensures that the augmented cost function does not increase at any step in the process and that the violation of constraints decreases at each step.

The bargaining process inherent in the above algorithm can be driven to an equilibrium solution that satisfies necessary conditions for Pareto optimality through the selection of the bargaining parameter, β_i. Unfortunately, the relationship between β_i and any specific solution is unclear, unlike the centralized case, where variation in the weighting vector, ω, results in Pareto optimal solutions that favor the more heavily weighted agent. Furthermore, we seek to ensure that the solution selected by the algorithm is "fair", that each agent receives an equal amount of the excess in the system, or incurs an equal amount of cost. The range of equilibrium solutions includes solutions where one agent ignores interconnected constraints while the other suffers dearly for it, and it is precisely these situations we wish to avoid by searching for the Nash Bargaining Solution.

4. Nash Bargaining Solution

Axiomatic Foundation Based on 4 axioms first defined by John Nash in 1950 [20], a unique optimal bargaining solution between two agents can be found if the set of feasible solutions is compact and convex. Let us define such a two-agent bargaining problem by $B = (V_1(x), V_2(x), d, S)$, where $x \in \mathbb{F} = [x_1, x_2]$, is as above with $p = 2$, $V_i : \mathbb{R}^{n_i} \to \mathbb{R}$ are the agents' Von Neumann-Morgenstern utility functions [28], $d = (d_1, d_2) \in \mathbb{R}^2$ is the disagreement point which defines the cost incurred by each agent if no agreement is reached and $S \subseteq \mathbb{R}^2$ is the compact, convex set of all feasible utility pairs that improve on d. We define $x_B^* \in \mathbb{F}$ to be the optimal bargaining solution with optimal utility $s_B^* \in S$.

Nash showed that a unique optimal solution exists which maximizes the product of the utility functions of both players if the following four axioms are satisfied. It was Nash who first chose to use the product of utilities to determine the Nash Bargaining Solution, and although there is no clear interpretation of this construct in relation to the bargaining problem, its simplicity has allowed for its wide adoption and varied uses (see [22] for an alternative formulation).

Axiom 4.1: **Axiom of Rationality**: Each agent prefers the locally optimal solution.

Axiom 4.2: **Axiom of Symmetry**: If S is symmetric about the line $V_1 = V_2$, then the optimal bargaining utility lies on that line.

Axiom 4.3: **Axiom of Linear Invariance**: Neither scaling nor offset of either utility function affects the resulting bargaining solution.

Axiom 4.4: **Axiom of Independence of Irrelevant Alternatives**: If we define $\tilde{B} = (V_1(x), V_2(x), d, \tilde{S})$, where $\tilde{S} \subseteq S$ and the optimal utility, $s_B^* \in \tilde{S}$, then $s_{\tilde{B}}^* = s_B^*$. If S is restricted and yet retains s_B^* of the original problem, then the original optimal bargaining solution remains optimal for the restricted problem.

Proof Outline (After Nash, [20]) To show existence and uniqueness of an optimal bargaining solution, we invoke properties of compactness and uniqueness of S, respectively. To show that the optimal solution maximizes the product of the utilities of both agents, the following elegant set of arguments was developed based on the four axioms. If both agents are rational they will try to maximize their local utility, V_i. If both utility functions are linearly invariant, then both can be scaled and offset such that $d = (0,0)$ and $s_B^* = (1,1)$. Let $B' = (V_1(x), V_2(x), d, S')$, where S' is augmented to include all points such that the sum of the two utilities is less than 2 (ie. let S' be the triangle formed by the points $\{(0,0),(2,0),(0,2)\}$). Since S' is symmetric, by Axiom 4.2, $s_{B'}^*$ must be on the line $V_1 = V_2$, and thus $s_{B'}^* = (1,1)$. By Axiom 4.4, we see that $s_{B'}^* \in S$, and so is also the optimal solution to the original problem. The final step is to see that s_B^* is the point of maximum product of utility improvements $(V_1(x) - d_1)$, $(V_2(x) - d_2)$, and hence that maximizing the product of utility improvements determines the unique optimal bargaining solution.□

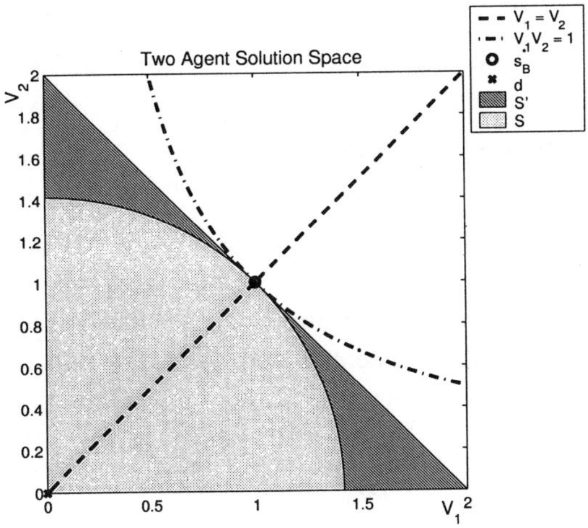

Fig. 1. Graphical representation of key elements of the Nash Bargaining Solution proof

A two dimensional representation of elements of the two-player bargaining problem can be seen in Figure 1.

Fact 4.1: The Nash Bargaining Solution is Pareto optimal.

As defined in Def. 2, the Pareto optimal solution requires that no other agent can improve their utility without decreasing the utility of another agent. By Axiom 4.1, both agents must select their locally optimal solution, and by convexity and compactness of the solution space, neither agent can improve their solution from this local optimum without decreasing the other agent's utility.

The same argument can be used for p agents, assuming that the solution space remains convex and compact and the same four axioms hold for all utility functions. The resulting central optimization for determining the p-agent Nash Bargaining Solution (NBS) is,

$$\max_{x \in \mathbb{F}} \left[\prod_{i=1}^{p} (V_i(x) - d_i) \right] \tag{10}$$

Reposing the formulation above as a minimization of cost functions, and adjoining problem constraints using the centralized Lagrangian method of Eq. 3, the NBS is found by minimizing,

$$\min_{x \in \mathbb{R}^n} \max_{\lambda \in \mathbb{R}^q_+, \mu \in \mathbb{R}^r} - \left[\prod_{i=1}^{p} (d_i - f_i(x_i)) \right] + \lambda^T g(x) + \mu^T h(x) \tag{11}$$

Likewise, in centralized Penalty Method form, Eq. 6 becomes

$$\lim_{\beta \to 0} \left(\min_{x \in \mathbb{X}} - \left[\prod_{i=1}^{p} (d_i - f_i(x_i)) \right] + \frac{1}{\beta} P(x) \right) \tag{12}$$

Necessary Conditions - Centralized Methods We now turn to a comparison of the necessary conditions for optimality [4], in order to determine a relationship between the decentralized penalty method and the NBS for a two agent problem. Using the centralized Lagrange multiplier formulation of Eq. 3, the resulting necessary conditions for optimality include,

$$\omega_i \frac{\partial f_i(x_i^*)}{\partial x_i} + \lambda^T \frac{\partial g(x^*)}{\partial x_i} + \mu^T \frac{\partial h(x^*)}{\partial x_i} = 0 \quad \forall i \in \mathbb{P} \tag{13}$$

By contrast, for the optimal solution x^* and the corresponding Lagrange multiplier values $\bar{\lambda}$, $\bar{\mu}$ the NBS necessary conditions can be written explicitly for each agent,

$$\left[\prod_{j \neq i} (d_j - f_j(x_j^*)) \right] \frac{\partial f_i(x_i^*)}{\partial x_i} + \bar{\lambda}^T \frac{\partial g(x^*)}{\partial x_i} + \bar{\mu}^T \frac{\partial h(x^*)}{\partial x_i} = 0 \quad \forall i \in \mathbb{P} \tag{14}$$

Dividing through by $\prod_{i=1}^{p}(d_i - f_i(x_i^*))$ yields,

$$\left[\frac{1}{(d_i - f_i(x_i^*))}\right]\frac{\partial f_i(x_i^*)}{\partial x_i} + \left[\frac{\bar{\lambda}^T}{\prod_{i=1}^{p}(d_i - f_i(x_i^*))}\right]\frac{\partial g(x^*)}{\partial x_i} + \left[\frac{\bar{\mu}^T}{\prod_{i=1}^{p}(d_i - f_i(x_i^*))}\right]\frac{\partial h(x^*)}{\partial x_i} = 0 \quad \forall i \in \mathbb{P} \tag{15}$$

If the weighting parameters ω_i in the centralized Lagrange multiplier formulation are chosen to be $\frac{1}{(d_i - f_i(x_i^*))}$, then the resulting Pareto optimal solution meets the necessary conditions for the NBS. Note that if $d_i = f_i(x_i^*)$, the problem is ill-posed as the optimal solution is disagreement.

Necessary Conditions - Decentralized Methods From the decentralized formulation of Eq. 9, the resultant necessary conditions become,

$$\beta_i \frac{\partial f_i(x_i^*)}{\partial x_i} + \frac{\partial P_i(x_i^* | \{x_j^*\}_i)}{\partial x_i} = 0 \quad \forall i \in \mathbb{P} \tag{16}$$

The NBS necessary conditions for the Penalty method formulation can be written for each agent as,

$$\left[\prod_{j \neq i}(d_j - f_j(x_j^*))\right]\frac{\partial f_i(x_i^*)}{\partial x_i} + \frac{1}{\beta}\frac{\partial P(x_i^* | \{x_j^*\}_i)}{\partial x_i} = 0 \quad \forall i \in \mathbb{P} \tag{17}$$

By Eqs. 5 and 7, the penalty function derivatives, $\frac{\partial P_i(x_i^* | \{x_j^*\}_i)}{\partial x_i}$ and $\frac{\partial P(x_i^* | \{x_j^*\}_i)}{\partial x_i}$ will appear identically in the two sets of necessary conditions, thus for the decentralized algorithm to meet the NBS necessary conditions for optimality, the bargaining parameters, β_i, must be chosen as,

$$\beta_i = \beta \cdot \prod_{j \neq i}(d_j - f_j(x_j^*)) \quad \forall i \in \mathbb{P} \tag{18}$$

Because the solution space, S, is compact and convex, the decentralized algorithm will converge to within ϵ of a Pareto optimal solution, as both necessary and sufficient conditions for Pareto optimality are satisfied if the solution converges. The optimal cost scaling factor for agent i, $\prod_{j \neq i} d_j - f_j(x_j^*)$, ensures the bargaining process converges to a solution that meets the necessary conditions of the NBS, and since the NBS must be unique, the decentralized algorithm with β_i as defined Eq. 18, will converge to the NBS.

Both centralized and decentralized results provide us with a method for determining the NBS, but are dependent on the optimal costs, and hence must be approximated for implementation. Immediately, the method of successive approximations [3] suggests itself as a means to approximate the desired coefficients. The disagreement point, d_j can be determined by first optimizing locally without interconnected constraints to find the ideal solution for each agent, and then to optimize locally with the ideal solutions for all other agents fixed, which results in a worst case non-cooperative solution for each agent. The NBS can now be found by setting β_i locally, at each iteration, k, of the optimization based on the intermediate optimization results x^{k-1} as follows,

$$\beta_i(k) = \beta(k) \prod_{j \neq i} \left(d_j - f_j(x_j^{k-1})\right) \quad \forall i \in \mathbb{P} \tag{19}$$

It is important to note the effect of defining bargaining parameters as in Eq. 19 on the communication network between agents. Up to this point, the decentralized framework required that only the current solution be passed by each agent to all others in its neighborhood. In addition, the new bargaining parameter definitions require that each agent receive the current best cost estimate x_j^{k-1} from all other agents in the system, and that each agent execute the update optimization using the same $\beta(k)$. These additional constraints on the communication structure may become restrictive with large numbers of agents, and remain an area for future investigation.

Implementation The algorithm, as modified by the above discussion, was implemented for the two vehicle collision avoidance problem. Vehicle 1 was located at the point $(6, 0)$ facing west, Vehicle 2 was located at $(0, -7)$ facing north, with desired trajectory defined as straight lines in the forward direction. A quadratic cost was associate with deviation from the desired trajectory, and a collision avoidance constraint required 5 m spacing between the vehicles. A simple kinematic model of an aircraft was used, with control inputs for velocity and turn rate, and a 5-step finite horizon lookahead policy was implemented.

A comparison was made between the original decentralized algorithm, as defined in [16], and the same algorithm with bargaining parameters selected as defined in 19. The following graph displays the evolution of the costs for each vehicle for both the original algorithm and the improved NBS inspired algorithm. The NBS method displays much faster convergence to the line between the greedy optimal point (0,0) and the feasible NBS, which will allow for future implementations to use fewer bargaining steps to arrive at the optimal solution. Calculation of the Pareto optimal front and the

NBS was performed in a centralized manner using the penalty method for reference.

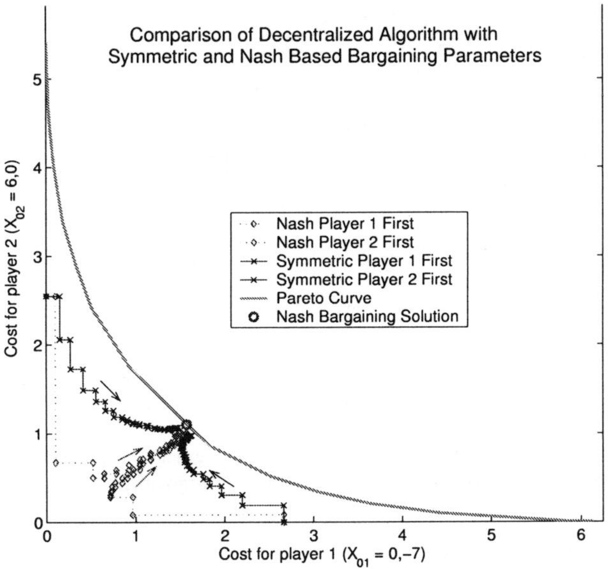

Fig. 2. Solution space and solution trajectories for NBS-based and symmetric decentralized algorithms. Arrows indicate direction of convergence of algorithm

As mentioned earlier, it is interesting to note that in a single solution thread, the advantage lies in not being the first vehicle in the process. With two vehicles, we can see that if vehicle 1 performs the first optimization given vehicle 2's desired trajectory, then it must select a trajectory that avoids vehicle 2, as required by the bargaining parameter β. Vehicle 2 then performs the next optimization based on vehicle 1's solution, and deviates slightly from its desired trajectory, due to a decrease in the value of β which increases the importance of satisfying the interconnected constraints. In Figure 2 and Figure 3 , both threads are displayed for each algorithm and it can be observed that the bargaining process must proceed for some time before this advantage is overcome, unless the Nash-inspired bargaining parameters are used.

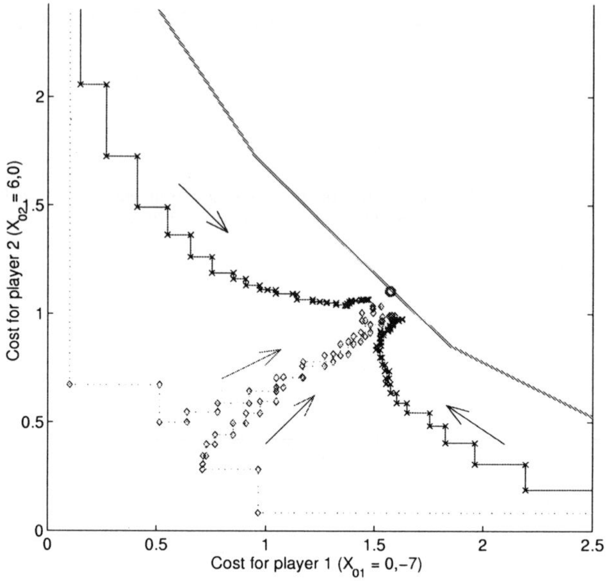

Fig. 3. Expanded view of the convergence of the algorithms to the Nash Bargaining Solution. Arrows indicate direction of convergence of algorithm

5. Complexity Analysis

The complexity of the decentralized algorithm is best compared with an equivalent centralized problem. For this analysis, let p be the number of agents, let A be the number of local control variables, and let B be the number of local constraints.

Nonlinear Program Complexity The nonlinear optimizations specified above are cast as standard nonlinear programs (NP), where one seeks to find a solution, x, to minimize the global cost function, $F(x)$. Since our problem makes no claim about convexity, we are constricted to finding local minima through an iterative process. The most common algorithm for solving NPs, used in Matlab functions fmincon, fminunc and others for medium scale problems is the sequential quadratic program (SQP), see [5], [11] and [23]. This method iteratively solves a quadratic approximation to the problem based on gradient and Hessian information. The Hessian of the Lagrangian is approximated using the BFGS update, and the quadratic program, (QP), is solved to find a search direction for the original problem. A standard line search is then performed in that direction and the process is repeated. The Broyden-Fletcher-Goldfarb-Shanno (BFGS) update requires the solution of a set of n linear equations ($O(n^3)$), unless sparsity can be exploited. The QP complexity can sometimes be bounded using self-concordant theory, [21] (when convex and self-concordant cost functions are used), but results in bounds that are orders of magnitude away from average numbers of iterations required. The line search is computationally trivial in comparison to the first two steps. The whole process must also be repeated an uncertain number of times to arrive at the NLP solution, but we assume a fixed problem complexity such that the number of SQP steps is relatively constant with respect to problem size.

Comparison In order to compare centralized and decentralized methods, first assume that the number of Newton steps required to solve any QP is reasonably constant and equal to K_{qp}, regardless of the order of the problem. Second, assume that the number of iterations needed to solve the NLP using SQP is equal to K_{sqp} and also does not depend on problem size. Furthermore, let us note that the relation between the number of iterations required to converge to a solution using the penalty method and the size of the optimization problem is not well understood, nor is the relation between the number of bargaining steps to converge to a solution in the decentralized problem and the number of vehicles bargaining. We therefore introduce variables K_b for the number of bargaining steps used in the decentralized problem and K_p for the number of penalty iterations in the centralized problem as parameters that can be varied in simulation.

The centralized approach with a fixed number of penalty method iterations results in a computational complexity of,

$$O(K_p \times p^3(A+B)^3 \times K_{qp} \times K_{sqp}) = O(K_p \times p^3(A+B)^3) \qquad (20)$$

Likewise, the decentralized approach solves

$$O((A+B)^3 \times K_{qp} \times K_{sqp})$$

at each of p vehicles for each of $p-1$ received solutions, and then repeats this process K_b times. The resulting algorithmic complexity is

$$O(K_b \times p^2(A+B)^3 \times K_{qp} \times K_{sqp}) = O(K_b \times p^2(A+B)^3) \qquad (21)$$

Hence, based on the assumptions made above and ignoring the effect of K_p and K_b on the quality of the solution, the result states that the decentralized approach outperforms the centralized approach as the number of agents grows, which is due to its ability to exploit the inherent problem structure.

Simulation Results In a multi-vehicle collision avoidance simulation, both algorithms were run with varying values for the number of vehicles, the number of bargaining steps/penalty steps and the number of control inputs and constraints. The simulation calculated finite horizon lookahead control policies for 30 time steps, based on quadratic costs for deviation from the desired straight-line trajectory, and 5 mile collision avoidance constraints for the entire horizon. The resulting simulation times are listed in Tbl. 1 below.

Table 1. Simulation Times (s) - Decentralized and Centralized Methods: 30-period collision avoidance problem

Decentralized Computation Times

Local Variables	10	14	20	10	14	20	10	14	20
Bargaining Steps	5	5	5	10	10	10	15	15	15
No. Vehicles									
2	65	127	267	106	178	377	136	265	407
3	139	292	745	197	494	1360	294	631	1812
4	318	773	1979	561	1224	2986	796	1586	4218

Centralized Computation Times

Local Variables	10	14	20	10	14	20	10	14	20
Penalty Steps	5	5	5	10	10	10	15	15	15
No. Vehicles									
2	147	315	719	220	458	1048	270	591	1296
3	466	1054	2948	682	1581	4414	787	2002	5020
4	980	2300	7090	1443	3554	9225	1716	4369	12637

For the decentralized algorithm, the computation times grew on the order of 0.7 with respect to K_b, which shows that the optimizations proceeded more quickly as K_b grew, which is most likely due to the fact that the number of steps required for convergence of the SQP algorithm reduced as the bargaining parameter, β converges to zero.

The centralized simulation results concur with the predicted complexity analysis, with the exception of the number of Penalty method iterations. Computation time varied as Q^3 and $\sqrt{K_p}$. The acceleration in the computation time for high number of iterates is due to the simplification of the problem as the iterations proceed, but at a faster rate than for the decentralized case. If the change in β is small, the optimization is nearly identical to the previous step, and so, with the solution of the previous iteration as initial estimate, almost no optimization is necessary.

The improvement in computation time of the decentralized algorithm over the centralized method was further investigated with a simplified problem of only one time step, such that initial conditions for each optimization were identical for both methods. The problems were posed such that significant optimization was necessary (the interconnected constraints were active in the optimal solution), and systems of 3-6 vehicles were simulated to get a better picture of the relation between the number of vehicles and computation time. The results, as displayed in Figure 5 showed p growth for the decentralized case, as predicted from the analysis, and p^3 for the centralized problem.

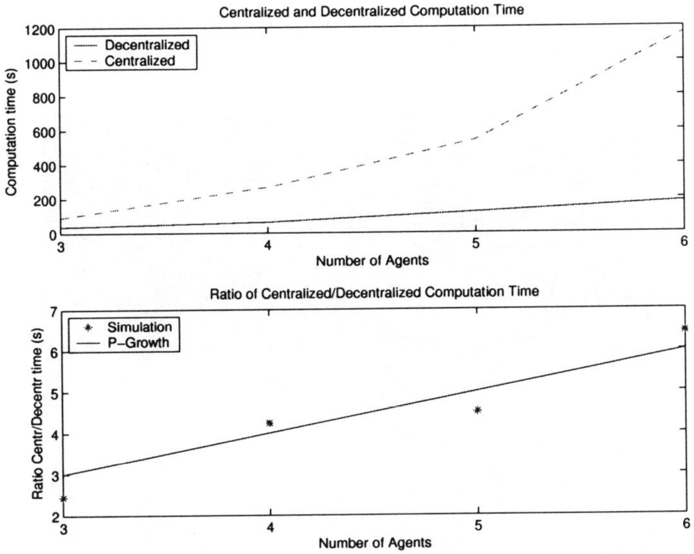

Fig. 4. Simulation time comparison of centralized and decentralized algorithms for 3-6 vehicles, 5 bargaining iterations and 10 step finite horizon lookahead control

We should note at this point that nonlinear optimization tools such as Stanford's SNOPT [10] can detect and exploit sparsity in any given optimization problem, and may be able to recover most or all of the gains in computation presented here. The decentralized algorithm is inherently designed around the problem structure, however, and so should maintain the advantage.

6. Testbed Validation

Working with the MIT Rover Testbed courtesy of Jonathan How and the MIT Aerospace Controls Laboratory [24], we implemented a three-vehicle collision avoidance scenario. The rovers are equipped with an indoor positioning system with cm-level accuracy and on board Sony Vaio laptops which communicate with a ground station via wireless ethernet, see Figure 6. The decentralized algorithm was implemented using 5 step, discretized, receding horizon control with 1 meter collision avoidance constraints between vehicles. The local optimizations were performed using Matlab's fmincon nonlinear optimization program, and new way points were passed to the vehicles at 2 second intervals. The results displayed in Figure 6 show the promise of implementing the proposed decentralized algorithm in real time on real hardware, and validates future extensions of the algorithm to multiple vehicle testbeds and real world applications.

Fig. 5. MIT Rover Testbed closeup with on board laptop and position sensor visible, courtesy of Jonathan How

References

[1] Barán, B., Kaszkurewicz, E., and Bhaya, A., Parallel asynchronous team algorithms: Convergence and performance analysis. *IEEE Transactions on Parallel and Distributed Systems*, 7(7):677–688, 1996.

[2] Benders, J. F., Partioning procedures for solving mixed-variables programming problems. *Numerische Mathematik*, 1962.

Fig. 6. MIT Rover Testbed in action performing 3 vehicle collision avoidance

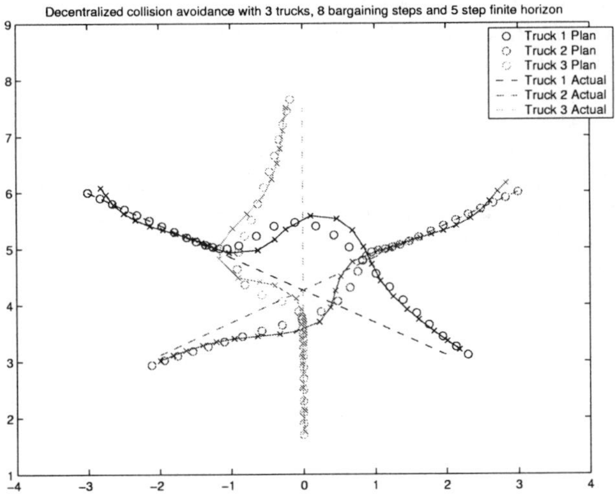

Fig. 7. MIT Rover Testbed Results, 3 vehicle traffic circle solution

[3] Bertsekas, D. P., *Dynamic Programming*, volume 1. Athena Scientific, Belmont, Mass., 2nd edition, 1993.

[4] Bertsekas, D. P., *Nonlinear Programming*. Athena Scientific, Belmont, Mass., 2nd edition, 1995

[5] Biggs, M., *Towards Global Optimization*, chapter Constrained Minimization

Using Recursive Quadratic Programming. North-Holland, 1975.

[6] Boyd, S. and Vandenberghe, L., *Convex Optimization.* Cambridge University Press, Cambridge, England, 2004.

[7] Coello, C. A. C., An updated survey of GA-based multiobjective optimization techniques. Technical Repot RD-98-08, Laboratorio Nacional de Informtica Avanzada (LANIA), Xalapa, Veracruz, Mexico, 1998.

[8] Conley, J. P. and Wilkie, S., An extension of the Nash bargaining solution to non convex problems. *Games and Economic Behavior*, 13(1):26–38, 1996.

[9] Geoffrion, A. M., Generalized benders decomposition. *Journal of Optimization Theory and Applications*, 10(4), 1972.

[10] Gill, P. E., Murray, W., and Saunders, M. A., *User's Guide for SNOPT Version 6', A Fortran Package for Large Scale Non Linear Programming*, 2002.

[11] Han, S., A globally convergent method for nonlinear programming. *J. Optimization Theory and Applications*, 22:297, 1977.

[12] Harsani, J. C., A simplified bargaining model for the n-person cooperative game. *International Economic Review*, 4(2):194–200, 1963.

[13] Heiskanen, P., Decentralized method for computing Pareto solutions in multi-party negotiations. *European Journal of Operational Research*, 117(3):578–590, 1999.

[14] Hillermeier, C., *Nonlinear Multiobjective Optimization: A generalized homotopy approach.* Birkhauser Verlag, Basel, 2001.

[15] Inalhan, G., Stipanovic, D. M., and Tomlin, C. J., Decentralized optimization, with application to multiple aircraft coordination. SUDAAR 759, Stanford, Palo Alto, CA, 2002a.

[16] Inalhan, G., Stipanovic, D. M., and Tomlin, C. J., Decentralized optimization, with application to multiple aircraft coordination. In *Proceedings of the 41st IEEE Conference on Decision and Control*, Las Vegas, 2002b.

[17] Kalai, E. and Smorodinsky, M., Other solutions to Nash's bargaining problem. *Econometrica*, 43(3):513–518, 1975.

[18] Klatte, D., Strong stability of stationary solutions and iterated local minimizations. In et. al., J. G., editor, *Parametric Optimization and Related Topics*, volume 35 of *Mathematical Research*, pages 119–136 Akademie-Verlag, 1987.

[19] Miettinen, K. M., *Nonlinear Multiobjective Optimization.* Kluwer Academic, 1999.

[20] Nash, J. F., The bargaining problem. *Econometrica*, volume 18(issue 2):pg 155–162, 1950.

[21] Nesterov, Y. and Nemirovskii, A., Self-concordant functions and polynomial time methods in convex programming. Moscow. USSR Academy of Science, Central Economic and Mathematical Institute, 1989.

[22] Osborne, M. J. and Rubinstein, A., *A Course in Game Theory.* MIT Press, Cambridge, Massachusetts, 1994.

[23] Powell, M., Fast algorithm for nonlinearly constrained optimization calculations. *Numerical Analysis*, 630. Lecture Notes in Mathematics, 1978.

[24] Richards, A., Kuwata, Y., and How, J., Experimental demonstrations of

real-time MILP control. In *Proceeding of the AIAA Guidance, Navigation, and Control Conference*, 2003.

[25] Tammer, K., The application of parametric optimization and imbedding to the foundation and realization of a generalized primal decomposition approach. In et. al., J. G., editor, *Parametric Optimization and Related Topics*, volume 35 of *Mathematical Research*, pages 376–386. Akademie-Verlag, 1987.

[26] Vazquez-Abad, F. J., Cassandras, C. G., and Julka, V., Centralized and decentralized asynchronous optimization of stochastic discrete-event systems. *IEEE Transactions on Automatic Control*, 43(5):631–655, 1998.

[27] Verkama, M., Ehtamo, E., and Hamalainen, R. P., On distributed computation of Pareto solutions in n-player games. Research Report A53, Helsinki University of Technology, Systems Analysis Laboratory, 1994.

[28] von Neumann, J. and Morgenstern, O., *Theory of Games and Economic Behavior*. John Wiley and Sons, New York, 1944.

[29] Šiljak, D. D., *Large-Scale Dynamic Systems: Stability and Structure*. North-Holland, New York, 1978.

